T0181424

Lecture Notes in Artificial Intelligence   12909

Subseries of Lecture Notes in Computer Science

Series Editors

Randy Goebel
University of Alberta, Edmonton, Canada

Yuzuru Tanaka
Hokkaido University, Sapporo, Japan

Wolfgang Wahlster
DFKI and Saarland University, Saarbrücken, Germany

Founding Editor

Jörg Siekmann
DFKI and Saarland University, Saarbrücken, Germany

More information about this subseries at http://www.springer.com/series/1244

Amrita Basu · Gem Stapleton ·
Sven Linker · Catherine Legg ·
Emmanuel Manalo · Petrucio Viana (Eds.)

Diagrammatic Representation and Inference

12th International Conference, Diagrams 2021
Virtual, September 28–30, 2021
Proceedings

Springer

Editors
Amrita Basu
Jadavpur University
Kolkata, India

Gem Stapleton ⓘ
University of Cambridge
Cambridge, UK

Sven Linker ⓘ
Lancaster University in Leipzig
Leipzig, Germany

Catherine Legg ⓘ
Deakin University
Burwood, VIC, Australia

Emmanuel Manalo ⓘ
Kyoto University
Kyoto, Japan

Petrucio Viana ⓘ
Universidade Federal Fluminense
Niterói, Brazil

ISSN 0302-9743   ISSN 1611-3349   (electronic)
Lecture Notes in Artificial Intelligence
ISBN 978-3-030-86061-5   ISBN 978-3-030-86062-2   (eBook)
https://doi.org/10.1007/978-3-030-86062-2

LNCS Sublibrary: SL7 – Artificial Intelligence

© Springer Nature Switzerland AG 2021
10 chapters are licensed under the terms of the Creative Commons Attribution 4.0 International License (http://creativecommons.org/licenses/by/4.0/). For further details see license information in the chapters.
This work is subject to copyright. All rights are reserved by the Publisher, whether the whole or part of the material is concerned, specifically the rights of translation, reprinting, reuse of illustrations, recitation, broadcasting, reproduction on microfilms or in any other physical way, and transmission or information storage and retrieval, electronic adaptation, computer software, or by similar or dissimilar methodology now known or hereafter developed.
The use of general descriptive names, registered names, trademarks, service marks, etc. in this publication does not imply, even in the absence of a specific statement, that such names are exempt from the relevant protective laws and regulations and therefore free for general use.
The publisher, the authors and the editors are safe to assume that the advice and information in this book are believed to be true and accurate at the date of publication. Neither the publisher nor the authors or the editors give a warranty, expressed or implied, with respect to the material contained herein or for any errors or omissions that may have been made. The publisher remains neutral with regard to jurisdictional claims in published maps and institutional affiliations.

This Springer imprint is published by the registered company Springer Nature Switzerland AG
The registered company address is: Gewerbestrasse 11, 6330 Cham, Switzerland

Preface

The 12th International Conference on the Theory and Application of Diagrams (Diagrams 2021) was hosted virtually during September 2021. For the first time, Diagrams ran as an annual event, representing a departure from its biennial history. The driver for this change was two-fold. Firstly, the COVID-19 pandemic disrupted the delivery of Diagrams 2020 and the Steering Committee felt it important to provide a virtual event that would bring the community together. Secondly, strong submission and attendance numbers in recent years served as motivation for a possible longer term move to an annual event. Diagrams 2021 allowed such a change to be tentatively explored. Diagrams 2021 provided an opportunity for our global community to respond to these challenges and needs in creative and innovative ways as, we believe, is represented in this volume.

Given this historical context, the organizers were keen to enable wide access to the conference from across the globe. As such, registration was free to all delegates, with running costs being absorbed by an underwriting fund. In addition, the virtual nature of the conference was reflected in the program schedule: Diagrams 2021 adopted a novel approach that scheduled talks across the full 24 hour period on each day, enabling fair access to the conference for delegates all over the world.

Submissions to Diagrams 2021 were solicited in the form of Long papers, Short papers, Posters, and non-archival Abstracts. All submissions received three reviews by members of the Program Committee or a nominated sub-reviewer. A rebuttal phase was included to ensure that authors had the opportunity to respond to reviewer concerns. The reviews and rebuttals led to a lively discussion involving the Program Committee and the conference chairs to ensure that only the highest quality submissions were accepted for presentation. The result was a strong technical program covering a broad range of topics, reflecting the multidisciplinary nature of the conference series.

We would like to thank the Program Committee members and the additional reviewers for their considerable contributions. The robust review process, in which they were so engaged, is a crucial part of delivering a major conference. A total of 94 submissions were received across the Main, Philosophy, and Psychology and Education tracks. Of these, 16 were accepted in the Long paper category. A further 25 were accepted as Short papers, 4 as Abstracts, and 22 as Posters, of which 6 are non-archival abstracts. These contributions were complemented by the inclusion of five tutorials, covering a diverse range of topics of interest to Diagrams delegates.

Diagrams 2021 had five outstanding keynote presenters, who delivered a wide variety talks:

- Shaaron Ainsworth, Professor of Learning Sciences at the University of Nottingham: Why and How Should we Draw to Learn.
- Daniel Rosenberg, Professor of History at the University of Oregon: Mapping Time.

- Katharina Scheiter, Head of the Multiple Representations Lab at the Leibniz-Institut für Wissensmedien and Full Professor for Empirical Research on Learning and Instruction at the University of Tübingen: Learning From Visual Displays: Processes and Interventions.
- Atsushi Shimojima, Professor at Doshisha University: A Philosophical View of Fundamental Properties of Diagrams.
- Frederik Stjernfelt, Full Professor of Semiotics, Intellectual History, and Philosophy of Science at Aalborg University: Diagrams and Dicisigns - the Interrelations of Peirce's Doctrines of Propositions and Diagrammatical Reasoning.

These keynotes were complemented by an Inspirational Early Career Researcher Invited Talk. The invitation to deliver this talk was reserved for an active Diagrams researcher, within approximately ten years of their PhD, who has demonstrable potential to be a major leadership force within the community. We were delighted that Francesco Bellucci, Assistant Professor at the University of Bologna, accepted our invitation and delivered a talk on *What is a Logical Diagram?* at the Graduate Symposium.

There are, of course, many people to whom we are indebted for their considerable assistance in making Diagrams 2021 a success. We thank Mohanad Alqadah, Graduate Symposium Chair; Mikkel Willum Johansen, Publicity Chair; Petrucio Viana, Proceedings Chair; Amirouche Moktefi, Finance Chair; and Daniel Raggi, Local Chair. We also thank Richard Burns, for his help producing the Diagrams 2021 website, and Reetu Bhattacharjee for her support with the technical delivery of the conference. Our institutions, Jadavpur University, the University of Cambridge, Lancaster University in Leipzig, Deakin University, and Kyoto University also provided support for our participation, for which we are grateful. Lastly, we thank the Diagrams Steering Committee for their continual support, advice and encouragement.

July 2021

Amrita Basu
Gem Stapleton
Sven Linker
Catherine Legg
Emmanuel Manalo

Organization

Program Committee

Eisa Alharbi	University of Brighton, UK
Mohanad Alqadah	Umm Al-Qura University, Saudi Arabia
Amrita Basu (Chair)	Jadavpur University, India
Francesco Bellucci	University of Bologna, Italy
Andrew Blake	University of Brighton, UK
Ben Blumson	National University of Singapore, Singapore
Leonie M. Bosveld-de Smet	University of Groningen, The Netherlands
Jean-Michel Boucheix	LEAD-CNRS, University of Burgundy, France
Richard Burns	West Chester University, USA
Jim Burton	University of Brighton, UK
Jessica Carter	Aarhus University, Denmark
Mihir Chakraborty	Jadavpur University, India
Marc Champagne	Kwantlen Polytechnic University, Canada
Peter Chapman	Edinburgh Napier University, UK
Peter Cheng	University of Sussex, UK
Daniele Chiffi	Politecnico di Milano, Italy
Lopamudra Choudhury	Jadavpur University, India
James Corter	Columbia University, USA
Gennaro Costagliola	Università di Salerno, Italy
Silvia De Toffoli	Princeton University, USA
Erica de Vries	Université Grenoble Alpes, France
Aidan Delaney	Bloomberg, UK
Lorenz Demey	KU Leuven, Belgium
Maria Giulia Dondero	Université de Liège, Belgium
George Englebretsen	Bishop's University, Canada
Judith Fan	University of California, San Diego, USA
Logan Fiorella	University of Georgia, USA
Jacques Fleuriot	The University of Edinburgh, UK
Amy Fox	University of California, San Diego, USA
Valeria Giardino	Archives Henri Poincaré, France
Mateja Jamnik	University of Cambridge, UK
Mikkel Willum Johansen	University of Copenhagen, Denmark
Yasuhiro Katagiri	Future University Hakodate, Japan
Vitaly Kiryushchenko	York University, Canada
John Kulvicki	Dartmouth College, USA
Brendan Larvor	University of Hertfordshire, UK
John Lee	The University of Edinburgh, UK
Catherine Legg	Deakin University, Australia

Javier Legris	CONICET - Universidad de Buenos Aires, Brazil
Jens Lemanski	FernUniversität in Hagen, Germany
Sven Linker	Lancaster University Leipzig, Germany
Emmanuel Manalo	Kyoto University, Japan
Kim Marriott	Monash University, Australia
Mark Minas	Universität der Bundeswehr München, Germany
Amirouche Moktefi	Tallinn University of Technology, Estonia
Martin Nöllenburg	Vienna University of Technology, Austria
Marco Panza	CNRS, France
Ahti Pietarinen	Tallinn University of Technology, Estonia
Margit Pohl	Vienna University of Technology, Austria
Uta Priss	Ostfalia University, Germany
João Queiroz	Federal University of Juiz de Fora, Brazil
Daniel Raggi	University of Cambridge, UK
Peter Rodgers	University of Kent, UK
Dirk Schlimm	McGill University, Canada
Stanislaw Schukajlow	University of Münster, Germany
Christoph Daniel Schulze	OTTO, Germany
Stephanie Schwartz	Millersville University, USA
Atsushi Shimojima	Doshisha University, Japan
Hans Smessaert	KU Leuven, Belgium
Pawel Sobocinski	Tallinn University of Technology, Estonia
Gem Stapleton (Co-chair)	University of Cambridge, UK
Takeshi Sugio	Doshisha University, Japan
Yuri Uesaka	The University of Tokyo, Japan
Jean Van Bendegem	Vrije Universiteit Brussel, Belgium
Peggy Van Meter	Pennsylvania State University, USA
Petrucio Viana	Federal Fluminense University, Brazil
Reinhard von Hanxleden	Christian-Albrechts-Universität zu Kiel, Germany
Michael Wybrow	Monash University, Australia
Chenmu Xing	Minot State University, USA

Additional Reviewers

De Rosa, Mattia	Schulz-Rosengarten, Alexander
Grimm, Lena	Smola, Filip
Morris, Imogen	Smyth, Steven
Papapanagiotou, Petros	Wallinger, Markus
Rentz, Niklas	Yang, Ying
Satriadi, Kadek	

Abstracts of Keynotes

A Philosophical View of Fundamental Properties of Diagrams

Atsushi Shimojima [ORCID]

Faculty of Culture and Information Science, Doshisha University, 1-3
Tatara-Miyakodani, Kyotanabe, 610-0394, Japan
ashimoji@mail.doshisha.ac.jp

I will discuss systems of diagrams that may seem ridiculously simple but, in my thought, have some of fundamental properties of more complex graphical systems. The properties in question are closely related to the potentials of free rides, over-specificity, and auto-consistency illustrated and analyzed in [1]. They were characterized roughly in the following way:

Free Ride: Expressing a set of information in diagrams can result in the expression of other, consequential information.
Over-Specificity: Expressing a set of information in diagrams can mandate the selective expression of other, often non-consequential pieces of information.
Auto-Consistency: It is not possible to express a certain range of inconsistent sets of information in diagrams.

In this presentation, I will offer a somewhat more general view of these properties, claiming that they all point to the existence of what may be called "proxy logics" in the diagrammatic systems in question. A proxy logic is a system of constraints that governs the arrangements of symbols and other elements in diagrams, to be distinguished from a "target logic" that governs the things represented by the diagrams. I will show that inference and comprehension that we perform with diagrams heavily depend on the soundness and completeness of the proxy logic relative to a part of the target logic.

This will lead us to the question how a diagrammatic system comes to be equipped with such a proxy logic. I will sketch an answer in the final part of my presentation. According to it, additional meaning relations hold in a diagrammatic system as logical consequence of its basic semantic conventions [1, 2]. Under these additional meaning relations, information is carried by properties of diagrams other than those designated in basic semantic conventions, and different ways in which these additional meaning carriers are related to basic meaning carriers are the basis of the proxy logic in that system and its correspondence with the target logic.

References

1. Shimojima, A.: Semantic Properties of Diagrams and Their Cognitive Potentials. CSLI Publications, Stanford (2015)
2. Shimojima, A., Barker-Plummer, D.: Channel-theoretic account of reification in representation systems. Logique Analyse **251**, 341–363 (2020)

Why and How Should We Draw to Learn

Shaaron Ainsworth 🆔

University of Nottingham, Nottingham NG95FH, UK
shaaron.ainsworth@nottingham.ac.uk

In recent years, there has been increasing interest in asking learners to draw diagrams for themselves. When learners pick up a pencil and paper or move a stylus on a screen to create a visual representation (i.e. one that uses position and space meaningfully), they can enhance this understanding. This is true whether they are learning chemistry, fashion design or medicine and at all stages of education. However, to date, most (but not all) studies have focussed on a narrow range of pedagogical practices based upon a predominantly cognitive approach. In this talk, I want to join with others to argue that to move the practice of drawing to learn forward, we must develop a synthetic theoretical framework that understands learning at multiple timescales (from the millisecond to millennium) and levels (from the neuron to the society).

Taking this approach leads us to recognise that drawing diagrams is not an optional "nice-to-have" but is fundamental to the way people learn. New knowledge emerges when we engage in representational practices such as drawing, as expressing what we currently know in external forms recruits cultural, cognitive, and sensory-motor resources that develop our own and others' understanding.

This also invites us to notice that drawing can serve many purposes: for example, we draw to prepare, to observe, to remember, to understand and to communicate. We can draw many sorts of things - varying from a quick back of the envelope sketch to a particular diagram whose form we may have struggled to learn. We can draw at different points of the learning process, and sometimes we draw for ourselves, our colleagues or our instructors.

In this talk, I illustrate these purposes of educational drawing using lots of examples from diverse domains, address what successful drawing looks like in each case and what support learners might need. I will also consider several open questions, such as whether everyone can draw to learn and if there are certain situations where we should avoid drawing diagrams.

You are warmly invited to draw your response to this talk.

Learning from Visual Displays: Processes and Interventions

Katharina Scheiter ⓘ

Leibniz-Institut für Wissensmedien, Schleichstraße 6,
72076 Tübingen, Germany
k.scheiter@iwm-tuebingen.de

In education, visual displays are ubiquitously used to teach students. Visual displays typically consist of multiple representations such as combinations of written explanations and illustrations (e.g., diagrams, pictures, animations, or simulations). I will refer to three potentials of visual displays for education, which are particularly relevant in many STEM domains: representing visuo-spatial information (visualization), enabling interaction with real-world phenomena (exploration), and augmenting phenomena beyond the observable (abstraction). To help students learn from visual displays, it is necessary to understand the learning processes that are linked to student achievement. In my talk, I will present studies that investigated said learning processes using eye tracking, log file analyses, and verbal protocols for different types of visual displays. A main finding of these studies is that learners often fail to apply effective learning processes spontaneously. Understanding these learning processes builds the basis for developing at least two types of support aimed at fostering their use: First, the design of the visual display can be optimized so that it will nudge students in applying helpful processes during learning. Second, trainings or processing prompts can be used to convey knowledge on learning processes and promote their application during learning from visual displays. In my talk, I will provide examples for both intervention approaches as regards their development and application in education.

Diagrams and Dicisigns: The Interrelations of Peirce's Doctrines of Propositions and Diagrammatical Reasoning

Frederik Stjernfelt

Aalborg University, Copenhagen, Denmark

Keywords: Diagrams · Propositions · Peirce

Famously, Peirce ascribed to diagrams a very central role in his philosophy and semiotics. Mathematics is possible only by the manipulation of diagrams and simultaneously, all deductive reasoning is taking place by means of diagrams. These ideas have a number of important implications:

1) Every investigation process involves a diagrammatic phase. In Peircean terminology, an abductive guess leads to an ideal hypothesis, more or less explicitly expressed in a diagram. [1] This diagram is subjected to manipulation, leading to a number of deductive implications of the hypothesis – theory development, if you so wish. Finally, these hypotheses are verified/falsified by the inductive sampling of evidence pertaining to them. Obviously, this general epistemology considerably generalizes the everyday notion of "diagram" [2].

2) This furthermore indicates that wherever, in the special sciences, in applied sciences or in everyday reasoning where deduction takes place, mathematical diagrams, simple or complex, are at work, more or less explicitly.

3) Some of such diagrams may remain implicit, in language, images, gesture, action, etc.

Reasoning, however, deductive or not, has to do with the truth-preserving derivation of *propositions*. How does that square with the claimed center role of diagrams? This paper makes the claim that diagrams form stylized iconic predicates of propositions. This is based on Peirce's less widespread theory of propositions – or "Dicisigns" – which is, importantly, multimodal. [3]) Much discussion of propositions in the analytic tradition is based on a tacit presumption that propositions are invariably linguistic, expressed in ordinary or formal languages. Peirce's doctrine of propositions differs here: it is purely functional, requiring of a proposition sign only that it fulfills the two functions of *denoting* some object and *describing* that same object. These functions may be satisfied by non-linguistic or partially linguistic expressions. Thus, when giving a basic example of a proposition, Peirce often picks "a painting with a label"; the painting serving the descriptive or predicative function and the label serving the denoting or referring function of the proposition.

In the light of this multimodal theory of propositions, diagrams in use are typically involved in propositions in the descriptive function. Stating some purported truth about some state-of-affairs, the denoting or referring function is indicated by the addition of subject indices to the naked diagram structure. In this analysis, diagrams are predicates, describing the detailed character of some structural property of the states-of-affairs under study. Doing so, diagrams may vastly transgress the linguistic limit of 3–4 subjects per sentence, and they add an indefinite increase in precision over merely linguistic predicates. [4] Simultaneously, they form the core of Peirce's version of what was later called "truth-maker" realism: real is that whose existence is presupposed by a true proposition. [5] Reality, then, is involved in three ways in true diagram propositions: 1) the diagram predicate describing some real structure; 2) the diagram subject indices pointing to the phenomenon possessing that structure; and 3) the diagram proposition as a whole depicting the real state-of-affairs described by the diagram and indicated by its indices.

This complex of ideas I shall present and discuss in the paper with a number of examples.

References

1. Peirce, C.: Collected Papers I-VIII. Belknap Press, Cambridge Mass (1931–1958)
2. Stjernfelt, F.: Diagrammatology: An Investigation on the Borderlines of Phenomenology, Ontology, and Semiotics. Springer, Dordrecht (2007). https://doi.org/10.1007/978-1-4020-5652-9
3. Stjernfelt, F.: Natural Propositions: The Actuality of Peirce's Doctrine of Dicisigns. Docent Press, Boston (2014)
4. Stjernfelt, F.: Co-localization as the syntax of multimodal propositions. In: Jappy, T. (ed.) The Bloomsbury Companion to Contemporary Peircean Semiotics., pp. 419–458, 482–485. Bloomsbury Academic, London (2019)
5. Stjernfelt, F.:: Peirce as a Truthmaker realist: propositional realism as backbone of Peircean metaphysics. Blityri. Studi di storia delle idee sui segni e le lingue, **XI**(2), 123–36 (2021)
6. http://www.springer.com/lncs. Accessed 21 Nov 2016

Contents

New Representation Systems

Analysis of Diagrams

Diagrams and Computation

xx   Contents

Cognitive Analysis

Diagrams as Structural Tools

Design of Concrete Diagrams

Aesthetics and Ordering in Stacked Area Charts

Steffen Strunge Mathiesen[ID] and Hans-Jörg Schulz$^{(\boxtimes)}$[ID]

Department of Computer Science, Aarhus University,
Åbogade 34, 8200 Aarhus, Denmark
hjschulz@cs.au.dk

Abstract. Stacked area charts are a common visualisation type for sets of time series. Yet, they are also known to be challenging to read, in particular if the time series exhibit much fluctuation or even abrupt changes. In this paper, we introduce a novel approach to improving the layout of stacked area charts by means of reordering the time series in the stack. This approach breaks down into two parts: First, we gather aesthetic criteria and define associated quality metrics for stacked area charts. Second, we use these quality metrics together with a new algorithm called UpwardsOpt to find orderings of the stacked time series that optimise a chart's aesthetic properties. The produced orderings guarantee optimality in the sense that no better result can be obtained by moving any individual time series to a different position in the stack. In a benchmark study, we show that our algorithm can increase the layout quality up to 25%–50% over the state-of-the-art approach at the expense of longer runtimes. All datasets and an open source implementation of our algorithm are provided to facilitate their reuse and the reproducibility of our results.

Keywords: Time series visualisation · Stacked graphs · Layout algorithm

1 Introduction

Stacked area charts are a common means to show the aggregate of numerical quantities over time and available in any serious spreadsheet or dashboard application. Yet academic literature on them is sparse. In most cases, stacked area charts are merely discussed as a by-product of their bigger brother, the stream graph. This paper aims to change this by providing a broader view on the aesthetic criteria of stacked area charts and how to algorithmically fulfil them.

Stacked area charts (sometimes also called *layer charts*, *strata charts*, or *band curves* [10]) show a set of n individual time series $f_0 \ldots f_{n-1}$ as horizontal layers that are piled on top of each other. As a result, the top contour of this stack of layers – the so-called *silhouette* g_n – conveys the sum of all time series: $g_n = \sum_{i=0}^{n-1} f_i$. In this stacked area charts differ from stream graphs, which do not stack on top of the horizontal time axis, but instead place layers above and below a curved baseline. Both chart types should only be used if the sum of the

© Springer Nature Switzerland AG 2021
A. Basu et al. (Eds.): Diagrams 2021, LNAI 12909, pp. 3–19, 2021.
https://doi.org/10.1007/978-3-030-86062-2_1

time series is a meaningful quantity and the silhouette carries useful information – e.g., for counts, percentages, or fractions [16, p. 27].

The expressiveness and effectiveness of stacked area charts are debated in the literature. Prominent studies have looked mainly into their usefulness for various analytic tasks as compared to other chart types [6,8,12,13]. One important observation is that any task performed with stacked area charts will be hindered by the accumulation of fluctuations across the layers – i.e., fluctuations of the lower layers affecting the look of the layers placed above them. In terms of expressiveness, this lets rather stable layers potentially look as if they are much more volatile than they actually are. In terms of effectiveness, this makes it very hard to read off concrete values for any time series but f_0 at the bottom of the stack and the overall silhouette g_n.

These apparent shortcomings stand in contrast to the unbroken popularity of stacked area charts for data journalism and reporting, with recent examples ranging from Donald Trump's tax returns[1] to the infection numbers of respiratory diseases including the COVID-19 virus[2]. These charts are made possible, as a clever ordering of the layers can reduce the readability problem of stacked area chart to some degree. This is done by finding an ordering of layers that minimises fluctuations, so that it eases the estimation of their thickness and does not convey a volatility that is not supported by the underlying data.

To achieve such an ordering that flattens the layers for better readability, existing algorithms focus on a quality metric called "wiggle" [3]. Generally speaking, wiggle is a measure of the fluctuations of a layer – i.e., its ups and downs. Different approaches for how to quantify a layer's wiggle have been proposed. They all have in common that they are based on the first derivative of a layer. In combination with an optimisation algorithm that aims to order the layers, so that the sum of their wiggle becomes minimal, the literature offers already good solutions to find suitable layer orderings for stacked area charts.

The first contribution of this paper is to show that there is more to the layout of stacked area charts than wiggle. We capture a set of aesthetic criteria for stacked area charts including means to quantify and combine them in one objective function: flatness, straightness, continuity, and significance. In their combination, these criteria yield a more balanced visual appearance for stacked area charts than it can be achieved by the mere optimisation of a chart's wiggle.

The second contribution of this paper is to give a novel optimisation algorithm with a stricter optimality guarantee than the state-of-the-art approach, which in turn yields better orderings for most charts. Where the state-of-the-art approach guarantees that no individual layer can be *swapped with a neighbouring layer* to improve the layout [5], our algorithm guarantees that no individual layer can be *moved to any other position* to improve the layout. This enables us to generate layouts that cannot be produced with the state-of-the-art algorithm – for example, where moving a layer to a better position would involve a

---

[1] https://www.nytimes.com/interactive/2020/09/27/us/donald-trump-taxes-timeline.html.

[2] https://publichealthinsider.com/2020/09/10/alongside-the-ongoing-transmission-of-covid-19-common-colds-are-on-the-rise-in-seattle-and-king-county/.

number of intermediate swaps that do not improve the overall layout. Together with other supplementary material, a Python implementation of our algorithm is available at https://vis-au.github.io/stackedcharts/.

2   Related Work

Stacked area charts have been around for a long time, making it impossible to pinpoint their concrete origin or inventor. Already Willard Briton mentions them in his 1914 book "Graphic Methods for Presenting Facts" under the heading *component parts shown by curves* [2, Chapter 8]. In 1938, the American Standards Association approved design guidelines for stacked area charts [1, pp. 64–65]. These guidelines were prepared by the Committee on Standards for Graphic Presentation of the American Society of Mechanical Engineers and they included advice on layout, gridding, scales, shading, and labelling. Some of these aspects still receive attention these days, as it is underlined by a novel labelling algorithm proposed in 2012 that seeks unused space to label layers [14]. Interestingly, the guidelines contain a list of circumstances under which stacked area charts should not be used:

– if accurate reading of values is desired, in the case of more than one layer,
– if irregular layers will unduly distort the contours of the others above it, or
– if changes in the series are abrupt, causing distortion of the layers' width.

In particular the latter two restrictions have ever since tickled the imagination of visualization designers, and a series of alternatives and improvements have been suggested to work around them. This resulted in an entirely new understanding of when and how stacked area charts can be used, as it is illustrated in Fig. 1. The most prominent of these layout alternatives is certainly the *ThemeRiver*™ technique [7] or its modern instance: the *stream graph* [3]. Though using different underlying algorithms, both yield very similar and appealing,

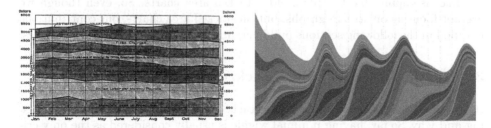

Fig. 1. The classic vs. the modern use of stacked area charts: On the left an example from 1914 that follows the then existing understanding that stacked area charts should only be used for mostly regular layers without any abrupt changes [2, p. 147]. On the right an example from 2016 that shows the much more relaxed understanding prevalent today that, given the proper aesthetic optimisations, stacked area charts can also be used for very irregular bands that do not even persist across the entire chart [13].

flow-like stacked diagrams for irregular layers. Using the terminology from the stream graphs algorithm, these techniques reduce the distortion introduced by irregular layers by means of a curved baseline instead of a flat x-axis. This curve allows for a better distribution of the irregularities by vertically stacking layers not only above it, but also below it.

An important consideration for stream graphs has always been the layout of grouped (e.g., clustered) layers [4,15,17]. In particular the idea of accommodating the display of "complementary evolution" of multiple layers has received some attention in the literature. Complementarity means in this case that the fluctuations of two or more layers cancel each other out at least in some part – i.e., when one layer gets thicker, another get thinner and vice versa. It is assumed that complementary layers are somehow related to each other – i.e., forming a group or cluster – and thus one wants to place them side-by-side in the chart. One way to address this layout challenge is to add a term capturing complementarity to the evaluation function of the chart's quality, so that it will be taken into account by the ordering algorithm [11].

The work most related to ours is the research on the ordering of layers to reduce distortions due to irregularities. Byron and Wattenberg [3] were the first to point out the ordering problem and to address it by introducing the notion of wiggle. They proposed to order the layers by a heuristic called OnSet – i.e., the first occurrence of a non-zero value for a time series f_i determined the vertical position of its layer in the stack. Bartolomeo and Hu [5] later argued that not all datasets exhibit this property of time series appearing at different time points, which limits the usefulness of OnSet for other data. They propose two new heuristics: BestFirst for finding an initial ordering and TwoOpt for iteratively improving that ordering. BestFirst uses a greedy approach, where the next layer to be added to a stack is the one which adds the least wiggle to the chart. Its output is then improved by TwoOpt, which iteratively swaps adjacent layers if this will improve the wiggle. Their algorithm is the current state of the art.

While both OnSet and the combination BestFirst+TwoOpt have been proposed for stream graphs, their principal approach as well as the notion of wiggle are likewise applicable to "plain-old" stacked area charts. So, even though we are not focusing on stream graphs, but on stacked area charts, our contributions detailed in the following sections build very much on top of these two approaches.

3   Aesthetic Criteria for Stacked Area Charts

It is not easy to pinpoint what makes a stacked area chart look aesthetically pleasing and why. So far, having minimal wiggle is usually considered as the only criterion in this regard. Yet in our layout experiments with different datasets,[3] we encountered a number of curious glitches and imperfections in the resulting charts that apparently are not captured (well) by wiggle alone. These observations have sparked our investigation into a broader set of aesthetic criteria to create more

---

[3] See https://vis-au.github.io/stackedcharts/ for a list of datasets used in this work.

balanced charts that trade some of that reduced wiggle for being a little less disagreeable in some other regards. In the following, we describe the four aesthetic criteria we found most useful in optimising the layout of stacked area charts.

As a layer can be geometrically captured in a variety of ways – e.g., through its centre line [3] or through its top and bottom outlines [5] – we introduce the quality metrics associated with the aesthetic criteria for a line l_i that is representative of layer f_i. Different options for choosing representative lines are discussed at the end of this section.

3.1 Flatness: Minimising Wiggle

The idea behind *flatness* is that a layer looks nicer and is easier to judge in terms of its vertical span if it is as horizontal as possible. The latter is rooted in the *line width illusion* [9], which is a bias that leads us to perceive sloped layers as thinner than they actually are. The smaller the absolute slope of a layer is, the more this perceptual bias will be reduced. Minimising the slope is exactly what Byron and Wattenberg [3] proposed with their wiggle metric: the less wiggle a layer exhibits, the flatter it is. Hence, we can capture flatness in the same way by measuring the absolute slope between time points t_{j-1} and t_j for a layer's representative line l_i:

$$wiggle_{i,j} = |l_i'(t_j)| \cdot (t_j - t_{j-1}) \quad \text{with} \quad l_i'(t_j) = \frac{l_i(t_j) - l_i(t_{j-1})}{t_j - t_{j-1}} \tag{1}$$

The effect of optimising a whole chart for flatness is illustrated in Fig. 2b, which reorders the layers from the chart in Fig. 2a to minimise the wiggle of all shown layers, flattening them out as best as possible. As one can see, the overall course of the layers is more horizontal with a less ragged and more orderly look to them as compared to the random order in Fig. 2a. This does not only help in tracing them from left to right, but also in estimating and comparing their respective vertical spans, which we call the *thickness* of a layer.

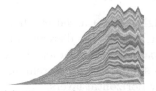 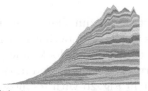 

(a) Chart without any opti- (b) Same chart as 2a, but (c) Same chart as 2a, but op-
misation (random order)   optimised for flatness   timised for straightness

Fig. 2. The visual effect of ordering the layers for flatness or straightness. Note that the colouring of layers is consistent across all three examples.

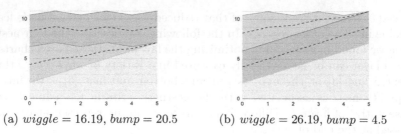

(a) *wiggle* = 16.19, *bump* = 20.5   (b) *wiggle* = 26.19, *bump* = 4.5

Fig. 3. Ordering the coloured layers for flatness (a) or straightness (b). (Color figure online)

3.2   Straightness: Minimising Bumps

Merely optimising for flatness can create charts that are made to be as horizontal as possible, no matter what. Yet in particular for charts that exhibit a large overall increase or decrease, that means for the optimisation to work against that overall trend of the chart. It does so by ordering the layers in a way that they meander between up and down to keep them as horizontal as possible. This results in a bumpy or wavy look for the individual layers that also shows in Fig. 2b. This bumpy look is an artefact of ordering the layers for flatness (i.e., minimal wiggle) and not inherent in the data.

To counter this visual effect, we introduce *straightness*, which is a criterion that aims to reduce those bumps and to keep the layers as steady and even as possible. This is schematically exemplified in Fig. 3b, where the order of the red and blue layers from Fig. 3a is reversed to maximise their straightness. As a result, the horizontal but bumpy look achieved by ordering for flatness is traded for a sloped but straight look – i.e., fewer, smaller bumps. The bumpiness of a layer can be captured using its second derivative as measure for how concave/convex it is. For a time point t_j and a representative line l_i, the second derivative can be computed using the first derivative from Eq. 1:

$$bump_{i,j} = |l_i''(t_j)| \quad \text{with} \quad l_i''(t_j) = l_i'(t_{j+1}) - l_i'(t_j) \tag{2}$$

Figure 2c shows the effects of maximising the straightness for an entire chart – the same chart that was optimised for flatness in Fig. 2b. By disregarding their flatness, the layers of this ordering look much calmer and steadier. They also better support the overall increasing trend of the silhouette, which remains somewhat of a baffling sight in Fig. 2b with its mostly horizontal layers.

3.3   Continuity: Minimising Broken Layers

In particular when layers appear or disappear suddenly, large singular jumps can sometimes occur in the chart which may "break" layers – i.e., have them end at time point t_j at a y-coordinate that is very different from the y-coordinate at which they continue at time point t_{j+1}. This is illustrated in Fig. 4, showing how this problem depends very much on the aspect ratio of the chart (left chart vs.

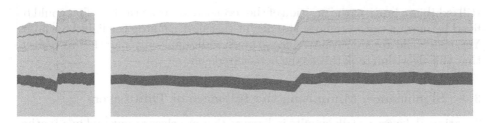

Fig. 4. Illustration of the effect of layer thickness and aspect ratio on the appearance of broken layers caused by sudden vertical shifts. Data: subset of *unempl* (Color figure online)

right chart) and on the width of the layer (red layer vs. blue layer). This is not only a cosmetic problem, but also a problem for a chart's effectiveness: breaks can make it tricky if not impossible to follow a layer – especially if there are many thin layers and the layers are shifted by many multiples of their thickness.

Hence, we introduce *continuity* as an additional aesthetic criterion. We capture continuity by measuring the vertical displacement of a layer in relation to its average thickness. We call this measure *break* and we define it for a time point t_j, a layer f_i, and a representative line l_i as:

$$break_{i,j} = \left| \frac{l_i'(t_j)}{f_i(t_j) + f_i(t_{j-1})} \right|^2 \cdot (t_j - t_{j-1}) \tag{3}$$

Using continuity by itself to generate an optimised layer ordering does not yield aesthetic results – it just removes the breaks. It is rather useful as an "add-on" to ordering by flatness or straightness to prevent breaks from appearing as optimal solutions when optimising for these criteria. This can be seen in Fig. 5, where we optimise solely for flatness in Fig. 5a, which introduces four breaks in the chart. Figure 5b shows the same chart, but this time also optimised to minimise breaks. It has less breaks, but they are not entirely gone – for example, the light blue layer

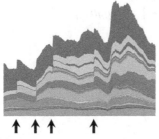

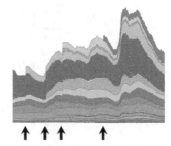

(a) Ordering solely for flatness.   (b) Ordering for flatness and continuity.

Fig. 5. The effect of introducing continuity as an additional criterion to reduce the sudden vertical shifts visible at the indicated time points. Data: *unempl* (Color figure online)

still exhibits a break at the position of the second arrow from the left. This could be fixed by moving the green layer starting at that position atop the light blue layer. Yet, this would apparently introduce so much additional *wiggle* for the green layer, that the algorithm took this as the better trade-off.

3.4   Significance: Minimising the Influence of Thin Layers

Treating all layers equally results in layouts that are close to optimal in a mathematical sense, but not by their visual appearance. The reason is that thin layers are optimised for their aesthetics in the same way as thick layers, even though thicker layers are visually much more salient. Sacrificing flatness or straightness of a prominently visible thick layer to improve the aesthetics of a barely visible thin layer is not a good visual trade-off. Whereas, getting the layout right for those thick layers contributes greatly towards an overall aesthetic chart and can make up for a number of thin layers which are not placed quite as optimally.

We propose the criterion of *significance*, which prioritises thicker layers in the layout process – or rather: thicker parts of the layers as thickness may vary over time and thus a layer maybe more important to the layout during one time interval than during another. Additionally, parts of the layers with zero thickness (i.e., a layer disappearing from the chart for some time) should not be counted at all at those time points, since they do not impact the aesthetics of the chart while they are not visible. To capture this, we weight each layer's aesthetic scores by its thickness. Already Byron and Wattenberg used a quadratic wiggle function to emphasise the importance of the thicker layers in the chart [3], and we extend this practice to all aesthetic measures in this section. Yet often a mere linear weighting does not give the thicker layers the prominence in the optimisation that they have in the visual appearance of a chart. Hence, we further introduce variable exponents to these weights to increase or decrease the effect of the weighting as needed. These exponents can be varied to adjust the weighting from chart to chart and from one aesthetic criterion to the next. This procedure yields the following three formulas for *wiggle*, *bump*, and *break* values as they are aggregated over all m time points for a layer f_i using line l_i and exponent s:

$$wiggle_i = \sum_{j=1}^{m-1} \left(\frac{f_i(t_j) + f_i(t_{j-1})}{2} \right)^s \cdot wiggle_{i,j}$$

$$bump_i = \sum_{j=1}^{m-2} f_i(t_j)^s \cdot bump_{i,j} \tag{4}$$

$$break_i = \sum_{j=1}^{m-1} \left(\frac{f_i(t_j) + f_i(t_{j-1})}{2} \right)^s \cdot break_{i,j}$$

Note that the weight for $bump_i$ is not an average, but depends on the thickness of a layer at the time point $f_i(t_j)$ where the bump is expressed. In other terms: $wiggle_i$ and $break_i$ use the first derivative defined between two time points t_{j-1} and t_j and we thus use the average thickness between these two time points.

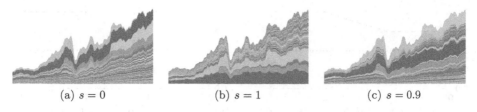

(a) $s = 0$ (b) $s = 1$ (c) $s = 0.9$

Fig. 6. Using different exponents for weighting the layout aesthetics – in this case straightness – by the layers' thickness. (a) employs no weighting and thus all thinner layers are placed at the bottom to "straighten out" as many layers as possible to reduce the overall *bump* measure. (b) employs linear weighting, which places all the thicker layers at the bottom as their straightness is prioritised. (c) uses an exponent in between 0 and 1 to get a compromise solution between (a) and (b). Data: *stocks*

But $bump_i$ uses the second derivative defined between three time points t_{j-1}, t_j, and t_{j+1} and we thus use the thickness $f_i(t_j)$ at t_j directly.

Using the exponent s, we can either remove the weighting by setting $s = 0$, use a linear weighting by setting $s = 1$, or increase s further to emphasise thicker layers even more. This can be seen in Fig. 6, where we optimise for straightness using different exponents for weighting the layers' thickness.

3.5 Choosing One or More Representative Lines

So far, Sects. 3.1, 3.2, 3.3 and 3.4 have not specified for which concrete line l_i of a layer to compute the aesthetics. Byron and Wattenberg [3] advocate the use of the centre line c_i of a layer – i.e., the line halfway between the bottom and the top of a layer. This line c_i can easily be determined by averaging bottom and top y-values of a layer at each time point t_j. Another option given by Bartolomeo and Hu [5] is to compute the aesthetics separately for the bottom line g_i and for the top line g_{i+1} (the top line of f_i being the bottom line of f_{i+1}) and then to average the resulting values to yield a single aesthetic quantity per layer. The three lines $l_i \in \{g_i, g_{i+1}, c_i\}$ of a layer f_i can be computed in a straightforward manner:

$$g_i = \sum_{k=0}^{i-1} f_k \qquad g_{i+1} = g_i + f_i \qquad c_i = \frac{g_i + g_{i+1}}{2} \qquad (5)$$

The main argument that speaks for using the outer lines g_i and g_{i+1} is that they are actually visible to the viewer. I.e., while the centre line is an "imaginary" line running in the middle of each layer, the outer lines are the borders to the respective next layers and thus shown in the chart. Hence, they are the ones that potentially produce a layer's wiggly or bumpy look, so it makes sense to optimise the layout with respect to them.

The main argument that speaks for using the centre line is its sensitivity to certain changes in a layer's thickness. This is illustrated in Fig. 7, where the centre line is able to capture the difference between the examples shown on the

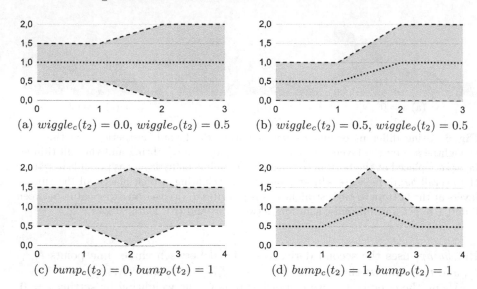

(a) $wiggle_c(t_2) = 0.0$, $wiggle_o(t_2) = 0.5$ (b) $wiggle_c(t_2) = 0.5$, $wiggle_o(t_2) = 0.5$

(c) $bump_c(t_2) = 0$, $bump_o(t_2) = 1$ (d) $bump_c(t_2) = 1$, $bump_o(t_2) = 1$

Fig. 7. Computing *wiggle* and *bump* values for centre line (dotted) and outer lines (dashed) of a layer. With the centre line, one can distinguish between left and right.

left and on the right. For example, in Fig. 7a, the *wiggle* values (i.e., absolute slopes) between t_1 and t_2 are 0.5 for both top and bottom line, so their average *wiggle* is 0.5. In Fig. 7b, outer lines again yield a *wiggle* value of 0.5 (the average of 0 for the bottom line and 1 for the top line). So, measured by their outer lines, left and right side would be considered equally good, even though the left example distributes the increase in thickness in a much nicer and more even way. Though, the *wiggle* values of the centre line reflect this nicer look of the left variant through a smaller value (0 on the left vs. 0.5 on the right). The same is true in Figs. 7c and 7d, which show a similar example for straightness and its associated *bump* values.

Hence, none of the two options is clearly superior to the other. Therefore, we propose a convex combination of all three lines with the weights α, β, and γ summing up to 1 for an aesthetic criterion $crit_i \in \{wiggle_i, bump_i, break_i\}$:

$$crit_i = \alpha \cdot crit_i(g_i) + \beta \cdot crit_i(c_i) + \gamma \cdot crit_i(g_{i+1}) \tag{6}$$

Setting $\alpha = 0, \beta = 1, \gamma = 0$ yields Byron and Wattenberg's procedure, while setting $\alpha = 0.5, \beta = 0, \gamma = 0.5$ yields Bartolomeo and Hu's procedure. Any other setting can be used to go beyond these two approaches, even combining all three lines with each other if desired. In the latter case, it has to be noted that the outer line at the top of one layer is at the same time the outer line at the bottom of the next layer and will thus be taken into account twice. Hence, weighing $\alpha = \beta = \gamma$ does not actually put all lines on par with each other, but will instead emphasise the outer lines over the centre line. Another more intricate option follows from the observations in Fig. 7 and would dynamically shift the weights so that when a layer's thickness changes the middle line gets higher weights, and when it is steady the outer lines get weighted more.

4   Ordering Layers

Being able to quantify a chart's aesthetics does not automatically yield a better chart. To that end, we need to combine the different criteria into an objective function and then search the space of all possible layer orderings for one that minimises that function.

4.1   Objective Function

Integrating the individual aesthetic criteria into one objective function is not quite straightforward due to the very different value ranges their respective quality measures produce. We chose an approach that multiplies the individual metrics with each other. To weight their influence on the overall outcome, we introduce the following weight function:

$$w(x, w_{metric}) = 1 - w_{metric} + x \cdot w_{metric} \qquad (7)$$

This function takes a value $x \in \mathbb{R}$ and a weight $w_{metric} \in [0 \ldots 1]$ with $metric \in \{wiggle, bump, break\}$. It produces a weighted output of x, returning 1 if $w_{metric} = 0$ and x if $w_{metric} = 1$, so it will have no effect on the multiplication in the objective function if set to 0 and full effect if set to 1.

Given that we aim to solve a minimisation problem, we formulate the objective function as a cost function, so as to reduce the values of $wiggle$, $bump$, and $break$. Its output depends on i, which specifies the layer and thus the time series f_i for which it shall be computed, as well as on its bottom line g_i produced by the underlying stack of layers on which layer i is placed and whose fluctuations add onto those of layer i itself:

$$cost_{layer}(i, g_i) =$$
$$w \left(\frac{wiggle_i}{\sum chart}, w_{wiggle} \right) \cdot w \left(\frac{bump_i}{\sum chart}, w_{bump} \right) \cdot w \left(\frac{break_i}{\sum chart}, w_{break} \right) \qquad (8)$$

The normalisation is done with $\sum chart = \sum_{i=0}^{n-1} \sum_{j=0}^{m-1} f_i(t_j)$. Without loss of generality, all other parameters (e.g., the significance exponents s or the weights α, β, γ for the convex combination of lines) are assumed as globally defined to keep the equations as well as the following algorithms readable. For all practical purposes, a reasonable default parameterisation is to set s, β, w_{wiggle}, w_{bump} to 1 and α, γ, w_{break} to 0. The resulting chart can then be used as a starting point to fine-tune individual parameters, like increasing the weight w_{break} if indeed breaks occur and need to be taken care of.

The cost for the whole chart is derived by summing its layers' costs:

$$cost_{chart}(stack) = \sum_{i=0}^{n-1} cost_{layer}(i, g_i) \tag{9}$$

Both variants, the cost per layer and the cost of the whole chart, are being used in the following when describing our optimisation algorithm.

4.2   Optimisation Procedure

Our ordering algorithm – called `UpwardsOpt` – improves the ordering of an initial stack of layers w.r.t. a given cost function $cost_{chart}$. `UpwardsOpt` iterates over all layers starting from the bottom and stopping at the top, hence its name. At each iteration i, `UpwardsOpt` (1) removes layer i from the stack, (2) calls a function *FindBestPosition* to determine the best position of layer i in the remaining stack, and (3) reinserts layer i there. After having gone through all layers, `UpwardsOpt` compares the cost values of the initial stack and of the resulting stack. If the cost improved by more than a given threshold, `UpwardsOpt` is run again on the resulting stack. Otherwise the algorithm terminates.

As for a suitable initial ordering, we recommend sorting the layers by their average thickness starting with the thickest layer at the bottom of the stack and then decreasing towards the top. Beginning with our incremental ordering algorithm from this sorted initial order can speed-up its runtime drastically, cutting down on the number of necessary executions of `UpwardsOpt` before passing the termination threshold. This is due to the fact that this particular ordering ensures an early consideration of the thick layers, whose impact on the overall layout is usually substantial. Once the thicker layers are moved to suitable positions, all the thinner layers can then fall in place around them. If the thicker layers were considered at a later point, many of the already well-positioned thinner layers would need to be repositioned again, requiring more executions of `UpwardsOpt` to converge on the final order. In our experiments, such an initial ordering by average thickness combined with a threshold of a minimum improvement of 1% resulted in 2 to 4 executions of `UpwardsOpt` before termination.

A naïve implementation of the algorithm described above would result in a cubic runtime complexity: For each execution of `UpwardsOpt`, we iterate over n layers, testing for each layer n different positions in the stack, and for each tested position we compute $cost_{chart}$ by summing the $cost_{layer}$ function for all n layers in the stack. Yet our implementation of the *FindBestPosition* function (shown as Algorithm 1) brings the computation down to quadratic complexity by eliminating the need to recompute $cost_{chart}$ each time a new position is tested for a layer. This is done by preprocessing and storing three types of costs:

Algorithm 1. FindBestPosition

1: **procedure** FINDBESTPOSITION($order, f_i$)
2:     $gBelow \leftarrow 0; gAbove \leftarrow f_i$ ▷ Preprocessing Stage
3:     $costBelow, costAbove, costLayer \leftarrow [\,]$
4:     **for** $pos = 0$ **to** $length(order) - 1$ **do**
5:         $costBelow.add(cost_{layer}(order[pos], gBelow))$
6:         $costAbove.add(cost_{layer}(order[pos], gAbove))$
7:         $costLayer.add(cost_{layer}(f_i, gBelow))$
8:         $gBelow \leftarrow gBelow + order[pos]$
9:         $gAbove \leftarrow gAbove + order[pos]$
10:     **end for**
11:     $costLayer.add(cost_{layer}(f_i, gBelow))$
12:
13:     $currentCost \leftarrow costLayer[0] + \sum_{l=0}^{j-2} costAbove[l]$ ▷ Testing Stage
14:     $bestIndex \leftarrow 0, bestCost \leftarrow currentCost$
15:     **for** $pos = 1$ **to** $length(order) - 1$ **do**
16:         $currentCost \mathrel{+}= costBelow[pos - 1] - costAbove[pos - 1]$
17:         $currentCost \mathrel{+}= costLayer[pos] - costLayer[pos - 1]$
18:         **if** $currentCost < bestCost$ **then**
19:             $bestIndex \leftarrow pos, bestCost \leftarrow currentCost$
20:         **end if**
21:     **end for**
22:     **return** $bestIndex$
23: **end procedure**

- *costBelow* (line 5) holds the layer cost under the assumption that the layer in question is below the layer i that is to be positioned – i.e., layer i does not add into the bottom line *gBelow* of that layer.
- *costAbove* (line 6) holds the layer cost under the assumption that the layer in question is above the layer i that is to be positioned – i.e., layer i adds into the bottom line *gAbove* of that layer.
- *costLayer* (line 7) holds the cost of layer i if being moved to position *pos*. This means at index *pos*, *costLayer*[*pos*] contains the cost of layer i sitting on the stack of layers 0 through *pos* − 1. The cost of layer i being positioned all the way at the top is added in line 11.

During testing, we then use these costs to determine the chart's overall cost if layer i is moved to position *pos* in the stack. This is done by combining the overall cost from *costBelow* for all layers below *pos*, from *costLayer* for the costs of layer i being placed at *pos*, and from *costAbove* for all layers above *pos*:

$$cost_{chart}(g_0) = \sum_{l=0}^{pos-1} costBelow[l] + costLayer[pos] + \sum_{l=pos}^{n-2} costAbove[l] \quad (10)$$

While this saves us from re-computing the cost function for all layers each time we try a new position for layer i, we can even get rid of the summations in Eq. 10. This is done by testing new positions for layer i from bottom to top, moving it up one position at a time. When moving layer i up to a new position pos, we do not need to recompute the cost for the whole stack, but only to adjust the cost value computed for the last position $pos - 1$ that we tested:

$$
\begin{aligned}
newCost = \ & currentCost \\
& + costBelow[pos - 1] - costAbove[pos - 1] \\
& + costLayer[pos] - costLayer[pos - 1]
\end{aligned}
\tag{11}
$$

This exact procedure can be found in Algorithm 1 in lines 16 and 17. This way, we only need to sum over the full stack once in the beginning for $pos = 0$ (line 16), i.e., testing the very bottom position for layer i. Afterwards, the above procedure can make use of the preprocessed cost values without having to run the cost function and without iterating over the full stack again.

5   Benchmarking

We implemented our ordering algorithm (UpwardsOpt) as well as the state-of-the-art algorithm (BestFirst+TwoOpt) as a Python 3.6 backends to a Tableau v.2019 chart. All benchmarks were run on a 2017 27" iMac 5K with a 3.4 GHz Intel Core i5 processor and 40 GB RAM. The datasets used in our benchmarks were chosen to span the different possibilities from only a few time series with many time points, all the way to many time series with only a few time points. They can be found at https://vis-au.github.io/stackedcharts/.

We restricted the benchmark to two cases: optimising only for flatness (minimising wiggle) and optimising only for straightness (minimising bumps). The significance exponent was set to $s = 1$. We used only outer lines – i.e., $\alpha = 0.5, \beta = 0.0, \gamma = 0.5$ – and a 1% threshold of minimum improvement. As a neutral reference point, we generated $100,000$ randomly ordered stacks for each dataset and averaged their $cost_{chart}$ values. We then computed the optimised orderings using BestFirst only, the combination of BestFirst+TwoOpt, as well as our algorithm UpwardsOpt. Their $cost_{chart}$ values were then set in relation to the averaged values to see how much each improves over the average random order. We further logged the runtimes of BestFirst+TwoOpt and of UpwardsOpt.

The results are documented in Table 1 and Fig. 8. In terms of quality and speed, we can observe that all trends persist for both, flatness and straightness. We can also observe that UpwardsOpt produces better, but slower outputs than BestFirst+TwoOpt throughout all datasets. The use of the BestFirst heuristic by itself produces very mixed results, from close to optimal orderings (e.g., for *movies*) to worse than the average random ordering (e.g., for *hotel*).

Quality-wise, UpwardsOpt performs only slightly better than BestFirst+textttTwoOpt for datasets with only few layers (e.g., for *unempl* or *sandy*), as well as for rather similar layers that do not exhibit much individual traits (e.g., for *liquor*). In both cases, the search space is not that large for both algorithms

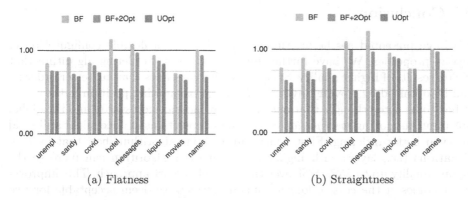

(a) Flatness   (b) Straightness

Fig. 8. Relative layout costs from Table 1. Lower is better.

to find much different solutions – either because there are only a few layers to reorder in the first place, or because no reordering would much affect the outcome. Yet, for some datasets that have a mix of longer and shorter layers (e.g., *messages*), UpwardsOpt improves significantly over BestFirst+TwoOpt. BestFirst will pick the shorter layers first, because they barely increase the overall cost. But in the end, only longer layers remain and will be placed on top of the shorter ones, creating a far from optimal starting point for TwoOpt.

Runtime-wise, we see that UpwardsOpt takes roughly about one order of magnitude more time to complete than BestFirst+TwoOpt. The only exception is the smallest dataset *unempl*, for which no significant differences could be observed. The measured runtimes increase mainly with the number of layers, but they are also dependent on the structure of the dataset itself. An example is the *movies* dataset with 881 layers, but its runtime is well below the *liquor* dataset with only 695 layers. This is due to the fact that the layers in the *movies* dataset only span rather short time intervals. As a result, reordering them disturbs only a small part of the chart, so that fewer iterations of UpwardsOpt are needed.

Table 1. Results of our benchmarking. Lower values are better. Costs are relative: $cost = 1.00$ denotes the quality of an average random order derived from $100,000$ random trials, $cost = 0.00$ denotes perfect quality with no wiggle or bumps.

Dataset	n	m	Relative costs, flatness			Relative costs, straightness			Times (secs), flatness		Times (secs), straightness	
			BF	BF+2Opt	UOpt	BF	BF+2Opt	UOpt	BF+2Opt	UOpt	BF+2Opt	UOpt
Unempl	28	443	1.06	0.82	0.81	1.04	0.73	0.67	3.79	3.58	2.05	4.29
Sandy	183	33	0.92	0.73	0.69	0.89	0.74	0.65	2.59	15.67	1.77	11.81
Covid	206	113	0.86	0.83	0.74	0.81	0.77	0.65	4.51	59.36	4.71	69.75
Hotel	334	115	1.13	0.90	0.59	1.10	1.00	0.54	16.47	214.02	12.85	196.42
Messages	604	135	1.08	0.98	0.58	1.23	0.97	0.49	45.50	640.39	58.64	1002.37
Liquor	695	240	0.95	0.88	0.84	0.96	0.92	0.89	167.38	1014.19	180.45	1264.28
Movies	881	51	0.73	0.71	0.64	0.77	0.76	0.60	28.55	504.54	31.11	589.37
Names	1000	135	1.01	0.94	0.69	1.00	0.98	0.74	165.47	2500.90	178.67	2024.34

6   Conclusion

We have presented aesthetic criteria that allow for a flexible configuration of layout properties. We have further introduced a novel ordering algorithm that yields results of higher quality. Specifically, it guarantees that no better ordering can be obtained by moving any *individual layer* to another position in the chart. It has to be noted though, that our algorithm does not necessarily find the global optimum, as moving *multiple layers* to different positions at once might still yield an even better ordering. First benchmarks of our algorithm show that in ideal situations (i.e., layers with high fluctuations) our algorithm can increase the layout quality up to 25%–50% over the state-of-the-art approach. This improvement comes at the cost of longer runtimes, which we deem acceptable for two reasons: First, stacked area charts are hardly ever used for thousands of time series, so that runtimes usually remain tolerable. Second, stacked area charts are usually generated for presentation purposes. It is hence sensible to spend some computation time to yield ready-to-print charts that look their best.

References

1. American Standards Association: Time Series Charts: A Manual of Design and Construction. ASME (1938). https://hdl.handle.net/2027/wu.89083916932
2. Brinton, W.C.: Graphic Methods for Presenting Facts. The Ronald Press Company, New York (1914). https://archive.org/details/cu31924032626792
3. Byron, L., Wattenberg, M.: Stacked graphs-geometry & aesthetics. IEEE TVCG **14**(6), 1245–1252 (2008). https://doi.org/10.1109/TVCG.2008.166
4. Cuenca, E., Sallaberry, A., Wang, F.Y., Poncelet, P.: MultiStream: a multiresolution streamgraph approach to explore hierarchical time series. IEEE TVCG **24**(12), 3160–3173 (2018). https://doi.org/10.1109/TVCG.2018.2796591
5. Di Bartolomeo, M., Hu, Y.: There is more to Streamgraphs than movies: better aesthetics via ordering and lassoing. Comput. Graph. Forum **35**(3), 341–350 (2016). https://doi.org/10.1111/cgf.12910
6. Harrison, L., Yang, F., Franconeri, S., Chang, R.: Ranking visualizations of correlation using Weber's law. IEEE TVCG **20**(12), 1943–1952 (2014). https://doi.org/10.1109/TVCG.2014.2346979
7. Havre, S., Hetzler, E., Whitney, P., Nowell, L.: ThemeRiver: visualizing thematic changes in large document collections. IEEE TVCG **8**(1), 9–20 (2002). https://doi.org/10.1109/2945.981848
8. Heuer, H., Polizzotto, A., Marx, F., Breiter, A.: Visualization needs in computational social sciences. In: Proceedings of MuC 2019, pp. 463–468. ACM (2019). https://doi.org/10.1145/3340764.3344440
9. Hofmann, H., Vendettuoli, M.: Common angle plots as perception-true visualizations of categorical associations. IEEE TVCG **19**(12), 2297–2305 (2013). https://doi.org/10.1109/TVCG.2013.140
10. Huggett, R.: Multiple Line Graphs (2), pp. 43–46. Palgrave Macmillan, London (1990). https://doi.org/10.1007/978-1-349-11245-6_10
11. Liu, S., et al.: TIARA: interactive, topic-based visual text summarization and analysis. ACM TIST **3**(2), 25:1–28 (2012). https://doi.org/10.1145/2089094.2089101

12. Rodrigues, A.M.B., Barbosa, G.D.J., Lopes, H., Barbosa, S.D.J.: Comparing the effectiveness of visualizations of different data distributions. In: Proceedings of SIBGRAPI 2019, pp. 84–91. IEEE (2019). https://doi.org/10.1109/SIBGRAPI. 2019.00020

13. Thudt, A., et al.: Assessing the readability of stacked graphs. In: Proceedings of GI 2016, pp. 167–174 (2016). https://doi.org/10.20380/GI2016.21

14. Toledo, A., Sookhanaphibarn, K., Thawonmas, R., Rinaldo, F.: Evolutionary computation for label layout on unused space of stacked graphs. Int. Scholarly Res. Not. **12**, 139603:1–10 (2012). https://doi.org/10.5402/2012/139603

15. Wang, Y., Wu, T., Chen, Z., Luo, Q., Qu, H.: STAC: enhancing stacked graphs for time series analysis. In: Proceedings of IEEE PacificVis 2016, pp. 234–238. IEEE (2016). https://doi.org/10.1109/PACIFICVIS.2016.7465277

16. Wills, G.: Visualizing Time: Designing Graphical Representations for Statistical Data. Springer, New York (2012). https://doi.org/10.1007/978-0-387-77907-2

17. Wu, T., Wu, Y., Shi, C., Qu, H., Cui, W.: PieceStack: toward better understanding of stacked graphs. IEEE TVCG **22**(6), 1640–1651 (2016). https://doi.org/10.1109/ TVCG.2016.2534518

Interactive, Orthogonal Hyperedge Routing in Schematic Diagrams Assisted by Layout Automatisms

Stefan Helmke$^{(\boxtimes)}$ ⓘ, Bernhard Goetze ⓘ, Robert Scheffler ⓘ, and Gregor Wrobel ⓘ

R&D Department of Graph-Based Engineering Systems, Society for the Advancement of
Applied Computer Science, Volmerstraße 3, 12489 Berlin, Germany
{Helmke,goetze,scheffler,wrobel}@gfai.de

Abstract. Schematic diagrams are used in graph-based engineering systems. They focus mainly on the structure of the design object. Graph-based engineering systems help to solve a concrete design task. This is primarily realized by the application of domain-specific languages. The layout of schematic diagrams is of particular importance, and a neat representation is desirable. But automatically generated layouts cannot always fully match the intention of a modeler. To improve automatic layouts and enable a user-specific representation, an algorithm that allows interactive changes of the orthogonal hyperedge geometry was implemented. In this paper, we present this algorithm and give an overview of such interactions. Additionally, several reductions of the hyperedge geometry are shown. Furthermore, a local, automatic routing considering interactions on the hyperedge geometry is presented. The consideration of domain-specific semantics and the possibility of interactive changes is a new approach. All algorithms were implemented in a self-developed software framework.

Keywords: Graph drawing · Orthogonal hyperedge routing · Interactive changes

1 Introduction

Schematic diagrams are used in graph-based engineering systems (GES). They focus mainly on the structure of the design object. GES help to solve a specific design task. This is primarily realized by the application of domain-specific languages. A GES is model-centered and uses a meta-model. The design model contains real or abstract objects. It is graphically represented in the GES and allows interactive editing.

The interactive treatment of these models, i.e., the design process, is a major concern of our research. We developed procedures that support the modeler in the design process, and we implemented these procedures in a framework for GES [1]. Graphical languages, i.e., schematic diagrams, and their layout are of particular importance. Principles for designing effective visual notation [2] for the depiction of vertices and edges are used to reach a neat representation. Furthermore, common methods of graph drawing like the placement of vertices and the routing of edges are applied. These representations need to be embedded in their specific problem domain to acquire good readability, i.e., semantic

© The Author(s) 2021
A. Basu et al. (Eds.): Diagrams 2021, LNAI 12909, pp. 20–27, 2021.
https://doi.org/10.1007/978-3-030-86062-2_2

transparency. This greatly improves the clarity, recognizability, and interpretability of complex models.

In the past, automatic routing algorithms were developed [3] to fulfill domain-specific requirements. But automatically generated layouts cannot always fully match the modeler's intention. Therefore, automatic layouts, e.g., activated by the movement, mirroring, or rotating of vertices, should consider interactive changes made by the modeler on the hyperedge geometry, while simultaneously preserving a neat layout.

The consideration of domain-specific semantics and the possibility of interactive changes is a novel approach. In this paper, we focus on an orthogonal hyperedge representation to ensure good readability, and we describe automatic routings that consider manual changes.

In the following sections, we refer to the internal data structure ELADO (Extended Layout Data Model) [1] to describe hypergraphs. In this data structure, vertices and hyperedges are not immediately connected, they are connected via so-called pins. The hyperedge geometry contains branch points and segments, while segments contain bends and (horizontal or vertical) segment parts. To avoid redundant information, segments contain no transit points. Transit points are pseudo-bends connecting two adjacent segment parts with the same orientation: either horizontal or vertical. Additionally, hyperedges are connected and acyclic, i.e., the hyperedge geometry is a tree. The vertices of this tree are bends, branch points, and pins; the edges of this tree are segment parts.

Section 3 describes and classifies possible, interactive changes of hyperedges by moving horizontal and vertical segment parts. This results in reductions of unwanted states. In Sect. 4, we explain how automatic routings executed after the movement of vertices consider the interactive changes mentioned in Sect. 3. Therefore we improved the geometrically stable routing PartRoute [4], which is a modification of the automatic routing OrthoRoute presented in [3]. Section 5 concludes the paper.

2 Related Work

Fundamental algorithms for drawing graphs were shown in [5], which is widely considered as a standard reference for graph drawing. A common set of aesthetic criteria was defined to improve the readability of diagrams: the minimization of crossings, bends, the length of the edges, or the area occupied by a drawing [6]. In [7] metrics were proposed to quantify the aesthetic quality of diagrams. In [8] the authors show an algorithm minimizing the number of hyperedge crossings. Additionally, in [9] the authors show an approach that dynamically reorders the vertices within the layers to further reduce the number of crossings. In [10] two crossing counting algorithms that predict the number of crossings between orthogonally routed hyperedges are presented. But besides that, there are only a few procedures for orthogonal hyperedge routing. An example is [11], where two automatic routings are given, and one semi-automatic routing, which improves the layout by local transformations, is added. A routing for directed hypergraphs with layers is shown by Sander [12].

3   Interactive Routing of Hyperedges

We now observe interactive changes of the hyperedge geometry. It is allowed only to move horizontal segment parts vertically and vertical segment parts horizontally. There are interactions that change the geometry of a hyperedge but not its structure. We call them trivial interactions (Fig. 1 (a)). These change only the length of segment parts, i.e., segment parts can contract or expand. There are also nontrivial interactions that are characterized by the transformation, genesis, or removal of bends, branch points, or segment parts (Fig. 1 (b)). Therefore, it is necessary to classify all the possible interactions.

First of all, we denote by sp the moved segment part, or more precisely the segment part that is about to be moved. Furthermore, we define the satellites s_1 and s_2 of sp. These are the vertices incident to sp. Satellites can be bends, branch points, or pins. We call the line segment between a satellite before an interaction and the same satellite after an interaction the satellite's trajectory.

We now study two kinds of interactions: interactions not causing collisions and interactions causing collisions with adjacent segment parts (Fig. 1 (a), (b)). These interactions depend completely on the satellites and the segment parts that are adjacent to sp. The type of a satellite is then given by the triple $(n_1, n_2, n_3) \in \{0, 1\}^3$ where n_1 is the number of adjacent segment parts in the direction of movement of sp, n_2 is the number of adjacent segment parts in the opposite direction of movement, and n_3 is the number of adjacent segment parts parallel to sp. It follows that there are 7 types of satellites. According to ELADO, the type $(0, 0, 1)$ (transit point) is not permitted.

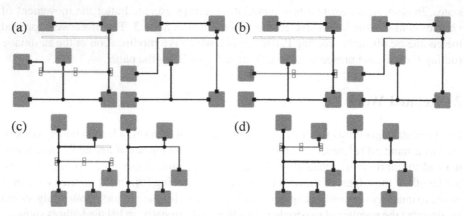

Fig. 1. (a) The right satellite collides with an adjacent segment part (trivial interaction). (b) The left satellite is a pin. Two bend points and a segment part are generated (nontrivial interaction). The right satellite collides with an adjacent segment part. (c) The left satellite collides with a branch point, sp collides with a bend. (d) The left satellite collides with a branch point. The right satellite collides with a segment part.

A segment part sp can also be moved so that sp or the satellites collide with other bends, branch points, or segment parts that are not adjacent to satellites (Fig. 1 (c), (d)). The following types of collisions are then allowed: a satellite collides with a bend or a branch point, a satellite collides with a segment part, or sp itself collides with a bend or a branch point.

First, we classify the bends or branch points colliding with a satellite depending on the segment parts that are incident to those bends or branch points. We define the type of such a colliding bend or branch point as the quadruplet $(n_1, n_2, n_3, n_4) \in \{0, 1\}^4$ where n_1 is the number of incident segment parts in the direction of movement of sp, n_2 is the number of incident segment parts in the opposite direction of movement, n_3 is the number of incident segment parts colliding with sp, and n_4 is the number of incident segment parts that are parallel to a potentially colliding segment part. There are 9 possible types of colliding bends or branch points. According to ELADO, the types $(0, 0, 0, 0)$ (isolated point); $(1, 0, 0, 0)$, $(0, 1, 0, 0)$, $(0, 0, 1, 0)$, $(0, 0, 0, 1)$ (pins); and $(1, 1, 0, 0)$, $(0, 0, 1, 1)$ (transit points) are not permitted.

Second, we describe the bends or branch points colliding with sp itself depending on the segment parts that are incident to them. The type of such a colliding bend or branch point is then given by the triple $(n_1, n_2, n_3) \in \{0, 1\}^2 \times \{0, 1, 2\}$ where n_1 is the number of incident segment parts in the direction of movement of sp, n_2 is the number of incident segment parts in the opposite direction of movement, and n_3 is the number of incident segment parts colliding with sp. There are 6 possible types. According to ELADO, the types $(0, 0, 0)$ (isolated point); $(1, 0, 0)$, $(0, 1, 0)$, $(0, 0, 1)$ (pins); and $(1, 1, 0)$, $(0, 0, 2)$ (transit points) are not permitted. If a bend or branch point collides with sp, we call it an explicit collision. A bend or branch point that is located on a satellite's trajectory can also collide with a segment part. We then call it an implicit collision.

Third, we class the segment parts colliding with a satellite depending on their position relative to sp. It is easy to see that there are only three cases. A colliding segment part and sp can be parallel or orthogonal. In a special case, a satellite can also be moved exactly onto a crossing.

Without a grid, it is not easy for users to force collisions, because it is nearly impossible to move a segment part exactly onto an object. In this case, the position of sp is automatically corrected if the distance between sp and a colliding object is smaller than a defined parameter ε. It is possible that multiple collisions appear simultaneously. Then the collision with the smallest distance to the original position of sp is chosen to correct sp.

In the first phase of the algorithm, the observation phase, the types of the satellites are determined. According to the aforementioned classification, the colliding objects and the bends and branch points on the trajectories of the satellites are determined. In the second phase, a canonical form is created. The first aim is to reduce the types of collisions so that satellites can only collide with bends, branch points, or temporary transit points. If a satellite collides with a segment part (Fig. 1 (d)), then a temporary transit point is generated on this segment part, and this segment part is divided into two. If a satellite is moved onto a crossing, then a branch point is generated on this crossing, and the two involved segment parts are divided into four. The position of these generated transit points and branch points equals that of the satellite. The second aim of this phase is to connect sp with bends or temporary transit points. If a satellite has the type $(0, 0, 0)$, i.e., the satellite is a pin (Fig. 1 (b)), a transit point is generated on sp depending on the exact position of the mouse click and the length of sp. This transit point divides sp into two segment parts. If the type is not $(1, 0, 0)$ or $(0, 1, 0)$, i.e., the satellite is not a bend, a transit point is generated on the satellite including a segment part with length 0 to maintain the orthogonal layout. In the next phase, the satellites are moved in the direction of movement, and the segment parts originally generated with length 0 are expanded. The aim of this phase is to consider and handle collisions. If a bend or branch point b is located on a satellite's trajectory (i.e., implicit collision), and there is a segment part in the opposite direction of movement, then b is transformed into a branch point. In addition, the colliding segment part is divided into two segment parts, and these are connected with b. If a satellite collides with a bend, a branch point (Fig. 1 (c), (d)), or a temporary transit point b (Fig. 1 (a), (b), (d)), then b is connected with sp. If the satellite is connected with a segment part in the opposite direction of movement, then this segment part is connected with b, too. If sp collides with a bend or branch point b (Fig. 1 (c)), then b is transformed to a branch point. In addition, the colliding segment part is divided into two segment parts, and these are connected with b. At the end of this phase, the unwanted structures in the hyperedge geometry are removed: Transit points, geometrically identical segment parts (to avoid interfering lines), and segment parts with length 0 are removed; branch points with valence 2 are transformed into bends.

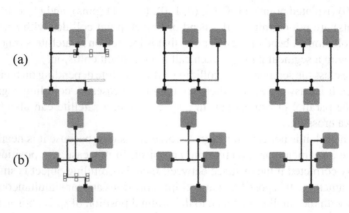

Fig. 2. (a) Automatic reduction of a cycle. (b) Automatic reduction of a stump.

In rare cases, some interactions can cause collisions that result in further unwanted states in the hyperedge geometry: cycles and stumps[1], which have to be removed automatically (Fig. 2). Of course, it is not predetermined how to reduce a cycle. The algorithm prefers to remove new, implicitly generated connections.

4   Automatic Routing Considering Manual Changes

In Sect. 3, we observed interactive changes of the hyperedge geometry to improve automatic routings. If another local, automatic routing is executed, e.g., by moving a vertex, users expect it to consider interactive changes. Otherwise, all interactive changes are lost. We improved the PartRoute method [4] to solve this task and explain the algorithm in the following.

---

[1] A stump is defined as a leaf, which is not a pin.

First, the PartRoute method is improved by choosing a flexible docking object; see Fig. 3. We define the critical segments as those segments that are incident to the pins of the moved vertex v, and we call the pins of v critical pins. The algorithm removes the critical segments and resulting transit points, while the rest of the hyperedge geometry remains. To find a suitable docking object, we search for the bend or branch point with the minimal Euclidean distance to the critical pin, and we denote it by b. Then a routing is executed in every cardinal direction. The type of the routing depends on the free directions of b. In the free directions, b is a potential docking object. But if a segment part is incident, i.e., this is not a free direction, a movable branch point is added on this segment part. This branch point is another potential docking object. Horizontal segment parts require two routings starting in this branch point with a northern or a southern segment part, and vertical segment parts require two routings starting in this branch point with an eastern or a western segment part. Therefore, there are up to 8 routings and 4 potential docking objects if b has degree 4 and no free directions. From these routings, the best one is chosen depending on the number of bends, the length of the critical segment, and the distance to other vertices. The segment part with the minimal Euclidean distance to the critical pin may also have a smaller distance to the critical pin than the nearest bend or branch point b, though this segment part is not incident to b. In this case, this segment part is an additional, potential docking object.

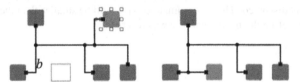

Fig. 3. PartRoute with flexible docking objects. The nearest bend or branch point is b. It has two free directions (east and south), in each of which a routing is performed. Additionally, two routings are executed on the western and northern segment part each time. The best routing is the one starting in b with an eastern segment part.

Having improved the PartRoute method by choosing flexible docking objects, the next task was to improve it in such a way that interactive changes of the hyperedge geometry are considered when a moved segment part sp is located on a critical segment. W.l.o.g., sp is vertical. If sp is horizontal the procedure is equivalent. The x-coordinate of sp should be fixed, while the length of sp is still flexible (Fig. 4 (a), (b)). The locked state of sp is dissolved when the routing is unsatisfying in terms of the length and the number of bends (Fig. 4 (c)). The specific approach depends on the position of the critical pin relative to sp.

We denote by s_1 the satellite with a smaller distance (in the tree) to the critical pin. The other satellite is denoted by s_2. Because sp is vertical the x-coordinate of s_1 is fixed, but the y-coordinate of s_1 is flexible. The coordinates of s_2 are also fixed. After a vertex v is moved, the critical pin is connected with s_1 in a way that this path minimizes the number of bends and the length of the path. This determines the y-coordinate of s_1. If the order of the y-coordinates of s_1 and s_2 does not change, then sp contracts, expands, or does not change its length (Fig. 4 (a), (b)). Otherwise, the locked state of *sp* is dissolved, and the improved PartRoute method is used (Fig. 4 (c)).

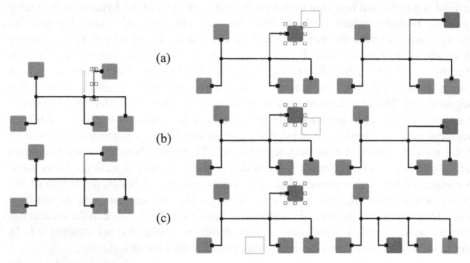

Fig. 4. PartRoute considering moved segment part *sp*. (a) *sp* expands. (b) *sp* contracts. (c) The locked state of *sp* is dissolved. The y-coordinate of s_1 would be smaller than the y-coordinate of s_2, while the order before the movement of v was reversed.

5 Conclusion

In Sect. 3, we show several possibilities of how users can manipulate an automatic routing by interactively changing the hyperedge geometry. A result of these manual changes is the automatic reduction of a hyperedge geometry to avoid unwanted states. In Sect. 4, we give several improvements to the PartRoute method for a more aesthetic routing minimizing the length and the number of bends. Furthermore, we present a modification of the PartRoute algorithm that considers manual interactions.

One challenge is that the PartRoute method and the OrthoRoute method differ significantly in the degree of geometric flexibility. Therefore, we want to improve the PartRoute algorithm so that it can automatically determine how geometrically flexible a routing can be depending on the specific situation. A future task is to create a geometrically flexible, automatic routing considering interactive changes—a dynamic, partial routing.

Acknowledgments. This R&D project of the Society for the Advancement of Applied Computer Science (GFaI) is supported by the funding program Innovation Competence (INNO-KOM) of the German Federal Ministry for Economic Affairs and Energy (BMWi), based on a resolution of the German Parliament.

References

1. Wrobel, G., Ebert, R.-E., Pleßow, M.: Graph-based engineering systems – a family of software applications and their underlying framework. Electronic Communications EASST Volume 6: Graph Transformation and Visual Modeling Techniques 2007, vol. 6 (2007). https://doi.org/10.14279/tuj.eceasst.6.50
2. Moody, D.: The "physics" of notations: toward a scientific basis for constructing visual notations in software engineering. IIEEE Trans. Software Eng. **35**, 756–779 (2009). https://doi.org/10.1109/TSE.2009.67
3. Goetze, B.: OrthoRoute – Ein gitterloses Verfahren zur Generierung orthogonaler Verbindungen in Schaltplänen. In: Roller, D., Opletal, S. (eds.) Tagungsband Workshop Elektrotechnik CAD 2007, pp. 13–32. Shaker, Aachen (2007)
4. Scheffler, R., Goetze, B.: PartRoute – Geometrisch stabiles Routing orthogonaler Verbindungen in Schaltplänen. In: Roller, D., Opletal, S. (eds.) Tagungsband Workshop Elektrotechnik CAD 2007, pp. 33–42. Shaker, Aachen (2007)
5. Di Battista, G., Eades, P., Tamassia, R., Tollis, I.G.: Graph Drawing. Algorithms for the Visualization of Graphs. Prentice Hall, Upper Saddle River (1999)
6. Tamassia, R., Di Battista, G., Batini, C.: Automatic graph drawing and readability of diagrams. IEEE Trans. Syst. Man Cybern. **18**, 61–79 (1988). https://doi.org/10.1109/21.87055
7. Purchase, H.C.: Metrics for graph drawing aesthetics. J. Vis. Lang. Comput. **13**, 501–516 (2002). https://doi.org/10.1006/jvlc.2002.0232
8. Eschbach, T., Günther, W., Becker, B.: Orthogonal hypergraph routing for improved visibility. In: Proceedings of the 14th ACM Great Lakes Symposium on VLSI 2004, Boston, MA, USA, 26–28 April 2004, pp. 385–388 (2004). https://doi.org/10.1145/988952.989045
9. Eschbach, T., Guenther, W., Becker, B.: Orthogonal hypergraph drawing for improved visibility. JGAA **10**, 141–157 (2006). https://doi.org/10.7155/jgaa.00122
10. Spönemann, M., Schulze, C.D., Rüegg, U., von Hanxleden, R.: Counting crossings for layered hypergraphs. In: Dwyer, T., Purchase, H., Delaney, A. (eds.) Diagrams 2014. LNCS (LNAI), vol. 8578, pp. 9–15. Springer, Heidelberg (2014). https://doi.org/10.1007/978-3-662-44043-8_2
11. Wybrow, M., Marriott, K., Stuckey, P.J.: Orthogonal hyperedge routing. In: Cox, P., Plimmer, B., Rodgers, P. (eds.) Diagrams 2012. LNCS (LNAI), vol. 7352, pp. 51–64. Springer, Heidelberg (2012). https://doi.org/10.1007/978-3-642-31223-6_10
12. Sander, G.: Layout of directed hypergraphs with orthogonal hyperedges. In: Liotta, G. (ed.) GD 2003. LNCS, vol. 2912, pp. 381–386. Springer, Heidelberg (2004). https://doi.org/10.1007/978-3-540-24595-7_35

Open Access This chapter is licensed under the terms of the Creative Commons Attribution 4.0 International License (http://creativecommons.org/licenses/by/4.0/), which permits use, sharing, adaptation, distribution and reproduction in any medium or format, as long as you give appropriate credit to the original author(s) and the source, provide a link to the Creative Commons license and indicate if changes were made.

The images or other third party material in this chapter are included in the chapter's Creative Commons license, unless indicated otherwise in a credit line to the material. If material is not included in the chapter's Creative Commons license and your intended use is not permitted by statutory regulation or exceeds the permitted use, you will need to obtain permission directly from the copyright holder.

Evidence of Chunking in a Simple Drawing Task

Yanze Liu(✉) ⓘ and Peter C-H. Cheng ⓘ

University of Sussex, Brighton BN1 9RH, UK
{yl680,p.c.h.cheng}@sussex.ac.uk

Abstract. Are perceptual chunks in memory central to the process of drawing? This study adopts a simple transcription drawing task in which patterns of dots are viewed and reproduced. Data from two experiments (Haladjian & Mathy, 2015) are re-analysed for evidence of chunking. Chunking was evident with long stimuli exposure time but not short ($\leq$200 ms). With more opportunity to chunk, various temporal and spatial signals suggest the occurrence chunking, including: actions are temporally clustering into groups with sizes of typical chunks; pauses between actions are longer with long exposure than with short exposure; spatial locations of responses are sometimes clustered as simple geometric shapes; clusters of responses are more likely to occur at the start or end of trials.

Keywords: Drawing diagrams · Perceptual chunks · Spatial memory

1 Introduction

There is large literature on how people perceive and reason with diagrams. In contrast, drawing diagrams has been rather neglected but its importance is shown in some cognitive science studies, including: van Sommers's work on low level cognition in drawing [7]; Cheng and colleagues' studies on higher level drawing strategies [5,6]. Therefore, our research aims to understand how people draw diagrams, with a particular focus on understanding the strategies they use.

Of particular interest is whether *chunking* has a central role in drawing. Chunking, a fundamental explanatory concept in cognitive science [3], is a mechanism that collects associated information into groups, which can greatly reduce our cognitive effort to hold and recall relatively long streams of information. Depending on the demands of the task, the size of chunks varies. In memory span tasks, Miller suggested the size is $\approx$7 items [3], Cowan proposed 4 as the size of working memory [2]. Additionally, individual differences may substantially impact the chunk size [1].

Drawing is the production of representations that have spatial formats (2D), which contrast with writing that is the production of linear (1D) sentential notations. Copying is one of several modes of production, which are orthogonal to the drawing-writing distinction. Cheng and collages have shown that chunking

© Springer Nature Switzerland AG 2021
A. Basu et al. (Eds.): Diagrams 2021, LNAI 12909, pp. 28–31, 2021.
https://doi.org/10.1007/978-3-030-86062-2_3

explains the strategies used in the copying of relatively complex geometric diagrams [5,6]. Might simpler strategies that do not exploit chunking be used in more basic drawing tasks? For instance, when quickly presented with an image to copy, people may simply hold it in their visuospatial working memory and then pick out elements to do the drawing. This will be inaccurate with large numbers of elements or when the image fades. If sufficient time is given, will people start to build up chunks in short term memory to enhance their memory for the target image? Moreover, if people build chunks, do they just pick out nearby items? Or will they encode the items with familiar geometric patterns to reduce the complexity of the stimulus?

Haladjian and Mathy [4] conducted a study on a localisation memorising task to investigate how people encode spatial information and whether stimulus exposure duration improves memorisation accuracy. Fortunately, their task and data are suitable for investigating our questions about when the chunking would happen in simple drawing tasks. In their experimental task participants were shown a stimulus consisting of a pattern of dots for a designated duration. After the stimulus disappeared the participants placed a marker on a blank computer screen for each object they saw using a mouse. Here, we take this dot pattern reproduction task to be a (very) simple form of drawing.

Haladjian and Mathy [4] hypothesised that if participants encode individual items or individual groups, then long stimulus exposure will allow rehearsal and so improve the memorisation accuracy of spatial information. However, their result suggested long exposure did not improve localisation accuracy substantively nor did grouped stimuli cancel out a global spatial compression effect. Thus, participants did not encode items or groups individually, but instead encoded stimuli as a complete 'snapshot'. In contrast to Haladjian and Mathy, we focus on lower-level patterns of dots – chunks – rather than the whole stimuli or their group level. In particular, is there evidence of chunking in this simple drawing task? Does the use of chunking change with exposure times?

Analyses of temporal and spatial signals were conducted, and the results suggested that chunking is important in this simple drawing task. For temporal signals: (1) the correlations of the pauses between actions and distances moved in trials which used ungrouped stimuli were weaker compared with trials which used grouped stimuli; (2) the thinking pause in long exposure trials were significantly longer than those in short exposure trials; (3) the mean values of thinking pauses of after-cluster responses were longer than means of within-cluster responses. For spatial signals: (4) spatial locations of responses were sometimes clustered as simple geometric shapes; (5) clusters of responses were more likely to occur at the start or end of trials. We will present one of the analyses as an example in this summary paper.

2   Sample Analysis: Inferring Potential Chunks from Pauses

This analysis attempts to identify clusters of responses from patterns of pause that may reflect the structure of chunks. As pauses at the beginning of chunks are substantially longer than pauses within chunks [5,6], we can attempt to identify chunks from the long pauses in a trial. There were two stimulus types in Haldajian and Mathy's experiments: ungrouped (UG) stimuli with dots in one cluster; grouped (G) stimuli with the dots split into two clusters. In this analysis, we only used ungrouped data to minimise the effect of mouse moving on the pause. Firstly, we should mention that the first response in a trial is particularly long, because it includes the initial reaction time prior to responding. The first responses have longer pauses (M = 976 ms, SD = 540) compared with the second to the fourth responses (M = 634, SD = 389), t(71678) = 91.9, p < .0001. In this analysis, we treated the first response as the beginning of a chunk. By identifying clusters, which are potentially chunks, from the long pauses and the successive short pauses, we can analyse the frequency of cluster of different sizes. Our hypothesis is that if participants did not use chunking the size of clusters would not exhibit distributions in sizes that are characteristic of chunks.

The procedure of cluster size frequency analysis is as follows. (1) Mark pauses whose z-scores are larger than 1 as long pauses. (2) Use the number of responses from one long pause to another long pause as the size of the cluster (for example, if 1st pause and 5th pause are long, then the size of this cluster is 4). (3) For each group of trials, count the numbers of clusters of each cluster size. (4) Normalise the numbers of clusters using corresponding numbers of trials to get the number of clusters per trial, because trial groups contain different numbers of trials.

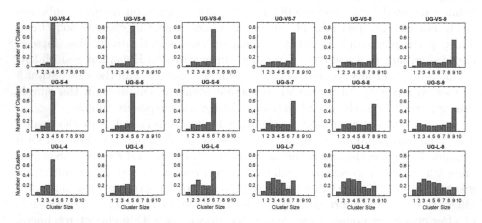

Fig. 1. Frequency of sizes of potential chunk for ungrouped trial data of EXP1 and EXP2. The titles of bar charts are the names of trial groups. The VS, V and L in the titles stand for three display lengths: very short, 50 ms; short, 200 ms; long for ≥ 1 s (1 s per dot). The x-axis is the cluster size, the y-axis is the number of clusters per trial.

Figure 1 shows that participants used two different strategies, snapshot (no chunking) and chunking, in short/very short and long exposure trials respectively. For all the very short and short duration trials (top and middle rows in Fig. 1), the cluster size that equals the total number of responses is the most frequent and the smaller size clusters are relatively uniformly distributed. This reveals that most of the participants took the whole pattern of dots as one snap-shot in the memorising stage, then did the responses in one uniform sequence using the snapshot in memory after the stimulus disappeared.

In marked contrast, for long trials L-6 to L-9 (bottom row), the small size clusters are more frequent than the corresponding VS or S trials, and the largest cluster size are less common. This shows that less participants used the snapshot strategy in long exposure trials compared with VS or S trials. Instead, the long exposure time allows the participants to split the whole stimulus into small groups and to store the groups separately in different chunks. In the response stage, participants reproduced the stimulus piece by piece using the chunks in memory. Notably, for L-7 to L-9, the two most common cluster sizes are 3 and 4, which is consistent with typical chunk sizes in everyday tasks [2]. This suggests that the clusters in L-7 to L-9 are actually chunks. With the increasing number of responses, there is more opportunity for chunking, which is reflected in the changing shape of the distributions (left to right, bottom row, Fig. 1).

3   Discussion

This study focuses on when people use perceptual chunks in a simple transcription drawing task and how they encode the spatial information into chunks. By examining temporal, spatial and combined spatiotemporal signals, we found chunking exists with long stimuli exposure time but not short, and participants tended to use regular geometric patterns to build up chunks. In very short and short exposure trials, participants held the stimulus as a mental image (snapshot). While, in long exposure trials, besides holding a mental image, participants were also likely to build chunks. Moreover, for the chunking strategies, participants spotted patterns where the dots are nodes in regular shapes.

Acknowledgment. We would like to thank Haladjian and Mathy for providing useful raw data of their experiments.

References

1. Cheng, P., Albehaijan, N.: Some determinants of chunk size in sequential behavior: individual differences in the transcription of alphanumeric strings. In: CogSci (2020)
2. Cowan, N.: The magical number 4 in short-term memory: a reconsideration of mental storage capacity. Behav. Brain Sci. **24**(1), 87–114 (2001)
3. Miller, G.A.: The magical number seven, plus or minus two: some limits on our capacity for processing information. Psychol. Rev. **101**(2), 343 (1994)
4. Haladjian, H.H., Mathy, F.: A snapshot is all it takes to encode object locations into spatial memory. Vision Res. **107**, 133–145 (2015)
5. Obaidellah, U.H., Cheng, P.C.: The role of chunking in drawing rey complex figure. Percept. Mot. Skills **120**(2), 535–555 (2015)
6. Roller, R., Cheng, P.: Observed strategies in the freehand drawing of complex hierarchical diagrams. In: Proceedings of the Annual Meeting of the Cognitive Science Society. vol. 36 (2014)
7. Van Sommers, P.: Drawing and Cognition: Descriptive and Experimental Studies of Graphic Production Processes. Cambridge University Press, Cambridge (1984)

Theory of Diagrams

Considerations in Representation Selection for Problem Solving: A Review

Aaron Stockdill[1]([⊠])([iD]), Daniel Raggi[1]([iD]), Mateja Jamnik[1]([iD]),
Grecia Garcia Garcia[2]([iD]), and Peter C.-H. Cheng[2]([iD])

[1] University of Cambridge, Cambridge, UK
{aaron.stockdill,daniel.raggi,mateja.jamnik}@cl.cam.ac.uk
[2] University of Sussex, Brighton, UK
{g.garcia-garcia,p.c.h.cheng}@sussex.ac.uk

Abstract. Choosing how to represent knowledge effectively is a long-standing open problem. Cognitive science has shed light on the taxonomisation of representational systems from the perspective of cognitive processes, but a similar analysis is absent from the perspective of *problem solving*, where the representations are employed. In this paper we review how representation choices are made for solving problems in the context of theorem proving from three perspectives: cognition, heterogeneity, and computational demands. We contrast the different factors that are most important for each perspective in the context of problem solving to produce a list of considerations for developers of problem solving tools regarding representations that are appropriate for particular users and effective for specific problem domains.

Keywords: Representations · Problem solving · Theorem proving

1 Introduction

Problem solving is a fundamental activity for intelligent agents – people included. The ability to solve problems is dependent on *expertise*, but what is expertise? Can we leverage expertise with one type of problem to solve others? And how can we support problem solvers in choosing the right representation to make their task easier? This paper reviews the current research on representations for problem solving, both from a human-centred perspective, and from a software-centred perspective. Our goal is to understand how we can support problem solvers, and use this knowledge to inform the design of problem solving software – specifically, theorem provers – by compiling a list of considerations for software developers.

We begin this paper in Sect. 2 by exploring the cognitive aspects of the human reasoning system: how people understand and solve problems, and how expertise affects the solving process. In Sect. 3 we consider how representations interplay with human reasoning: the different modalities, their effectiveness, and how human reasoners choose representations for their problems. Diagrammatic representations in particular exhibit many of the aspects identified. From the software

ⓒ Springer Nature Switzerland AG 2021
A. Basu et al. (Eds.): Diagrams 2021, LNAI 12909, pp. 35–51, 2021.
https://doi.org/10.1007/978-3-030-86062-2_4

angle, we consider how problems are solved by automated and interactive theorem provers; Sect. 4 explores different types of theorem provers, and how some incorporate multiple representations to varying degrees. In Sect. 5 we analyse the cognitive aspects of these systems, and conclude with a list of considerations for software designers to better align their software with the cognitive needs of the user.

2 Cognitive Factors in Problem Solving

Whilst the importance of choosing the right representation for a problem and a person attempting to solve the problem has long been established [28], computationally choosing the right representation remains an open problem. Much taxonomisation of knowledge representations has been done before, but little in the context of problem solving. Some people are able to solve problems more effectively than others, exhibiting expertise in particular domains. But changing representations, adapting and updating problem solving strategies remains hard for people – it seems these are the skills of experts. We focus on *problem solving* because of its general nature: there is an initial state, some way of identify goal states, and actions that can be taken that modify the state. A wide array of tasks can be modelled as problem solving, so we wish to understand how human experts model problems in order to solve them, and how expertise is related.

2.1 Problem Solving

Solving a problem is conjectured to be a tight loop of understanding, planning, executing, and evaluating progress until a condition is met [28]. Pólya's influential work on problem solving, *How to Solve It: A New Aspect of Mathematical Method*, lays out these four steps clearly, presents many varied examples of each step, and exemplifies the loop in its entirety. A more formal treatment of problem solving comes from Simon and Newell, where they introduce the *problem space* [33]. The problem space is modelled as a (possibly infinite) graph, where nodes are the problem state and the edges are the actions that allow movement between them; a walk in this graph originating from the initial state and ending on a goal state is a solution to the problem. The nature of the problem, and the representation of the problem, determine the problem space. The person solving the problem must traverse the problem space.

In this paper, we shall work with Simon and Newell's model of problem solving as our grounding. We choose this model because it maps cleanly to common models of theorem proving, making our discussion more direct. Other ways to frame problem solving (such as Zhang's distributed cognition [47], or Johnson-Laird's mental models [18]) may also function as a suitable model for this discussion, but are beyond the scope of this paper.

When a person is traversing the problem space, some factors are fixed: the fundamentally serial information processing, small-capacity[1] but rapid-recall

---

[1] Famously, seven plus or minus two *chunks* [22].

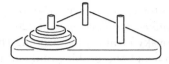

Fig. 1. A typical Tower of Hanoi puzzle. The goal is to move all three discs from the starting peg to another peg, without placing a larger disc atop a smaller disc.

short-term memory, and effectively infinite slow-recall long-term memory. But other factors are mutable, such as *how* the space is traversed. Kotovsky et al. presented participants with variations on the 'Towers of Hanoi' problem, recording how they interacted with the problem and made progress towards (and away from) the solution [19]. The Towers of Hanoi puzzle, depicted in Fig. 1, involves three discs with holes stacked atop one peg – a small disc on top, then a medium sized disc, then a large disc at the bottom. Alongside, there are two pegs without any discs. The goal is to move all three discs from the first peg to either of the remaining pegs such that they all end up in one single peg, with the condition that at no point in the process should a larger disc be on top of a smaller disc. Kotovsky et al. analysed how people perform when presented with isomorphic variants of the Towers of Hanoi puzzle, such as monsters-and-globes, boxes-and-dots, or acrobats-and-flagpoles, and with different types of action: either *moving* objects (as in moving a disk from one peg to another) or *changing* their size. Notably, representations involving unfamiliar scenarios (e.g., monsters rather than acrobats), and representations involving changing rather than moving, strongly hindered problem-solving performance, in spite of the problem being isomorphic. Moreover, when facing an unfamiliar problem, participants tended to *probe* the problem space: they would perform a short sequence of actions with minimal deviation from the planned sequence before returning to the initial state. After these probes had been completed, and the participants were satisfied with their ability to traverse the space, they applied short leaps of two actions chained together, rapidly converging on the goal state. These leaps achieved sub-goals, unblocking the next action [19].

Kotovsky et al.'s work has two implications: first, there are two distinct methods of traversing the problem space (the probing back-and-forth approach, and the rapid sub-goal chaining approach); and second, changing the representation of the problem without changing its nature made the Towers of Hanoi-like problems easier or harder. These two results are tightly coupled: the representation of the problem impacted how the participants were able to traverse the problem space, and the participants' relative expertise in the problem space affected how difficult they found the task. To better understand this, we must understand expertise.

2.2   Space Traversal and Expertise

Algorithmically, there are many ways to traverse a graph: breadth first search, depth first search, A* heuristic search, etc. While people are less procedural,

Larkin et al. identify two strategies that solvers use to traverse the problem space: *means-ends analysis* and *knowledge development* [20]. The former is similar to the behaviour seen by Kotovsky et al., using probing then sub-goal unblocking; the latter uses heuristics to avoid the probing and sub-goal analysis to immediately start chaining actions. Further, the solvers who use each strategy can be identified: means-ends analysis is indicative of *novices* in the problem domain, while *experts* employ knowledge development [20].

The strategy of *means-ends analysis*, which is employed by novices, is a type of 'working backwards'. The solver must identify what necessary conditions must be met to move towards the goal, and then work towards this new sub-goal [19]. Thus the novice begins to probe the problem space, understanding what effect their actions have, and then can begin to achieve their sub-goals. Maintaining this internal sub-goal chain is cognitively demanding, using working memory that could otherwise be devoted to the problem itself, not the 'traversal state'; even small problem spaces overwhelm human working memory [19].[2] Worse, the high cognitive load required to employ means-ends analysis can inhibit *schema acquisition*, a method of becoming an expert [40].

Expert problem solving is best modelled through *knowledge development*, in which powerful heuristics guide the expert through the problem space [40]. Because experts are familiar with the domain – and thus the problem space – there is little to no 'probing' phase; they have seen and solved similar problems in the past. Instead, experts can immediately begin applying *schemas*, which are patterns that the expert can recognise in the new problem space, and so immediately apply actions [40]. Not only does this approach eliminate the probing and sub-goal creation, this approach induces less cognitive load – the utilisation of working memory – than means-ends analysis [40]; experts will be faster *and* more cognitively efficient.

2.3   Cognitively Effective Representations

A representation is a view of a problem: the problem is expressed using some representation. The representation itself belongs to some representational *system*: a collection of syntax and semantics that generate some agreed-upon notation and interpretation. This is sometimes called an *external* representation because it exists outside the mind; there is a corresponding *internal* representation that exists within the mind of the problem solver [30]. Cheng links internal and external representations in two directions: an appropriate external representation can induce an effective internal representation, while an effective internal representation encourages external representation generation [7].

With a diverse range of representational systems at our disposal, some with more diagrammatic aspects than others, we must consider: what makes a representation *effective*? In the context of problem solving, there are quantifiable

---

[2] By analogy to computers, we devote 'registers' that would otherwise be used on the problem to maintaining the 'call stack', but the human brain's 'call stack' capacity is small, and – due to the nature of graph search – easy to overflow.

results we might be interested in: lower cognitive load, shorter times to generate a solution, shorter solution paths. But in this subsection we look at the representations themselves, not the results they generate: in order to achieve these results, what properties do our representations have?

We consider effective external representations in relation to the internal representations they induce. Green and Blackwell created the 'Cognitive Dimensions' framework as a guide on creating representations, but note that it is not intended for a deep analysis of existing representations [13].[3] Instead we here consider the 5 general criteria containing the more specific 19 criteria identified by Cheng [7] for effective representations. These are: direct encoding, low-cost inference, conceptual transparency, generality, and conceptual-syntactic compatibility. In the next section, we will compare them to diagrammatic aspects of representations.

Direct Encoding. Cheng's first criterion for effective representations is that it *directly encodes* the types, structures, and relations of the problem [7]. This is the same benefit that diagrammatic representational systems provide. But *why* is this necessary for a representation to be effective? Consider, for example, Duncker's 'candle problem': given a box of tacks, some matches, and a candle, attach the candle to the wall [45]. Participants will attempt to tack the candle to the wall, or melt some wax to use as glue, neither being effective; rarely do they consider they could pin the tack box to the wall and sit the candle in the box [45]. Condell et al. call this inability to re-contextualise the tack box *functional fixedness*: the 'type' of the box is wrong, since in the 'representation' people have, the box is a container for tacks, not a container for candles as required for the solution [8]. Tversky highlights a similar point in regards to structure: people have a mental hierarchy to categorise their environment, and benefit when the representation follows the same hierarchy [41].[4] Thus a representation that more directly encodes a problem is likely more effective than those that encode the problem indirectly.

Low-Cost Inference. The cost of inference in representations is a combination of factors: while the inferential actions themselves should be low-cost to perform, they must also be low-cost to identify [7]. In diagrammatic representational systems, this is a mixture of geometric and spatial aspects, syntactic constraints, and syntactic plasticity. One notable variety of low-cost inference is the *free ride* – an inference that can be made without specifically taking steps to make that inference [31]. Stapleton et al. generalise this to *observational advantages*: some representations allow information to be observed 'for free' that would require purposeful inference in other representations [35]. 'Free' is certainly low-cost; representations exhibiting observational advantages are likely to be more effective than their disadvantaged counterparts.

---

[3] Although work that builds upon these dimensions (e.g., [4]) often includes concepts very close to those we are about to discuss.

[4] In this case, the hierarchy is that the box is restricted to tacks, and there is no hierarchical relationship to the candle; the *necessary* hierarchy has box restricted to *objects*, which includes tacks and candles.

Conceptual Transparency. More difficult to define, conceptual transparency is the ability to 'see through' the representation to its underlying meaning. Cheng decomposes conceptual transparency into five aspects: coherence and unambiguity; small conceptual gulf; integration of conceptual perspectives; integration of granularity scales; and the comparing and contrasting of typical, special, and extreme cases [6]. These are themselves difficult to resolve, and are beyond the scope of this review.

Generality. It is desirable that a representational system allows us to represent a variety of situations (generality) and, equally important, that there are available a variety of syntactic operations that allow us to traverse the space (generativity) towards desirable goals. This relates to the syntactic power of the systemdiscussed next.

Conceptual-Syntactic Compatibility. Cheng's final criterion for effective representations is conceptual-syntactic compatibility. In diagrammatic representational systems, this is related to the idea of syntactic constraints: a close relationship between 'expressible' and 'valid' results in a more effective system [7]. By analogy, in computer science we discuss making illegal states unrepresentable [23] for the same effect: if you *cannot* say something incorrect, then you have reduced the ways in which you can make a mistake. Other mechanisms through which representations have conceptual-syntactic compatibility is by having a construction process which mirrors the problem solving process, and by allowing for distinct phases in encoding, interpreting, and making inferences [7].

In the problem solving context, we can consider an effective representation to be one that provides a problem space in which the solver is sufficiently expert: they already have access to low-cost inferences and powerful schemas. All the above criteria contribute to making the space easier to traverse for the solver. However, it is worth noting that they are not independent. For example, while a general representational system might be more likely to guarantee the existence of a solution, the problem space will often be more 'branchy', making correct inferences more costly. This compromise between generality and low-cost inference is captured by Cheng's concept of *syntactic plasticity* [7]. This sort of compromise indicates that choosing an effective representation is a complex optimisation problem where many competing factors have to be weighted. Furthermore, often they cannot be weighted equally for solvers with differing levels of expertise.

Novices and experts alike solve problems by traversing a problem space, applying actions to change state within the space such that they eventually reach a goal state. But their traversal methods are very different: novices have a costly, means-ends analysis approach to searching the problem space; experts apply powerful heuristics called schemas to efficiently work from the start to the goal. Clearly, being an expert is advantageous: can we somehow transfer these advantages to a novice? Or perhaps, can we change the problem space so that our novice is already expert?

3 Heterogeneity of Representations

As Kotovsky et al. discovered, the way a problem is represented can significantly impact how difficult the problem is to solve [19]. But why is this, and what exactly is involved in the representation of a problem? In this section we consider different modalities of representations, and what it means for a problem to be represented effectively.

3.1 Diagrammatic Aspects of Representations

Restricting ourselves to external representations, we can attempt to classify representations: a common distinction is between 'sentential' – a sequence of characters composed only through concatenation [37] – and 'diagrammatic' representations.[5] Despite their apparent value of '10 000 words' [21], diagrammatic representations are often second-class in mathematics, even in highly visual domains such as graph theory. Only in the last 25 years have diagrammatic representational systems begun to be treated formally to address this gap. Informally, diagrammatic representations are widely used by mathematicians; formally, diagrams are often stripped from the discussion, because mathematicians consider them unsuitable for proof [16]. Even educational materials such as textbooks often present only sentential solutions to problems, obscuring any intuition that a diagram can provide [46]. Perhaps it is because diagrammatic systems are difficult to define: what makes a diagrammatic representation *diagrammatic*?

Taken in the extremes, there is obvious consensus around which representations are 'sentential' and which are 'diagrammatic': in mathematics, standard propositional logic notation is sentential, while Euler diagrams are diagrammatic. But as we drift away from these extremes, the boundary becomes indistinct: positioning limits on a summation is not concatenative, and hints towards some vertical-positioning relationship; a table filled with words uses space and positioning to encode information, but uses strings extensively. The distinction is difficult because, as Giardino observes, there is no sharp distinction to be made [10]. Representations exist on a continuum, some with more diagrammatic aspects than others; when we discuss diagrammatic representations we are referring to representations exhibiting four diagrammatic aspects: direct encoding, syntactic constraints, syntactic plasticity, and heavy use of geometric and spatial attributes and relations. Let us consider each of these in more detail.

Direct Encoding. Diagrammatic representations directly encode the types, structures, and relations of the problem, rather than using some indirect association as in sentential representations [37]. Consider a relation 'to the right of': we can easily state an instance of this sententially:

$$a \text{ is to the right of } b$$

---

[5] We consider only *visual* representations; representations that are audial or tactile, for example, are beyond the scope of this review.

while observing that a is visibly *left* of b. By comparison,

$$b \qquad a$$

is a more *direct* encoding: a is literally to the right of b. This extends to all levels: rather than using the word 'square', diagrams can include squares; rather than explaining how nodes and edges form a graph, we can draw the graph. But this can also enforce specificity: we can sententially state that 'a zebra has some stripes', without making any claim to how many stripes, but any particular drawing of a zebra has a *fixed number* of stripes. This makes the representation easier to process at the cost of reducing generality [38].

Syntactic Constraints. Shimojima observed that rules of a representational system come in two broad classes: intrinsic, and extrinsic[6] [32]. An *intrinsic* (or *syntactic*) constraint is imposed by the syntax of the representational system: the geometry, topology, or physics of the representation enforces the rules. An *extrinsic* constraint is imposed by the problem solver: the representational system allows for statements that the solver wishes to avoid. Going back to our 'to the right of' example, let us assume a system where we can write $a >_r b$, meaning a is to the right of b.[7] Then we can state the following three facts:

$$a >_r b, \; b >_r c, \text{ and } c >_r a$$

Now, if we try to represent this in our 'positional' notation from earlier, we hit an *intrinsic* constraint: we cannot arrange the letters on the page such that this is true! The representational system has prevented us from representing some state. On a plane, the sentential notation is too permissive: we failed to apply the *extrinsic* constraints necessary to identify a nonsense statement. But on a sphere, the positional representation is overly restrictive: the intrinsic constraints are preventing us from encoding a valid state.

Geometry and Space. Finally, diagrammatic representations make use of *geometry* and *space* [37]. The benefit of this is that it exploits the human visuo-spatial reasoning system – the Towers of Hanoi variants presented by Kotovsky et al. to participants consistently demonstrated that participants more efficiently solved the 'physically plausible' variants [19]. Humans evolved in a physical world that obeys particular rules: we are well-adapted to manage systems that follow these rules. But geometry and space are limiting; just as we identified in direct encoding and syntactic constraints, we forfeit *abstraction* and *generality* by following the physical rules.

Characterising diagrammatic representations by these properties exhibits why diagrams fit various criteria from Sect. 2.3 for effective representations.

---

[6] [32] identified variations on this divide, but all are sufficiently similar for our discussion.

[7] Note again that, visually, a is to the *left* of b.

3.2   Recommending a Representation

We have seen that representations can affect how difficult a problem is to solve and argued that diagrammatic representations often exhibit favourable aspects for problem solving. Undeniably, changing to an effective representation is useful [1,6,12] – the problem is that students do not necessarily *change* to a more effective representation [39]. Ideally, we want to support students to change representation to one that is appropriate for the problem and for their expertise. But how should they be guided to change towards effective representations?

In the restricted domain of extracting information from a database, Grawemeyer's External Representation Selection Tutor (ERST) was able to recommend an information visualisation to users to answer queries [12]. The visualisations were scatter plots, sector graphs, pie charts, bar charts, tables, and Euler diagrams; when supported by ERST in choosing an effective representation, participants were more effective at answering the queries [12]. But to consider tasks beyond information extraction, the literature on representation recommendation becomes scarce. To *solve* a problem, we explore the representation *design* literature: what factors are important when designing representational systems, which we may consider for representation recommendation?

Representation design recommendation is a product of three factors: what is the problem, who is approaching it, and why are they working on it? This combination of factors determines the *cognitive fit* of a representation [24,44]. Vessey introduces cognitive fit as the combination of the specific problem under consideration, and the overarching task and context in which the problem is encountered, which together influence the internal representation a person constructs [44]. But implicit in Vessey's discussion is that the *person* influences the internal representation; as we saw earlier, an expert and a novice will be operating with different internal representations: the novice's internal representation is tuned for search, and the expert's for heuristics [20]. Moody makes this explicit: cognitive fit is the interaction between the problem, the person, and the task [24].

A representation is a complex thing: it is an encoding of information into the real world, which induces an *internal* representation in people. A range of factors determine representational efficacy, and diagrammatic aspects of representations align to allow for effective representations. By considering cognitive fit, we can begin to understand how to recommend a representation based on the problem being solved, the person solving the problem, and the task and context in which the problem was encountered.

4   Computational Considerations of Representation

In the previous two sections we considered why and how representations are evaluated and recommended. In this section, we explore the use of representations in computational systems. While artificial intelligence researchers have attempted to build general problem solvers for a long time – consider the aptly named 'General Problem-Solving Program' [25] – most success has been had in solvers specialised to particular domains. We focus on interactive and automated theorem provers,

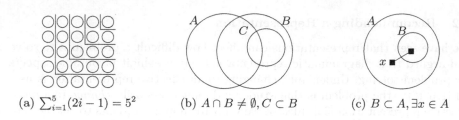

(a) $\sum_{i=1}^{5}(2i - 1) = 5^2$ (b) $A \cap B \neq \emptyset, C \subset B$ (c) $B \subset A, \exists x \in A$

Fig. 2. (a) Dot diagrams, (b) Euler diagrams, and (c) Spider diagrams.

as this class of software is forced to consider concerns similar to ours: solving problems, representing them effectively, and considering their users.

4.1   Homogeneous Systems

Theorem provers are used by people to solve a very specific type of problem: given some assumptions, derive a specific conclusion. To make progress, the set of assumptions is updated using already-proved theorems (or axioms) through inference mechanisms. This maps directly to the problems space we discussed: the current state is the current set of assumptions, a goal state is any set of statements which contains the desired conclusion, and the actions to move between states are the inference mechanisms. So the difference between the theorem provers is the state space they model – and so the representations they exploit.

Most theorem provers are *homogeneous*: that is, they use a single representational system. This single representational system is usually sentential, but the details vary. One family of theorem provers are those based on type theory: two notable members are Coq [15], and Nuprl [9]. These systems use Martin-Löf type theory as their foundational system, and proofs are constructions of a value that has the type which is an encoding of the theorem to prove. A second notable family of theorem provers are those with HOL/LCF[8] ancestry [11]: HOL4 [34], HOL Light [14], and Isabelle/HOL [26].[9] These systems use a small core of actions that is intended to be easy to verify, and all other actions must be built on top of this core. Both families use a syntax that is programming-language-like, and purely sentential.

Equally homogeneous, but no longer sentential, are the family of *diagrammatic* theorem provers. DIAMOND focuses on diagrammatic proofs of arithmetic using grids of dots (Fig. 2a), and ways of partitioning the grid [17]. The high-level approach of DIAMOND is different to that of the sentential provers

---

[8] HOL stands for higher-order logic, an extension of predicate/first-order logic. LCF stands for the logic for computable functions, a theorem prover based on the logic *of* computable functions.

[9] Isabelle (without 'HOL') is a *meta*-logic system: a developer tailors Isabelle to work in their particular system. For example, Isabelle/ZF allows people to use Zermelo-Fraenkel (ZF) set theory rather than higher-order logic. This is an interesting step towards heterogeneity, but the different logics are inaccessible to each other.

mentioned earlier: it works with instances of a proof and generates a generalised version automatically, rather than expecting the person proving the theorem to work in the most general case at all times. Edith, and its successor Speedith, focus on Euler diagrams (Fig. 2b) and Spider diagrams (Fig. 2c), respectively [36, 43]. Their proof structure more closely resembles that of the sentential systems: from some diagrams you can construct a new diagram; analogously, from some assumptions you derive a new conclusion. These systems show that software is capable of supporting diagrammatic representational systems, yet they do not yet push the bounds to *heterogeneous* reasoning: exploiting multiple representations.

4.2 Semi-heterogeneity

Homogeneous theorem provers have continued to grow in sophistication and power, but their generality comes at the cost of speed: certain problems are best left to dedicated tools that have a better representation for that problem. In the HOL/LCF tradition, these tools are integrated as *hammers* [5]. For example, Isabelle/HOL uses 'Sledgehammer' to transform a higher-order logic problem into a first-order logic problem before passing the transformed problem (along with a set of relevant lemmas) to automated first-order logic provers; the proof is returned to Isabelle/HOL, and validated in the verified core like any other proof [27]. While not obviously heterogeneous – every representational system involved is sentential – Isabelle/HOL exploits a system with a more effective problem space by transforming the problem.

Isabelle has a second means of semi-heterogeneous reasoning: the *Transfer* package. The Transfer package was designed as a general tool for code generation, and for the development of quotient types and subtypes. Raggi et al. used Transfer as the basis for a tool for heterogeneous reasoning [29], so that statements are transferred across a network of theories that formalise natural numbers in various ways. With this tool, theorems about natural numbers can be proved under their representation as either successors of zero, multisets of primes, classes of finite sets, and others. The one heuristic used to select between potential representations is the size of sentences. Otherwise, the task of selection is left to the user.

We consider this approach semi-heterogeneous, as it enables the transformation of sentences across different mathematical theories, while remaining in purely sentential representational systems.

4.3 Fully Heterogeneous Theorem Provers

We move now from homogeneous or purely sentential systems to heterogeneous reasoning systems. An early and notable heterogeneous system, Hyperproof, was an educational tool for first-order logic that used a three-dimensional chessboard environment alongside a more typical sentential representational system [3]. The actions available in the two representational systems were different, as would be expected; proofs in the sentential first-order logic system are often more verbose than their chessboard counterparts [3]. Barker-Plummer et al. generalised Hyperproof to Openproof, a framework allowing heterogeneous reasoning with many different representational systems [2]. The framework avoids an *inter-lingua* (common language) but maintains a common proof state; this avoids some 'lowest-common-denominator' expressiveness concerns while maintaining a valid proof. But as a result, there is a tight coupling between the representational systems in Openproof: there is a one-to-one correspondence between the objects and relations in each representation, and formal translations between them.

MixR is a heterogeneous theorem proving framework that grew out of a desire to integrate Speedith, the spider diagram reasoner, with Isabelle [42]. The MixR framework consisted of two parts: one theorem prover that 'owned' the proof state, and many 'working' theorem provers that could modify the proof state. MixR aimed to reuse existing theorem provers, rather than develop specialist heterogeneous theorem provers. Moreover, MixR allowed for unsound transformations between the representational systems used by each of the 'working' theorem provers. MixR also introduced heterogeneous statements – using multiple representational systems simultaneously – through *placeholders*.[10] MixR, like all the heterogeneous systems we have discussed, provides the *option* for heterogeneous reasoning. But it does not *encourage* or *guide* heterogeneous reasoning: representation selection is a human-driven process.

This section explored how current software, designed to work towards solving problems alongside a person, manages the issue of representation. For the most part, software systems maintain a single representational system, whether sentential or diagrammatic. Some software is heterogeneous, notably MixR: it allows the user to combine multiple representational systems together to solve a single problem, in particular, allowing for informal transformations between representations. But the decision on which representation system to use at any given point is driven by the user – while multiple representations may be available, the user is *not* helped to use them.

5   Cognitive Analysis of Computational Systems

Each of these computational systems comes with compromises: their representational systems exist at different points along each of the cognitive factors we have looked at, with varying degrees of heterogeneity. This in turn interacts with their generality. We would encourage the developers of existing and future problem solving software to be cognisant of the compromises they are making for the cognitive benefit of their users.

Many of the systems we explored are sentential, meaning they can be incredibly general tools at the cost of abstraction. By being able to encode effectively all of mathematics, the encoding becomes less *direct*, with less *conceptual transparency*, and potentially more effort is required by the user to map the notation to the problem. Exceptions to this are tools like DIAMOND, Edith, and Speedith, which trade *generality* for directness: DIAMOND encodes natural numbers as arrangements of dots, Edith encodes sets as Euler diagrams, and Speedith directly encodes sets with existential elements as spider diagrams. They restrict their domain to afford a more direct representation of said domain.

---

[10] For example, $(x = \square) \rightarrow (\text{shape}(x) = \texttt{square})$ places the square in the statement.

Similarly, we see these diagrammatic tools lower the *inference cost* for the user in their domains, again trading generality for lowered cognitive cost. The user can often exploit the observational advantages of the diagrammatic representations over more general sentential representations. In contrast, sentential tools like Isabelle or Coq typically exhibit fewer observational advantages, meaning even simple inferences still require effort.

This impacts the accessibility of these systems. On the one hand we have domain-specific diagrammatic tools which exhibit all the cognitive benefits of diagrams and, on the other hand, we have very general tools that require high skill from the user. Systems such as Hyperproof were designed with the goal of education from the very beginning: the creators understood the value of re-representation and diagrammatic aspects as a powerful tool for novices to reach for. Conversely, Isabelle and Coq are designed for research-level mathematics: learning to use these tools is *hard*, but their generality and power reward the effort. Neither is 'better' – these are different classes of tools with different audiences.

The heterogeneous systems we examined (MixR and Hyperproof) have seen most of their success limited to research, with only limited deployment or applications in the real world (Openproof). As we have argued in this paper, cognitive considerations are of utmost importance for the selection of effective representations for problem solving. Thus, we contend that the limited success of these systems is due to their lack of flexibility in regards to the variety of users with different levels of expertise and familiarity. This flexibility could be harnessed through intelligent recommendation by estimating the cognitive fit of representation using the criteria presented in this paper. As we argued, this is necessarily a complex assessment that involves the calculation of trade-offs between competing criteria.

In summary, we recommend software designers consider the following:

Consideration	Explanation
Direct encoding	Information is usually easier to extract, but the representation is typically less general
Low-cost inference	Make each step simple, and the next step easy to identify. These are relative to your target users. Exploit free rides
Generality	Be general enough to be useful, but not so general as to be confusing. More expert users can typically handle more general representations
Conceptual-syntactic compatibility	Make it 'easy' to do the right thing, but 'hard' to do the wrong thing. Easy and hard depend on the target users
Syntactic constraints	The representation imposes rules, rather than forcing users to remember them. More constraints typically limit generality
Geometry and space	Favour physically plausible interactions. This typically limits the ability to generalise

No one representation will be perfect for every user and every problem, so we encourage heterogeneous (but coherent) systems: use direct representations with strong syntactic constraints when available, but more general representations when flexibility is needed. We hope that software designers will be more open to different modalities and develop new, more effective tools to solve problems.

One final recommendation is for the software to *guide* heterogeneous reasoning by making the right representation available at the right time. Consider: theorem provers that suggest illuminating diagrams; spreadsheets that encourage explanatory charts; video editing software that intelligently switches between timelines and nodes. We wish to encourage software that dynamically switches between – not just allows – varied representations. Research in this direction could open up new opportunities in human-computer collaboration.

6   Conclusion

We examined in this paper how human problem solving can be modelled as traversing a problem space: the solver is attempting to reach goal states by applying actions to the current state. The expertise of the solver impacts their ability to navigate the problem space, but by selecting an effective representation we can induce a problem space in which the solver is already expert. The nature of the representation determines its effectiveness, and specific aspects – each with trade-offs – are generally agreed to be better; conveniently, these align with diagrammatic aspects of representational systems. We also discussed how representations are used in software, specifically theorem proving software: few support heterogeneous reasoning, and those that do, fail to support the user in selecting an appropriate representational system. We encourage developers of these problem solving tools to be aware of how the design decisions they are making impact the cognitive aspects of their tools, and what effect this will have upon their users. This review identifies and illuminates some of the factors impacting those decisions.

Acknowledgements. Aaron Stockdill is supported by the Hamilton Cambridge International Scholarship. This work was supported by the EPSRC grants EP/R030650/1, EP/T019603/1, EP/R030642/1, and EP/T019034/1.

References

1. Ainsworth, S.: The educational value of multiple-representations when learning complex scientific concepts. In: Gilbert, J.K., Reiner, M., Nakhleh, M. (eds.) Visualization: Theory and Practice in Science Education, pp. 191–208. Springer, Dordrecht (2008). https://doi.org/10.1007/978-1-4020-5267-5_9
2. Barker-Plummer, D., Etchemendy, J., Liu, A., Murray, M., Swoboda, N.: Openproof - a flexible framework for heterogeneous reasoning. In: Stapleton, G., Howse, J., Lee, J. (eds.) Diagrams 2008. LNCS (LNAI), vol. 5223, pp. 347–349. Springer, Heidelberg (2008). https://doi.org/10.1007/978-3-540-87730-1_32

3. Barwise, J., Etchemendy, J.: Visual information and valid reasoning. In: Allwein, G., Barwise, J. (eds.) Logical Reasoning with Diagrams, pp. 3–25. Oxford University Press Inc, New York (1996)
4. Blackwell, A.F., Whitley, K.N., Good, J., Petre, M.: Cognitive factors in programming with diagrams. Artif. Intell. Rev. 15(1), 95–114 (2001)
5. Blanchette, J.C., Kaliszyk, C., Paulson, L.C., Urban, J.: Hammering towards QED. J. Formaliz. Reason. 9, 101–148 (2016)
6. Cheng, P.C.H.: Electrifying diagrams for learning: principles for complex representational systems. Cogn. Sci. 26(6), 685–736 (2002)
7. Cheng, P.C.-H.: What constitutes an effective representation? In: Jamnik, M., Uesaka, Y., Elzer Schwartz, S. (eds.) Diagrams 2016. LNCS (LNAI), vol. 9781, pp. 17–31. Springer, Cham (2016). https://doi.org/10.1007/978-3-319-42333-3_2
8. Condell, J., et al.: Problem solving techniques in cognitive science. Artif. Intell. Rev. 34(3), 221–234 (2010)
9. Constable, R.L., et al.: Implementing Mathematics with the Nuprl Proof Development System. Prentice-Hall Inc, Hoboken (1986)
10. Giardino, V.: Towards a diagrammatic classification. Knowl. Eng. Rev. 28(3), 237–248 (2013)
11. Gordon, M.J., Milner, A.J., Wadsworth, C.P.: Edinburgh LCF. LNCS, vol. 78. Springer, Heidelberg (1979). https://doi.org/10.1007/3-540-09724-4
12. Grawemeyer, B.: Evaluation of ERST – an external representation selection tutor. In: Barker-Plummer, D., Cox, R., Swoboda, N. (eds.) Diagrams 2006. LNCS (LNAI), vol. 4045, pp. 154–167. Springer, Heidelberg (2006). https://doi.org/10.1007/11783183_21
13. Green, T., Blackwell, A.: Cognitive dimensions of information artefacts: a tutorial. In: BCS HCI Conference, vol. 98 (1998)
14. Harrison, J.: HOL light: an overview. In: Berghofer, S., Nipkow, T., Urban, C., Wenzel, M. (eds.) TPHOLs 2009. LNCS, vol. 5674, pp. 60–66. Springer, Heidelberg (2009). https://doi.org/10.1007/978-3-642-03359-9_4
15. Huet, G., Kahn, G., Paulin-Mohring, C.: The Coq proof assistant: a tutorial. The Coq Project (2004)
16. Inglis, M., Mejía-Ramos, J.P.: On the persuasiveness of visual arguments in mathematics. Found. Sci. 14(1), 97–110 (2009)
17. Jamnik, M., Bundy, A., Green, I.: On automating diagrammatic proofs of arithmetic arguments. J. Logic Lang. Inform. 8(3), 297–321 (1999)
18. Johnson-Laird, P.N.: Models and heterogeneous reasoning. J. Exp. Theor. Artif. Intell. 18(2), 121–148 (2006)
19. Kotovsky, K., Hayes, J., Simon, H.: Why are some problems hard? Evidence from tower of Hanoi. Cogn. Psychol. 17(2), 248–294 (1985)
20. Larkin, J.H., McDermott, J., Simon, D.P., Simon, H.A.: Models of competence in solving physics problems. Cogn. Sci. 4(4), 317–345 (1980)
21. Larkin, J.H., Simon, H.A.: Why a diagram is (sometimes) worth ten thousand words. Cogn. Sci. 11(1), 65–100 (1987)
22. Miller, G.A.: The magical number seven, plus or minus two: some limits on our capacity for processing information. Psychol. Rev. 63(2), 81–97 (1956)
23. Minsky, Y.: Effective ML revisited (2011). Blog Post. https://blog.janestreet.com/effective-ml-revisited/. Accessed 28 Nov 2020
24. Moody, D.: The "physics" of notations: toward a scientific basis for constructing visual notations in software engineering. IEEE Trans. Softw. Eng. 35(6), 756–779 (2009)

25. Newell, A., Shaw, J., Simon, H.A.: Report on a general problem-solving program. In: Proceedings of the International Conference on Information Processing, pp. 256–264 (1959)
26. Paulson, L.C.: The foundation of a generic theorem prover. J. Autom. Reason. **5**(3), 363–397 (1989)
27. Paulson, L.C., Blanchette, C.: Three years of experience with sledgehammer, a practical link between automatic and interactive theorem provers (2010)
28. Pólya, G.: How to Solve It: A New Aspect of Mathematical Method. Princeton Science Library, Princeton University Press (1957)
29. Raggi, D., Bundy, A., Grov, G., Pease, A.: Automating change of representation for proofs in discrete mathematics (extended version). Math. Comput. Sci. **10**(4), 429–457 (2016)
30. Scaife, M., Rogers, Y.: External cognition: how do graphical representations work? Int. J. Hum. Comput. Stud. **45**(2), 185–213 (1996)
31. Shimojima, A.: Operational constraints in diagrammatic reasoning. In: Allwein, G., Barwise, J. (eds.) Logical Reasoning with Diagrams, chap. 2, pp. 27–48. Oxford University Press, New York (1996)
32. Shimojima, A.: The graphic-linguistic distinction. In: Blackwell, A.F. (ed.) Thinking with Diagrams, pp. 5–27. Springer, Dordrecht (2001). https://doi.org/10.1007/978-94-017-3524-7_2
33. Simon, H.A., Newell, A.: Human problem solving: the state of the theory in 1970. Am. Psychol. **26**(2), 145–159 (1971)
34. Slind, K., Norrish, M.: A brief overview of HOL4. In: Mohamed, O.A., Muñoz, C., Tahar, S. (eds.) TPHOLs 2008. LNCS, vol. 5170, pp. 28–32. Springer, Heidelberg (2008). https://doi.org/10.1007/978-3-540-71067-7_6
35. Stapleton, G., Jamnik, M., Shimojima, A.: What makes an effective representation of information: a formal account of observational advantages. J. Logic Lang. Inform. **26**(2), 143–177 (2017)
36. Stapleton, G., Masthoff, J., Flower, J., Fish, A., Southern, J.: Automated theorem proving in Euler diagram systems. J. Autom. Reason. **39**(4), 431–470 (2007)
37. Stenning, K., Lemon, O.: Aligning logical and psychological perspectives on diagrammatic reasoning. Artif. Intell. Rev. **15**(1), 29–62 (2001)
38. Stenning, K., Oberlander, J.: A cognitive theory of graphical and linguistic reasoning: logic and implementation. Cogn. Sci. **19**(1), 97–140 (1995)
39. Superfine, A.C., Canty, R.S., Marshall, A.M.: Translation between external representation systems in mathematics: all-or-none or skill conglomerate? J. Math. Behav. **28**(4), 217–236 (2009)
40. Sweller, J.: Cognitive load during problem solving: effects on learning. Cogn. Sci. **12**(2), 257–285 (1988)
41. Tversky, B.: Visualizing thought. topiCS **3**(3), 499–535 (2011)
42. Urbas, M., Jamnik, M.: A framework for heterogeneous reasoning in formal and informal domains. In: Dwyer, T., Purchase, H., Delaney, A. (eds.) Diagrams 2014. LNCS (LNAI), vol. 8578, pp. 277–292. Springer, Heidelberg (2014). https://doi.org/10.1007/978-3-662-44043-8_28
43. Urbas, M., Jamnik, M., Stapleton, G., Flower, J.: Speedith: a diagrammatic reasoner for spider diagrams. In: Cox, P., Plimmer, B., Rodgers, P. (eds.) Diagrams 2012. LNCS (LNAI), vol. 7352, pp. 163–177. Springer, Heidelberg (2012). https://doi.org/10.1007/978-3-642-31223-6_19
44. Vessey, I.: Cognitive fit: a theory-based analysis of the graphs versus tables literature. Decis. Sci. **22**(2), 219–240 (1991)

45. Weisberg, R., Suls, J.M.: An information-processing model of Duncker's candle problem. Cogn. Psychol. **4**(2), 255–276 (1973)
46. Zazkis, D., Weber, K., Mejía-Ramos, J.P.: Bridging the gap between graphical arguments and verbal-symbolic proofs in a real analysis context. Educ. Stud. Math. **93**(2), 155–173 (2016). https://doi.org/10.1007/s10649-016-9698-3
47. Zhang, J.: The nature of external representations in problem solving. Cogn. Sci. **21**(2), 179–217 (1997)

Diagrams as Part of Physical Theories: A Representational Conception

Javier Anta(✉) ⓘ

Departament de Lògica, University of Barcelona, Història i Filosofia de Ciència, C/ Montalegre 6-8, 4º Floor, Office 4013, 08001 Barcelona, Spain

Abstract. Throughout the history of the philosophy of science, theories have been linked to formulas as a privileged representational format. In this paper, following [8], I defend a semantic-representational conception of theories, where theories are identified with sets of scientific re-presentations by virtue of their epistemic potential and independently of their format. To show the potential of this proposal, I analyze as a case study the use of phase diagrams in statistical mechanics to convey in a semantically consistent and syntactically correct way theoretical principles such as Liouville's theorem. I conclude by defending this philosophical position as a tool to show the enormous representational richness underlying scientific practices.

Keywords: Phase diagrams · Semantic conception · Physical representations

1 Introduction

Throughout the history of the philosophy of science in the twentieth century, scientific theories have been constantly linked to sets of symbolic formulas (whether logical or mathematical) as the privileged representational format to convey their theoretical content or to reconstruct them rationally. Here I will defend the recent proposal of [8], who argue for a semantic-representational (in contrast to the semantic-structural) conception of scientific theories, where these are constitutively identified with sets of scientific representations by virtue of their capacity to provide scientific knowledge and independently of their format (i.e. formulaic, diagrammatic, iconic, etc.). To show the potential applicability of this philosophical proposal, we will analyze in detail as a case study the use of phase diagrams in statistical mechanics. From this analysis we will conclude that these phase diagrams could be satisfactorily understood as vehicles capable of conveying in a semantically consistent and syntactically correct way theoretical principles of statistical mechanics such as Liouville's theorem. We will conclude by defending this position as a tool to show the enormous representational richness underlying scientific practices.

2 Conceptions of Scientific Theories

During the twentieth and twenty-first century philosophy of science, scientific theories have been characterized in a variety of ways [4]. Initially, in logical positivism (whose

© Springer Nature Switzerland AG 2021
A. Basu et al. (Eds.): Diagrams 2021, LNAI 12909, pp. 52–59, 2021.
https://doi.org/10.1007/978-3-030-86062-2_5

history extends from the 1920s to the late 1950s) they employed the formal logical tools developed by Frege, Russell and Wittgenstein to characterize or 'rationally reconstruct' scientific theories as sets of symbolic formulas generated by means of axiomatically articulated logical languages. The logical architecture of scientific theories was determined 'syntactically' by a set of axioms expressed as primitive symbols, assigning empirical content by means of correspondence rules. In the context of this syntactic conception of theories, the relationship between diagrammatic vehicles (and of course also other non-syntactically defined iconographic elements) and scientific theories was relegated to the background, as merely accessory elements. Had a logical equivalence between diagrammatic systems and formulaic systems been demonstrated in this period (e.g. [2]), we may question whether diagrammatic resources would have constituted acceptable tools for the rational reconstruction of scientific theories by the advocates of the syntactic conception.

From the late 1950s onwards, the syntactic conception of logical positivism gave way to the first proposals (e.g. [9]) to characterize or 'rationally reconstruct' scientific theories not as sets of symbolic formulas but as sets of mathematical structures (set-theoretical, model-theoretical, etc.). In this direction certain physical theories such as Newtonian mechanics would be reconstructed by sets of set-theoretical objects {P, T, s, m, f} where P corresponds to the set of particles, T (real-valued) sets of time intervals, s(p, t) the position of each particle, m(p) the (real-valued) mass of the particle and f(p, q, t) the force that a particle q exerts on p at time t (ibid.). These strategies of characterizing scientific theories as mathematical structures are referred to in the literature as the 'semantic-structural conception', identifying formal theories and models. However, the transition between the syntactic and the semantic-structural conception did not really constitute an advance in the consideration of new representational formats (since both are based on identifying theories as symbolic formulas, either logical or mathematical, respectively), but an advance in the consideration of more expressive formal structures.

It was from the 1980s onwards that philosophers of science such as [5] began to pay attention to the fundamental role of non-formulaic elements (e.g. conceptual schemes, detector images, phase diagrams, etc.) in obtaining scientific knowledge through the use of certain models. This gave impetus to the so-called 'pragmatic conception' of scientific theories and models [3], where models were conceived as epistemically active and intermediary elements between theories and modeled phenomena. On the other hand, in the 1990s there was also a growing philosophical interest in the problem of scientific representation (i.e., what are they, how do they work, etc.) led by [5] and by [7] deflationary proposal, where a representation is characterized by its capacity to (i) 'denote' a phenomenon, (ii) 'demonstrate' certain properties and (iii) 'interpret' its content. In this sense, the pragmatic conception of model-theories and the problem of scientific representations allowed to question the idea that theories should be characterized exclusively as sets of symbolic formulas, what we can here call 'formulaic dogma' of scientific theories.

3 Semantic-Representational View on Theories

However, according to the pragmatic conception à la [3], only scientific models can be considered as format-independent sources of scientific knowledge. According to this conception, what from the formulaic dogma are assumed as 'peripheral representations' (e.g. diagrams, icons, images, etc.) would belong to the broad domain of modeling tools but not properly to scientific theories, as formulations of natural laws. In this sense the pragmatic conception presupposes a hierarchy of scientific elements according to their epistemic potential: formulaic theories-laws (epistemically inactive) proto-formulaic theoretical models (epistemically quasi-active) format-independent phenomenological models (epistemically active). The main problem with this hierarchy lies in assuming that diagrammatic representations cannot constitute vehicles of the theoretical principles of the disciplines, but merely tools for the application of these principles. In my proposal I argue against this idea that diagrammatic representations constitute adequate semantic vehicles for encoding theoretical principles. To defend this thesis, I will adopt a semantic conception that directly rejects the 'formulaic dogma' about scientific theories. Against the semantic-structural conceptions of [8,9] recently defended a proposal to character-ize scientific theories as sets not of logico-mathematical structures (i.e. syntactic and semantic-structural conceptions) but properly of representational models or representa-tions. By 'representation' I mean any structure that can consistently encode information about a target phenomenon. Based as a premise on what these authors call the 'Hughes-Giere-Suarez thesis' (i.e. all scientific models are representations, regardless of their representational format), [8] argued that scientific theories are composed precisely of sets of representations, or properly a particular subset of all possible representations associated with that theory:

(P1) Theories are composed of a subset M of all models.

(Semantic conception of theories)

(P2) All models are format-independent representations.

(Hughes-Giere-Suarez thesis)

(SR) Theories are composed of a subset of all representations.

(Semantic representational conception)

As can be extracted from the argument above, this semantic-representational (SR) conception of scientific theories presents important theoretical advantages for defending our main thesis (i.e. certain diagrammatic representations are part of physical theories) as opposed to the syntactic and semantic-structural conception. On the one hand, this con-ception of scientific theories undermines (i) the "criterion of rational reconstruction" of the structuralists, according to which theories can only be identified and, therefore, ana-lyzed by means of logical (syntactic conception) or mathematical (semantic-structural conception) formulas; and (ii) the formulaic-symbolic dogma, according to which the only "rationally legitimate" vehicles of theoretical content are certain types of privileged symbolic formulas. Note that the main difference (as far as our aim here is concerned) between SR and the semantic-structuralist conception (SS, below), is that whereas the latter identifies scientific theories with a subset of models associated with a partic-ular format (formulaic dogma), for SR this subset is completely independent of the representational format of the model:

(P3) Only a subset of formulaic models can be identified with theories.
(Formulaic dogma)

(SS) Theories are composed of a subset of formulaic models.
(Semantic structural conception)

The main advantage of SR for assessing the enormous representational richness of science lies in its ability to get rid of format-centric biases (e.g., historically predominant among syntactic and semantic-structural conceptions) when analyzing the epistemic potential of any model to convey theoretical principles. This does not mean that any representation is valid to be constitutively part of a scientific theory. At this point in our analysis, we argue precisely that the criterion for including a representation in the subset of representations identifiable with a theory is that it contributes to a (i) semantically consistent and (ii) syntactically correct encoding of the theoretical content. Illustratively, for SR particle physics would not be identified exclusively with sets of models or formulaic structures such as {P, T, s, m, f} (i.e. as with SS, above) but with any representation regardless of its format that contributes significantly to the attainment of knowledge by conveying such theoretical content, as in the case of the semantically consistent encoding of 'electric charge' by the direction of the white lines (i.e. up charge-negative, down charge-positive) in a bubble chamber picture (Fig. 1).

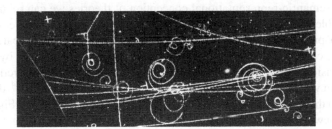

Fig. 1. Bubble chamber picture as a part of particle physics.

4   Case Study: Phase Space Diagrams in Statistical Mechanics

We will now explore phase diagrams in statistical mechanics as a case study (extended in [1]) to demonstrate the potential of the semantic-representational conception of theories in the evaluation of the epistemic potential of diagrams in science. Statistical mechanics is the discipline that studies certain macroscopic behaviors (i.e. an expanding gas inside a container V, see Fig. 2) from the dynamics of its microscopic components of material systems. To achieve this goal, the position and velocity of all the molecules that make up a gas are encoded in what is technically known as the phase space Γ of this system, whose exact values for each moment (or 'microstate') are represented by points x in this abstract 6n-dimensional Γ-space (wherein n is the number of particles, usually $n = 10^{24}$ per mole of substance). This abstract space is usually represented by 'portraits' or 'phase diagrams', which are nothing more than two-dimensional simplifications that remove large amounts of redundant and irrelevant information from the phase space of a system. Within this phase-diagrammatic apparatus, the set of microstates compatible with an observable property of the system (i.e. the volume 1/2V of a gas before it expands) is represented as bidimensional areas or 'macrostates' Γ_m in this phase space. As the gas progressively expands through the total physical volume V of the vessel, the microstates contained in the initial macrostate Γ_m (i.e. phase volume of the macrostate) progressively move through the phase space Γ.

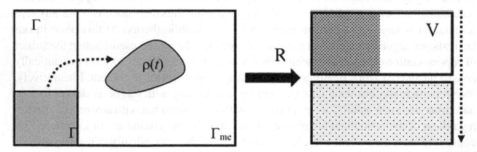

Fig. 2. Phase diagram (left) representing a gas approaching equilibrium (right).

From our semantic-representational conception of theories we can robustly defend that it is possible to employ diagrammatic resources as representational vehicles of certain theoretical principles of statistical mechanics. In particular, the statistical mechanical meaning of the celebrated 'Liouville theorem' would be visually encoded by preserving the two-dimensional dark gray area from Γ_m to $\rho(t)$, wherein the theoretical content of that theorem would be semantically-consistently preserved from a 10^{24}-dimensional phase space to the two-dimensional graphical representation viewable in Fig. 2. In fact, from the framework of the representational semantic conception, certain semantically consistent and syntactically correct devices such as these kind of phase diagrams would no longer be considered as marginal representational practices (as it would make sense to claim from the formulaic dogma), but would be included in the body of the theory itself as a constitutive part of it. Thus, statistical mechanics as a theory would not be constituted by a set of statements set out in a technical textbook, but by a set of valid representational practices with which to obtain meaningful information about a physical field.

To check whether the type of phase diagrams represented in Fig. 2 allow us to convey theoretical content in a semantically consistent and syntactically correct way, let us explore the scenario in which we represent the evolution of an expanding gas in a vessel and also perform a macroscopic measurement during this process. In this case, the phase space of the expanding gas under consideration is divided into three different macroscopic macrostates (Γ_m, $\Gamma_{m'}$ and Γ_{meq}), each associated with a particular value of a macroscopic observational variable such as the volume V, where the macrostate having a larger volume (Γ_{meq}), represented graphically as a two-dimensional area of the macrostates, corresponds appropriately to the thermal equilibrium state of the system. Being a "dynamic" phase diagram, the phase structures it contains represent synchronously three different moments in the evolution of the target system. At the initial time (t_0) the macrostate Γ_m, the farthest from thermal equilibrium, is considered to have a positive and uniform probability measure (light gray) that statistically describes the system at time t_0; this means that the current microstate of the system is found with equal probability at any of the points it contains.

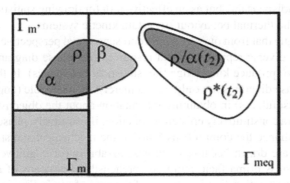

Fig. 3. Phase space representation of the dynamic evolution of a physical system (encoded via a density ρ) in the theoretical context of classical statistical mechanics.

At the beginning of the dynamic evolution of the system, all the microstates contained in Γ_m move along the phase space generating (let say at time t_1) what Shenker and Hemmo call a 'dynamic blob' ρ having a uniform distribution of probability defined over different regions in Γ. At this precise moment, a measurement would be carried out on the macrovariable associated with the macrostates, dividing the dynamic blob into two parts α and β depending on the particular macrostate ($\Gamma_{m'}$ or Γ_{meq}, respectively) in which they are found. Note that the partition of ρ determines the probabilistic results of such measurement. Finally, let us imagine that the macrostatistical measurement results in the value of the macrovariable associated to $\Gamma_{m'}$; then we take into consideration ρ/α (that is, the part of contained in $\Gamma_{m'}$, visually highlighted in dark grey) and let it evolve dynamically until at the moment t_2 we would obtain $\rho/\alpha(t2)$. On the other hand, we take ρ/α and carry out a phase averaging of its probability values along macrostate $\Gamma_{m'}$ (namely, coarse-graining procedure as detailed in Sect. 2) generating a new probability distribution $\rho*$ that will dynamically evolve into $\rho*(t_2)$.

Firstly, we can point out how various graphic resources of the diagram serve to encode in a formal-syntactically correct and semantically consistent way statistical mechanical content. For example, the fact that the area contained in macrostate Γ_m is uniformly light-grey colored can be considered as a graphic-chromatic resource used to encode that the probability distribution defined on that very macrostate will be uniform. Since any agent competent with (i) the syntactic-semantic functioning of this type of phase diagrams and (ii) with the basics concepts of statistical mechanics will be perfect able to access such graphically encoded theoretical content to draw inferences about the target system (e.g. concluding that any two possible microscopic configurations contained in this macrostate will be equally likely to be the actual one), such a diagrammatic element (the uniform grey colouring of Γ_m) may be considered as a valid representational vehicle.

In the same way, other graphic resources such as the area invariance between ρ/α at time t_1 and ρ/α at time t_2 would correctly and consistently encode the meaning of the Liouville principle as a mechanical statistical content (notice that if such an area were to change from t_1 to t_2, then the content of the Liouville theorem would be encoded in an incorrect and inconsistent way). From our semantic-representational conception of statistical mechanics, a well-formed phase diagrams like the one in Fig. 3 would properly render not as an indispensable but as an effective tool for drawing statistical mechanical inferences about the thermal behaviour of certain kinetic systems.

Finally, we argue that from our semantic-representational perspective we can delimit the way in which diverse representational resources (i.e. phase diagrams, in our case study) contribute to produce knowledge (i.e. statistical mechanics). In this sense it can be shown how phase diagrams not only have a merely ancillary role (connected to their greater or lesser usefulness) in obtaining information about the objective thermal phenomenon, but also a constitutively epistemic function. Illustratively, phase diagrammatic representations enhance the comprehensibility of the mechanical statistical content on which inferences are drawn because of its visualizability (or cognitive accessibility). Liouville's theorem is prima facie more comprehensible by means of its visualizable diagrammatic representation (i.e. invariance of the area of ρ/α in Fig. 3) than by means of the symbolic-analytical formula $|\rho/\alpha (t_1)| = |\rho/\alpha (t_1 + \Delta t)|$, since any agent will require more technical and conceptual skills (as well as cognitive processing resources) to access the same statistical mechanical content from the formulaic than from the diagrammatic representation.

5   Conclusion

We conclude by defending that from a conception of scientific theories such as the semantic-representational one of [8] we can evaluate the epistemic potential of certain diagrammatic representations (semantically consistent and syntactically well-defined) as theoretical vehicles. An example of this can be found precisely in the case study analyzed in the previous section, where the semantic-syntactic manipulation of diagrammatic resources allows us to obtain theoretical knowledge in certain statistical mechanical scenarios at the same level as from equivalent formulaic representations. For example, we can satisfactorily explain what a macroscopic measurement consists of through the valid representational resources contained in Fig. 3, explicating that when performing a measurement at t_1 of an observable property (e.g. volume V) of the target system, the whole pre-measurement density ρ collapses in either ρ/α or ρ/β with a degree of probability proportional to the graphically encoded area of each of these post-measurement distributions. We have shown how the capacity to exploit inferential a phase diagram has a direct impact on the possibility of generating mechanical statistical explanations, for example, by making explicit how graphically separating the area of ρ into two non-overlapping regions associated with macrostates $\Gamma_{m'}$ and Γ_{meq} (respectively) at time t_1 could constitute a robust explanation of a macroscopic measurement process in statistical mechanics. In short, our results should be taken as a modest vindication of the enormous representational richness that underlies science, disregarded for decades by the philosophy of science.

References

1. Anta, J.: Integrating Inferentialism about physical theories and representations: a case for phase space diagrams. Critica **52**(156) (2021)
2. Barwise, J., Etchemendy, J.: Heterogeneous logic. In: Allwein, G., Barwise, J., (eds.) Logical Reasoning with Diagrams. Oxford University Press, Oxford (1996)
3. Cartwright, N.: The Dappled World: A Study of the Boundaries of Science. Cambridge University Press, Cambridge (1999). https://doi.org/10.1017/CBO9781139167093
4. Frigg, R., Hartmann, S.: Models in science. In: Zalta, E.N., (ed.) The Stanford Encyclopedia of Philosophy. https://plato.stanford.edu/archives/spr2020/entries/models-science/ (2020)
5. Giere, R.: Explaining Science: A Cognitive Approach. University of Chicago Press, Chicago, IL (1988)
6. Hacking, I.: Representing and Intervening: Introductory Topics in the Philosophy of Natural Science. Cambridge University Press, Cambridge (1983)
7. Hughes, R.: Models and representation. Philos. Sci. **64**(4), 336 (1997)
8. Suárez, M., Pero, F.: The representational semantic conception. Philos. Sci. **86**(2), 344–365 (2019)
9. Suppes, P.: Introduction to Logic. Princeton, D. Van Nostrand Co (1957)

Diagrams and Mathematics

Beyond Counting: Measuring Diagram Intensity in Mathematical Research Papers

Henrik Kragh Sørensen$^{(\boxtimes)}$ (iD)

Section for History and Philosophy of Science, Department of Science Education,
University of Copenhagen, Copenhagen, Denmark
henrik.kragh@ind.ku.dk
http://www.dh4pmp.dk/

Abstract. In the Diagrams 2020 conference, we (Mikkel Willum Johansen and myself) reported on our first successes with machine-learning agents identifying and counting diagrams in mathematical papers. One year later, we have progressed, and in this paper I present and discuss ways of creating evidence on the use of diagrams in mathematical publications. Studying a corpus of mathematical journals from the early 21st century and focusing on the intensity of their reliance on diagrams, I explore different means of measuring the use of diagrams as a precursor to further studying their integration into the mathematical argument.

1 Introduction

Despite what may be called the *formalist ethos*, diagrams and other visual representations play important parts in mathematical research. Although philosophers and mathematicians, such as David Hilbert, have claimed that appeals to visual representations are risky, they enter into the heuristic as well as the justificatory practices of research mathematics (see e.g. Johansen and Misfeldt 2018). Therefore, it has become an interesting and important program within philosophy of mathematical practice to understand the actual roles of diagrams and visual representations in general (Giaquinto 2020).

To study how mathematicians think and reach their insights is a complicated philosophical endeavour. When confronted with the additonal condition that philosophy of mathematical practice be empirically informed, the situation is even worse. We can perform interviews with practitioners or observe their practices, but such studies remain constrained to a small number of cases. If we want to attempt to access the *products* of mathematical practice, we are mainly served by studying mathematical research papers.

Until quite recently, the preferred and almost exclusive method for studying mathematical products has been through qualitative methods, in particular what we may call 'close reading' (Mizrahi 2020). Yet, by combining digital humanities and machine learning, we are beginning to explore other, quantitative perspectives on the content and structure of mathematical publications. Of course,

© Springer Nature Switzerland AG 2021
A. Basu et al. (Eds.): Diagrams 2021, LNAI 12909, pp. 63–70, 2021.
https://doi.org/10.1007/978-3-030-86062-2_6

scientometry has already produced insights into the *network* of mathematical publications, but to get at their individual epistemic aspects, new tools are required.

Among the limited set of such tools, we have developed a machine-learning agent which can detect and count diagrams in mathematical papers (Sørensen and Johansen 2020). When first reported, the detector was quite successful, but also restricted in its use. Counting the number of visual representations in a mathematical paper does not, per se, say very much about the epistemic and communicative roles that such representations play. Therefore, additional triangulation with qualitative methods is required before we can say anything substantial and detailed about the roles of diagrams from a quantitative perspective.

This paper thus forms part of our research agenda of using machine learning and text mining to 1) extend our corpus of mathematical diagram, 2) aid in classifying mathematical diagrams, 3) study the spread of mathematical diagrams between subfields and over time, and 4) suggest interesting research questions and remarkable diagrams for closer, qualitative analyses.

In this article, I report on an improved detector for mathematical diagrams which not only counts but also measures the size of diagrams. Similar approaches have been adopted in studying the use of visual elements in other disciplines (see e.g. Cleveland 1984; Arsenault, Smith, and Beauchamp 2006). In the present context, this allows us to develop different metrics for diagram use, which may say more (or at least something else) about diagram use. When combined with simple text extraction, we begin to have the means to study the integration of visual reasoning in mathematical research papers.

To illustrate this potential, I have chosen a sample of five international mathematical research journals from the past 25 years. By focusing on a short and contemporary time span, I lose the ability to perform long-duree historical comparisons. What I gain, though, is support for the assumption that diagram usage as well as technical means of producing diagrams remain relatively stable throughout the corpus.

2 Detecting and Measuring Diagrams

In order to deploy a machine-learning agent to the automated detection of mathematical diagrams, it is required to set up a training pipeline. Our detector pipeline is a reimplementation of the one reported in Sørensen and Johansen (2020), which achieved an F1-score on the *counting* of diagrams of 0.90777, which was found to be quite satisfactory.

Our new detector is based on the Detectron2 framework developed by Facebook and released as open source (Wu et al. 2019). Among the extensive model zoo provided with Detectron2, we have, as before, chosen a Faster RCNN architecture, which is pretrained on a standard set of objects. This model is then fine-tuned by training it on our previous training set of 11.500 images of diagrams verified by human identification from a subset of articles in the *Journal für die reine und angewandte Mathematik*, colloquially known as *Crelle's Journal* (Sørensen and Johansen 2020). In addition, the training included 4700 'negatives,' i.e. images without diagrams.

By training the detector in such a way, it could be said that it is based on an inductive, extensional definition of a diagram. And as reported in Sørensen and Johansen (2020) such a definition faces quite a lot of choices and demarcations. It is part of our research agenda to further explore the variation of diagrams in mathematical practice, but for the present purpose it suffices to notice that we take diagrams to be 2-dimensional pictorial objects in mathematical texts which serve functions of cognitive offloading and procedural reading (Larkin and Simon 1987; Johansen, Misfeldt, and Pallavicini 2018). Of particular notice is that 'commutative diagrams' *are included* in this concept of a diagram, although expert mathematicians sometimes seem to work with such diagrams in ways similar to equations, which we do not consider to be diagrams.

Another difficulty in training a detector is whether a visual representation should count as one or multiple diagrams when different parts each could also be encountered in isolation. This difficulty raises some problems for counting the *number* of diagrams as a good means to assess diagram reliance and intensity. Luckily, the detector can now be deployed also to measure the *area* of the diagrams that it detects, and by normalizing this by the print area of the page, we obtain a different measure from the *diagrams-per-page* (d/p).

3   Measuring Textual Proxies for Diagrams

A different means of assessing diagram usage comes through text extraction and various linguistic analyses. The most basic such method is counting the occurances of certain words, that serve as proxies for the interesting feature, in this case diagram usage.

Thus, I devised a list of 'diagram words', which would signal reference to visual representations in the articles:

diagram, drawing, figure, illustration, picture.

Obviously the words from the list of 'diagram words' can also occur in derived forms, and therefore I simply counted all words beginning with the given string of characters.

Initially, I had also included "graph" in the list, but this word was found to be much more prevalent than the others as it would often also refer to a mathematical object without referring to a visual representation of it. Therefore it was dropped from the list until it is possible to distinguish these references by more advanced linguistic tools. Clearly, other words could also be included, but this list seems to encapsulate the essential indicators for pictorial and diagrammatic elements in mathematical publications.

The counting of 'diagram words' is normalized by the total number of 'regular words' in the text, where so-called stop-words are filtered out. This was done to eliminate short words which could derive from processing mathematical formulas and less relevant, frequent words of the English language such as prepositions. Clearly, again, this choice could be different, but since it is applied to all texts, the measures produced are comparable over different journals and over time.

4 Contemporary Use of Diagrams in Mathematical Research

After training the detector, it was time to deploy it in detecting (actually *predicting*) diagrams in a corpus different from the one on which it was trained.

As mentioned above, I chose to focus on a set of prominent journals through the past 25 years, since these would presumably represent mathematical practice—both in general and in subfields—and live up to uniform technical standards in using and preparing visual representations for print. Thus, it seems likely that recent journals would not be subject to huge variations in typography or layout.

The selection was based on Scimago Journal Rankings (SJR) but constrained by ease of access to the pdf files on which the detector would run (Guerrero-Bote and Moya-Anegón 2012). The final selection is presented in Fig. 1 and can be seen to include some of the most impactful and prestigious mathematical journals as well as a few more specialized journals from main fields such as analysis, algebra and number theory, and geometry and topology.

Abbreviation	Name	SJR	Categories
ihes	Publications Mathématiques de l'IHÉS	8.890	Misc
inventiones	Inventiones Mathematicae	5.848	Misc
jdg	Journal of Differential Geometry	3.623	ANT; Ana; GT
geofuncanalysis	Geometric and Functional Analysis	3.508	Ana; GT
arma	Archive for Rational Mechanics and Analysis	3.420	Ana; Misc

Fig. 1. The selection of journals for quantitative study. The SJR is the Scimago Journal Ranking, and the categories are as given in Scimago.

Figure 2 lists the different metrics for journals in the sample. It shows considerable variation between the journals, not least in the new area metric, da/pa. Figure 3 further illustrates the variation of this particular metric across the journals included in the sample.

It is also interesting to investigate the variation over time to substantiate the assumption that this would be small when considering a "short", contemporary period of mathematical publication. Figure 4 shows the combined variations of the three metrics considered here over time for journals in the sample. The figure serves also to illustrate that the different metrics seem to be well aligned, but also that the show (small) deviations from each other over time. In particular, the textual metric dw/krw seems to drop more drastically over the past decade than the two other metrics, thus illustrating the kind of refined analysis multiple metrics allow for.

Finally, by comparing the different metrics for the same papers, the graphs in Fig. 5 are produced. These plots show that the number of diagrams and the

journal	number	w/diagram	d/p	da/pa	dw/krw
arma	530	46.8%	0.0737	0.0458	0.0298
geofuncanalysis	1168	47.2%	0.1184	0.0684	0.1025
ihes	165	77.0%	0.2677	0.1229	0.2559
inventiones	1893	69.8%	0.2156	0.0748	0.1963
jdg	110	53.6%	0.1562	0.1006	0.1037

Fig. 2. Different measures of diagram usage for different journals in the sample. As metadata, the number of articles included in the sample is given along with the percentage of articles with at least one diagram. Here, and throughout, d/p means diagrams per page, da/pa means diagram area per page area, and dw/krw means diagram words per 1000 regular words.

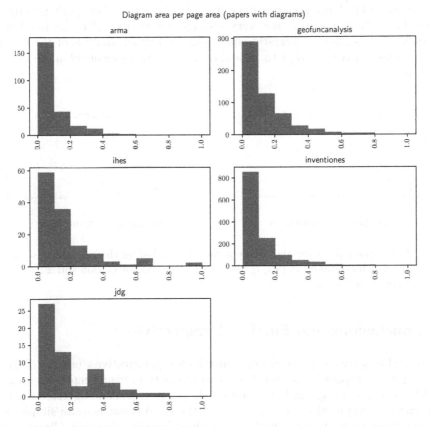

Fig. 3. Variation of the metric da/pa over the different journals. Here, only articles with at least one diagram are included.

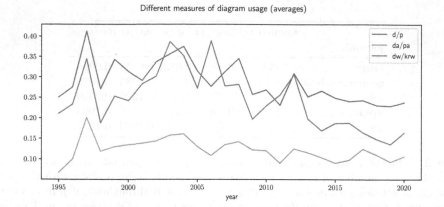

Fig. 4. Proportion of the printed area covered by diagrams.

area taken up by diagrams are rather strongly correlated. But perhaps surprisingly, the textual metric is not very strongly correlated with the number of diagrams. This suggests interesting questions for further analysis of the interrelation between textual and pictorial and diagrammatic elements of mathematical publications.

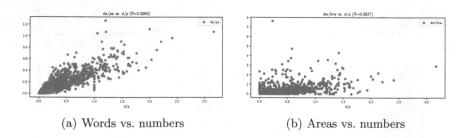

(a) Words vs. numbers   (b) Areas vs. numbers

Fig. 5. Correlation plots for the different metrics, including the relevant Pearson R-coefficients which are values in $[-1, 1]$ where 0 indicates no correlation and ± 1 indicates perfect linear correlation.

5   Conclusions and Further Perspectives

Based on the motivation of developing metrics for quantitative studies of diagram usage that encompass both prevalence and intensity of visual representations, I considered area per page and 'diagram words' per page as new suggested metrics. Both can be computed with relative ease—the word count is quite simple, and the area measurements come with the machine-learning detector. Clearly, even higher precision can be obtained by further training the detector against a larger training set.

Once the metrics and the measuring tools are further developed, they will open the way to new questions of philosophical interest. For instance, it will become possible to *locate* the reference to visual representations within each article, thus opening the way to testing whether such references occur mostly in the first, middle or last parts of the article, as hypotheses about the framing, argument and perspective of the paper might suggest. A preliminary analysis based on the current sample is shown in Fig. 6 which indicates where diagrams mostly occur in articles by plotting the average proportion of the text block occupied by diagrams for 5% intervals of the text. When combined with further *structural* information mined from other corpora which include e.g. LaTeX markup, it will even be possible to distinguish between diagram usage in different structural contexts.

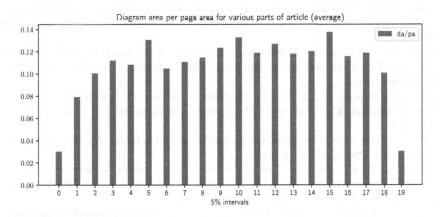

Fig. 6. Location of diagrams in mathematical papers, measured as the average da/pa in various parts of the paper when divided into 5% intervals.

Thus, the construction and integration of different metrics holds the potential of fine-tuning our quantitative 'telescope' from the "distant reading" suggested here. This makes it possible to identify particular journals or certain papers for further, qualitative investigation.

Moreover, by combining more sophisticated linguistic tools and, in particular, looking for references between text and visual representations, it will be possible to further ascertain the role that diagrams play in the mathematical argument. Both of these further perspectives hold great promise for the philosophy of mathematical practice aided by digital humanities and machine learning.

References

Arsenault, D.J., Smith, L.D., Beauchamp, E.A.: Visual inscriptions in the scientific hierarchy. Sci. Commun. **27**(3), 376–428 (2006). https://doi.org/10.1177/1075547005285030

Cleveland, W.S.: Graphs in scientific publications. Am. Stat. **38**(4), 261 (1984). https://doi.org/10.2307/2683400

Giaquinto, M.: The epistemology of visual thinking in mathe- matics. In: Zalta, E.N. (ed.) The Stanford Encyclopedia of Philosophy. Spring 2020. Metaphysics Research Lab, Stanford University (2020). https://plato.stanford.edu/archives/spr2020/entries/epistemology-visual-thinking/

Guerrero-Bote, V.P., Félix, M.-A.: A further step forward in measuring journals' scientific prestige. The SJR2 indicator. J. Inform. **6**(4), 674–688 (2012). https://doi.org/10.1016/j.joi.2012.07.001

Johansen, M.W., Misfeldt, M.: Material representations in mathematical research practice. Synthese **197**(9), 3721–3741 (2018). https://doi.org/10.1007/s11229-018-02033-4

Chapman, P., Stapleton, G., Moktefi, A., Perez-Kriz, S., Bellucci, F. (eds.): Diagrams 2018. LNCS (LNAI), vol. 10871. Springer, Cham (2018). https://doi.org/10.1007/978-3-319-91376-6

Larkin, J.H., Simon, H.A.: Why a diagram is (sometimes) worth ten thousand words. Cogn. Sci. **11**(1), 65–100 (1987). https://doi.org/10.1111/j.1551-6708.1987.tb00863.x

Mizrahi, M.: The case study method in philosophy of science. An empirical study. Perspect. Sci. **28**(1), 63–88 (2020)

Pietarinen, A.-V., Chapman, P., Bosveld-de Smet, L., Giardino, V., Corter, J., Linker, S. (eds.): Diagrams 2020. LNCS (LNAI), vol. 12169. Springer, Cham (2020). https://doi.org/10.1007/978-3-030-54249-8

Wu, Y., et al.: Detectron2 (2019). https://github.com/facebookresearch/detectron2

On the Relationship Between Geometric Objects and Figures in Euclidean Geometry

Mario Bacelar Valente[✉] [iD]

Pablo de Olavide University, Seville, Spain
mar.bacelar@gmail.com

Abstract. In this paper, we will make explicit the relationship existing between geometric objects and figures in planar Euclidean geometry. Geometric objects are defined in terms of idealizations of the corresponding figures of practical geometry. We name the relationship between them as a relation of idealization. It corresponds to a resemblance-like relationship between objects and figures. This relation is what enables figures to have a role in pure and applied geometry. That is, we can use a figure in pure geometry as a representation of geometric objects because of this relation. Moving beyond pure geometry, we will defend that there are two other 'layers' of representation at play in applied geometry.

Keywords: Figures · Diagrams · Geometry · Geometric objects

1 Introduction

The role of diagrams in geometry has been the subject of many philosophical inquires. Here, we endeavor to determine what kind of relationship exists between geometric objects and geometric figures in planar Euclidean geometry.

The rationale behind this work is the following. If there is a clear relation existing between geometric objects and geometric figures, then, this might condition or even determine what role geometric figures and diagrams (composite figures) can have when used in the context of pure or applied geometry.

The work unfolds as follows. First, in Sect. 2, we will address geometric figures. In Sect. 3, we will consider the treatment of geometric objects in Euclid's *Elements*. This will enable us to bring to light the relationship that geometric objects have with geometric figures. In Sect. 4, we will determine basic features that geometric figures or diagrams have due to this relationship when used in the context of pure or applied planar Euclidean geometry.

2 Geometric Figures in Practical Geometry

We can have geometric figures even without a clear indication of how they are conceptualized (see, e.g., [1, pp. 45–46]). A conceptualization proper of geometric figures arises in the context of a practical geometry where there are clear geometrical practices, and,

© Springer Nature Switzerland AG 2021
A. Basu et al. (Eds.): Diagrams 2021, LNAI 12909, pp. 71–78, 2021.
https://doi.org/10.1007/978-3-030-86062-2_7

importantly, the figures are named. This is already the case during the Old Babylonian period [2].

A good example is that of the rectangle. Each side is given a name that refers to agricultural field plots. They are named the 'long side' and the 'front'. The side called the 'front' is one of the small sides that is parallel to an irrigation channel [3, p. 34].

That Mesopotamian practical geometry arises in the context of field measurements has important implications regarding how the geometric figures were conceptualized. The rectangle, be it an actual field plot or a drawing (for example, a field plan), is conceptualized in terms of the boundary that establishes an inner space separated from the outside by it. The figure proper is what is inside the boundary [1, p. 64].

In Mesopotamia, the area of a field plot was calculated from the measurement of its boundary. Land surveyors could only rely on length measurements. For that purpose, they could use, e.g., ropes whose lengths were given in terms of a metrological length unit [4, pp. 296–297]. To calculate the area of quadrilateral field plots, surveyors applied the so-called surveyors' formula. This formula enables us to calculate what for us is the approximate value of a quadrilateral figure [5, pp. 106–107].

This boundary-oriented conceptualization of space lasted in Mesopotamian mathematics. In geometrical problems from the Old Babylonian period, one still finds "the assumption that the area of a quadrilateral is determined by the surveyors' formula" [5, p. 117]. We can say that the notion of area of geometric figures derives from the notion of practical geometry [5, pp. 115–117]. We see an example of this in the conception of circle in Mesopotamian mathematics. Like in the case of a rectangular figure, a circle is conceptualized in terms of its boundary: "a circle was the shape contained within an equidistant circumference" [2, p. 20]. The circle and its boundary were given the same name, something like "thing that curves" [2, p. 20]. Like in the case of quadrilateral figures the area of the circle is calculated, using a formula, from the length of its boundary (which can be measured) [2, p. 18].

Circle figures are well-attested in ancient Mesopotamian mathematical problems. The drawings can be very sketchy but also quite precise. In one example, there is a drawing of an equilateral triangle inscribed in a circle [6, pp. 207 and 488]. Not only the sides of the triangle are quite rectilinear, but also the circle is very precise since it was drawn using a compass [6, p. 207].

3   Relating Geometric Objects to Geometric Figures

In pure planar geometry as developed in Euclid's *Elements,* the geometric object called circle, like other geometric objects, is explicitly defined in the definitions.[1] At this point, it might be useful to make a silly question: Why do we name this geometric object with the same name as that of a figure?

To help to answer our silly question we will first consider another one: Why not take the definition of a circle as a geometric object as a definition of a circle figure? As it is, the definition of circle in the *Elements* seems to correspond to the practice of drawing a

---

[1] "A circle is a plane figure contained by one line such that all the straight lines falling upon it from one point among those lying within the figure are equal to one another; and the point is called the center of the circle" [7, pp. 153–4].

circle figure using a compass. The center of the circle is the needle point of the compass, and all points of the circumference drawn with the compass lead are at the same distance from this point as measured using, e.g., a ruler.[2] How can this be possible, using the same definition for a drawn figure and for a geometric object? For us, it comes down to semantics; in particular, the meaning of a few terms, which must be understood in the context of a particular geometrical practice.

A circle as a geometric object is instantiated in an idealized plane [9, p. 208] – an abstraction from a real physical plane [10, p. 19], like a dusted surface or a wax tablet [11, pp. 14–16]. As defined, the circle as a geometric object has all radii equal to one another. Here, 'equal' does not mean the same as 'equal' in practical geometry. In the latter case, the equality of different radii is a practical one; we simply neglect whatever small differences in lengths there are. In pure geometry, it is made an idealization of this practical approach and instead of conceiving of practically equal radii, these are conceived as exactly equal. We have what we might call an exactification of the equality of lengths. The relationship between the geometric circle and the circle figure is what we might call a relation of idealization: the abstract object is defined in terms of an idealization of the concrete figure.

This relation of idealization is made clearer by considering similar relations for lines and points [12]. As defined in the *Elements*, "a point is that which has no part" [7, p. 153]. This definition can be seen as arising from an idealization of the practice of practical geometry: "a point is characterized as a non-measurable entity, as it has no parts that can measure it" [13, p. 18]. In the same way, a "line is breadthless length" [7, p. 153]. Drawn lines have small breadths whose lengths are disregarded in practical terms. We idealize the concrete line as a geometric object that has an exact length and is breadthless. We can say that both the geometric line and the geometric point are in a relation of idealization with lines and points from practical geometry.

The relationship between geometric objects and figures is manifested very clearly in the definition of geometric line. We can see that the definition of geometric line is made by reference to an idealization at play. The geometric line is defined in relation to what is being implicitly idealized: a line from practical geometry. It only makes sense a definition in terms of breadthless length, in relation to something that has breadth. The definitions of geometric objects are dependent on the figures and result from an idealization of these. This enables us to say that between geometric objects and figures we have a relation of idealization.

---

[2] To the best of our knowledge, there is no extant text containing a practical definition of circle corresponding to that of geometric circle in the *Elements*. However, there are records of a conceptualization that approaches that in terms of radius [2, p. 20]. That this conceptualization can be ascribed to circle figures independently of having been adopted in the context of pure geometry is suggested, e.g., by a Greek third-century BC papyrus containing practical geometrical problems among others [8, pp. 70–2].

4 Basic Features of the Role of Diagrams in Pure and Applied Geometry

In the previous section, we have determined what we called the relation of idealization between geometric objects and figures. We expect that this relationship determines basic features regarding the role of figures or diagrams when used in the context of pure or applied geometry. To show this, we will consider two propositions, one from pure geometry and another from applied geometry. We will start with pure geometry.

In proposition 1 of book 1 of Euclid's *Elements* (proposition I.1), one constructs a geometric object – an equilateral triangle – by a particular procedure where one uses two circles that intersect each other. The text is accompanied by a lettered diagram (see Fig. 1), and with the letters one refers in the text to parts of the diagram.[3]

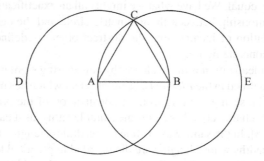

Fig. 1. The diagram in proposition I.1 of the *Elements*

The basic point we want to make here is to question how come that in the demonstration of a result in pure geometry we use a diagram which is a drawing consisting of several figures? The evident answer is that we take the diagram to represent the geometric objects. But what justifies using figures from practical geometry as a representation of geometric objects? More generally, we have to know what enables something to be a representation of something else. There are two features related to representation that are relevant here: intentionality and resemblance (see, e.g., [15–17]).

Regarding the intentionality in the adoption of a representation, what is relevant for us here is not so much that the intention of the author that adopts a particular representation is usually relevant in the interpretation of the representation, but that 'intentionality' underlies the possibility of choosing quite freely what we take to be the representation of something else. For example, we might decide that a hand-drawn line represents a segment drawn using a straightedge. That is, we intentionally take the sketchy line to represent the practical segment or segment figure.

The intentionality enables us to choose whatever we want as a symbol (representation) of something else. With an ad hoc representation, we would not go very far in the case under consideration. So, we rely on another concept related to representation. That of resemblance. We would go from just a symbol to a symbol that has iconic properties.

---

[3] We can expect that early versions of the *Elements* included at least unlettered diagrams [14].

That is, to a symbol that in some way resembles what it is symbolizing. But here we face a major problem. A geometric object is not something that we can see. It is instantiated in an idealized plane not in the space of our experience. There is no way in which we might say that a circle figure resembles a geometric circle. How do we overcome this difficulty?

A circle figure does not resemble a geometric circle; this simply has no meaning, unless we twist considerably the semantics of the word 'resemblance'. However, we have another kind of relationship between geometric objects and figures. We can adopt the relation of idealization, e.g., between a geometric circle and a circle figure to take the second as a representation of the first. The relation of idealization works as a resemblance-like relation. While the circle figure does not resemble the geometric figure, we can nevertheless establish a simulacrum of a resemblance between them. The circle figure has as its center the needle point of the compass. To this concrete point corresponds the geometric point as the center of the geometric circle. To the drawn circumference corresponds a breadthless line. While the radii of the circle figure are equal only within a particular practice of practical geometry where we neglect small measurement differences, the radii of a geometric circle are equal exactly as if corresponding to an idealized measurement in which all lengths are exactly equal.

The relation of idealization is a sort of resemblance-like relation that enables us to take the circle figure to be a representation of the geometric circle. Since we have one-to-one resemblance-like relations between all relevant elements of the geometric circle and the circle figure (center, circumference, radii, diameter, etc.), the circle figure works as an avatar of the geometric object in the diagram. When in the text we refer to aspects of the diagram we can take these as referring to the corresponding aspects of geometric objects.

This is a very basic characteristic of the use of diagrams in pure geometry. We suggest that any account of the role of diagrams in pure geometry should be compatible with this feature and how it arises.

Let us now address the issue of the role of diagrams in applied geometry. We will consider Proposition 1 of Euclid's *Optics*. What we want to determine here is in what way, if any, do we move beyond the representational role that a figure has in pure geometry. In that case, as we have just seen, we can establish a resemblance-like relationship between geometric object and figure.

When applying geometry like in the *Optics*, we take geometric objects to represent physical phenomena. The basic idea developed in Euclid's *Optics* is that the eyes emit 'visual fire'. It is the 'visual fire' that enables us to see the world around us. For example, the incidence of 'visual fire' in objects is what enables us to see them. 'Visual fire' is represented in the *Optics* by geometric segments [18, p. 8]. Proposition 1 of the *Optics* is as follows:

Proposition 1: No observed magnitude is seen simultaneously as a whole.

Call AD the observed magnitude, and B the eye from which the visual rays BA, BG, BK, BD fall. Since the visual rays diverge, they do not fall on the magnitude AD in a contiguous manner; so that there are intervals of this magnitude on which the visual rays do not fall. Consequently, the entire magnitude is not seen simultaneously. However, as the visual rays move rapidly, it is as if we saw [the entire magnitude] simultaneously. (Cited in [18, p. 10]) (Fig. 2).

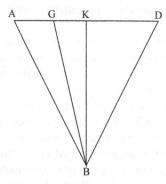

Fig. 2. The diagram in Proposition 1 of the *Optics*

Here, we are going to make some magic. We will use the above diagram to help us to interpret the text. We will take advantage of the representational roles of the diagram even if we have not clarified what these are. The geometric segment AD represents the physical object that we see. The geometric segments BA, BG, BK, BD represent the physical 'visual fire' emitted by an eye of the observer. The geometric point B represents an eye. Here, we are using the diagram to help us clarify the representational role of geometric figures. In fact, we are ascribing to the diagram these features. When we look, for example, at the drawn line BA, we take it to represent the 'visual fire'. The point is that since we have a resemblance-like relation established between the figure and its corresponding geometric object, and we take the geometric object to represent a physical entity, we can intentionally ascribe to the figure the representational role of its corresponding geometric object. We put another 'layer' of representation on top of the first one (see also [19]).

At this point, we can say that in applied geometry, the geometric figure has a double representational role. The geometric figure (or diagram) represents the geometric object, and this represents a physical entity. In this way, the geometric figure represents the physical entity, via the geometric object represented by the figure.[4]

There is in our view a third 'layer' of representation in the diagrams of Euclid's *Optics*. The geometric objects are given a representational character in the context of several assumptions. For example, it is assumed that "the straight lines drawn from the eye diverge to embrace the magnitudes seen" (cited in [18, p. 9]). How the visual rays 'diverge to embrace' is further specified in Proposition 1. There, the visual rays are taken to "move rapidly" [18, p. 10]. This corresponds to ascribing to the diagram a new 'layer'

---

[4] One might ask what justifies taking a geometric object to represent physical phenomena in the.first place. Again, it is due to the relation of idealization that we have between geometric objects and concrete objects. For instance, a geometric segment is in a relation of idealization not only with, e.g., a practically drawn segment but also, e.g., with a rod, a stretched rope, or with the 'visual fire' taken to be a sort of light beam [12].

of representation of the optical phenomena. We have the assumption that there is a sort of scanning of magnitudes by emitting successively the visual rays BA, BG, BK, and BD. The diagram as a whole is a static representation of a dynamic situation (see also [19]).

While this layer of representation relates, as the second one, to the geometric objects, it is only meaningful when taking into account the whole diagram. Like with the second layer of representation (where we can regard it as implemented directly on the figures), we can see this further 'layer' of representation as implemented directly on the diagram. An important difference with the second 'layer' is that it is not implemented so much on the figures that form the diagram but on the diagram as a whole.

For applied geometry, the situation is then as follows. The figures represent geometric objects due to the relation of idealization existing between them. Since we take the geometric objects to represent physical phenomena like, e.g., 'visual rays', we take the corresponding figures to represent the physical phenomena. This is a second 'layer' of representation that we ascribe to the figures. Besides this, the geometric objects on a whole have a dynamic relationship between them since they represent not only physical entities but also their dynamics. We must take into account that the visual rays 'move rapidly'. This corresponds to ascribing a third 'layer' of representation not to each figure individually but to the diagram as a whole since it is only at this 'level' that we can represent the dynamics. With this third 'layer' of representation, the diagram represents the dynamics of the physical phenomena.

These are basic aspects of the role of diagrams in applied geometry that follow from the relation of idealization that exists between geometric objects and figures and from taking geometric objects to represent physical phenomena. We suggest that any account of the role of diagrams in applied geometry should be compatible with this view.

5 Conclusions

In this work, we have tried to determine what kind of relationship there is between geometric figures of practical geometry and geometric objects of pure geometry. We have established that there is a relationship between geometric objects and figures. Geometric objects in the *Elements* are defined in terms of idealizations of the corresponding figures of practical geometry. We have named the relationship between them as a relation of idealization.

This relation existing between objects and figures is what in our view enables figures to have a role in pure and applied geometry. That is, we can use a figure or diagram as a representation of geometric objects or composite geometric objects because the relation of idealization corresponds to a resemblance-like relationship between objects and figures.

Moving beyond pure geometry we have defended that there are two other 'layers' of representation at play in applied geometry: 1) geometric figures can be ascribed as representing physical phenomena when we give this representational role to their corresponding geometric objects due to the relation of idealization existing between them; 2) The diagram as a whole can be taken to represent dynamical features of the physical phenomena also for the same reason.

References

1. Robson, E.: Mathematics in Ancient Iraq. Princeton University Press, Princeton (2008)
2. Robson, E.: Words and pictures: new light on Plimpton 322. In: Anderson, M., Katz, V., Wilson, R. (eds.) Sherlock Holmes in Babylon and other tales of mathematical history, pp. 14–26. Mathematical Association of America, Washington (2004)
3. Høyrup, J.: Lengths, Widths, Surfaces: a Portrait of Old Babylonian Algebra and its Kin. Springer, New York (2002). https://doi.org/10.1007/978-1-4757-3685-4
4. Baker, H.D.: Babylonian land survey in socio-political context. In: Selz, G.J., Wagensonner, K. (eds.) The Empirical Dimension of Ancient Near Eastern Studies, pp. 293–323. LIT, Wien (2011)
5. Damerow, P.: The impact of notation systems: from the practical knowledge of surveyors to babylonian geometry. In: Schemmel, M. (ed.) Spatial Thinking and External Representation: Towards a Historical Epistemology of Space, pp. 93–119. Edition Open Access, Berlin (2016)
6. Friberg, J.: A Remarkable Collection of Babylonian Mathematical Texts: Manuscripts in the Schøyen Collection Cuneiform Texts I. Springer, New York (2007). https://doi.org/10.1007/978-0-387-48977-3
7. Euclid: The thirteen Books of the Elements, vol. I–III, edited by T.L. Heath. Dover Publications, New York (1956)
8. Cuomo, S.: Ancient Mathematics. Routledge, London (2001)
9. Mueller, I.: Philosophy of Mathematics and Deductive Structure in Euclid's Elements. MIT Press, Cambridge (1981)
10. Taisbak, C. M.: ΔΕΔΟΜΕΝΑ. Euclid's Data or the Importance of Being Given. Museum Tusculanum Press, Copenhagen (2003)
11. Netz, R.: The Shaping of Deduction in Greek Mathematics: a Study in Cognitive History. Cambridge University Press, Cambridge (1999)
12. Valente, M.B.: From practical to pure geometry and back. Rev. Brasil. Hist. Mat. **20**, 13–33 (2020)
13. Harari, O.: The concept of existence and the role of constructions in Euclid's *Elements*. Arch. Hist. Exact Sci. **57**, 1–23 (2003)
14. Saito, K.: Diagrams and traces of oral teaching in Euclid's *Elements*: labels and references. Zentralbl. Didaktik Math. (ZDM) **50**, 921–936 (2018)
15. Abell, C.: Canny resemblance. Philos. Rev. **118**, 183–223 (2009)
16. Kulvicki, J.V.: Images. Routledge, London (2006)
17. Blumson, B.: Resemblance and representation. An essay in the philosophy of pictures. Open Book Publishers, Cambridge (2014)
18. Darrigol, O.: A History of Optics: from Greek Antiquity to the Nineteenth Century. Oxford University Press, Oxford (2012)
19. Valente, M.B.: Geometrical objects and figures in practical, pure, and applied geometry. Disputatio. Philos. Res. Bull. **9**, 33–51 (2020)

What Diagrams Are Considered Useful for Solving Mathematical Word Problems in Japan?

Hiroaki Ayabe[1]([⊠]), Emmanuel Manalo[2], Mari Fukuda[3], and Norihiro Sadato[1]

[1] National Institute for Physiological Sciences, Aichi, Japan
{ayabe,sadato}@nips.ac.jp
[2] Graduate School of Education, Kyoto University, Kyoto, Japan
manalo.emmanuel.3z@kyoto-u.ac.jp
[3] Graduate School of Education, The University of Tokyo, Tokyo, Japan
mari_fukuda@p.u-tokyo.ac.jp

Abstract. Previous studies have shown that diagram use is effective in mathematical word problem solving. However, they have also revealed that students manifest many problems in using diagrams for such purposes. A possible reason is an inadequacy in students' understanding of variations in types of problems and the corresponding kinds of diagrams appropriate to use. In the present study, a preliminary investigation was undertaken of how such correspondences between problem types and kinds of diagrams are represented in textbooks. One government-approved textbook series for elementary school level in Japan was examined for the types of mathematical word problems, and the kinds of diagrams presented with those problems. The analyses revealed significant differences in association between kinds of diagrams and types of problems. More concrete diagrams were included with problems involving change, combination, variation, and visualization of quantities; while number lines were more often used with comparison and variation problems. Tables and graphs corresponded to problems requiring organization of quantities; and more concrete diagrams and graphs to problems involving quantity visualization. These findings are considered in relation to the crucial role of textbooks and other teaching materials in facilitating strategy knowledge acquisition in students.

Keywords: Math education · Diagram use · Textbook research

1 Introduction

Mathematical word problem solving is commonly used in schools to cultivate students' abilities in applying mathematical knowledge and skills to everyday life. However, the complexity of cognitive processes demanded by this activity leads to a prevalence of difficulties in students [1]. Diagram use is generally considered one of the most effective strategies for promoting success in this kind of problem solving [2]. However, students tend not to use diagrams spontaneously, and even when they do use them, such use does not always lead to obtaining the correct solutions.

© The Author(s) 2021
A. Basu et al. (Eds.): Diagrams 2021, LNAI 12909, pp. 79–83, 2021.
https://doi.org/10.1007/978-3-030-86062-2_8

One reason for the failure to solve despite the use of diagrams is a lack of correspondence between the problem schema/type (i.e., its structure and requirements) and the kind of diagram that students construct [3]. Word problems are included in the school curriculum to facilitate understanding of target learning contents, and diagrams are used as a way of "scaffolding" to make solving easier. However, inadequate attention is placed on teaching "appropriate diagram use" to match the problems given because the curriculum focuses on teaching "mathematical contents". Thus, much of the knowledge that students develop about diagram use remains implicit.

To address this problem, ways to effectively facilitate understanding of the correspondence between problem schema/type and appropriate diagrams need to be investigated. While it would be difficult to capture and evaluate the wide range of methods that teachers use to promote such understanding, examining how the textbooks might facilitate that understanding is much more manageable. In Japan, all schools use textbooks certified by the Ministry of Education in line with the national guidelines (issuance law of Japan). Examining such textbooks could provide valuable insights into how those correspondences are portrayed, which had not been examined in previous research. Therefore, this study aimed to clarify: (i) the types of word problems used in math textbooks, (ii) the kinds of diagrams deemed useful for solving those problems, and (iii) the correspondence between problems and diagrams.

2   Method

The textbooks used were one of six mathematics textbook series eligible for use in Japan at the elementary school level (6 books corresponding to Grades 1 to 6, each containing 13–19 chapters) [4]. Each chapter contains example problems with detailed explanations. The main purpose of those example problems is to teach the learning contents set by the national standard, assuming that children would be learning them for the first time. The present study focused on all the example problems that take the form of word problems, defined here as problems comprising two or more sentences with a background story, and which requires a mathematical solution. The word problems were examined and categorized according to their problem schema (i.e., type of word problem), and the kind of diagram deemed appropriate for solving them.

In line with previous studies [5], the word problems found were classified into one of six types: Change (e.g., Ken had 3 candies. Naomi gave him 5 more candies. How many candies does Ken have now?), Combine (e.g., Ken has 4 candies. Naomi has 3 candies. How many candies do they have altogether?), Compare (e.g., Ken has 4 candies. Naomi has 12 candies. How many times are Naomi's candies greater than Ken's?), Vary (e.g., Ken packs 5 candies in each box. How many boxes will he need to pack to fit 40 candies?), Organize (i.e., organizing data given to find an answer by using tables or graphs; e.g., The following data shows students' reading times in a month. What number appears the most often?), and Visualize (visualizing the conditions given in the problem with figures, graphs, tables, etc., to facilitate search for rules difficult to find based only on superficial details; e.g., There are 3 children in front of Ken, and 4 behind him. How many children are there in total?).

Diagrams were categorized into seven kinds: Pictures (images relating to the problem but conveying no quantitative information), Concrete diagrams (illustrations/pictures

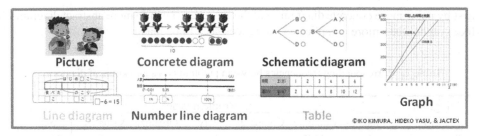

Fig. 1. Kinds of diagrams (with examples) used to categorize those found in the textbooks

with quantitative information, including semi-concrete representations of quantities using circles and other counters), Schematic diagram (arrows, lines, figures showing procedures and quantitative/functional relationships, including tree diagrams), Line diagrams (line or tape diagrams, segments of which indicate quantities to facilitate visual comparison), Number line diagrams (one or more lines arranged to show relationships between quantities, including proportional relationships), Tables (arrays of numbers, symbols, and words/letters), and Graphs (bar graphs, line graphs, statistical graphs such as histograms and dot plots, function graphs) (see Fig. 1). Coding for each textbook was independently undertaken by combinations of two of four school teachers, none of whom had any vested interest in the outcomes.

The reliability statistic between inter-raters indicated almost perfect agreement (Cohen's kappa = 0.84). Differences were settled through discussion.

3   Results and Discussion

There were 246 word problems found in the six textbooks (mean = 41.0 ± 9.76), and diagrams were included with all of them. The textbooks therefore clearly convey the importance of diagram use in problem solving. The numbers of word problems varied depending on grade level ($\chi^2_{(5)} = 11.61$, $p = .041$): lowest at Grade 1 ($n = 25$, $p = .003$), and highest at Grade 5 ($n = 54$, $p = .013$). The Grade 1 textbook contains mostly simpler one-sentence problems, while the concept of proportionality is introduced in the Grade 5 textbook, which may partially explain those differences.

An analysis of residuals after Chi-squared tests identified proportional differences. Table 1 shows the frequencies with which each kind of diagram was included with each type of problem, with asterisks indicating those proportionally significantly higher or lower. In total, Concrete diagrams and Number line diagrams were found to be the most frequently used diagrams ($p < .001$, for both). Concrete diagrams were often used with Change, Combine, Vary, and Visualize problems, suggesting that they are helpful in facilitating the four arithmetic operations and understanding of the problem scenario and contents. Number line diagrams were often used with Compare and Vary problems, suggesting that they are helpful when comparing measurement quantities and performing proportional calculations. More abstract diagrams (Tables and Graphs) were used with Organize problems. Concrete diagrams and Graph were used with Visualize problems to support visual search. These results suggest the possibility of scaffolding in diagram use, first by using more concrete diagrams and then progressing onto more abstract diagrams like graphs when more complicated inferences need to be made.

Table 1. Corresponding frequencies between problem types and kinds of diagrams

Diagram kind	Problem type						Total
	Change	Combine	Compare	Vary	Organize	Visualize	
Picture	3	**11****	0*	10	2	0*	26*
Concrete	**25****	**13****	3	20*	0**	**8****	**69****
Schematic	4	3	0*	8	5	0*	20**
Line	3	9	2	9	0**	1	24*
Number line	0**	1**	**15****	**44****	0**	2	**62****
Table	0**	0**	1	2***	**14****	4	21**
Graph	0**	0**	2	0***	**15****	**7****	24*
Total	35	37	23	93	36	22	246
Mean	5.00	5.29	3.29	13.29	5.14	3.14	35.14
SD	8.98	5.56	5.28	15.00	6.64	3.29	20.93
$\chi^2_{(6)}$	96.80	35.08	50.96	101.57	51.50	20.64	74.81
p (adjusted)	0.00	0.00	0.00	0.00	0.00	0.00	0.00

p (adjusted): p-value adjusted by Bonferroni method. $* p < .05$, $** p < .01$, $*** p < .001$. The asterisks indicate significantly higher (with bolded values) or lower proportions.

These results clarify the correspondence between problem types and kinds of diagrams as portrayed in Japanese textbooks. As this examination was conducted using only one of the book series for elementary schools, it needs to be conducted also for the other series/levels of education, and for textbooks in other countries to promote a better understanding of how textbooks contribute to the cultivation of competencies in diagram use. Future research will also need to scrutinize the efficacy of the kinds of diagrams indicated in textbooks for solving different types of problems.

Acknowledgment. This research was supported by a grant-in-aid (20K20516) received from the Japan Society for the Promotion of Science.

References

1. Daroczy, G., Wolska, M., Meurers, W.D., Nuerk, H.-C.: Word problems: a review of linguistic and numerical factors contributing to their difficulty. Front. Psychol. **06**, 1–13 (2015)
2. Uesaka, Y., Manalo, E., Ichikawa, S.: What kinds of perceptions and daily learning behaviors promote students' use of diagrams in mathematics problem solving? Learn. Instr. **17**, 322–335 (2007)
3. Ayabe, H., Manalo, E., Hanaki, N.: Teaching diagram knowledge that is useful for math word problem solving. In: EAPRIL 2019 Conference Proceedings, pp. 388–399 (2020)

4. Souma,K., et al.: Tanoshii Sansuu 6 Nen (Fun Math Grade 6), 1st edn. Dainippon Tosho Publishing, Tokyo (2020)
5. Kintsch, W., Greeno, J.G.: Understanding and solving word arithmetic problems. Psychol. Rev. **92**, 109–129 (1985)

Open Access This chapter is licensed under the terms of the Creative Commons Attribution 4.0 International License (http://creativecommons.org/licenses/by/4.0/), which permits use, sharing, adaptation, distribution and reproduction in any medium or format, as long as you give appropriate credit to the original author(s) and the source, provide a link to the Creative Commons license and indicate if changes were made.

The images or other third party material in this chapter are included in the chapter's Creative Commons license, unless indicated otherwise in a credit line to the material. If material is not included in the chapter's Creative Commons license and your intended use is not permitted by statutory regulation or exceeds the permitted use, you will need to obtain permission directly from the copyright holder.

Diagrams and Logic

The Search for Symmetry in Hohfeldian Modalities

Matteo Pascucci[1(✉)] and Giovanni Sileno[2]

[1] Slovak Academy of Sciences, Bratislava, Slovakia
matteo.pascucci@savba.sk
[2] University of Amsterdam, Amsterdam, The Netherlands
g.sileno@uva.nl

Abstract. In this work we provide an analysis of some issues arising with geometrical representations of a family of deontic and potestative relations that can be classified as Hohfeldian modalities, traditionally illustrated on two diagrams, the Hohfeldian squares. Our main target is the lack of symmetry to be found in various formal accounts by drawing analogies with the square of opposition for alethic modalities. We argue that one should rather rely on an analogy with the alethic hexagon of opposition and exploit the notions of contingency and absoluteness in order to restore the symmetry of Hohfeldian modalities in accordance to the diagrams presented by Hohfeld. Interestingly, the investigation unveils three potestative squares defined at different levels of granularity (force, outcome and change) and allows us to further elaborate on the connections between deontic and potestative relations.

Keywords: Hohfeldian modalities · Modal hexagon of opposition ·
Deontic relations · Potestative relations · Symmetry · Power ·
Hohfeld's cube

1 Introduction

Diagrams are powerful tools for conceptual modelling, and powerful didactic tools. In the early XX century the legal scholar W. N. Hohfeld offered a systematic analysis of different uses of the word 'right' in the context of legal and judicial reasoning [5,6], illustrating the resulting framework by two diagrams, consequently named Hohfeldian squares. Hohfeld's theory of rights is based on a conceptual distinction between two families of normative relations: first-order relations (duty, claim, liberty, and no-claim), which can be also called *deontic relations*, and second-order relations (power, liability, disability, and immunity), also called *potestative relations*, which specify how first-order relations can be modified. Because of the focus on judiciary settings, Hohfeld's investigation is about *subjective rights*: all the uses of the word 'right'[1] express relations between two normative parties and a certain behaviour (a normative party can be taken

---

[1] We will not enter here into the debate whether power is a "proper" right, or rather an accessory construct necessary for the functioning of norms, see e.g. [9].

© Springer Nature Switzerland AG 2021
A. Basu et al. (Eds.): Diagrams 2021, LNAI 12909, pp. 87–102, 2021.
https://doi.org/10.1007/978-3-030-86062-2_9

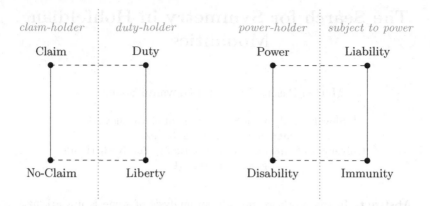

Fig. 1. The two Hohfeldian squares: (left) the deontic square, (also obligative or of the first-order), and (right) the potestative square (or of the second-order).

to be either an individual or a collective; a behaviour can be equated with a type of action). The eight normative relations were illustrated by Hohfeld on two squares: the obligative (or deontic) square and the potestative square (Fig. 1). In a broader perspective, these relations can be conceived of as ternary modalities; for this reason we will speak of *Hohfeldian modalities*. A natural question is then the following: how can the diagrammatic representation offered by Hohfeld's theory be transformed into a standard geometrical representation of modalities, such as an Aristotelian polygon of opposition?

Hohfeld's squares have an important role in legal education; Aristotelian squares are instead used in linguistic, literary and semiotic studies, and have attracted a renewed interest in logic. The convergence to geometric constructs is plausibly not by chance: cognitive studies show that symmetries facilitate perception of structure, memorization and thus recall. Unveiling an underlying connection between the two representations would have in principle both a theoretical and practical value. Indeed, diagrams help in understanding relations between norms. Suppose a legal code includes two norms N1 and N2. N1 speaks of the duties of a normative party x, N2 speaks of the liberties of a party y and it is intended that N1 and N2 jointly describe the normative relation between x and y. Diagrams (and, in particular, Aristotelian polygons of opposition) provide hints on how to translate, e.g., statements about duties into statements about liberties (and vice versa). In the present work we will point out that the task of translating Hohfeldian squares into Aristotelian polygons is very challenging, due to an overlap of perspectives from which some fundamental notions, such as power and liberty, can be analysed.

The paper proceeds as follows. Section 2 provides the background, with a brief overview of related works. Section 3 presents the notation upon which our proposal will be constructed, and introduces to the most common formalization of the normative concepts illustrated in Hohfeld's framework. Section 4 investigates those concepts through the framing of squares of opposition, reorganizing, integrating, and extending several contributions presented in the literature.

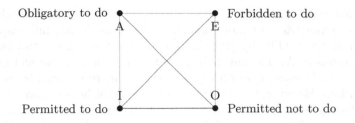

Fig. 2. Aristotelian square of opposition for basic deontic modalities, with subaltern (green), contrary (orange), sub-contrary (blue) and contradictory (red) bindings, and the traditional labels associated to each node (A, E, I, O). (Color figure online)

In particular, we unveil three squares of opposition associated to the potestative dimension, defined at different level of abstraction (force, outcome, and change). Section 5 argues that, in order to obtain a proper symmetric treatment of the two families of Hohfeldian modalities (in particular with respect to the notions of liberty and of power), one has to move from squares to hexagons of opposition. Finally, Sect. 6 elaborates shortly on the causal and logical connections between the deontic and potestative relations.

2 Background and Related Works

A good starting point to build a geometrical representation of Hohfeldian modalities is the Aristotelian square of opposition for *basic* deontic modalities, that is, obligation, permission and forbiddance (prohibition),[2] reproduced in Fig. 2.[3] Hohfeld did not refer to Aristotelian squares in his work; however, his diagrams were meant to clarify the logic of the relations at the basis of his theory.[4]

While Hohfeld's notions have been long since analysed via formal logic (see, e.g., the seminal works of Lindahl [8] or Makinson [10], or, more recently, the work of Markovich [11] or our own proposal [17]), a full understanding of the meaning of his two squares has been regarded as problematic. No standard

---

[2] We stress the difference between basic deontic modalities and Hohfeldian deontic modalities, since the former do not make reference to normative parties, while the latter do. Basic deontic modalities have occupied a central role in the development of formal systems of deontic logic. For an historical survey, see [2].

[3] The Aristotelian square of opposition for basic deontic modalities is construed following the parallel between deontic and alethic modalities usually attributed to Leibniz ('Elementa Juris Naturalis', 1669-71), but later also independently acknowledged by Jeremy Bentham ('Of Laws in General', 1782). According to Leibniz's definition, an obligation is "the necessity which constrains the wise to do good".

[4] An interesting historical remark is that, before studying law in the footsteps of Bentham and Austin, Hohfeld started his formal education in chemistry. In hindsight, one can see in Hohfeld's account traces of both the jurisprudential analytical tradition, as well as an attempt to support legal practitionners just as Mendeleev's periodic table (first presented in 1869) supports chemists.

formalization exists today. Hohfeld's framework suggests that the relations at its core are symmetric, but most used formalizations do not fully capture this feature. In contrast, O'Reilly [12] elaborated on reframing the two Hohfeldian squares in terms of Aristotelian polygons (focusing on what we will name here "change-centered" power), in order to provide a more systematic encoding of logical relations. However, the Aristotelian squares that he gets also encode problematic relations between, e.g., liberties and duties due to a lack of symmetry. He points out that a different sort of Aristotelian polygon is called for. Sileno [16], following Blanché's insights [4], proposed to use triangles of opposition (focusing on a "force-centered" power). However, both works presented a semi-formal conceptualization only.

The present contribution aligns and extends both diagrammatic and formal characterizations. Hohfeldian squares are mapped to Aristotelian hexagons, to restore the symmetries on various levels. Additionally, we formalize change-, force-, and outcome-centered squares of power, visualizing their mutual relations.

3   Formalization

According to Lindahl [8], all relations in each of the deontic and potestative families of concepts analysed by Hohfeld are interdefinable, in the sense that one could take a single deontic relation and a single potestative relation as primitives and introduce all the others via logical operations. This idea will serve as a guide through our formal transposition of Hohfeld's theory in the present section.

3.1   Language

In order to analyse the notions at stake, we can conveniently introduce a language of first-order logic. We will use two categories of variables: x, y etc. to denote normative parties and α, β, etc. to denote action types. We will also have *constants* p, q, etc. for normative parties and constants A, B, etc. for action types. The symbol $-$ (overline) will denote complementation on action types. Complementation will be the only operation that allows one to build *complex* action types: given an action type A, $\overline{A}$ will denote the complement of A, that is, the type of any action that does not instantiate A. We will work under the assumption that the Law of Double Complementation holds ($\overline{\overline{A}} = A$). Hohfeldian modalities will be represented via n-ary predicates (relations) and will be given an explicit name throughout the presentation. We will use a different font for relations not corresponding to Hohfeldian modalities (the only relation that is not a Hohfeldian modality in the rest of the exposition is *Ability*). In some cases the argument of a relation can be a statement involving another relation. However, no quantification on such statements will be employed; therefore, the language will remain at the level of first-order logic. Finally, we will employ standard symbols for logical connectives: $\neg$ to denote Boolean negation, $\rightarrow$ to denote material implication, $\equiv$ to denote material equivalence, $\exists$ and $\forall$ as quantifiers, etc. For the sake of brevity, we will omit quantification over variables for normative parties, interpreting a formula of the form $\phi(x, y, ...)$ as implicitly having the form $\forall x \forall y ... \phi(x, y, ...)$. Thus, while we will read an expression of the form Claim(x, y, A) as "for all x, for all y: x has a claim that A be performed by y", we will read an expression of the form Claim(p, q, A) as "p has a claim that A be performed by q". We divide our analysis into two parts, respectively dealing with first-order and second-order Hohfeldian relations.

3.2   First-Order Hohfeldian Relations

The formal renderings of the fundamental deontic relations identified in Hohfeld's framework, for two normative parties p and q and an action type A, are the following: Claim(p, q, A), Liberty(p, q, A), Duty(p, q, A) and NoClaim(p, q, A). One can immediately notice that the last Hohfelidian modality in this list is, due to its name, just the negation of the first. Therefore, if one wants to take Claim as the primitive deontic concept, the definability of NoClaim with respect to a given action type A turns out to be obvious, thanks to Boolean negation:

$$\text{NoClaim}(x, y, A) \equiv \neg\text{Claim}(x, y, A)$$

Furthermore, one can treat Claim and Duty as *correlative* notions, in the sense that they are two faces of the same modality, seen from the points of view of the two normative parties involved, (whence, Duty just results from a permutation of the two parties):

$$\text{Duty}(y, x, A) \equiv \text{Claim}(x, y, A)$$

Finally, one can define Liberty in terms of Claim, Boolean negation, a permutation of normative parties and action complementation:

$$\text{Liberty}(y, x, A) \equiv \neg\text{Claim}(x, y, \overline{A})$$

Note that the last two equations, together with the Law of Double Complementation, entail that:

$$\neg\text{Duty}(y, x, A) \equiv \text{Liberty}(y, x, \overline{A})$$

that is, the negation of a duty of performance corresponds to the liberty of non-performance.

3.3   Second-Order Hohfeldian Relations

Relations of the potestative family concern *actions* that trigger changes of first-order or even second-order relations (although for most legal scholars legal power concerns only first-order relations), such as, for instance, an action B creating a duty for a party q to perform an action A to the advantage of a party p. A possible way of writing that p has such a power would be by means of a predicate expression *Ability*(p, B, R) (cf. the predicate *has_ability* investigated in [17]), where R is a Hohfeldian relation issued at B's performance by p; for instance, *Ability*$(p, B, \text{Claim}(p, q, A))$. Indeed, to simplify the notation, we may abstract the triggering action B, and focus on a common relation, e.g. $R = \text{Claim}(p, q, A))$.

In the following we denote this canonic[5] power construct with reference to a given action type A by means of the expression $\mathsf{Power}(p, q, A)$, whose definition involves an existential quantification on the set of action types:

$$\mathsf{Power}(x, y, A) \equiv \exists \beta : Ability(x, \beta, \mathsf{Claim}(x, y, A))$$

The four fundamental potestative relations identified by Hohfeld can then be encoded following the same syntactic pattern used in the case of first-order relations, namely as ternary relations whose first and second argument is a normative party and whose third argument is an action type (that is, as the equivalence above indicates, the action type mentioned in the relation affected): $\mathsf{Power}(p, q, A)$, $\mathsf{Liability}(p, q, A)$, $\mathsf{Disability}(p, q, A)$, $\mathsf{Immunity}(p, q, A)$.

Now, suppose we take Power as the primitive potestative notion. First, one can treat $\mathsf{Disability}$ as the negation of Power and thus define it with reference to a given action type A as follows:

$$\mathsf{Disability}(x, y, A) \equiv \neg \mathsf{Power}(x, y, A)$$

Then, also in this case, one can identify correlative statements involving $\mathsf{Liability}$ and Power via permutations of the normative parties:

$$\mathsf{Liability}(y, x, A) \equiv \mathsf{Power}(x, y, A)$$

Finally, one can define $\mathsf{Immunity}$ in terms of Power, Boolean negation and a permutation of the normative parties:

$$\mathsf{Immunity}(y, x, A) \equiv \neg \mathsf{Power}(x, y, A)$$

In this case, the last two equations entail a structurally different template:

$$\neg \mathsf{Liability}(x, y, A) \equiv \mathsf{Immunity}(x, y, A)$$

that is, the negation of a liability (correlatively, power) towards performance corresponds to the immunity (disability) towards performance (whereas for the first square it was towards non-performance). Therefore, according to this formalization (or analogous proposals by most subsequent authors, e.g. [10,11,14]), the two Hohfeldian squares lose the symmetry suggested in Hohfeld's diagrams.

4   Hohfeldian Squares and Aristotelian squares

The previous formalization makes clear that any of the four relations on each of the Hohfeldian squares can be defined in terms of any other relation belonging to the same family. Some authors, as e.g. O'Reilly [12] (in turn extending Sumner's

---

[5] In legal scholarship, synonymous terms for power like *legal ability*, *legal capability* or *legal competence* are generally used only when the target of change constrains the conduct of agents: "[..] power (*Konnen*) is a legal concept only in-so-far as it includes within its ambit, claims or duties" [7].

analysis [18]), observe that, for any choice of a primitive deontic and potestative modality, one can build two Aristotelian squares of opposition whose corners are labelled by a formula where only the primitive modality of the relevant family is mentioned. The four formulas needed as labels for a square are obtained by the possible combinations of Boolean negation and action complementation (either both present, or both absent, or one present and one absent).

4.1   Deontic Square of Opposition

Let us choose Claim as primitive for the square of opposition extracted from the first-order Hohfeldian diagram. This choice leads to the following set of labels DR (for "deontic relations") with respect to a given action type A:

$$DR = \{\mathsf{Claim}(p, q, A), \mathsf{Claim}(p, q, \overline{A}), \neg\mathsf{Claim}(p, q, \overline{A}), \neg\mathsf{Claim}(p, q, A)\}$$

We will say that this is a Claim-based description of DR; exploiting the correlativity principles one could write, equivalently, a Duty-based description. The same holds, relying on other principles, for Right- and NoClaim-based descriptions of DR. The set DR, together with the meaning of Boolean negation and action complementation, naturally gives rise to a deontic square of opposition. The only additional principle needed is the following, used to characterize *subalternate statements*:

$$\mathsf{Claim}(x, y, A) \rightarrow \neg\mathsf{Claim}(x, y, \overline{A})$$

By substituting the (implicitly) quantified variables for normative parties x and y with constants p and q, this can be read as saying that if p has a claim towards q about the performance of A, then p does not have a claim towards q about its non-performance.

4.2   O'Reilly's (or Change-Centered) Potestative Square of Opposition

O'Reilly applies a similar approach to the potestative relations. He first considers power as the ability of p to *affect* q with respect to a relation R. This can be rephrased in our formal setting by saying that there are triggering actions that produce a *change* w.r.t. R. More precisely, a change can occur when either R or its contrary or its contradictory is created. Therefore, this is a *change-centered* notion of power. Focusing on $R = \mathsf{Claim}(p, q, A)$, we can take a triggering action B and write the definitional equivalence for this O'Reillian notion of power as:

$$\mathsf{Power}_{\mathsf{OReilly}}(x, y, B, A) \equiv Ability(x, B, \mathsf{Claim}(x, y, A))$$
$$\vee\ Ability(x, B, \mathsf{Claim}(x, y, \overline{A}))$$
$$\vee\ Ability(x, B, \neg\mathsf{Claim}(x, y, A))$$

We can then use quantification over the set of possible action types and the O'Reillian notion of power to define a form of *positive-change power* (Power^+), with respect to a given action type A, as below:

$$\mathsf{Power}^+(x, y, A) \equiv \exists\beta : \mathsf{Power}_{\mathsf{OReilly}}(x, y, \beta, A)$$

Positive-change power corresponds to the ability of affecting (in any sense) a relation. O'Reilly then refers to a distinct form of internal negation, capturing the ability of a party p to *not* affect a party q with respect to a relation R. This will be said to represent a form of *negative-change* or *no-change power* (Power^-). In other words, this means that p may choose an action that does not produce any change. In our formalization the resulting definitional equivalence would be:

$$\mathsf{Power}^-(x, y, A) \equiv \exists \beta : \neg\mathsf{Power}_{\mathsf{OReilly}}(x, y, \beta, A)$$

Starting from these concepts, we can define a set $\mathsf{PR}^\pm$ of potestative (change-centered) relations with the aim of building a square of opposition for second-order Hohfeldian relations. Here the four formulas needed are obtained via possible combinations of Boolean negation and of positive- vs. negative-change power:

$$\mathsf{PR}^\pm = \{\mathsf{Power}^+(p, q, A), \mathsf{Power}^-(p, q, A), \neg\mathsf{Power}^-(p, q, A), \neg\mathsf{Power}^+(p, q, A)\}$$

To build a square of opposition upon $\mathsf{PR}^\pm$, and follow O'Reilly's approach, one has to add the principle below, which captures subalternation:

$$\neg\mathsf{Power}^-(x, y, A) \rightarrow \mathsf{Power}^+(x, y, A)$$

However, in this case such a principle is not independent from the rest; the formalization proposed here, together with the plausible assumption that the set of action types is non-empty, already entails this principle. Note also that:

$$\mathsf{Power}(x, y, A) \rightarrow \mathsf{Power}^+(x, y, A) \qquad \mathsf{Power}^-(x, y, A) \not\rightarrow \neg\mathsf{Power}(x, y, A)$$

The first implication above indicates that the canonic notion of power introduced in Sect. 3.3 has a narrower scope than the notion of positive-change power.

4.3   Force-Centered Potestative Square of Opposition

The notion of power considered by O'Reilly is rather complex: one may then wonder whether a square of opposition may be constructed starting instead from more primitive forms of power. As observed in [16], [15, Ch.4], power relations can be put in analogical correspondence to physical phenomena as attraction, repulsion, and absence of those (independence). To express such physical metaphor of "force", we need to separate the *stimulus* component (a particular type of action, such as a verbal command) and the consequent target *manifestation* (a type of action that is due on the basis of the stimulus). If the latter is denoted by the action type symbol A, then, the former can be here conveniently represented via the symbol "A", rather than with a generic symbol for an action type B. In this way, one emphasizes the connection between stimulus and target manifestation. Relevant scenarios can be then identified on, e.g., whether stimulus and manifestation converge (A is always performed in correspondence to its stimulus) or diverge (A is never performed in correspondence to its stimulus). Using our

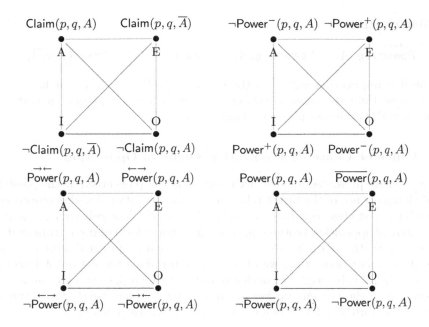

Fig. 3. Deontic and potestative (change-centered, force-centered, and outcome-centered) Aristotelian squares of opposition constructed from the two Hohfeldian diagrams. The usual convention on the colour of bindings apply.

notation, the definition of force-centered notions of power is:

$$\overset{\rightarrow\leftarrow}{\mathsf{Power}}(x, y, A) \equiv Ability(x, \text{"A"}, \mathsf{Claim}(x, y, A))$$

$$\overset{\leftarrow\rightarrow}{\mathsf{Power}}(x, y, A) \equiv Ability(x, \text{"A"}, \mathsf{Claim}(x, y, \overline{A}))$$

We will say that $\overset{\rightarrow\leftarrow}{\mathsf{Power}}$ represents positive-force power and that $\overset{\leftarrow\rightarrow}{\mathsf{Power}}$ represents negative-force power. (As an empirical confirmation, see e.g. the negative-force liability position found in the Dutch Act of Abjuration [15].)

From these concepts we can define a new set of potestative relations $\mathsf{PR}^{\overset{\leftarrow}{\rightarrow}}$ as labels for a force-centered potestative square of opposition. More precisely, here the four formulas needed for the square are obtained by taking into account all possible combinations of positive- vs. negative-force power and Boolean negation:

$$\mathsf{PR}^{\overset{\leftarrow}{\rightarrow}} = \{\overset{\rightarrow\leftarrow}{\mathsf{Power}}(p, q, A), \overset{\leftarrow\rightarrow}{\mathsf{Power}}(p, q, A), \neg\overset{\leftarrow\rightarrow}{\mathsf{Power}}(p, q, A), \neg\overset{\rightarrow\leftarrow}{\mathsf{Power}}(p, q, A)\}$$

The subalternity is here captured by the logical principle:

$$\overset{\rightarrow\leftarrow}{\mathsf{Power}}(x, y, A) \rightarrow \neg\overset{\leftarrow\rightarrow}{\mathsf{Power}}(x, y, A)$$

which is acceptable because otherwise the same stimulus "A" could generate two conflicting first-order relations.

Note also that:

$$\overrightarrow{\overleftarrow{\mathsf{Power}}}(x, y, A) \to \mathsf{Power}(x, y, A) \qquad \overleftarrow{\overrightarrow{\mathsf{Power}}}(x, y, A) \to \mathsf{Power}(x, y, \overline{A})$$

As the first implication indicates, the notion of positive-force power has a narrower scope than the canonic notion of power, whence *a fortiori*, a narrower scope than the notion of positive-change power.

4.4   Outcome-Centered Potestative Square of Opposition

So far, we have presented an abstract notion of power (concerned by the possibility of change or not of the target relation), and an operational notion (concerned by the interaction between directive and performance). Strangely enough, we lack of a square of opposition centered around the canonic form of power captured by $\mathsf{Power}(p, q, A)$ (the power to issue a duty to A), that is, power centered around the outcome. For doing this, we also need to introduce the notion of *power to release a duty*. In this way we can distinguish between a *positive-outcome* notion of power (i.e., the canonic notion of power) and a *negative-outcome* notion of power. More formally, let us define the power to release a command as:

$$\overline{\mathsf{Power}}(x, y, A) \equiv \exists \beta : \mathit{Ability}(x, \beta, \neg \mathsf{Claim}(x, y, A))$$

Thus, with respect to any action type A we can form a set PR of four powers:

- the power to issue a duty to A, or $\mathsf{Power}(p, q, A)$
- the power to issue a prohibition to A, or $\mathsf{Power}(p, q, \overline{A})$
- the power to release a duty to A, or $\overline{\mathsf{Power}}(p, q, A)$
- the power to release a prohibition to A, or $\overline{\mathsf{Power}}(p, q, \overline{A})$

Furthermore, we adopt the following logical principles, which provide conditions for the truth of statements involving outcome-centered notions of power:

$$\mathsf{Power}(x, y, A) \to \neg \mathsf{Claim}(x, y, A)$$
$$\overline{\mathsf{Power}}(x, y, A) \to \mathsf{Claim}(x, y, A)$$

The rationale behind these formulas is that power captures a potential of a manifestation, and so the manifestation must not hold, for the potentiality to hold. Thus, having a power to impose A on q entails that one does not already have a claim that A be performed by q. Analogously, having a power to release q from the performance of A entails that one has (until the power at issue will be exercised) a claim that A be performed by q. These principles shed light on the way in which an Aristotelian square of opposition for outcome-centered modalities should be built.

In fact, at a first glance one might be inclined to consider the statement $\mathsf{Power}(p, q, \overline{A})$ as contrary to the statement $\mathsf{Power}(p, q, A)$ (namely, the power to forbid A as contrary to the power to impose A), but this is not a valid choice: it may well be the case that p has the power to impose the duty to A,

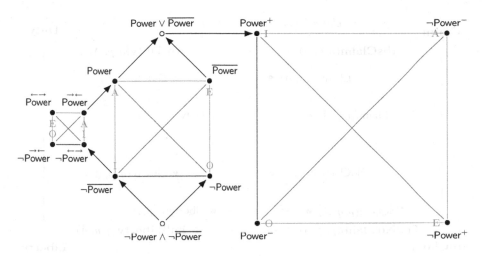

Fig. 4. Map of potestative relations defined in terms of triggering action (force-centered square of opposition, the left one), in terms of outcome (middle square), in terms of change or affecting outcomes (O'Reilly's square of opposition, the right one). Notice that the leftmost square is vertically mirrored and the rightmost square underwent a 90° clockwise rotation. The usual convention on the colour of bindings apply.

as well as to impose the prohibition to A. This intuition can be confirmed by analyzing their truth-conditions: $\mathsf{Power}(p, q, A) \wedge \mathsf{Power}(p, q, \overline{A})$ entails (according to the principles stated above and the Law of Double Complementation) $\neg\mathsf{Claim}(p, q, A) \wedge \neg\mathsf{Claim}(p, q, \overline{A})$. Looking at the Claim-based deontic square of opposition in Fig. 3, the latter is a conjunction of subcontrary statements, whence it can be true.

By contrast, in order to find the appropriate contrary to the statement $\mathsf{Power}(p, q, A)$, one has to rely on the observation that, according to the logical principles on outcome-centered notions of power, for the same normative party it is not possible to have the power to create a claim that A be performed by q *and* the power to release q from the duty of performing A. Thus, it is not possible that $\overline{\mathsf{Power}}(p, q, A)$ is true at the same time of $\mathsf{Power}(p, q, A)$. This observation provides us with the sub-alternation principle

$$\mathsf{Power}(p, q, A) \rightarrow \neg\overline{\mathsf{Power}}(p, q, A)$$

to construct a square of opposition by means of the set PR.

Comparison. It is interesting to check which corners of squares of oppositions are occupied by the notions of power discussed thus far. We will make reference to the four corners in a square with the labels A (upper left corner), E (upper right corner), I (lower left corner) and O (lower right corner) as in the tradition (see e.g. [3]). Looking at Fig. 4, one sees that $\mathsf{Power}^{\rightarrow\leftarrow}$ and Power occupy the A position in their respective squares, whereas $\mathsf{Power}^{\leftarrow\rightarrow}$ and $\overline{\mathsf{Power}}$ occupy the E position, Power^+ occupies the I position and Power^- occupies the O position.

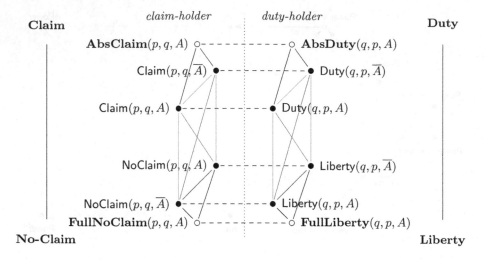

Fig. 5. The two correlative hexagons of opposition for Hohfeld's deontic relations.

5   Of Lost Symmetries

Here we return on the issue of the asymmetry between the analysis of first-order and second-order Hohfeldian diagrams. Symmetry is a desired property not only among relations belonging to the same square (given that these should be all interdefinable), but also between pairs of relations belonging to the deontic and to the potestative square, respectively. The reason for this will be clarified below.

5.1   Half-Liberties and Full-Liberties

According to the description of the deontic square that we have provided in Fig. 3, the formula Liberty(q, p, A) is logically equivalent to ¬Claim$(p, q, \overline{A})$. However, this captures a notion that does not match the ordinary meaning of 'liberty' (see, on similar lines, [12,15]). The two statements ¬Claim$(p, q, \overline{A})$ and ¬Claim(p, q, A)—the latter is logically equivalent to Liberty$(q, p, \overline{A})$—are sub-contraries, which means that they cannot be both false. But then, it may be that only one of Liberty$(q, p, \overline{A})$ and Liberty(q, p, A) is true. In this case, speaking of a 'liberty' is misleading. In fact, suppose, without loss of generalization, that Liberty(q, p, A) is true and that Liberty$(q, p, \overline{A})$ is false: then, ¬Liberty$(q, p, \overline{A})$ is true and so is Claim(p, q, A). The latter, in turn, is equivalent to Duty(q, p, A). Therefore, one gets that q is free to perform A with respect to p and, at the same time, has a duty to p to perform A. This is only a *half-liberty*, rather than a genuine one. By contrast, a *full-liberty* for q with respect to the performance of A obtains only when both formulas Liberty(q, p, A) and Liberty$(q, p, \overline{A})$ are true:

$$\text{FullLiberty}(y, x, A) \equiv \text{Liberty}(y, x, A) \land \text{Liberty}(y, x, \overline{A})$$

We can define correlatively a full no-claim relation: FullNoClaim$(x, y, A) \equiv$ ¬Claim$(x, y, A) \land$ ¬Claim$(x, y, \overline{A})$.

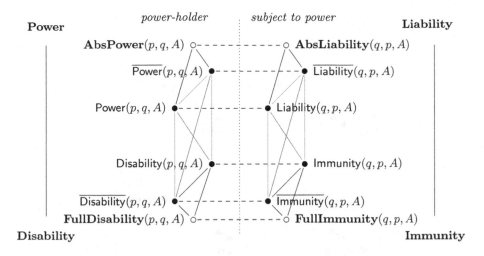

Fig. 6. The two correlative hexagons of opposition for Hohfeld's potestative relations (outcome-centered).

5.2 Disjoint or Absolute Duty

Being moved by the aim of an overall symmetry of the geometrical construction of deontic modalities, one could argue that there must be a notion of duty associated with the combination of the two formulas $\mathsf{Duty}(q, p, A)$ and $\mathsf{Duty}(q, p, \overline{A})$, which correspond to $\mathsf{Claim}(p, q, A)$ and $\mathsf{Claim}(p, q, \overline{A})$ in the Claim-based deontic square. However, as it is acknowledged by O'Reilly, such a combination cannot correspond to the joint truth of the two formulas. Indeed, if q were required both to perform A and to forbear from A, then there would be a conflict between norms, since q could not avoid doing something regarded as wrong. Here, we rather propose to further exploit the analogy between deontic and alethic modalities in order to find a more plausible solution. In fact, the notion of liberty is associated with the alethic notion of *possibility*; by contrast, the notion of claim and the correlative notion of duty are associated with the alethic notion of *necessity*. The square of opposition can thus be expanded to an *hexagon of opposition*, following the ideas in [4], in order to make room for two notions that respectively correspond with two-sided possibility and two-sided necessity. In the alethic case, the former notion is also known as *contingency*, the latter notion as non-contingency or *absoluteness* (see, e.g., [3], and [13].)

We can define, accordingly, an absolute duty, and a correlative absolute claim, confirming its duality with the full no-claim (see Fig. 5):

$$\mathsf{AbsDuty}(y, x, A) \equiv \mathsf{Duty}(y, x, A) \vee \mathsf{Duty}(y, x, \overline{A})$$
$$\mathsf{AbsClaim}(x, y, A) \equiv \mathsf{Claim}(x, y, A) \vee \mathsf{Claim}(x, y, \overline{A})$$
$$\mathsf{FullNoClaim}(x, y, A) \equiv \neg\mathsf{AbsClaim}(x, y, A)$$

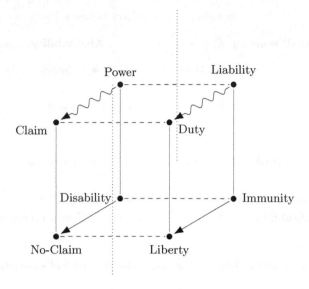

Fig. 7. Hohfeld's cube: relations connecting the second-order Hohfeldian square constructed upon the canonic power $\mathsf{Power}(p,q,A)$ with relations of the first-order Hohfeldian square. The wavy arrows are causal relationships.

Similar considerations can be applied to the correspondent potestative concepts, resulting in the construction of full-disabilities, or absolute powers (see Fig. 6). As the two figures show, taking as descriptions of normative relations the top and bottom points of the hexagons, concepts of the first and second Hohfeldian square follow the same structure. Another possibility, that diagrammatically maintains the core structure, would be to construct Hohfeldian prisms made of triangles of opposition, i.e. the positions A, E, and $I \wedge O$ of each hexagons [15].

6   Prototypical Relations Between the Two Squares

As a second-order relation, power reifies the possibility of an agent to modify some normative relation (of the first-, or of the second-order). The most prototypical power is the one that, once enacted by an agent, creates a duty upon another agent, and thus a correlative claim upon the first one towards the second (e.g., a commander w.r.t. a soldier). Abstracting the internal workings of power (for a possible formalization, see [17]), we can still observe that *power generally leads to a claim*, or (using $\leadsto$ for the notion of a dynamic/causal entailment):

$$\mathsf{Power}(x,y,A) \leadsto \mathsf{Claim}(x,y,A)$$

As observed by Andrews [1], one can construct, following the same idea, similar patterns between the other notions, resulting into a Hohfeld's cube (Andrews however does not consider the temporal/causal aspect we suggest here and reduces all relations to variations of composite deontic structures). Proceeding along this path, we observe that *liability generally leads to a duty*:

$$\text{Liability}(x, y, A) \rightsquigarrow \text{Duty}(x, y, A)$$

This approach has however a different interpretation on the other relations: "immunity generally leads to a liberty" and "disability generally leads to a no-claim". Superficially, these sentences capture that a liberty becomes explicit only in the moment in which immunity is utilized against a supposed power. However, from a logical point of view, it is the very presence of immunity that makes the agent free of behaving how she prefers with respect to the holder of a disability, whereas the manifestation is only an epiphenomenon.

$$\text{Immunity}(x, y, A) \rightarrow \text{Liberty}(x, y, A)$$
$$\text{Disability}(x, y, A) \rightarrow \text{NoClaim}(x, y, A)$$

These four relations are illustrated in Fig. 7.

7 Conclusion

The paper revisits, reorganizes and extends several distinct contributions developed around the primitive normative relations expressed in the framework of Hohfeld, with the purpose of capturing underlying patterns. In contrast to most papers on this topic, we gave here precedence to the systematization of views that are generally lost when we look at more general logical constructs. The primary focus on a canonic form of power (the one creating a claim/duty) allowed us to make explicit three distinct levels of abstraction on which power can be defined (force, outcome and change). We showed in what sense contemporary formalizations were losing part of the appeal of Hohfeld's proposal, and how this can be solved by making reference to concepts expressed on the deontic hexagon, as e.g. full-liberty and absolute duty. Finally, we illustrated the prototypical connections (causal, or logical) between the second-order (potestative) and first-order (deontic) relations by means of a cube. In this effort of systematization, diagrams have proven to us to be an effective method to discover gaps in the theoretical framework that were not evident from the syntactic view.

Furthermore, in a normative context, diagrams may be used to create user-friendly interfaces for the analysis of legal/contractual constructs. Rather than inspecting hundreds of sentences in the text of a contract, a subject may more easily figure out her normative relations (duties, rights, etc.) with the other parties by navigating or exploring a diagram-construed model.

Acknowledgments. Matteo Pascucci was supported by the Štefan Schwarz Fund for the project "A fine-grained analysis of Hohfeldian concepts" (2020–2022). Giovanni Sileno was supported by NWO for the DL4LD project (628.009.001) and for the HUMAINER AI project under contract KIVI.2019.006. This article is the result of a joint research work of the two authors.

References

1. Andrews, M.: Hohfeld's cube. Akron Law Rev. **16**, 471–485 (1983)
2. Åqvist, L.: Deontic logic. In: Gabbay, D., Guenthner, F. (eds.) Handbook of Philosophical Logic, vol. 8, pp. 147–264 (2002)
3. Béziau, J.Y.: The power of the hexagon. Log. Univers. **6**(1–2), 1–43 (2012)
4. Blanché, R.: Sur l'opposition des concepts. Theoria **19**(3), 89–130 (1953)
5. Hohfeld, W.N.: Some fundamental legal conceptions as applied in judicial reasoning. Yale Law J. **23**(1), 16–59 (1913)
6. Hohfeld, W.N.: Fundamental legal conceptions as applied in judicial reasoning. Yale Law J. **26**(8), 710–770 (1917)
7. Kocourek, A.: The Hohfeld system of fundamental legal concepts. Ill. Law Rev. **15**(1), 24–39 (1920)
8. Lindahl, L.: Position and Change: A Study in Law and Logic, vol. Synthese Library. Springer, Netherlands (1977). https://doi.org/10.1007/978-94-010-1202-7
9. MacCormick, N., Raz, J.: Voluntary obligations and normative powers. Proc. Aristotelian Soc. **46**, 59–102 (1972)
10. Makinson, D.: On the formal representation of rights relations. J. Philos. Logic **15**, 403–425 (1986)
11. Markovich, R.: Understanding Hohfeld and formalizing legal rights: the Hohfeldian conceptions and their conditional consequences. Studia Logica **108**(1), 129–158 (2020)
12. O'Reilly, D.T.: Using the square of opposition to illustrate the deontic and alethic relations. Univ. Toronto Law J. **45**(3), 279–310 (1995)
13. Pizzi, C.: Contingency logics and modal squares of opposition. In: Béziau, J., Gam-Krywoszynska, K. (eds.) New Dimensions of the Square of Opposition, pp. 201–222 (2012)
14. Sartor, G.: Fundamental legal concepts: a formal and teleological characterisation. Artif. Intell. Law **14**(1), 101–142 (2006)
15. Sileno, G.: Aligning Law and Action. Ph.D. thesis, University of Amsterdam (2016)
16. Sileno, G., Boer, A., van Engers, T.: On the interactional meaning of fundamental legal concepts. In: Hoekstra, R. (ed.) Proceedings of the 27th International Conference on Legal Knowledge and Information Systems (JURIX 2014), vol. FAIA 271, pp. 39–48 (2014)
17. Sileno, G., Pascucci, M.: Disentangling deontic positions and abilities: a modal analysis. In: Calimeri, F., Perri, S., Zumpano, E. (eds.) Proceedings of the 35th Edition of the Italian Conference on Computational Logic (CILC 2020), vol. 2710, pp. 36–50 (2020)
18. Sumner, L.W.: The Moral Foundation of Rights. Clarendon Press, Oxford (1987)

Wittgenstein's Picture-Investigations

Michael A. R. Biggs$^{(\boxtimes)}$ ⓘ

University of Hertfordshire, Hatfield AL10 9AB, UK
m.a.biggs@herts.ac.uk

Abstract. This paper reports on Wittgenstein's use of pictures and diagrams undertaken through an analysis of the surrounding co-text in the published works. It is part of a larger project to develop tools for the integrated semantic analysis of images and text in Wittgenstein's original manuscript and typescript sources. The textual analysis took keywords, phrases and punctuation as possible indicators of definitive samples and rules in propositions and non-propositions. For reasons argued in the paper we focused on non-propositions and differentiated those that functioned descriptively from those that functioned definitively. Finally, from the range of definitive statements we investigated those that functioned according to Wittgenstein's concept of a rule. In all cases we focused on collocation of indicative text with images. We concluded that Wittgenstein's practice accorded with his early statements about images needing accompanying words to activate their propositional status, but that images could function independently as non-propositional descriptive or definitive samples. As definitive samples, many images also had the capability to function as rules, or independently as proofs. Since the picture-sentences rely on iconicity to communicate rules that may otherwise be hidden in our language practice, we speculate that the iconic relationship may belong to hinge epistemology. This is proposed as a strand for future research.

Keywords: Wittgenstein · Corpus analysis · Picturing · Propositions · Rules · Semantics

1 Introduction

This paper describes a survey of the use of images in Wittgenstein's published works and is part of a feasibility study for the development of a semantic tool with which to investigate the range of meaning and use of literal pictures in the digital corpus of Wittgenstein's manuscripts known as the Bergen Nachlass Edition[1], and how that might be integrated with more established text-based tools. As an initial interpretative tool, corpus analysis attempts to complement any deficits in existing semantic information by offering collocation analysis, etc., i.e., syntactic and pragmatic tools rather than semantic tools. In the case of image-rich texts such as Wittgenstein's, there is a stage at which the interpretative analyses of the texts need to be integrated with analyses of the images, establishing

---

[1] Wittgenstein, L. (2015-). Bergen Nachlass Edition. Edited by the Wittgenstein Archives at the University of Bergen under the direction of Alois Pichler. In: Wittgenstein Source, curated by Alois Pichler (2009–) and Joseph Wang-Kathrein (2020–). http://www.wittgensteinsource.org.

© Springer Nature Switzerland AG 2021
A. Basu et al. (Eds.): Diagrams 2021, LNAI 12909, pp. 103–117, 2021.
https://doi.org/10.1007/978-3-030-86062-2_10

any mutual relationships and the impact they have on semantic interpretation. Previous studies of Wittgenstein's use of images and diagrams have focused on their semantic contribution to the co-text. However, the images also contribute to, and impact on, the syntax of the sentences in which they occur. In particular, we note that Wittgenstein warns us that neither words nor pictures make sense on their own but require a context of practice in which they have meaning (PI-I §23) We therefore undertook a pilot study to see how propositions and non-propositions, rules and proofs, were functioning when they contained literal pictures. We claim that such picture-sentence investigations are distinct from previous text-image interpretations because the latter have overlooked the difference between an image that illustrates what is said in the surrounding text, and an image that functions syntactically as part of a sentence. The current revision of the Bergen Nachlass Edition and the Wittgenstein Ontology Explorer that is under development, offer a platform in which the outcomes can be implemented as tools for complex text-image analysis[2].

1.1 The Presence of Literal Pictures

There are about 500 images in the published works, which were mostly produced posthumously from a collection of handwritten and typewritten documents known as Wittgenstein's Nachlass. There are 2000–3000 images in the Nachlass, depending on how one counts them, although this figure is somewhat misleading because the contents are quite repetitive owing to their status as preparatory studies and reworkings of what is now well known through the edited published works. In the digital publications, many image files are generated as a response to problems of notational representation rather than being intrinsically pictorial or diagrammatic. The presence of literal pictures causes presentational problems for anyone involved in publishing these texts, and they cause search problems in the digital environment owing to the different file types[3].

1.2 The Function of Images

Given that images are rare in works of philosophy, a relevant question is "what is the function of these images"; for example, what do they contribute to the text, and what do they contribute that would be missing if, like most philosophical writings, they were not there? By approaching the problem from this novel angle, i.e., what would be missing if they were not there, we could separate many of the larger, eye-catching images such as the duck-rabbit because they sit between sentences and show or illustrate what is said in the text. We call these inter-sentential images. This contrasts with many of the smaller, more easily overlooked images that lie within a sentence and if they were to be removed the sentence would no longer make any sense. We call these intra-sentential images. This highlighting of intra-sentential images pointed us towards the possibility of using text analysis tools and methods that might reveal the function of these literal pictures in picture-sentences. We acknowledge that it is still the case that the interpretation of

---

[2] http://wab.uib.no/sfb/.

[3] Owing to the different copyright rules and permissions for images and text, it has been necessary to either redraw or use the scanned manuscript images for this paper.

Wittgenstein's use of images in his writings needs to be undertaken in context, but it does provide a syntactic view of that use in addition to the semantic view that has been undertaken previously by researchers.

1.3 Propositions, Non-propositions and Pictures

When Wittgenstein says that a proposition is a picture, the term "picture" refers to an underlying model *[Bild]* theory of representation in which language functions like a picture because it shares the morphology of what it represents. This is a very restricted correspondence theory of language, but it perhaps served Wittgenstein's early purpose to focus on a specific class of language acts, i.e., propositions. In TLP[4], an early work, Wittgenstein develops a sophisticated model of this language-world relationship that shows propositions to be either true or false. The model-theory gave rise to what has become known as his "picture theory of meaning" (cf. TLP §§2.16–2.2). The picture theory of meaning is not a theory of how pictures convey meaning (cf. Mitchell 1986, p. 20), but rather it is a theory of how language conveys meaning based on a comparison with how pictures convey meaning. In TLP, Wittgenstein approaches this issue from a structural point of view, identifying that there needs to be a correspondence between the logical dimensions of the representation and what it represents, and there is little mention of the appearance or iconicity of the picture itself.

Anscombe, in her introduction to TLP, objects that a picture cannot function on its own as a proposition because it cannot assert that the things depicted can actually be found somewhere in the world (Anscombe 1959, p. 64). Anscombe's objection is that a picture is truth-functionally neutral because it belongs to a non-shared language that simply resembles a state of affairs rather than asserting "this is how things are in the world". The issue for us here is whether a picture can "say" (assert) anything, and if so, in what sense? Shier proposes that this difficulty can only be resolved within the constructs of TLP and the early picture theory by adding something external to the picture, e.g., by adding "some account of what we do with pictures" (Shier 1997, p. 73). Gregory also adopts a similarly Fregeian reservation about the ability of pictures to be propositions, "given that the notion of a proposition has its natural home in thought about language" (Gregory 2020, p. 155). He thinks that it is "a little odd" to talk about the truth conditions of pictures, and it is perhaps better to speak of their accuracy-conditions i.e., that the depiction can be seen in this way from this viewpoint (2020, p. 157). We regarded the implication that images rely on the surrounding co-text in order to function as propositions, as an indicator of a strong thesis that integrated text-image tools are necessary for the semantic analysis of Wittgenstein's published works and the broader Nachlass.

In the middle period, especially in the early 1930s, the number of literal pictures in Wittgenstein's writings increases, and his focus shifts to the difference between propositions and non-propositions. Propositions are important in philosophy because they tell us something about the world, and for this reason they can be either true or false. On the other hand, non-propositions are neither true nor false, but instead make a statement about the language game in which they occur. Such statements are often grammatical

---

[4] Tractatus Logico-Philosophicus, translated by Ogden and Ramsey (1922).

rules about how words or practices are to be used rather than moves within the practice itself. However, in the example of correctly counting $2 + 2 = 4$, the implied proposition "it is asserted that $2 + 2 = 4$ is the case" does not tell us something intrinsic about numbers but does tell us about our practice and what we do or do not describe as correctly counting (cf. PI-I §22[5]). Wittgenstein would call this a pseudo-proposition because one cannot imagine it to be otherwise (PG[6] p. 129) and claims that rather than telling us about the world it tells us about the rules or grammar of our language-game of counting, i.e., the correct use of words or "operating with signs" (BBB[7] p. 6) rather than about the fundamentals of logic and arithmetic. In discussing picture-sentences, we might therefore need to differentiate between propositions and non-propositions, and then consider what can be said about the contribution of literal pictures in each case.

1.4   Problem Statement and Present Aims

We have discussed that the early Wittgenstein has a variable but sustained concept of the proposition as a picture or model of the world. This is sometimes expressed as the assertion "the proposition is a picture." Sometimes such claims are accompanied by literal pictures that may occur between sentences (inter-sentential) or within a sentence (intra-sentential). According to Frege and Anscombe, images cannot "say" anything in isolation from the co-text, and this context-dependent semantics is the way in which they have hitherto been analyzed. However, it seemed to us that the claim that what an image "says-shows" is context dependent, is altered depending on whether one is considering inter- or intra-sentential images. The alteration arises because the inter-sentential context of the former affects the interpretation of the image whereas the intra-sentential context of the latter affects the interpretation of the text. Furthermore, what the picture-sentence "says-shows" in the case of a proposition is different from what the picture-sentence "says-shows" in the case of a non-proposition. In the former the sentence makes an assertion which may or may not be true, whereas in the latter we have a description of how the language game is to be played. Such non-propositional picture-sentences may be demonstrations or proofs, for example.

Undertaking an analysis of the distribution and function of images in these cases will shed light on both Wittgenstein's practice, and on the instrumentality of the images in relation to the concepts in the texts, i.e., semantic categories that have been developed independently for words and images may need to be harmonized. In addition to contributing to Wittgenstein scholarship, such insights will also contribute to the presentation of Wittgenstein's works by providing a decision-making tool for the representation of images by normalized types, e.g., Unicode. It will also contribute to the design of integrated image-text search and content analysis tools such as Wittgenstein Source (image-text representation) and the Wittgenstein Ontology Explorer (image-text search and semantic analysis). We believe such tools are necessary to facilitate scholarly exploration and discussion of Wittgenstein's "picture-investigations" advocated by Baker (2001, p. 21).

---

[5] Philosophical Investigations (1953).

[6] Philosophical Grammar (1974).

[7] Blue and Brown Books, second edition (1969).

2 Method

We have used tools from corpus linguistics, semiotic analysis, and textual analysis, to locate and interpret instances of intra-sentential literal pictures that may relate to the concepts of propositions, rules, and hinges. The corpus assumed to be "Wittgenstein's published works" consisted of 14 books in digital format by Intelex[8] plus a digitized copy of TLP1922. This c.925,000-word corpus was encoded so that the images became locatable within the texts, and parsed for examples of image-text collocations, and syntactic forms indicated by punctuation. These examples were then used to inform our hypotheses and proposals for further study.

In terms of images, there are 409 instances in the corpus according to the strict definition of Biggs and Pichler (1993, p. 92). However, according to their criteria, a number of significant works are omitted from the corpus because they do not have an origin in Wittgenstein's Nachlass, i.e., a manuscript source for the publications in Wittgenstein's own handwriting, e.g., BBB. Since it is also discretionary whether some series of images are counted as one image or several, it is safer to make the more general claim that there are c.400 images in the corpus. We therefore adopted the approach that the statistics in this paper are indicative rather than definitive, since the boundaries and conditions of the quantitative analysis are insufficiently rigid. However, our method could generate quantitative data if there were to be consensus about the scope of the sample. Since the present study reports on a pilot for a larger study of the Nachlass, the issue of scope is not critical at this stage.

Given the focus of the present conference on diagrams as a specific form of visual communication, we need to defend our use of the terms image, picture and diagram. We have adopted a technical use of both image and picture. An image is taken to be anything non-textual, i.e., that is not part of the Unicode basic multilingual plane (0000-04FF) and therefore needs to be represented via an image file-type. Although Unicode can be used to construct quite complex forms, we avoid using it "creatively" or combinatorially, e.g., by using box drawing elements. Adopting this approach results in some mathematical and musical notation being classified as images, owing solely to presentational issues, and overlooking some tabular layouts of letters that are functioning diagrammatically (e.g., PG p. 328). Where necessary we note such cases. We follow Wittgenstein's occasional use of the term "picture", for example in "picture-face" (PI-II p. 194) if the iconicity of the image is relevant for the co-text. Following Giardino and Greenberg (2014) we agree on some pragmatic differences between picture and diagram, in that pictures imply a viewpoint whereas diagrams do not, and diagrams tend to require text whereas pictures do not. We also found a linguistic difference between the adjective "pictorial" which suggested iconic resemblance, and "diagrammatic" which suggested something with more of a structural resemblance. However, owing to the very sketchy quality of Wittgenstein's images, and the almost universal requirement that the signification of Wittgenstein's images need to be clarified by the text, not least owing to the context of philosophical discussion, neither of these differentiations is very reliable.

We use the terms "picture-sentence" and "intra-sentential" to describe a sentence in which an image occurs, in contrast to the term "inter-sentential" to describe an image that

---

[8] BBB, CV, LW1&2, NB, OC, PG, PI, PR, RFM, ROC, RPP1&2, Z.

occurs after a full-stop (actual or implied) and before an initial capital letter.[9] Finally, on this issue of "location", it should be noted that Wittgenstein frequently draws his images on a new line and then starts the text again after another line-break. However, although these look like "inter-paragraph" images, and are frequently represented as such in the published works, we gave sentence punctuation priority over page layout. We therefore adopted a structural encoding of the corpus, and our descriptions of inter- and intra-sentential images relates to structural rather than presentational features, i.e., intra-sentential images were regarded as part of a continuous line of text. We differentiated two classes of sentence: propositions, non-propositions; and two classes of image-text collocation: inter-sentential and intra-sentential. In response to the objections implied by Anscombe (above) we focused on intra-sentential images, although we comment on some inter-sentential examples where they seem to shed light on Wittgenstein's use of images. We were interested in three functions within the class of non-propositions: examples, rules and proofs. In other words, our investigation could not be undertaken simply using corpus analysis tools because it involved the interpretation of what Wittgenstein had written in terms of his philosophy.

We hypothesized that Wittgenstein's use of quotation marks in connection with diagrams was probably indicative of assertion, i.e., propositional content, and demonstration, i.e., non-propositional samples, respectively. Terms such as "thus:", "like this:", and "example"; we regarded as possibly indicative of description by example, but also as possibly indicating a notation for ostensive definition. We associated the words "rule" and "proof" with our textual searches for images that may be acting as rules.

We experimented with various indicators of the way in which the text-diagram collocation may be functioning in the corpus. We first divided the total corpus images (n = 409) between inter- (n = 122, 30%) and intra-sentential cases (n = 287, 70%). We then searched the intra-sentential cases for collocation of the images with keywords that might indicate the sentential functions of examples, including the term "example"[10], the abbreviation "e.g." and the word-punctuation collocation "thus:". We hypothesized that certain punctuation, such as the colon, was instrumental in creating a syntactic role for an image as a word-substitute. We also used contextual reading to identify intra-sentential uses in connection with rules and proofs. In order to focus on the active role played by the images we also searched using Wittgenstein's term "perspicuous representation" and its cognates, together with "surveyability, synoptic view, bird's-eye view" in accordance with the translator's note by Rush Rhees in PR[11]. These were by no means comprehensive

---

[9] This may seem like a rather laborious description of a sentence, but in manuscripts one often encounters orthographic errors in which sentences are not "properly" concluded by a full stop or begun by an initial capital. Wittgenstein rarely puts a full stop after an image at the end of a sentence. Sentence termini may need to be inferred.

[10] It should be noted that this initial research was conducted on a digital text of the published works in English (Intelex 2000). To this collection was added a digital version of TLP (1961). Most of the original manuscripts in the Nachlass are written in German. Where there are known to be special difficulties owing to translation, we will include a brief discussion of the impact of the issue on the current research. The published works were selected in preference to the Nachlass for reasons of textual determinacy and feasibility. At the time of writing, we did not have access to a digital corpus of the published works in German.

[11] Philosophical Remarks (1975).

search terms but provided a basic set to test the utility of the approach. Finally, further segmented cases between images that arose owing to presentational complexities of representing the notation in the text, e.g., conventional musical or mathematical notation, from images that were essentially pictorial or diagrammatic, or creatively "misused" notational conventions.

2.1 Discussion

Broadly speaking, we took a proposition to be a subset of sentences which make claims that are either true or false. This is a technical description adopted by Wittgenstein in his very earliest notebooks. Whether it is feasible to determine the truth or falsehood of a proposition is irrelevant to making the classification. On this basis it is clear that many well-formed sentences should not be called propositions, a point not always observed by translators. Nonetheless, propositions are normally of special interest in philosophy because they tell us something new about the world. One of the novelties of the present research is that we focus on non-propositions because they may describe our practices, and Wittgenstein's use of images in non-propositions may be related to his therapeutic method of showing us a way out of our linguistic entrapment.

By the middle period of Wittgenstein's philosophical development, in the 1930s, he was focusing on a form of non-propositional sentence that he sometimes called a "rule". Such sentences tell us about how words and concepts should be used within a language game. Some of these rules sound as though they are making claims about the external world, but this is misleading. An example of a rule would be "an object cannot be red and green at the same time". This sounds like a proposition that could be tested in practice and might involve a search for a simultaneously red-and-green object. However, Wittgenstein would regard this was a waste of time because the statement that "an object cannot be red and green at the same time" tells us about how we use colour words and concepts rather than telling us about phenomena. It belongs to the grammar of our concepts that we cannot use the words red and green in this way. The possibility of making a well-formed, grammatical sentence encourages us to view it as a proposition, but this possibility is misleading and reinforces why, in the present research, it is not sufficient to simply employ corpus analysis tools independently from an understanding of the philosophical text to which they are being applied.

Wittgenstein's rules function like the rules of the game of chess. Rules are discretionary, in the sense that although they are arbitrary, having been established they are non-negotiable. If I do not follow the rules of the game of chess, then I am not playing chess but some other game. The rules of chess lie outside the game itself and form a framework within which the game is played. It is equally so with language; the rules and grammar of language determine how one goes about using the language; grammar in this case meaning conceptual grammar rather than English grammar. Using the words red and green within the language game of English involves the rule that red excludes green even though there is no corresponding rule that not-red implies green. Rules, according to Wittgenstein, are "nonsense", not because they are useless but because they do not tell us anything meaningful within the language game, i.e., they are not propositions.

3   Case Studies

The intra-sentential images were analyzed in a variety of ways as discussed above, differentiated by collocations of words and punctuation. In the following sub-sections, we discuss some paradigmatic cases of intra-sentential images collocated with text matching these criteria. These cases therefore serve only as examples, and we do not claim that these are the best or only cases.

3.1   Propositions as Pictures and Pictures as Propositions

The familiar claim that a proposition is a picture is mainly found in Wittgenstein's early work (e.g., TLP§4.01). However, we wanted to examine the concept from the point of view of the picture, i.e., whether a picture is a proposition. The high frequency lemmas "pict*" (n = 1949) and "proposit*" (n = 3911) were searched for collocation with one another within the span R5 (n = 53), and further refined as the 2-g assertion "the/a proposition is... picture (n = 18)", "proposition as... picture" (n = 1). There was one assertion that "the picture can serve as a proposition" (NB p. 33), and one of inference "I have... in a picture in front of me; then this enables me to form a proposition, which I as it were read off from this picture" (RFM-IV §49 p. 249). Finally, there was one explicit statement about the picture as proposition:

> *"Can one negate a picture? No. And in this lies the difference between picture and proposition. The picture can serve as a proposition. But in that case something gets added to it which brings it about that now it says something. In short: I can only deny that the picture is right, but the picture I cannot deny" (NB p. 33).*

An alternative possibility, of the neutral picture as a "proposition-radical", occurs in PI-§22 and, similar to the example in NB[12] p. 7, describes a picture that depicts people. The textual context, including the notation of enclosing the picture in quotation marks, seems to reinforce his early view influenced by Frege that the surrounding textual cues are part of transforming the proposition-neutral picture into an assertion. A further possibility "between picture and proposition" occurs in a context discussing negation and whether one can negate a picture (cf. NB p. 33 above, no image). Both alternative possibilities reflect Frege's theory that an image can only communicate when it is supplemented by words. The picture itself says nothing and needs to be contextualized and turned into a proposition. However, a semantic interpretation of the contribution made by the image to the meaning of the sentence already requires that the sentence is seen-as a proposition. We therefore attributed some agency to the role that the quotation marks have in turning the picture-sentence into a proposition, and we added this syntactic indicator to our search list but noted that it was serving only to clarify that the sentence may be functioning as a proposition and not what was being claimed by that proposition. This is the inherent semantic ambiguity, according to Wittgenstein, of each picture-proposition, e.g., the stance of the fencer (NB p. 7) and the stance of the boxer (PI-§22, no image): that there is some content, for example an internal relationship, that is the content of the image

---

[12] Notebooks 1914–1916, second edition (1979).

but that its external relationships or the assertion that is made about the state of affairs in the picture, is ambiguous.

3.2 Images Within Quotation Marks (Picture-Assertions)

Wittgenstein frequently introduces an interlocutor into his texts. The interlocutor is the voice that asks difficult questions or puts forward an expected response that Wittgenstein wishes to refute. The interlocutor's comments are frequently enclosed within quotation marks, and even when there are no other indicators, the reader soon learns to expect that she should disagree with the stance the interlocutor is taking. Therefore, there is a high frequency of the occurrence of quotation marks in the corpus as a whole (n > 13,000 pairs).

There are only 12 instances of images immediately enclosed within quotation marks. This includes a set of four symbols that is repeated in two places (PG p.188 & PI §495). There are an additional 7 images in ≤5-g lexical phrases within quotation marks. Of most interest, owing to what is said in the collocated text, are the cases of the stick-men (NB p. 7 mentioned above), the musical example (BBB p. 84), the cartographic sign (RPP-I p. 42) and the case of "seeing-as" (PI-II p. 206).

The musical example (BBB p. 84) is a somewhat marginal case because the image is only necessary to present musical notation, and it is part of a phrase that is encompassed in the quotation marks[13].

Compare these cases: a) Someone says "I whistled..." (whistling a tune); b) Someone writes, "I whistled ♪♪♪♪ "

However, it represents an interesting contrast between an action and the notation for that action. The image itself, is taking a role that would normally be fulfilled by the written word. To that extent it is not especially interesting. On the other hand, the voice of person a) says the words "I whistled" and then whistles (action). This is converted into notation in which the words are written but the whistling remains as a description of the action instead of being annotated in musical notation. This is contrasted with the voice of person b) who also says the words "I whistled" and then whistles, but this is converted into notation in which the words are written, and the whistling is annotated in musical notation. The difference between a) and b) is clarified by the use of quotation marks, because the action is the same whereas the annotation is different. This example occurs in a textual context within BBB in which Wittgenstein is drawing attention to the way in which there are many different ways of representing or annotating an action and furthermore many different ways of interpreting and acting upon that interpretation.

The cartographic sign "⌂" enclosed in quotation marks (RPP-I p. 42) serves a similar function[14]. The textual context discusses whether there is a significant difference between the sign for a house on a map, and the house itself, and whether the house could stand as a sign for itself if it could be put on a map. As with the musical example, this is an issue regarding the relationship of a thing to the notation or representation of that

---

[13] As a result, we here use a normalised (typographic) representation of the image.

[14] This is stated as a conventional sign which we represent with a normalised typographic symbol.

thing; representations that include the spoken word, the written word, drawings, etc. It is possible that the function of the quotation marks is to make the whole phrase into an assertion or proposition of the form "there is a △ here".

The stick-men " 大大 " enclosed in quotation marks (NB p. 7) also seem to be acting as an assertion or proposition. The preceding text explicitly states that "actual pictures of situations can be right and wrong". The subsequent text also states that "the proposition in picture-writing can be true and false". The content in this case is that the image acts as a proto-proposition and has content independent of its truth or falsehood. We do not believe this image is acting as a definition. The term "assert" is here used to claim the semantic content of the image, owing to some iconicity, is that "A is fencing with B". An example of a definitive version of a similar situation would be the sculpture Discobolus by Myron. There was a time when this sculpture was regarded as definitive of how to stand and throw the discus.

The "seeing-as" picture-sentence "I see △ as ↘" (PI p. 206) has a structure similar to the musical example, except that there is a notational problem that is resolved twice over by the inclusion of an image. In the musical example there was a contrast between the musical sound itself and music-as-notation. In the seeing-as example there are two forms of notation, the first of a triangle, the second of an arrow, both functioning as images of a direction (towards the bottom right). As is the case with the musical notation, the broader textual context is about the complexity of the way in which we both annotate and interpret actions and objects, and the way in which one might be blind to alternative aspects and uses, such as seeing the triangle as pointing towards the bottom left.

We concluded that the enclosure of an image within quotation marks implies that the image is acting as a proposition by making an assertion that may be true or false. However, in the early image of the stick-men there is an ambiguity regarding what is being asserted; whether the men are fencing, or demonstrating a stance in fencing, etc. On the other hand, the later image of seeing-as consists in whether an image can be interpreted in one way or another in the sense that the aspect that is seen is demonstrated by the way in which someone acts in relation to the image. The assertion "I see x as y" asserts the aspect even if that aspect is surprising.

3.3   Images Collocated with "example, thus:, like this:" (Picture-Samples)

As one might expect, pictures frequently occur as illustrations of what has been said in the text, in other words, as examples. Searches were made for combinations of intra-sentential images with the term "example:" (n = 963, collocation n = 5), "e.g.:" (n = 888, collocated n = 5) and constructions "like this:" (n = 471, collocated n = 31), and "thus:" (n = 465, collocated n = 9).

The most discussed image in the secondary literature of this kind is the "eye/visual field" image at NB p.80. The image is collocated with "like this:" and has been critiqued by Bazzocchi (2013) in terms of its graphical appearance in published texts but not in terms of its propositional or rule functionality.

The visual field has not, e.g., a form like this: ⟨image⟩ *(NB p. 80, cf. also TLP §5.6331)*

For our purposes, the syntax "like this:" suggests its use as a descriptive sample. Bazzocchi argues from this point too, that the importance and the common misrepresentation is to show the eye within the visual field instead of completely outside it. There are several occurrences of the form "is it not like this", but there are no cases of picture sentences that show how things "do not stand". Indeed, at Z[15] §249: Wittgenstein's interlocutor has quite a dialogue with herself about an un-picturable "four-dimensional cube". In the published work an image has been added by the editors, but in the original manuscript Wittgenstein simply inserted a row of dots "......". At Z §699, Wittgenstein experiments with another boundary case of number representation, by writing the recurring decimals of *pi* on top of one another (repeated at RPP-I p. 66).

To summarize: descriptive samples can be used to demonstrate both what can and what cannot be represented by images. This cannot be said of definitive images because they cannot be negative in isolation, i.e., denying the form they describe, as is the case at NB p.80 above. Examples of definitive samples include instruction tables at BBB pp.123f. and the use of novel ciphers such as ⌐ʃ used as a sign for the smell of coffee (RPP-I p. 103, and one of the 5-g lexical phrases within quotation marks). What differentiates definitive from descriptive samples is that it makes no sense to deny the assertion of the definition because it is functioning as a rule within a particular language game (PG p. 129). As a result, we concluded that images in propositions acted as descriptive samples owing to them also being assertions of a state of affairs whereas non-propositions were probably acting as definitive samples or, conversely, the presence of images in some non-propositions caused them to be definitive samples.

3.4   Images Collocated with "Rule" (Picture-Rules)

The concept of non-propositions as rules of our grammar of concepts occurs principally in texts from the 1930s ("rule*" n > 1000). We comment on just a small number of examples of "rule" collocated with an image. Three examples (PG pp. 98, 202, 203) make the connection between intention and following a rule. The issue at stake in these remarks is how or under what [additional] conditions would one say someone was following a rule. Such additional requirements do not remove the difficulty, e.g., RFM[16]-VI §29 p. 330 on rule-following and how to go on:

> When I have been taught the rule of repeating the ornament ⌐⌐ and now I have been told "Go on like that": how do I know what I have to do the next time? (RFM-VI §29 p. 330)

Sometimes, despite following a rule/prediction, the outcome results in surprise, suggesting that the outcome is obscured in the form of the notation, i.e., is not perspicuous:

> Before I have followed the two arrows ↘ like this ↘ I don't know how the route or the result will look. I do not know what face I shall see... But why wasn't this a genuine prediction: "If you follow the rule, you will produce this"? Whereas

---

[15] Zettel, second edition (1981).

[16] Remarks on the Foundations of Mathematics, third edition (1978).

the following is certainly a genuine prediction: "If you follow the rule as best you can, you will..." The answer is: the first is not a prediction because I might also have said: "If you follow the rule, you must produce this." It is not a prediction if the concept of following the rule is so determined, that the result is the criterion for whether the rule was followed. (RFM p. 316f.)

We concluded that picture-rules were closely related to practices and embodied them as memorable representations of those practices through iconicity. The image reveals what is either hidden or only implied by the textual "rule", e.g., assumptions about adding and not counting any elements twice, and in particular that rules that we recognize as such, e.g., lexical rules, can be misconstrued in a wide range of often surprising ways. The surprise seems to come from aspect-blindness that the rule could be misinterpreted in the way that the image reveals. This seems to be related to the very fundamental nature of the rules under discussion, e.g., of basic arithmetic, which we erroneously believe have only one aspect and one way of following the rule.

3.5   Images Collocated with "Proof" (Picture-Proofs as Picture-Acts)

We initially interpreted the collocation of the term "proof" with an image as an assertion of the possibility of the image as a paradigmatic sample, i.e., the same as a picture-rule. The concept of an image as a proof is absent from the early works but is widely distributed in Wittgenstein's middle period. It arises typically in connection with mathematical proof or foundational concepts in arithmetic. The most explicit example is the correlation of five strokes, known as a hand, with the five points of a star as paradigmatic of counting the number of points on the star (RFM-I §40 p. 53). Wittgenstein introduces the aspect of a correct correspondence of the hand to the star in contrast with an incorrect one.

 (RFM-I §25 p. 47 & §40 p. 53).

There is also the mapping of one star onto another which epitomizes the apparently obvious 1:1 relationship (RFM-I §41 p. 54). These are contrasted with the graphical demonstration of an alternative way of seeing the aspect of counting:

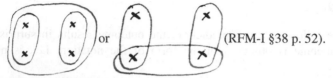 (RFM-I §38 p. 52).

These images show that there is more than one possible procedure but that only one procedure is regarded as correct. This reveals the first example to be a rule. Therefore, following our initial assumption that the term "proof" indicated a paradigmatic case of the image as a sample, we revisited it as a possible indicator of an image as a rule. RFM-I §50 p. 57 offers the clearest case owing to the syntax of the accompanying text:

A rectangle can be made of two parallelograms and two triangles. Proof:

A child would find it difficult to hit on the composition of a rectangle with these parts, and would be surprised by the fact that two sides of the parallelograms make a straight line, when the parallelograms are, after all, askew. (RFM-I §50 p. 57)

What we mean by a proof in this context is that it shows that it can be done, whereas a rule legitimizes one of the possible practices. The rectangle is the surprising result of a physical procedure of rearranging two parallelograms and two triangles. A movie that followed the actions of somebody rearranging paper triangles and parallelograms would perhaps be more convincing. The key point is that there is a practice that embodies the meaning of "a rectangle can be made of..." In this sentence the word "made" reinforces the idea that there is a constructive practice that could be undertaken or observed.

We concluded that "proofs" were often being used as antitheses to "rules", where a diagrammatic proof embodied an alternative, surprising, or scarcely credible case of rule-following. Diagrammatic proofs can therefore be demonstrations, i.e., speech-acts. Proofs could be said to be diagram-acts or picture-acts.

4 Conclusions and Future Research

If one could identify examples of pictures as propositions, it would provide a new or alternative critical framework for the discussion of the so-called picture theory, i.e., one might propose a language theory of pictures rather than a picture theory of language. Similarly, if one could identify cases of non-propositional pictures as rules it would give a new tool for discriminating when sentences that appear to be propositions are not acting as propositions, i.e., what are the syntactic indicators of rules in the textual context and how are these modified or supplemented in the diagrammatic context. Finally, the concept of "hinge epistemology" may or may not be illuminated by consideration of whether non-textual elements might also function in this foundational way, i.e., it might be symptomatic of what we call a *picture-hinge* that few image theorists have been able to describe the way in which an iconic image resembles its subject[17]. Our conclusions consist of two main claims, the first regarding the non-possibility of pictures as propositions, the second regarding the possibility of picture-sentences as rules; followed by a proposal for future research into picture-hinges.

---

[17] For an excellent survey of this problem see Hyman and Bantinaki "Depiction". In: Zalta, E.N. (ed.), (2021) *The Stanford Encyclopaedia of Philosophy* https://plato.stanford.edu/archives/sum 2021/entries/depiction/.

4.1 Picture-sentences as Propositions

In 1942, in the late period, Wittgenstein writes explicitly that "a proposition describes a picture", accompanied by an image (RFM-IV §2 p. 223). We take this to mean that behind the proposition is a practice that gives it meaning. In agreement with Frege, the picture alone is not acting as a proposition: the picture is a proposition-radical. Pictures can be asserted, as sometimes indicated by quotation marks, but a co-text is required to clarify a *picture-sentence* as a proposition. We are skeptical that any of the picture-sentences meet Wittgenstein's [variable] requirements for a proposition, but we do claim that the images in picture-sentences are doing useful work as *descriptive samples*.

4.2 Picture-sentences as Rules

Many of Wittgenstein's picture-sentences are non-propositions. Some are *definitive samples,* and a subset of those are functioning as rules (§3.4). We found the highest frequency of images occurred during the middle period, especially in conjunction with the discussion of rules. The notions of proof and of convincing procedures were textual markers of picture-sentences that functioned as conceptual grammar and therefore as non-propositions. Images collocated with such textual elements revealed underlying practices, principally on the theme of the foundations of arithmetic. We suggested that proofs could be regarded as visual equivalents to speech-acts, which we called *picture-acts*. *Picture-rules* and *picture-proofs* were used to reveal correct and incorrect practices by virtue of having iconic relationships with concepts, i.e., the images critiqued the textual representation which obscured the practices. This ability to show rather than to say belongs to the iconicity of the images, as mentioned in §2, however, the iconic relationship remains ineffable, i.e., the resemblance problem in depiction. We speculated that its ineffability may be a consequence of its foundational relationship of the type described in Wittgenstein's final writings on certainty, as "hinges", for example:

> *And if it were not like this the ground would be cut away from under the whole proof. For we decide to use the proof-picture instead of correlating the groups; we do not correlate them, but instead compare the groups with those of the proof (in which indeed two groups are correlated with one another). (RFM-I §31 p. 49 referring back to the hand and star correlation mentioned above).*

4.3 Future Research: The Potential for Pictures as Hinges

Picture-rules, and conventionalized picture-acts, are epistemic and part of our conceptual understanding. Hinges, on the other hand, are incontestable and are non-epistemic.

> *the mathematical proposition has, as it were officially, been given the stamp of incontestability. I.e.: "Dispute about other things; this is immovable--it is a hinge on which your dispute can turn." (OC§655).*

Hinge certainty (e.g. that an arrow points, that white is lighter than black) is more fundamental than merely seeing an aspect, which is merely particular to my pathology. Moyal-Sharrock (2021, p. 140) lists the characteristics of hinges as:

(1) non-epistemic: they are not known; not justified (2) indubitable: doubt and mistake are logically meaningless as regards them (3) foundational: they are the unfounded foundation of thought (4) non-empirical: they are not conclusions derived from experience (5) grammatical: they are rules of grammar (6) non-propositional: they are not propositions (7) ineffable: they are, qua certainties, ineffable (8) enacted: they can only show themselves in what we say and do..

Some non-propositional picture-sentences meet these criteria. For example, once we have learned to see the pointing gesture of arrows, or picture-faces as faces, they become embedded in our practices of image-based communication. In the current research we speculated that *picture-hinges* would be an additional set of non-propositional picture-sentences that embody the certainties underpinning our practices.

Acknowledgements. We would like to acknowledge the advice and resources received from Prof Alois Pichler and the Wittgenstein Archives at the University of Bergen. The scanned [diplomatic] images in this paper are taken from the facsimiles of Wittgenstein's handwritten manuscripts held at the Wren Library at Trinity College Cambridge, which are made available under the Creative Commons license CC BY-NC 4.0 (Attribution-Noncommercial 4.0 International) since February 19, 2017 via the Bergen Nachlass Edition (cf. "about BNE" at https://www.wittgensteinsource.org accessed on 17 June 2021).

References

Anscombe, G.E.M.: An Introduction to Wittgenstein's Tractatus. Hutchinson University Library (1959)

Baker, G.P.: Wittgenstein: concepts or conceptions? Harvard Rev. Philos. **9**, 7–23 (2001). https://doi.org/10.5840/harvardreview2001912

Bazzocchi, L.: A significant 'false perception' of Wittgenstein's draft on mind's eye. Acta Anal. **29**(2), 255–266 (2013). https://doi.org/10.1007/s12136-013-0197-1

Biggs, M.A.R., Pichler, A. (eds.): Wittgenstein: Two source catalogues and a bibliography: catalogues of the published texts and of the published diagrams, each related to its sources. Wittgensteinarkived ved Univ. i Bergen (1993). http://www.wittgensteinrepository.org/ojs/index.php/agora-wab/article/download/3232/3930

Giardino, V., Greenberg, G.: Introduction: varieties of iconicity. Rev. Philos. Psychol. **6**(1), 1–25 (2014). https://doi.org/10.1007/s13164-014-0210-7

Gregory, D.: Pictures, propositions, and predicates. Am. Philos. Q. (oxf.) **57**(2), 155–170 (2020). https://doi.org/10.2307/48570845

Mitchell, W.J.T.: Iconology: Image, Text, Ideology. University of Chicago Press, Chicago (1986)

Moyal-Sharrock, D.: Certainty in Action: Wittgenstein on Language, Mind and Epistemology. Bloomsbury, London (2021)

Shier, D.: How can pictures be propositions? Ratio (oxford) **10**(1), 65–75 (1997). https://doi.org/10.1111/1467-9329.00027

What Kind of Opposition-Forming Operator is Privation?

José David García Cruz[(✉)] [iD]

Pontifical Catholic University of Chile, Santiago, Chile
jdgarcia2@uc.cl

Abstract. In this paper, a new kind of opposition relations is presented. Taking privation as a main negative operation on predicates, in this paper is presented a relative opposition theory, *i.e.*, a sub-theory of oppositions.

Keywords: Privation · Negation · Opposition

1 Introduction

In his commentary on Aristotle's *De Interpretatione*, Ammonius Hermiae [2] presents a hexagon of oppositions whereby he analyses the logical relations between simple, indefinite, and privative propositions. Interpreting a passage from Aristotle, he concludes that privation is a more specific form of canceling a property, while negation is the more general one. Ammonius further draws on several passages from the Metaphysics and Categories to emphasize this idea, although he does not specify whether privation produces a contrariety, a sub-contrariety, or a contradiction. In this regard, Aristotle *(Met. 1055 a 34 - 1055 b 10)* mentions that privation satisfies characteristics of contrariety since it is possible that privation and its opposite can be false simultaneously; but furthermore, privation satisfies characteristics of contradiction in the sense that, being a specific negation, it is a particular case of contradictory negation relative to a subject that must possess the property of which it is deprived.

Considering this scenario, the question that motivates this paper is what kind of opposition-forming operator is privation? This question is relevant for two reasons. On the one hand, from the point of view of the theory of oppositions, analyzing an operation such as privation is interesting because this notion presents a hybrid case of oppositions. This may imply that we are faced with a new form of opposition with dynamic characteristics, which changes its truth conditions concerning the context of predication. To exemplify this case of dynamism, even and odd are not opposed in the same way as just and unjust, even though, in both pairs of opposites, we find similar properties. On the other hand, this question is relevant to the general study of privation from a logical point of view. Privation as a matter of logical scrutiny arises, as far as we know, in the passage

Supported by ANID-Chile. Special thanks to the reviewers, without their valuable comments this work would not have been possible.

© Springer Nature Switzerland AG 2021
A. Basu et al. (Eds.): Diagrams 2021, LNAI 12909, pp. 118–131, 2021.
https://doi.org/10.1007/978-3-030-86062-2_11

of Aristotle's *De Interpretatione 19 b 22 - 24*, in which the mode of opposition of propositions with this type of predicates is studied. Therefore, answering the question that concerns us may offer a new matter of analysis for this discussion.

The thesis to be defended in this paper is the following: privation is a dynamic opposition-forming operator, a relative negation, which, depending on the properties of the opposing propositions, establishes specific truth conditions of a contradiction or a contrariety. In other words, privation produces relative opposition relations.

To justify this thesis, we propose the following strategy. We will begin with a definition of what we will call the term logic of privation. This logic shares syntactic features with other known systems of logic, from the traditional formalization of Łukasiewicz to the term logic of Fred Sommers and George Englebretsen. The novelty in this respect is the introduction of operators on terms, one for producing indefinite terms and another for privative terms. On the other hand, semantics is the most original part of this system. We draw inspiration from Igor Sédlar and Karel Sebela's definition of "range of applicability" of a term to define the semantics for privative terms in our system[1]. We propose a variation of Sédlar and Sebela's notion of range to define privation as the complement of the extension of a term, relative to its range of predication[2].

In this sense, privation is an operation derived from negation, relative to the possible predicates for a given subject. That is to say, if "man" has as a range of predication all that we can predicate to men, then, to deprive a man of a property X, means that, X could belong to the range of predication of "man" and such a man does not have the property X. Based on this characterization, we will study the properties of privation in the context of opposition relations, whereas that it is required to propose a specific definition of opposition for privation. We end the paper with this definition, which for lack of a better name, we will call *r-opposition*.

The plan of the paper is as follows. Section two presents in detail the term logic of privation, the syntax is analyzed, the semantics and the consequence are defined. Subsequently, in section three, the textual evidence necessary to support this logic and to justify the proposal of the paper is taken up. Finally, section four presents the question of what kind of opposition-forming operator is privation and solves the issue by defining *r-oppositions*.

---

[1] "The range $\hat{P}$ of a predicate P can be seen as *the range of applicability* of P, denoting things that may be meaningfully– even if not truly–described by P". [19, pp. 266].

[2] This characterization is close to the one proposed by Seuren & Jaspers in [20], specifically related to the concept of *morphological negation* that they develop. In addition, Seuren & Jaspers' approach can be compared with many-valued analysis of privation in [13]. Due to lack of space, more in-depth comparisons are not elaborated here, but this will be left for future work.

2 TL_σ: The Term Logic of Privation

2.1 Syntax of TL_σ

In this section are presented all the syntactic elements for term logic of privation. The inspiration for the system proposed here coming from [1,6,7,11,12,14–18, 21–24].

Definition 1 *(Alphabet of TL_σ). The alphabet of TL_σ is the union of the following collections:*

1. *A collection of singular-term variables: $S = \{t_i, t_j, t_k, ...\}$,*
2. *A collection of universal-term variables: $U = \{T_i, T_j, T_k, ...\}$,*
3. *A collection of quantifiers: $Q = \{[...], \langle...\rangle, (...)\}$,*
4. *A collection of negations: $N = \{|, -, *\}$.*

For term variables, in practice we use upper and lower case letters for singular and universal terms, treating them as meta-variables ranging over S and U respectively. The quantifiers are the same as in first-order logic, universal and particular ($[...], \langle...\rangle$, respectively), but there is a new kind of quantifier, the undetermined one ($(...)$), which could be read as "the...", "a...", as in "the man is animal", or "a man is animal". Now, let us proceed to the definition of complex term and formula.

Definition 2 *(Term collection). The collection $\mathbb{T}$ of terms is given by the following two conditions:*

1. *$S \cup U \subset \mathbb{T}$,*
2. *$\forall T \in U; \overline{T}, T^* \in \mathbb{T}$.*

Terms of the first class are called *singular* terms and *universal* terms as we have said, and terms of the second class are of two kinds: *indefinite* and *privative* terms, respectively. Indefinite and privative terms are formed only by the concatenation of a negative particle and a universal term. Singular terms are the only terms that cannot be predicates and the formulas in which they appear are called singular formulas. Universal terms can be subject or predicate on a formula. The following definition states all these intuitions.

Definition 3 *(Formula). The collection FOR of term formulas is the smallest collection of sequences satisfying the following conditions:*

1. *if $t_i \in S$ and $T_1, T_j \in \mathbb{T} - S$, then $(t_i T_i), (t_i | T_i), (T_i T_j), (T_i | T_j) \in FOR$,*
2. *if $T_i, T_j \in \mathbb{T} - S$, then $[T_i T_j], [T_i | T_j], \langle T_i T_j \rangle, \langle T_i | T_j \rangle \in FOR$.*

In specific this collection will be called FOR_{TL_σ}, but for short in the following will be called only FOR, and its elements will be called *formulas*.

Example 1 (Formula). The following are examples of formulas:

1. $[MW]$ "Every man is wise"
2. $\langle A | \overline{J} \rangle$ "Some animal is not non-just"

3. (AW^*) "The (an) animal is in-wise (deprived of wisdom)"
4. (sJ) "Socrates is just"

Example 2 (non-Formula). The following are examples of non-Formulas

1. $[sA]$ "Every Sócrates is animal"
2. $\langle c|W \rangle$ "Some Calias is not wise"
3. (s^*A) "in-Socrates is animal"

2.2 Semantics and Consequence

Now, lets turn to the semantic aspect of this logic. We will define the models and the consequence relation.

Definition 4 *(Term-Frame). A Term-Frame is a tuple* $\mathbb{F}_T = \langle D, \sqsubseteq, \eta \rangle$, *where:*

1. $D = \{a, b, c, ...\}$,
2. $\sqsubseteq \subseteq P(D) \times P(D)$,
3. $\eta : \mathbb{T} \longrightarrow P(D)$.

The first element of the Term-Frame is a non-empty collection of atoms, from this collection we obtain the *domain* by its power-set $P(D)$. The second element $\sqsubseteq$, is a transitive, non-symmetric and non-reflexive relation on the domain; and finally, the function η is an operation that assigns sub-collections of D to the terms of the language, intuitively we say that η assigns to each term its extension. Another way to consider those elements is as the *complete lattice* $(P(D), \sqsubseteq)$ in which top and bottom elements are D and $\emptyset$, respectively. Before going to consequence we need some definitions related to the properties of above elements. First, we have to talk about *range of predication*, and semantics for indefinite and privative terms.

Definition 5 *(Range of predication). Let* $T_i, T_j \in \mathbb{T}$, *and let* $\mathbb{F}_T = \langle D, \sqsubseteq, \eta \rangle$ *be a Term-Frame, we will say that* $\mathbb{R}_{T_i} = \{T_j : \eta(T_i) \sqsubseteq \eta(T_j) \text{ or } \eta(T_i) \cap \eta(T_j) \neq \emptyset\}$ *is the range of predication of the term* T_i.

Intuitively talking, the range of predication of a term T_i is a collection of terms such that, its extensions are related by $\sqsubseteq$ with the extension of T_i or the infimum between $\eta(T_i)$ and $\eta(T_j)$ (for all $T_j \in \mathbb{T}$) is non-empty[3].

Definition 6 *(Indefinite Term semantics). Let* $\overline{T} \in \mathbb{T}$, *we say that the extension of* $\overline{T}$ *is the collection* $\eta(\overline{T}) = P(D) - T$.

Definition 7 *(Privative term semantics). Let* $T \in U$, *we say that* $\forall X \in \mathbb{T}$ *the extension of* T^* *is the collection* $\eta(T^*) = \mathbb{R}_X - T$.

---

[3] This definition is a variation of the notion of range of applicability defined in [19, 271].

The next definition is the concept of model, that concept will allow us to define the notion of truth in a model and therefore, the concept of consequence.

Definition 8 *(Model for TL_σ). A model $\mathcal{M}_{TL_\sigma}$ for TL_σ will be the structure $\mathcal{M}_{TL_\sigma} = (\mathbb{F}_\mathbb{T}, V, v)$, where:*

- $\mathbb{F}_\mathbb{T} = \langle D, \sqsubseteq, \eta \rangle$ *is a Term-Frame,*
- $V = \{F, T\}$ *is a collection of truth values,*
- $v : FOR \longrightarrow V$ *is a valuation function.*

Definition 9 *(Truth in a model). Let $\mathcal{M}_{TL_\sigma}$ be a model for TL_σ, let be $\varphi \in FOR$, we say that $v(\varphi) = T$ in model $\mathcal{M}_{TL_\sigma}$ (for short $\mathcal{M}_{TL_\sigma} \Vdash \varphi$), if the conditions below are met:*

1. $\mathcal{M}_{TL_\sigma} \Vdash (SP)$, iff $\eta(S) \sqsubseteq \eta(P)$,
2. $\mathcal{M}_{TL_\sigma} \Vdash (S|P)$, iff $\eta(S) \not\sqsubseteq \eta(P)$,
3. $\mathcal{M}_{TL_\sigma} \Vdash \langle SP \rangle$, iff $\exists T \in \mathbb{T}$ such that, $\eta(T) \sqsubseteq \eta(S)$ and $\eta(T) \sqsubseteq \eta(P)$,
4. $\mathcal{M}_{TL_\sigma} \Vdash \langle S|P \rangle$, iff $\exists T \in \mathbb{T}$ such that, $\eta(T) \sqsubseteq \eta(S)$ and $\eta(T) \not\sqsubseteq \eta(P)$,
5. $\mathcal{M}_{TL_\sigma} \Vdash [SP]$, iff $\forall T \in \mathbb{T}$, if $\eta(T) \sqsubseteq \eta(S)$, then $\eta(T) \sqsubseteq \eta(P)$,
6. $\mathcal{M}_{TL_\sigma} \Vdash [S|P]$, iff $\forall T \in \mathbb{T}$, if $\eta(T) \sqsubseteq \eta(S)$, then $\eta(T) \not\sqsubseteq \eta(P)$.

Given a formula $\varphi \in FOR$, we say that $v(\varphi) = F$ in model $\mathcal{M}_{TL_\sigma}$ (for short $\mathcal{M}_{TL_\sigma} \nVdash \varphi$) if non of the above conditions are met. Let $X \subseteq FOR$, we say that X is true in a model $\mathcal{M}_{TL_\sigma}$ (for short $\mathcal{M}_{TL_\sigma} \Vdash X$) iff $\forall \varphi \in FOR, \mathcal{M}_{TL_\sigma} \Vdash \varphi$. On the other side X is false in a model $\mathcal{M}_{TL_\sigma}$ (for short $\mathcal{M}_{TL_\sigma} \nVdash X$) iff $\forall \varphi \in FOR, \mathcal{M}_{TL_\sigma} \nVdash \varphi$

Finally, semantic consequence relation is defined as follows.

Definition 10 *(Semantic Consequence). Let $X \subseteq FOR$ and $\varphi \in FOR$ and let $\mathbb{M}_{TL_\sigma}$ a collection of models, we say that φ follows from X (for short $X \Vdash \varphi$) iff, for all $\mathcal{M}_{TL_\sigma} \in \mathbb{M}_{TL_\sigma}$, if $\mathcal{M}_{TL_\sigma} \Vdash X$ then $\mathcal{M}_{TL_\sigma} \Vdash \varphi$. The relation $\Vdash$ is called* semantic consequence *of TL_σ. When this conditions is not met we say that φ do not follows from X, and we write $X \nVdash \varphi$.*

3   Some Textual Evidence for Semantics of Privative Terms

In this section we will present the general discussion about the private terms from Aristotle's commentators perspective. The first part presents an informal outline on the private and indefinite terms, and its links with the negation in Aristotle. We follow the analysis of Manuel Correia in his work *Is being non-just the same as being unjust? Aristotle and his commentators* [8][4]. We will present a synthesis of this work highlighting the author's conclusion.

---

[4] In Spanish: "¿Es lo mismo ser no-justo que ser injusto? Aristóteles y sus comentaristas".

3.1   Two-Term and Tree-Term Propositions

Manuel Correia's text is divided into three parts, initially he presents the problem and the passage of Aristotle, secondly he presents the expositions of the passage that some commentators elaborate, and finally he discusses the expositions of the commentators. The discussion begins with the distinction between two-term and three-term propositions and their mode of opposition. We therefore start with a brief outline of the distinction between these terms and a basic presentation of the square of oppositions.

In the first four chapters of the *De Interpretatione* the definition of noun, verb, proposition is presented. A noun is a meaningful sound by convention without reference to time, and its parts are not separately meaningful (*De In. 16 a 19*), a verb is like a noun but in addition contains a reference to time and means what is said about something else. Further on in (*17 a 36*), he proposes an analogous distinction between universal and particular terms. These types of propositions can be combined positively or negatively to form affirmations or negations, and if a quantifier is also added, four types of propositions are obtained, the well-known propositions of the square of opposition.

This combination of propositions can occur in two ways, with a noun and a verb, or with a noun a verb and the verb to be, that is, forming a two-term or a three-term proposition, respectively. An example of a two-term proposition is "Socrates walks", while a three-term proposition is "Socrates is white", where, the verb to be is added as a third element [8, pp. 42], (*19 b 19*). Thus, the main distinction is the presence or absence of the verb to be. This distinction becomes meaningful as soon as we ask ourselves what is the mode of opposition between two-term and three-term propositions, and what is the number of propositions opposed to a two-term or three-term proposition.

Aristotle presents several squares of oppositions, as Correia states in [10]. The main one, and the one to which most attention has been paid, is the square containing quantified propositions, the *traditional square*. Our reformulation of this square is the one presented in Fig. 1[5].

Having established these definitions, we can present the mode of opposition of simple propositions of two and three terms. Taking into account that oppositions are formed by adding the negation to the verb, in the case of two-term propositions we only have one line of contradiction, and in the case of three-term propositions the number is doubled because the negation can be positioned in front of the verb to be or in front of the predicate. As Correia mentions [8, pp. 42]. With this context we can move on to Correia's presentation.

3.2   Correia's Analysis

The problem here is how three-term propositions, with simple, indefinite and privative predicates oppose? Manuel Correia presents an analysis of the inter-

---

[5] It is worth mentioning that Aristotle's formulation only included contradiction, contrariety and subcontrariety; subalternation was added by the commentators.

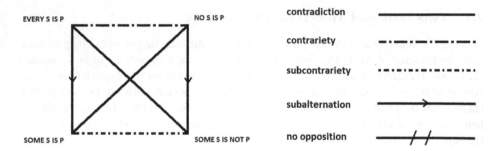

Fig. 1. Traditional square of oppositions

pretation given by commentators of the (*De Interpretatione* 19 b 22 - 24) passage, which we will reproduce in Table 1, and we will explain in detail.

Table 1. Comparison of commentators

	Prior Analytics	Unjust ≠ non-just	Indefinite subject	Syntactic interpretation	Semantic interpretation	Negative to Positive
Herminus	√	√	√	×	×	×
Pophyry	√	√	×	×	√	√
Alexander	√	√	×	√	×	×
Ammonius	√	×	×	×	√	√

The exposition takes Boethius and Ammonius as its main sources. The former is in charge of presenting the interpretations of Herminus, Alexander and Porphyry, the latter recovers the oral tradition of his teacher Proclus, and proposes the interpretation of deprivation that interests us.

First, Herminus' interpretation has the following characteristics: 1) he maintains that privative terms are equivalent to indefinite terms, and 2) he introduces indefinite terms for the subject into the discussion. This interpretation is the one that interests us the least.

Boethius, in agreement with Correia's report, maintains that Herminus loses direction in the last association by introducing indefinite subjects. With this formalization we can appreciate the loss of symmetry between the distribution of indefinite and privative negation over subjects and predicates, which affects Herminus' interpretation, for, as we see in Fig. 2, the last two formulas of the hexagon are affirmative, while the arrangement of the four upper formulas is similar to the arrangement of the traditional opposition square, the left formulas affirmative and the right formulas negative.

This same lack of symmetry makes us reconsider what relationship is maintained by the formulas introduced by Herminus at the end of the diagram, with indefinite subject, since both are affirmative, and therefore are not in an affirmation/negation relationship. Assuming that Herminus has arranged them in the

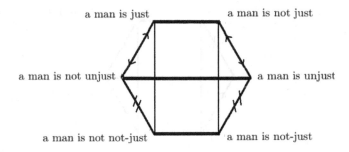

a man is just   a man is not just

a man is not unjust   a man is unjust

a man is not not-just   a man is not-just

Fig. 2. Herminus' Hexagon

way the diagram shows, we can assume that Herminus considers one to be the negation of the other, even though they are both affirmations. Accordingly, for Herminus the indefinite terms produce negations, even though the proposition is an affirmation. Therefore, these would be contradictory for him. On the other hand, transitivity also allows us to suppose other relations.

Alexander's interpretation is similar to Herminus' except 1) he does not introduce indefinite subjects, and 2) he interprets Aristotle's passage syntactically, Fig. 3 shows this interpretation. In Correia's words Alexander explains Aristotle's passage as follows:

> Since the privative affirmation is equivalent with the indefinite affirmation and, in the other column, the privative negation is equivalent with the indefinite negation, and this does not occur with the respective simple affirmation and simple negation, it follows that: "two propositions" (namely, the affirmative and the negative with indefinite predicate), "are related, in order of sequence (are similar), according to affirmation and negation, in the way that privations are related, while two are not" (namely, simple affirmation and simple negation).[6] [8, pp. 49]

Our interpretation is the following. Two propositions, "man is unjust" and "man is not unjust", are related in order of sequence, *i.e.*, they have a syntactic resemblance (they have indefinite predicates although one is affirmative and the other negative); "just like privations"; that is, the propositions "man is unjust" and "man is not unjust" have a syntactic resemblance analogous to the previous pair, namely, that their predicate is "negative" in a certain way, and that one is affirmative and the other negative. On the other hand, there is a pair that is not similar to these two pairs, the pair of propositions "man is just" and "man

---

[6] In Spanish: Puesto que la afirmación privativa es equivalente con la afirmación indefinida y, en la otra columna, la negación privativa es equivalente con la negación indefinida, y ello no ocurre con las respectivas afirmación simple y negación simple, se tiene que: "dos proposiciones" (a saber: la afirmativa y la negativa con predicado indefinido), "se relacionan, en orden de secuencia (sc. son similares), según la afirmación y la negación, del modo como las privaciones se relacionan, mientras que dos no" (a saber, la afirmación y la negación simples).

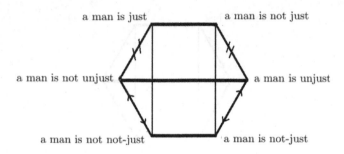

Fig. 3. Alexander's Hexagon

is not just", for their predicate is neither indefinite nor privative, there is no affection on the predicate. For this reason Alexander arranges the propositions as in the hexagon, the affirmative ones in the left column and the negative ones on the right side, the ones with simple terms at the top, the privative ones in the middle and the indefinite ones at the bottom.

Alexander takes a step beyond Herminus' analysis, since he manages to focus the discussion on the simple, privative and indefinite predication. His syntactic interpretation prevents him from moving forward, since the relations displayed in the hexagon separate negation from privation and indefinite negation. The problem of identifying privation and indefiniteness is that semantically there is no contribution in their introduction in the hexagon, since they are equivalent, they are related in the same way with the rest of the formulas. In general, if they are equivalent, there is also no contribution in distinguishing them in everyday or scientific discourse, but this is not the case.

Porphyry, on the other hand, differs from Alexander, only in 1) not interpreting the passage syntactically, but semantically, and modifying the relations between propositions, allowing the negative ones to follow from the positive ones.

According to Correia, Porphyry gives a semantic reading of the passage and relies on (*An. Pr. I 46*). The interpretation is semantic because the similarity that Alexander emphasized before, now becomes a similarity in truth value. Porphyry exchanges the negative propositions with indefinite and privative predicate to the right side, because of the similarity they have with respect to the truth value of the simple affirmative proposition. Thus we obtain Porphyry's hexagon, shown in Fig. 4, with the respective relations. This interpretation has the same difficulties as that of Alexander, with respect to equivalence. The originality comes with Ammonius' interpretation.

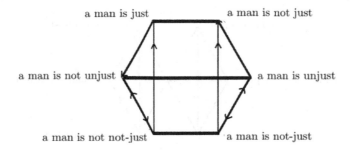

Fig. 4. Porphyry's Hexagon

Finally, Ammonius, differs from all except 1) interpreting the passage semantically (like Porphyry), and 2) allowing negatives to follow from positives. The main point of difference between Ammonius and the others is that privative terms are not equivalent to indefinite terms unless a condition is satisfied between the opposites.

The basis of Ammonius' interpretation rests on several facts that we will detail below. The first fact to highlight is that according to Correia, Ammonius takes from his teacher Proclus the idea that propositions with privative predicate are not equivalent to propositions with indefinite predicate. Moreover, Ammonius takes evidence, to justify this idea, from texts such as (*Physics A, 7 189 b 30*) and (*Categories 11 b 38 - 12 a 5*). And consequently, Ammonius takes up the doctrine of contraries, which in synthesis maintains that, "contraries are of two kinds, necessary to the substance in question or accidental to it" [8, pp. 49]. Examples of necessaries are "even/odd" for numbers, "health/disease" for animal. For these cases privative and indefinite do coincide, e.g. not-even is equivalent to odd. Examples of accidental contraries are "white/black" for bodies, "just/unjust" for men. In these cases the privative and the indefinite do differ, since black is not equivalent to not-white, and not-just is not equivalent to unjust.

Bearing this doctrine in mind, and assuming that justice is accidental to man, Correia's explanation is that Ammonius is going to order the types of negations, first of all "(a) 'to be unjust', said of a man is more specific than (b) 'to be unjust', and this is more specific than (c)'not to be just', since the number of beings to whom 'to be unjust' applies is less than the number of beings to whom 'to be unjust' applies, and this more specific than 'not to be just'. " [8, p. 49] That is, these three types of negative predicates form a strict order of logical consequence as shown in Fig. 5. As can be seen in the diagram, there are several missing relations, and because of that it is necessary to analyze what kind of relations are missing, *i.e.*, between a negative proposition with a negative predicate (MJ^*) and a negative one ($M|J$) what kind of relation could be? Therefore, to sum up, we turn to the presentation of privation as opposition-forming operator.

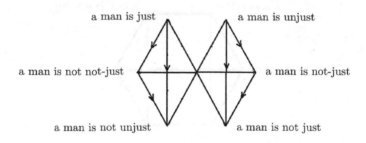

a man is just

a man is unjust

a man is not not-just

a man is not-just

a man is not unjust

a man is not just

Fig. 5. Ammonius' Hexagon

4   Privation as Opposition-Forming Operator

As we mention on the introduction, Aristotle himself suggest[7] that privation have some properties of contradiction and contrariety, despite not being the same as contradiction or contrariety. Following this idea, in this final part, we elaborate a formal analysis of oppositions from the standpoint of what Manuel Correia have been called *The Proto-Exposition of Aristotelian Categorical Logic* [9].

In that paper, Manuel Correia presents [9, pp. 25] a very useful arrangement of the way in which propositions relate. We will reformulate this description using our formal language, and introducing some specific definitions for opposition relations.

Definition 11 *(Meta-logical properties of propositions and terms). The following are some operations on formulas and terms. For all $t \in \mathbb{T}, \varphi, \psi \in FOR$:*

1. $\tau : FOR \longrightarrow \mathbb{T} \times \mathbb{T}$
2. $Sub : FOR \longrightarrow \mathbb{T}$
3. $Pre : FOR \longrightarrow \mathbb{T}$
4. $K : FOR \longrightarrow \{Univ, Par\}$
5. $Q : FOR \longrightarrow \{Aff, neg\}$
6. $ty : \mathbb{T} \longrightarrow \{sim, ind, priv\}$

The first one, is a function that takes a formula and gives its terms. The second one, is a function that takes a formula as input and its output is the subject of the formula, *i.e.*, the first term. The next one is a similar function as the previous, but in this case, the output is the predicate of the formula. Function number four is the quantity function, this function takes as input a formula and as output gives us its quantity. Function number five is the quality function, as in the previous case, takes as input a formula and as output gives it quality. Finally, the last function is the type function, that takes a term as input and its output is one of the three kinds of term, *i.e.*, simple, indefinite, or privative. These operations are the basic elements to present our definition of r-oppositions. First,

---

[7] *(Met. 1055 a 34 – 1055 b 10).*

we will define the notion of *opposite pair* following Correia's Proto-Exposition of Aristotelian Categorical Logic. As Correia explain, the theory of oppositions is a theory of relations between propositions with the same terms and the terms are in the same order. The corresponding opposition relations emerge considering the kind of quantifier and the negation in each case, that is to say, taking into account the quantity and quality of the propositions. In our terms we define opposites as a pairs of formulas holding the mentioned conditions. The following definition states this idea[8].

Definition 12 *(Opposites). $Opp \subseteq FOR \times FOR$, such that:*
$Opp = \{<\varphi, \psi> \in FOR \times FOR : \tau(\varphi) = \tau(\psi) \& (Sub(\varphi) = Sub(\psi) \& Pre(\varphi) = Pre(\psi))\}$

This definition lead us to define opposition as follows.

Definition 13 *(Opposition relations). $\forall < \varphi, \psi > \in Opp$, we have that:*

1. *φ, ψ are contradictory ($\mathbb{K}\varphi\psi$) iff $K(\varphi) \neq K(\psi)$ and $Q(\varphi) \neq Q(\psi)$*
2. *φ, ψ are contraries ($\mathbb{C}\varphi\psi$) iff $K(\varphi) = K(\psi) = Univ$ and $Q(\varphi) \neq Q(\psi)$*
3. *φ, ψ are subcontraries ($\mathbb{SC}\varphi\psi$) iff $K(\varphi) = K(\psi) = Par$ and $Q(\varphi) \neq Q(\psi)$:*
4. *φ, ψ are subalterns ($\mathbb{SA}\varphi\psi$) iff $K(\varphi) \neq K(\psi)$ and $Q(\varphi) = Q(\psi)$*

That definition can be seen as a formal reconstruction of Aristotle's one presented in *(De In. 17b 17–26)*, with the extra condition that we have added subalternation. Now we will present our specific definition for propositions with private terms. The first step is to define the general notion of *r-opposition*, as the pairs of formulas that are opposites and the predicate is part of the range of the subject. Subsequently, we will define the specific opposition relations for privative propositions, and to conclude we present a diagram that generalizes and complete Ammonius' hexagon.

Definition 14 *(r-Opposites). $r\text{-}Opp \subseteq Opp$, such that:*
$r\text{-}Opp = \{< \varphi, \psi > \in Opp : \eta(Pre(\varphi)) \in \mathbb{R}_{Sub(\varphi)}\}$

Then, r-opposition relations are defined as follows.

Definition 15 *(r-Opposition relations). $\forall < \varphi, \psi > \in r\text{-}Opp$, we have that:*

1. *φ, ψ are r-contradictory ($\mathbb{K}_r\varphi\psi$) iff $K(\varphi) \neq K(\psi)$ and $Q(\varphi) = Q(\psi)$ and $ty(Pre(\psi)) = priv$*
2. *φ, ψ are r-contraries ($\mathbb{C}_r\varphi\psi$) iff $K(\varphi) = K(\psi) = Univ$ and $Q(\varphi) = Q(\psi)$ and $ty(Pre(\psi)) = priv$*
3. *φ, ψ are r-subcontraries ($\mathbb{SC}_r\varphi\psi$) iff $K(\varphi) = K(\psi) = Par$ and $Q(\varphi) = Q(\psi)$ and $ty(Pre(\psi)) = priv$*
4. *φ, ψ are r-subalterns ($\mathbb{SA}_r\varphi\psi$) iff $\mathbb{K}\varphi\psi$ and $ty(Pre(\psi)) = priv$; or $\mathbb{C}\varphi\psi$ and $ty(Pre(\psi)) = priv$*

---

[8] This definition may seem very redundant, and can possibly be omitted. It is explicitly included here only to keep us aligned to Proto-exposition.

Finally, we only want to mention that this theory is a sub-theory of opposition theory defined following the Proto-Exposition of Aristotelian Categorial Logic. In the same sense, we are able to define other sub-theories as r-Syllogistic, or theory of r-Conversion. We now conclude the diagram by presenting the remaining relationships. In this case there are two r-contrariety relations, one between (MJ) and (MJ^*), and the other between $(M|\bar{J})$ and (MJ^*). Also there are two r-subcontrariety relations, one between $(M|J^*)$ and $(M\bar{J})$, and the other between $(M|J^*)$ and $(M|J)$. To represent the corresponding relative relationship we use the same code of bridged lines together with a black line[9].

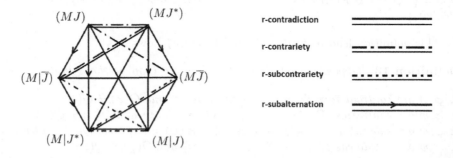

Fig. 6. Ammonius' hexagon 2 in TL_σ

5   Conclusion

In this paper we have presented a term logic of privative terms, this system is capable of representing in an explicit way propositions with privative terms, its semantics is equipped with a notion that lead us to define a clear semantics for privative terms and to define opposition relations for propositions with that terms. The main result is the definition of r-oppositions as a sub-collection of relations of opposition with the reference to the range of predication of terms. The paper begins with a question that this definition answer, privation is a relative opposition-forming operator.

References

1. Alvarez, E., Correia, M.: Syllogistic with indefinite terms. History Philosophy Logic **33**(4), 297–306 (2012)
2. Busse, A. (ed.): Ammonius. Ammonii In Aristotelis De Interpretatione Comentarius. Reimer (1985)
3. Barnes, J.: The Complete Works of Aristotle: The Revised Oxford Translation, vol. 1. Princeton University Press, Princeton (1984)
4. Bochenski, I.M.: Ancient Formal Logic. North Holland Publishing Company, Amsterdam (1968)

---

[9] All these relations are valid by Definition 9.

5. Boecio. Los tratados silogísticos: De syllogismo Categorico y Introductio Ad Syllogismos Categoricos. Traducción, estudio preliminar y notas de Manuel Correia. Ediciones Universidad Católica de Chile (2011)
6. Corcoran, J.: Completeness of an ancient logic. J. Symbol. Logic **37**(4), 696–702 (1972)
7. Corcoran, J.: Aristotle's natural deduction system. In: Corcoran, J. (eds.) Ancient Logic and Its Modern Interpretations. Synthese Historical Library (Texts and Studies in the History of Logic and Philosophy), vol 9. Springer, Dordrecht (1974). https://doi.org/10.1007/978-94-010-2130-2_6
8. Correia, M.: "¿Es lo mismo ser no-justo que ser injusto? Aristóteles y sus comentaristas", Méthexis XIX, 41–51 (2006)
9. Correia, M.: The Proto-exposition of Aristotelian Categorical Logic. In: Béziau, J.-Y., Basti, G. (eds.) The Square of Opposition: A Cornerstone of Thought. SUL, pp. 21–34. Springer, Cham (2017). https://doi.org/10.1007/978-3-319-45062-9_3
10. Correia, M.: Aristotle's squares of opposition. South Am. J. Logic **3**(2), 313–326 (2017)
11. Englebretsen, G.: The New Syllogistic. P. Lang (1987)
12. Englebretsen, G.: Something to Reckon with: The Logic of Terms. Books collection. Canadian electronic library: University of Ottawa Press (1996)
13. García Cruz, J.D.: A Useful Four-Valued Logic with Indefinite and Privative Negations: Ammonius and Belnap on Term Negations. To appear on The Logica Yearbook (2020)
14. Jarmuzek, T.: Tableau Methods for Propositional Logic and Term Logic. Peter Lang, Berlin (2020)
15. Lear, J.: Aristotle and Logical Theory. Cambridge University Press, Cambridge (1980)
16. Łukasiewicz, J.: La silogística de Aristóteles. Desde el punto de vista de la lógica formal moderna. Editorial Tecnos, Madrid (1977)
17. Martin, J.: Proclus and the neoplatonic syllogistic. J. Philos. Log. **30**(3), 187–240 (2001)
18. Matrin, J.: All brutes are subhuman: Aristotle and Ockham on private negation. Synthese **134**(3), 429–461 (2003)
19. Sedlár, I., Sebela, K.: Term negation in first order logic. Logique et Anal. (N.S.) **247**, 265–284 (2019)
20. Seuren, P.A.M., Jaspers, D.: Logico-cognitive structure in the lexicon. Language **90**(3), 607–643 (2014)
21. Smiley, T.J.: Syllogism and quantification. J. Symbol. Logic (1962)
22. Smiley, T.J.: What is a syllogism. J. Philosophical Logic (1972)
23. Sommers, F.: The Logic of Natural Language. Clarendon Library of Logic and Philosophy. Clarendon Press, Oxford: New York: Oxford University Press (1982)
24. Sommers, F., Englebretsen, G.: The Logic of Terms. Ashgate, An Invitation to Formal Reasoning (2000)

Presenting Basic Graph Logic

Márcia R. Cerioli[1], Leandro Suguitani[2]([✉]), and Petrucio Viana[3]

[1] Programa de Engenharia de Sistemas e Computação-COPPE e Instituto de Matemática, UFRJ, Rio de Janeiro, RJ, Brazil
marcia@cos.ufrj.br
[2] Instituto de Matemática, UFBA, Salvador, BA, Brazil
leandro.suguitani@ufba.br
[3] Instituto de Matemática e Estatística, UFF, Niterói, RJ, Brazil
petrucio_viana@id.uff.br

Abstract. In this paper, we present and exemplify the Basic Graph Logic (BGL), this is an initial formalism which can be extended to provide a diagrammatic representation within which Set Theory and, hence, the whole of mathematics can be diagrammatic developed. We present the syntax, semantics, and inference engine of BGL. We introduce and exemplify the BGL-inference rules by showing, throughout diagrammatic proofs, that the BGL-operators satisfy the analogous of the relation algebra axioms, i.e., BGL is build as a distributive complemented lattice with operators satisfying the involutive monoid axioms and the De Morgan's Theorem K.

Keywords: Diagrammatic reasoning · Binary relations · Basic Graph Logic · Relation algebra

1 Introduction

This paper is a gentle introduction to *Basic Graphic Logic* (BGL). Its main contribution is a presentation of the syntax, semantics, and inference engine of BGL through the exam of simple examples. With this work we hope to contribute to the dissemination of knowledge about this beautiful diagrammatic system.

BGL is designed to express and reason with information based on binary relations. Given an universe of discourse U, a **binary relation** defined on U is a subset of $U \times U$. Binary relations provide ways to "organize the universes" on which they are defined. This can be easily seen, at least, for the finite universes.

Example 1. If we want to organize a finite universe U in layers, it suffices to define a relation on U satisfying the three well known conditions: reflexivity, antisymmetry, and transitivity. By satisfying conditions above a relation R is called a *partial order* on U. As the reader will be able to check later on, after going through Sect. 3, the BGL-inclusion:

$$- \underset{R^\smile}{\overset{R}{\rightrightarrows}} + \ \sqsubseteq \ - \overset{\iota}{\longrightarrow} + \tag{1}$$

© Springer Nature Switzerland AG 2021
A. Basu et al. (Eds.): Diagrams 2021, LNAI 12909, pp. 132–148, 2021.
https://doi.org/10.1007/978-3-030-86062-2_12

expresses "R is antisymmetric" in the BGL-language.

Given a partial order R on U and elements $a \neq b \in U$, we say that a is *before* b or that b is *after* a when $(a, b) \in R$. Now, it is easy to see that a partial order on U produces a layering on U. Indeed, we can classify the elements of U in layers as follows: first those elements which have no elements before them; second those elements which have exactly one element before them; third those elements which have exactly two elements before them; and so on, until those elements which have no elements after them. Elements for which there are no elements before them are called *minimal*. As the reader will be able to check, after going through Sect. 3, the BGL-diagram:

$$[(R^\vee \sqcap I^-) \circ E]^-$$
$$- \underset{I}{\overset{\longrightarrow}{\rightleftharpoons}} + \tag{2}$$

expresses in BGL-language that "x is an element for which there is no element z such that x is related to and different from z", that is "there is no element before x", that is "x is a minimal element".

Binary relations also provide ways to differentiate the universes in which they are defined, according to the existence or not of certain elements that satisfy special conditions regarding these relations.

Example 2. Comparing the usual orders on $\mathbb{N}$ and $\mathbb{Z}$, the first has an element which is below all the others while and the later does not have such an element.

The (necessarily unique) element which is below all the others is called *minimum*. As the reader will be able to check, after going through Sect. 3, the BGL-diagram:

$$(R^- \circ E)^-$$
$$- \underset{I}{\overset{\longrightarrow}{\rightleftharpoons}} + \tag{3}$$

expresses "x is an element for which there is no element z such that x is not before z", that is "x is a minimum element", in the BGL-language.

Finally, we want to highlight that reasoning about binary relations provides ways to investigate the relationships between the various universes on which they are defined, as well as the logical relationships between the various concepts defined through these binary relations.

Example 3. If an element is the minimum according to a certain order, it seems to be fair to conjecture that it is also a minimal element. As the reader will be able to check, after reading this paper, the following sequence of BGL-diagrams, showing how to transform (3) into (2) throughout (1) is a proof of this fact:

$$(R^- \circ E)^- \qquad (R^- \circ E)^- \qquad (R^- \circ E)^-$$

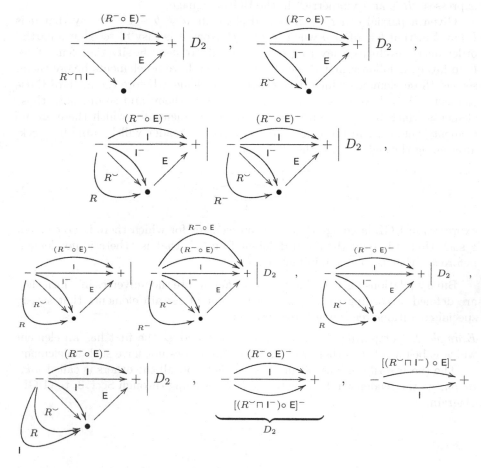

We believe that BGL has two main characteristics pointing to its acceptance as a visual diagrammatic system for the treatment of mathematical information: (1) its language "forces" the user fragmenting the information into more or less disjoint parts—represented as diagrams—whose formal treatment can be carried out without interference from the other parts and (2) its inference mechanism acts strictly on the only diagram under scrutiny, forcing, so to speak, the user to operate step-wisely in the visual search for patterns that can lead to the next diagram in the diagrams chain which constitutes—in its entirety—the reasoning being performed.

Systems based on directed multi graphs having two singled out nodes and arcs labelled with terms from the calculus of binary relations were pioneered in [1,2]. Our system is a melange of ideas from these works and from [3,4]. In this last works, the graphical systems were formally presented and studied as proper diagrammatic systems. There are other approaches for using directed multi graphs in the study of binary relations as, for example [5,6]. We left a comparison between the work reported here and these works for another occasion.

In Sects. 2, 3 and 4 we review the syntax and the semantics of BGL. In Sect. 5, we present the inference rules of BGL, one by one, over a series of examples, where the need for each rule is felt and met.

2   Syntax of Basic Graph Logic

The BGL-syntax is given by defining diagrams of four different kinds: labels, components, graphs, and inclusions.

Labels. The BGL-**labels** are built from **relation symbols** $X_1, X_2, \ldots, X_n, \ldots$, and **relation constants** E (the *symbol for universal relation*), O (the *symbol for empty relation*), I (the *symbol for identity relation*) by applying the **relation operators** $\sqcap$ (the *symbol for intersection*), $\sqcup$ (the *symbol for union*), $\circ$ (the *symbol for composition*), $^-$ (the *symbol for implementation*), $^\smile$ (the *symbol for conversion*), according to the following grammar:

$$L ::= X \mid \mathsf{O} \mid \mathsf{E} \mid \mathsf{I} \mid (L)^- \mid (L)^\smile \mid (L \sqcap L) \mid (L \sqcup L) \mid (L \circ L).$$

Relation symbols are generically denoted by X, Y, Z; labels are generically denoted by L, L_1, L_2, ...; as usual, we omit some occurrences of the auxiliary symbols (and) when no ambiguity seems to arise; for enhance the reading of the labels, we also use the symbols $\langle, \{, [,], \}, \text{and} \rangle$ as auxiliary symbols.

Example 4. The following are eight labels:

$$\mathsf{E} \,,\, \mathsf{I} \,,\, X^\smile \,,\, X^- \,,\, X^- \circ \mathsf{E} \,,\, X^\smile \sqcap \mathsf{I}^- \,,\, (X^\smile \sqcap \mathsf{I}^-) \circ \mathsf{E} \,,\, [(X^\smile \sqcap \mathsf{I}^-) \circ \mathsf{E}]^- \quad (4)$$

Components. The BGL-**components** are *multidigraphs*—i.e., directed graphs with possibly multiple arcs and loops—having two, not necessarily distinct, singled out nodes, and arcs labeled with labels. That is, the components are structures $\langle N, A, i, o \rangle$, where N is a set of **nodes**, $A \subseteq N \times \text{Labels} \times N$ is a set of **arcs** with labels, and $i, o \in N$, where i is the **input node** and o is the **output node**. Components are generically denoted by C, C_1, C_2, ...

We see the components as diagrams that can be drawn respecting—whenever possible—the following conventions: the i node is represented by a mark $-$ and is drawn further to the left; the node o is represented by a mark $+$ and is drawn further to the right; the other nodes are represented by marks, $\bullet$, each occupying a specific position inbetween the marks $-$ and $+$; each arc (n_1, L, n_2) is drawn in such a way that an arrow labeled with L starts from the mark that represents n_1 and arrives at the mark that represents n_2, as, e.g.:

$$- \xrightarrow{L} + \quad , \quad - \xrightarrow{L} \bullet \quad , \quad \bullet \xrightarrow{L} + \quad , \quad \bullet \xrightarrow{L} \bullet$$

according to n_1, n_2 being $-$, $+$, $\bullet$. When n_1 is the same as n_2, the arc (n_1, L, n_2) is drawn as a loop from n_1 to n_2 labelled with L, as, e.g.:

$$L \,\circlearrowright\, - \quad , \quad + \,\circlearrowright\, L \quad , \quad \bullet \,\circlearrowright\, L$$

according to n_1 (and, consequently, n_2) being $-$, $+$, $\bullet$.

The reader can see samples of components in the Introduction, for instance, (2) and (3) are two components.

Graphs. The BGL-**graphs** are finite sets of components, $\{C_i : 1 \leq i \leq n\}$. Observe that neither order nor repetition of components inside a graph is taken into consideration. Graphs are generically denoted by G, G_1, G_2, ... Graphs are drawn by depicting their components—in some order from left to right— separated by the symbol $|$. According to this, when convenient, we denote a generic graph $G = \{C_i : 1 \leq i \leq n\}$ by $C_1 \mid \cdots \mid C_n$.

The reader can see samples of BGL-graphs in Example 3. More specifically, the first diagram, all the diagrams between commas, and the last diagram displayed in the proof are BGL-graphs.

Inclusions. The BGL-**inclusions** are the expressions of the form $G_1 \sqsubseteq G_2$, where G_1 and G_2 are graphs. As usual, we say that $G_2 \sqsubseteq G_1$ is the converse of $G_1 \sqsubseteq G_2$.

The reader can see a sample of inclusion in Example 1.

3   Semantics of Basic Graph Logic

The BGL-semantics is given by defining the denotations of labels, components, and graphs, as well as the truth of inclusions in a given universe of discourse.

Denotation of a Label. Given an universe of discourse U, the **denotation** of a BGL-label L is a binary relation on U, defined as follows. The denotation of a relation symbol X is a binary relation which depends on a given function which assigns binary relations on U to the relation symbols. Fixed the denotations of the relation symbols, the denotations of the remaining labels, are defined as: the denotation of O is $\emptyset$, the empty relation on U; the denotation of E is $U \times U$, the universal relation on U; the denotation of I is $\{(a, a) : a \in U\}$, the identity relation on U; the denotation of L^- is the complement (with respect to $U \times U$) of the denotation of L; the denotation of $L^\smile$ is the converse of the denotation of L; the denotation of $L_1 \sqcap L_2$ is the intersection of the denotation of L_1 with the denotation of L_2; the denotation of $L_1 \sqcup L_2$ is the union of the denotation of L_1 with the denotation of L_2; the denotation of $L_1 \circ L_2$ is the composition of the denotation of L_1 with the denotation of L_2.

Denotation of a Component. Given a universe of discourse U, the **denotation** of a BGL-component $C = \langle N, A, -, + \rangle$ is a binary relation on U, defined as follows.

For all $a, b \in U$, we have that (a, b) belongs to the denotation of C iff there is an *assignment function* from nodes of C to elements of U, being $-$ labelled with a and $+$ labelled with b, such that for every arc (n_1, L, n_2), including n_1, n_2 as $-$ or $+$, if $c \in U$ labels the "left node" n_1 and d labels the "right node" n_2, then (c, d) belongs to the denotation of L. That is, a pair (a, b) of elements of

U belongs to the relation denoted by a component C iff it is possible mapping each node of C with an element of U in a manner that the input node is mapped with a, the output node is mapped with b, and the extremes n_1 and n_2 of any arc (n_1, L, n_2) are mapped with elements which form a pair that belongs to the relation denoted by the label L.

Example 5. Given an universe of discourse U in which the denotation of the relation symbol X is a fixed binary relation R, whose converse is denoted by R^{-1}, and $a, b \in U$: (a, b) belongs to the denotation of $- \xleftarrow{\quad X \quad} \bullet \xrightarrow{\quad X \quad} +$ iff one can map $-$ to a, $+$ to b and some $c \in U$ to $\bullet$, in such a way that $(c, a) \in R$ and $(c, b) \in R$ iff $\exists c \in U : (a, c) \in R^{-1}$ and $(c, b) \in R$. Hence, the denotation of $- \xleftarrow{\quad X \quad} \bullet \xrightarrow{\quad X \quad} +$ is the composition of R^{-1} with R.

Denotation of a Graph. Given a universe of discourse U, the **denotation** of a BGL-graph $C_1 \mid \cdots \mid C_n$ is the binary relation obtained by the union of the denotations of its components.

Truth of an Inclusion. Given a BGL-inclusion $G_1 \sqsubseteq G_2$, an universe of discourse U, and an assignment of binary relations on U to the relations symbols occurring in G_1 or G_2, the *truth* (or *falsity*) of $G_1 \sqsubseteq G_2$ is defined as follows. We say that $G_1 \sqsubseteq G_2$ is **true** (in this universe under this assignment) iff the denotation of G_1 is a subrelation of the denotation of G_2. And we say that $G_1 \sqsubseteq G_2$ is **false** iff it is not true.

Example 6. Given an universe of discourse U in which the denotation of the relation symbol X is a fixed binary relation R, the inclusion displayed in (1) is true iff R is an antisymmetric relation, in the sense that for any $a, b \in U$, if $(a, b), (b, a) \in R$, then $a = b$.

4 Validity and Consequence in Basic Graph Logic

In this section, we define the validity of inclusions and the consequence of an inclusion from a set of inclusions taken as hypotheses in the Basic Graph Logic.

Validity. Validity is universal truth. A BGL-inclusion $G_1 \sqsubseteq G_2$ is BGL-**valid**, denoted by $\models G_1 \sqsubseteq G_2$, if it is true in all universes of discourse U under all assignments of binary relations on U to the relation symbols occurring in any one of G_1 and G_2.

As usual, equalities between relations are proved as double inclusions between the graphs denoting them.

Consequence from Hypothesis. Consequence from hypothesis is preservation of truth. A BGL-inclusion $G_1 \sqsubseteq G_2$ is a BGL-**consequence** of a set $\Gamma = \{G_1^i \sqsubseteq G_2^i : i \in I\}$ of BGL-inclusions, denoted $\Gamma \models G_1 \sqsubseteq G_2$, if $G_1 \sqsubseteq G_2$ is true in all universes of discourse under all assignments of binary relations on U to the relation symbols occurring in any one of G_1, G_2, and G_i, for every $i \in I$, where all the inclusions in Γ are simultaneously true.

5   Diagrammatic Proofs in Basic Graph Logic

In this section, we present and exemplify the inference engine of Basic Graph Logic. Each BGL-inference rule transform a graph into another, preserving truth. We present the BGL-rules throughout examples in two circumstances: proving that the operators $\sqcap$, $\sqcup$, $\bar{}$ and constants O, E satisfy the Boolean algebra axioms; studing the effect of the Boolean operators $\sqcap$, $\sqcup$ on functional relations.

Proofs of Validities. The proof of a BGL-validity $\models G_1 \sqsubseteq G_2$, conduced inside BGL, have two major steps:

1. Display the graph G_1, that is, the LHS of the inclusion.
2. Apply truth preserving rules which transform a graph into another, to transform G_1 into G_2, the RHS graph of the inclusion.

We illustrate this procedure proving that the valid inclusions analogous to the usual axioms for Boolean algebras hold for the operators $\sqcap$, $\sqcup$, $\bar{}$, and constants O, E; and that the valid inclusions analogous to the usual axioms for involutive monoids hold for the operators $\circ$, $\smile$ and I. Along the way, we introduce the inference rules associated to these operators. In what follows, a component having one arc $n_1 L n_2$ singled out, is denoted by $\Gamma, n_1 L n_2, \Delta$ and G denotes a general graph. A similar notation applies for any number of components.

Example 7. To prove $\sqcap$ is associative, we need to prove the validity

$$- \xrightarrow{(X\sqcap Y)\sqcap Z} + \ \sqsubseteq\ - \xrightarrow{X\sqcap(Y\sqcap Z)} + \tag{5}$$

and its converse. To do this, we need a rule that allows us to transform a graph, based on an occurrence of a label of the form $L_1 \sqcap L_2$ in one of its arcs.

Rule to Parallelize Arcs. A pair (a, b) belongs to the denotation of $L_1 \sqcap L_2$ iff it belongs to the denotation of L_1 and to the denotation of L_2. Thus, the following rule is sound: for any graph, any component C having an arc $n_1 L_1 \sqcap L_2 n_2$ can be transformed in one component, obtained from C, by adding two arcs $n_1 L_1 n_2$ and $n_1 L_2 n_2$, that is,

$$\mathsf{PAR}\quad \frac{G \mid \Gamma, x L_1 \sqcap L_2 y, \Delta}{G \mid \Gamma, x L_1 \sqcap L_2 y, x L_1 y, x L_2 y, \Delta}$$

When applying PAR, $n_1 L_1 \sqcap L_2 n_2$ may or may not be erased in the new component. PAR can be applied from bottom to up.

To prove (5), we apply PAR obtaining the sequence of five graphs:

$$- \xrightarrow{(X\sqcap Y)\sqcap Z} + \ , \quad - \underset{Z}{\overset{X\sqcap Y}{\rightrightarrows}} + \ , \quad - \overset{X}{\underset{Z}{\rightrightarrows}}\!\!\!\overset{}{\underset{}{Y}}\, + \ , \quad - \overset{X}{\underset{Y\sqcap Z}{\rightrightarrows}} + \ , \quad - \xrightarrow{X\sqcap(Y\sqcap Z)} +$$

whose first and last terms are, respectively, the LHS and the RHS of the inclusion we want to prove. We prove the converse of (5) applying PAR (bottom-up), reversing the sequence above.

Example 8. To prove $\sqcup$ is associative, we prove the validity:

$$\vDash \; - \xrightarrow{\;(X \sqcup Y) \sqcup Z\;} + \; \sqsubseteq \; - \xrightarrow{\;X \sqcup (Y \sqcup Z)\;} + \tag{6}$$

and its converse. To do this, we need a rule that allows us to transform a graph, based on an occurrence of a label of the form $L_1 \sqcup L_2$ in one of its arcs.

Rule to Split Arcs. A pair (a, b) belongs to the denotation of $L_1 \sqcup L_2$ iff it belongs to the denotation of L_1 or to the denotation of L_2. Thus, the following rule is sound: for any graph, any component C having an arc $n_1 L_1 \sqcup L_2 n_2$ can be transformed in two alternative components, both obtained from C, one by adding an arc $n_1 L n_2$ and the other by adding an arc $n_1 L_2 n_2$, that is,

$$\text{SPL} \quad \frac{G \mid \Gamma, n_1 L_1 \sqcup L_2 n_2, \Delta}{G \mid \Gamma, n_1 L_1 \sqcup L_2 n_2, n_1 L_1 n_2, \Delta \mid \Gamma', n_1 L_1 \sqcup L_2 n_2, n_1 L_2 n_1, \Delta}$$

When applying SPL, $n_1 L_1 \sqcup L_2 n_2$ may or may not be erased in one or both of the new components. SPL can be applied from bottom to up.

To prove (6), we apply SPL, obtaining the sequence of five graphs:

$$- \xrightarrow{\;(X \sqcup Y) \sqcup Z\;} + \; , \quad - \xrightarrow{\;X \sqcup Y\;} + \left| - \xrightarrow{\;Z\;} + \; , \quad - \xrightarrow{\;X\;} + \right| - \xrightarrow{\;Y\;} + \left| - \xrightarrow{\;Z\;} + \right.$$

$$- \xrightarrow{\;X\;} + \left| - \xrightarrow{\;Y \sqcup Z\;} + \; , \quad - \xrightarrow{\;X \sqcup (Y \sqcup Z)\;} + \right.$$

We prove the converse of (6), applying SPL (bottom-up), reversing the sequence above. We prove that $\sqcap$ and $\sqcup$ are commutative, applying PAR and SPL.

Example 9. To prove $\sqcup$ absorbes $\sqcap$, we prove the validity:

$$\vDash \; - \xrightarrow{\;(X \sqcap Y) \sqcup Y\;} + \; \sqsubseteq \; - \xrightarrow{\;Y\;} + \tag{7}$$

and its converse. We start proving (7) applying SPL and PAR, obtaining the sequence:

$$- \xrightarrow{\;(X \sqcap Y) \sqcup Y\;} + \; , \quad - \xrightarrow{\;X \sqcap Y\;} + \left| - \xrightarrow{\;Y\;} + \; , \quad - \underset{Y}{\overset{X}{\rightarrowtail}} + \right| - \xrightarrow{\;Y\;} +$$

Now, we need a rule to transform $- \underset{Y}{\overset{X}{\rightarrowtail}} + \left| - \xrightarrow{\;Y\;} + \text{ in } - \xrightarrow{\;Y\;} + \right.$.

Rule for Existential Homomorphism. A component C represents certain conditions which must be satisfied in order to assure that the pair (a, b), whose terms are assigned to nodes $-$ and $+$, respectively, belong to the denotation of

C. Since each component of $- \underset{Y}{\overset{X}{\rightarrowtail}} + \left| - \xrightarrow{\;Y\;} + \text{ has a copy of } - \xrightarrow{\;Y\;} + \right.$

inside it, all the conditions imposed by the later are already present in the former. Then, if the conditions expressed by $-\underset{Y}{\overset{X}{\rightrightarrows}}+\ \Big|\ -\overset{Y}{\longrightarrow}+$ are fulfilled, the conditions expressed by $-\overset{Y}{\longrightarrow}+$ are automatically fulfilled.

To state this idea as a general inference rule, we need two new related concepts, one for components and the other for graphs.

Let $C_1 = \langle N_1, A_1, s_1, t_1 \rangle$ and $C_2 = \langle N_2, A_2, s_2, t_2 \rangle$ be BGL-components. An **homomorphism from C_2 to C_1** (observe the order) is a function $h : N_2 \to N_1$ such that:

1. h **preserves input and output**, i.e., $g(s_2) = g(s_1)$ and $g(t_2) = g(t_1)$.

2. h **preserves labels**, i.e., for all $m, n \in N_2$ and label L, *if* $m \overset{L}{\longrightarrow} n \in A_2$, then $g(m) \overset{L}{\longrightarrow} g(n) \in A_1$ (observe that the label does not change).

If there is a homomorphism from C_2 to C_1, then we can say that all the restrictions expressed in C_2 are already, in some way, expressed in $h(C_2)$, the image of C_2 under h, inside C_1:

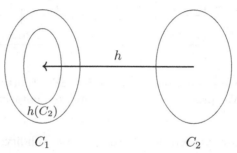

So, we can conclude that, given an universe of discourse U, it must be the case that the denotation of C_1 can not have more elements than those belonging to the denotation of C_2. But, maybe, C_1 has still more restrictions than those expressed in C_2. Hence, if there is homomorphism from C_2 to C_1, we can conclude that the denotation of C_1 is a subrelation of the denotation of C_2; but we are not allowed to conclude that later is a subrelation of the former.

The notion of homomorphism for graphs in general is defined as follows.

Let $G_1 = C_1^1 \mid \cdots \mid C_m^1$ and $G_2 = C_1^2 \mid \cdots \mid C_n^2$ be BGL-graphs. We say that G_2 **is a homomorphic image of** G_1, denoted by $G_1 \overset{\mathsf{H}}{\leftarrow} G_2$, iff for every C_i^1, $1 \leq i \leq m$, there exists a C_2^j, $1 \leq j \leq n$ and a homomorphism $h : C_2^j \to C_1^i$.

Thus, the following rules is sound:

HOM. For any graphs G_1 and G_2, if G_2 is a homomorphic image of G_1, we can transform G_1 in G_2:

$$\frac{G_1}{G_2}, G_1 \overset{\mathsf{H}}{\leftarrow} G_2$$

We finish the proof of (7), applying HOM to $-\overset{X}{\underset{Y}{\rightrightarrows}}+\Big|-\xrightarrow{\;Y\;}+$ obtain-

ing $-\xrightarrow{\;Y\;}+$ which is the RHS of the inclusion we want to prove. To prove the converse of (7) we apply HOM and SPL (bottom-up), obtaining the following sequence:

$$-\xrightarrow{\;Y\;}+\quad,\quad -\xrightarrow{X\sqcap Y}+\Big|-\xrightarrow{\;Y\;}+\quad,\quad -\xrightarrow{(X\sqcap Y)\sqcup Y}+$$

As noted in Example 9, a nice consequence of HOM is that from any graph G we can deduce a graph of the form $G\mid C$, were C is a new component.

Example 10. To prove that $\sqcap$ absorves $\sqcup$, we need to prove the validity

$$-\xrightarrow{(X\sqcup Y)\sqcap Y}+\;\sqsubseteq\;-\xrightarrow{\;Y\;}+ \tag{8}$$

and its converse. We prove (8) applying PAR, SPL, and HOM. We prove the converse of (8) applying HOM and PAR (bottom-up).

Example 11. To prove $^-$ is a complement, we need to prove the validities

$$-\xrightarrow{X\sqcap X^-}+\;\sqsubseteq\;-\xrightarrow{\;O\;}+\;\text{ and }\;-\xrightarrow{X\sqcup X^-}+\;\sqsubseteq\;-\xrightarrow{\;E\;}+ \tag{9}$$

and their converses.

We start proving (9a) applying PAR, obtaining the sequence:

$$-\xrightarrow{X\sqcap X^-}+\quad,\quad -\overset{X}{\underset{X^-}{\rightrightarrows}}+$$

Now, we need a rule to transform graphs having occurrences of parallel arcs of the form $-\overset{L}{\underset{L^-}{\rightrightarrows}}+$.

Rules for Inconsistent Components. A pair (a,b) can not belong to both the denotation of L and L^-. Thus, the following rule is sound: for any graph, any component C having arcs n_1Ln_1 and $n_1L^-n_2$ can be transformed in one component, obtained from C, by adding the arc n_1On_2, that is,

$$\mathsf{INC}\quad \frac{G\mid \Gamma, n_1Ln_2, n_1L^-n_2, \Delta}{G\mid \Gamma, n_1Ln_2, n_1L^-n_2, n_1On_2\Delta}$$

or be replaced by the component $-\xrightarrow{\;O\;}+$ (when what remains is still a graph):

$$\frac{G\mid \Gamma, n_1Ln_2, n_1L^-n_2, \Delta}{G\mid\;-\xrightarrow{\;O\;}+}$$

or, simply, be erased (when what remains is still a graph): :

$$\frac{G \mid \Gamma, n_1 L n_2, n_1 L^- n_2, \Delta}{G}$$

When applying the first part of INC, $n_1 L n_2, n_1 L^- n_2$ may or may not be erased in the new component. All parts of INC work in both directions. The bottom-up application of the third part of INC is a consequence of HOM.

Back to the proof of (9), now, we apply INC, increasing the sequence above to:

$$- \xrightarrow{X \sqcap X^-} + \quad , \quad - \overset{X}{\underset{X^-}{\rightleftarrows}} + \quad , \quad - \xrightarrow{\text{O}} +$$

We prove the converse of (9a) applying INC and PAR (both bottom-up), reversing the sequence above. To prove (9b), we need a rule to manage the display of arcs $- \xrightarrow{\text{E}} +$ inside components.

Rule for Universal Arcs. Any pair (a, b) belongs to the denotation of E. Thus, the following rule is sound: for any graph and any of its components, any nodes n_1, n_2 belonging to this component can be decorated with an arc labeled E, that is,

$$\text{UNI} \quad \frac{G \mid \Gamma, \Delta}{G \mid \Gamma, n_1 E n_2, \Delta}$$

UNI can be applied from bottom to up.

We prove (9b), applying UNI and HOM, obtaining the sequence:

$$- \xrightarrow{X \sqcup X^-} + \quad , \quad - \underset{\text{E}}{\overset{X \sqcup X^-}{\rightrightarrows}} + \quad , \quad - \xrightarrow{\text{E}} +$$

To prove the converse of (9b), we need a rule to manage the display of arcs $n_1 \xrightarrow{L} n_2$ and $n_1 \xrightarrow{L^-} n_2$ in different components of a graph.

Rule for Alternatives. Given any label L and pair (a, b) it is always true that (a, b) belongs to the meaning of L or (a, b) does not belong to the meaning of L—that is, (a, b) belongs to the meaning of L^-. Thus, the following rule is sound: for any graph, any component C, and any nodes n_1, n_2 belonging to this component, the component can be transformed in two others obtained from C, one having an arc $n_1 L n_2$ and the other having an arc $n_1 L^- n_2$, that is,

$$\text{ALT} \quad \frac{G \mid \Gamma, \Delta}{G \mid \Gamma, n_1 L n_2, \Delta \mid \Gamma, n_1 L^- n_2, \Delta}$$

ALT can be applied from bottom to up.

We prove the converse of (9b), applying ALT, HOM and SPL (bottom-up), obtaining the sequence:

$$- \xrightarrow{\;E\;} + \quad , \quad - \underset{X}{\overset{E}{\rightrightarrows}} + \; \Big| - \underset{X^-}{\overset{E}{\rightrightarrows}} +$$

$$- \xrightarrow{\;X\;} + \; \Big| - \xrightarrow{\;X^-\;} + \quad , \quad - \xrightarrow{\;X \sqcup X^-\;} +$$

Example 12. To prove $\circ$ is associative, we need to prove the validity

$$- \xrightarrow{\;(X \circ Y) \circ Z\;} + \; \sqsubseteq \; - \xrightarrow{\;X \circ (Y \circ Z)\;} + \tag{10}$$

and its converse. To do this, we need a rule that allows us to transform a graph, based on an occurrence of a label of the form $L_1 \circ L_2$ in one of its arcs.

Rule for Intermediate Nodes. A pair (a, b) belongs to the denotation of $L_1 \circ L_2$ iff there exists an element c such that (a, c) belongs to the denotation of L_1 and (c, b) belongs to the denotation of L_2. Thus, the following rule is sound: for any graph, in any component, any arc $n_1 L_1 \circ L_2 n_2$ can be replaced by two arcs $n_1 L_1 m$ and $m L_2 n_2$, that is,

$$\mathsf{INT} \quad \frac{G \mid \Gamma, n_1 L_1 \circ L_2 n_2, \Delta}{G \mid \Gamma, n_1 L_1 m, m L_2 n_2, \Delta}, m \text{ is new.}$$

INT can be applied bottom-up, provided that m is a node not occurring in $\Gamma, n_1 L_1 \circ L_2 n_2, \Delta$, and there are no arcs having m as an extremity, besides $n_1 L_1 m$ and $m L_2 n_2$.

To prove (5), we apply INT obtaining the sequence:

$$- \xrightarrow{\;(X \circ Y) \circ Z\;} + \quad , \quad - \xrightarrow{\;X \circ Y\;} \bullet \xrightarrow{\;Z\;} + \quad , \quad - \xrightarrow{\;X\;} \bullet \xrightarrow{\;Y\;} \bullet \xrightarrow{\;Z\;} +$$

$$- \xrightarrow{\;X\;} \bullet \xrightarrow{\;Y \circ Z\;} + \quad , \quad - \xrightarrow{\;X \circ (Y \circ Z)\;} +$$

We prove the converse of (10) applying INT (bottom-up), reversing the sequence above.

Example 13. To prove I is a neutral element for $\circ$, we need to prove the validities

$$- \xrightarrow{\;X \circ I\;} + \; \sqsubseteq \; - \xrightarrow{\;I\;} + \quad \text{and} \quad - \xrightarrow{\;I \circ X\;} + \; \sqsubseteq \; - \xrightarrow{\;I\;} + \tag{11}$$

and their converses. We start proving (11a) applying INT to the LHS of (11a), obtaining:

$$- \xrightarrow{\;X\;} \bullet \xrightarrow{\;I\;} +$$

Now, we need a rule that allows us to transform a graph, based on an occurrence of a label I in one of its arcs (which is not a loop).

Rule for Identify Nodes. An ordered pair (a, b) belongs to the denotation of I iff $a = b$. Thus, the following rule is sound: for any graph, inside any component, the arcs $n_1 L_1 n$, ..., $n_k L_k n$, $n \mathsf{I} m$, $m L_1 m_1$, ..., $m L_l m_l$ can be replaced by the arcs $n_1 L_1 n$, ..., $n_k L_k n$, $n \mathsf{I} n$, $n L_1 m_1$, ..., $n L_l m_l$, that is,

$$\mathsf{IDE} \quad \frac{G \mid \Gamma, n_1 L_1 n, \ldots, n_k L_k n, n \mathsf{I} m, m L_1' m_1, \ldots, m L_l' m_l, \Delta}{G \mid \Gamma, n_1 L_1 n, \ldots, n_k L_k n, n \mathsf{I} n, n L_1' m_1, \ldots, n L_l' m_l, \Delta}$$

or, alternatively, by the arcs $n_1 L_1 m$, ..., $n_k L_k m$, $m \mathsf{I} m$, $n L_1' m_1$, ..., $n L_l' m_l$:

$$\mathsf{IDE} \quad \frac{G \mid \Gamma, n_1 L_1 n, \ldots, n_k L_k n, n \mathsf{I} m, m L_1' m_1, \ldots, m L_l' m_l, \Delta}{G \mid \Gamma, n_1 L_1 m, \ldots, n_k L_k m, m \mathsf{I} m, m L_1' m_1, \ldots, m L_l' m_l, \Delta}$$

where $n \neq m$, $n_1, \ldots, n_k$ are all the nodes adjacent to n and $m_1, \ldots, m_l$ are all the nodes adjacent to m. IDE can be applied from bottom to up. When applying IDE, we have to preserve $-$ or $+$, when there is a choice between them and another node different from $-$ or $+$ (as it is done next).

Hence, continuing the proof, we apply IDE to $- \xrightarrow{X} \bullet \xrightarrow{\mathsf{I}} +$, obtaining $- \xrightarrow{X} + \, \bigcirc \, \mathsf{I}$. Now, we need a rule that allows us to transform a graph, based on an occurrence of a label I in one of its loops.

Rule for Erase Identity. An ordered pair (a, b) belongs to the denotation of I iff $a = b$. Thus, the following rule is sound:

for any graph, inside any component, any loop $x \mathsf{I} x$ can be erased (provided what remais is a graph), and vice-versa, that is,

$$\mathsf{ERA} \quad \frac{G \mid \Gamma, x \mathsf{I} x, \Delta}{G \mid \Gamma, \Delta}$$

ERA can be applied from bottom to up.

We finish the proof, applying ERA to $- \xrightarrow{X} + \, \bigcirc \, \mathsf{I}$, obtaining $- \xrightarrow{X} +$.

We also prove (11b) applying INT, IDE, and ERA.

Example 14. To prove $^\smile$ is an involution w.r.t $\circ$, we need to prove the validities

$$- \xrightarrow{X^{\smile\smile}} + \ \sqsubseteq \ - \xrightarrow{X} + \quad \text{and} \quad - \xrightarrow{(X \circ Y)^\smile} + \ \sqsubseteq \ - \xrightarrow{Y^\smile \circ X^\smile} + \quad (12)$$

and their converses. To do this, we need a rule that allows us to transform a graph, based on an occurrence of a label of the form $L^\smile$ in one of its arcs.

Rule to converse Arcs. A pair (a, b) belongs to the denotation of $L^\smile$ iff the pair (b, a) belongs to the denotation of L. Thus, the following rule is sound: for any graph, any component C having an arc $n_1 L^\smile n_2$ can be transformed in a new component obtained from C by adding one arc $n_2 L n_1$:

$$\mathsf{CON} \quad \frac{G \mid \Gamma, n_1 L^\smile n_2, \Delta}{G \mid \Gamma, n_1 L^\smile n_2, n_2 L n_1, \Delta}$$

When applying CON, $n_1 L^\smile n_2$ may or may not be erased in the new component. CON can be applied from bottom to up.

We prove (12a) applying CON obtaining:

$$- \xrightarrow{X^{\smile\smile}} + \quad , \quad - \xleftarrow{X^\smile} + \quad , \quad - \xrightarrow{X} +$$

We prove the converse of (12a) applying CON (bottom-up), reversing the sequence above. We prove (12b) we apply SPL, CON and SPL to the LHS of (12b), obtaining:

$$- \xleftarrow{X \circ Y} + \quad , \quad - \xleftarrow{Y} \bullet \xleftarrow{X} + \quad , \quad - \xrightarrow{Y^\smile} \bullet \xrightarrow{X^\smile} + \quad , \quad - \xrightarrow{Y^\smile \circ X^\smile} +$$

To prove the converse of (12b) we apply CON and SPL (bottom-up), reversing the same sequence of graphs.

Proofs of Consequences. The proof of a BGL-consequence $\{G_1^i \sqsubseteq G_2^i : i \in I\} \models G_1 \sqsubseteq G_2$, conduced inside BGL, has three major steps:

1. Suppose the inclusions $G_1^i \sqsubseteq G_2^i$ as premises.
2. Consider the graph G_1, that is, the LHS of the conclusion that we want to proof as a consequence of the premises.
3. Apply truth preserving rules which transform a graph into another, to transform G_1 into G_2, the RHS graph of the conclusion. In particular, there is an inference rule that allows us to transform parts of the graphs already obtained using the premises $G_1^i \sqsubseteq G_2^i$, $i \in I$.

We exemplify this procedure by proving that the valid consequences analogous to the De Morgan's Theorem K [4]. Along the way, we introduced the rule associated to the transformation of graphs from hypothesis.

Example 15. To prove that $^-$, $\circ$, and $^\smile$ satisfy the analogous of Theorem K, we prove the consequences

$$- \xrightarrow{X \circ Y} + \ \sqsubseteq \ - \xrightarrow{Z} + \ \models \ - \xrightarrow{X^\smile \circ Z^-} + \ \sqsubseteq \ - \xrightarrow{Y^-} + \quad (13)$$

$$- \xrightarrow{X^\smile \circ Z^-} + \ \sqsubseteq \ - \xrightarrow{Y^-} + \ \models \ - \xrightarrow{Z^\smile \circ Y^-} + \ \sqsubseteq \ - \xrightarrow{X^-} + \quad (14)$$

$$- \xrightarrow{Z^\smile \circ Y^-} + \ \sqsubseteq \ - \xrightarrow{X^-} + \ \models \ - \xrightarrow{X \circ Y} + \ \sqsubseteq \ - \xrightarrow{Z} + \quad (15)$$

To prove (13), we proceed as follows. First, we suppose the hypothesis:

$$- \xrightarrow{X \circ Y} + \ \sqsubseteq \ - \xrightarrow{Z} +$$

Second, we display the LHS of the conclusion we want to prove:

$$- \xrightarrow{\;X^{\smile}\circ Z^-\;} +$$

Now, we transform the graph above applying INT, and CON, obtaining:

$$- \xleftarrow{\;X\;} \bullet \xrightarrow{\;Z^-\;} +$$

Now, we transform the graph above applying ALT, obtaining:

$$- \overset{\displaystyle X}{\underset{\displaystyle Y}{\overleftrightarrow{\bullet}}} \xrightarrow{Z^-} + \quad \Big| \quad - \xleftarrow{\;X\;} \bullet \underset{Y^-}{\xrightarrow{\;Z^-\;}} +$$

Now, we look to the hypothesis $- \xrightarrow{\;X\circ Y\;} + \sqsubseteq - \xrightarrow{\;Z\;} +$. According to the semantics of $\circ$, it can be rewritten as $- \xleftarrow{\;X\;} \bullet \xrightarrow{\;Y\;} + \sqsubseteq - \xrightarrow{\;Z\;} +$. So, according to the semantics of $\sqsubseteq$, it can be interpreted as "whenever we find a pair (a,b) which belongs to the denotation of a graph as $n_1 \xleftarrow{\;X\;} \bullet \xrightarrow{\;Y\;} n_2$, we can conclude that (a,b) also belongs to the denotation of the graph $n_1 \xrightarrow{\;Z\;} n_2$." In diagrammatic terms, this means that for all graph G, component C of G, and nodes n_1, n_2 of C, whenever we find inside C a diagram as $n_1 \xleftarrow{\;X\;} \bullet \xrightarrow{\;Y\;} n_2$, we are allowed to add a diagram as $n_1 \xrightarrow{\;Z\;} n_2$ between the two nodes n_1 and n_2. We apply this idea to the graph above, obtaining:

$$- \overset{\displaystyle Z}{\overset{\curvearrowright}{\underset{\displaystyle Y}{\overleftrightarrow{\;\bullet\;}}}} \xrightarrow{Z^-} + \quad \Big| \quad - \xleftarrow{\;X\;} \bullet \underset{Y^-}{\xrightarrow{\;Z^-\;}} +$$

Now, we transform the graph above, applying INC, obtaining:

$$- \underset{Y^-}{\overset{\displaystyle X}{\overleftrightarrow{\;\bullet\;}}} \xrightarrow{Z^-} +$$

Finally, we transform the graph above, applying HOM, obtaining:

$$- \xrightarrow{\;Y^-\;} +$$

the RHS of the conclusion.

The proofs of (14) and (15) are entirely analogous.

Rule for Hypothesis. The major step given in the proof of (13) is the transformation of $- \overset{\displaystyle X}{\underset{\displaystyle Y}{\overleftrightarrow{\;\bullet\;}}} \xrightarrow{Z^-} +$ into $- \overset{\displaystyle Z}{\overset{\curvearrowright}{\underset{\displaystyle Y}{\overleftrightarrow{\;\bullet\;}}}} \xrightarrow{Z^-} +$ based on the hypothesis

$- \xleftarrow{\quad X \quad} \bullet \xrightarrow{\quad Y \quad} + \sqsubseteq - \xrightarrow{\quad Z \quad} +$. To state this idea as a general inference rule, we need two new concepts.

We say that a function f **displays a copy of** a component C_j **inside** a component C_i if $f : N_{C_j} \to N_{C_i}$, is injective, and preserves labels.

Let C_i, C_j, C_k be components and $f : N_{C_j} \to N_{C_i}$ be a function that displays a copy of C_j inside C_i. The component obtained from C_i by **append a copy** of C_k **in parallel** to C_j (according to f), denoted by $C_i[C_j \leftarrow C_k]$ is defined by:

1. The set of nodes of $C_i[C_j \leftarrow C_k]$ is the disjoint union of the following sets: N_{C_i} of the nodes of C_i; $N_{C_k} \setminus \{i_{C_k}, o_{C_k}\}$ of the nodes of C_k which are neither the input nor the output nodes of C_k.
2. The input and output nodes of $C_i[C_j \leftarrow C_k]$ are the input and output nodes of C_i.
3. The set of arcs of $C_i[C_j \leftarrow C_k]$ is the disjoint union of the following sets: A_{C_i} of the arcs of C_i; $A_{C_k}[i_{C_i}, o_{C_i}]$ of the arcs of C_k which the extremities i_{C_k}, o_{C_k} of C_k substituted by the extremities i_{C_i}, o_{C_i} of C_i.

Thus, the following rule is sound:

HYP. For any graph G and inclusion $G_1 \sqsubseteq G_2$, if there are components C_i, C_j, C_k of G, G_1, G_2, respectively, and a function f such that f displays a copy of C_j inside C_i, then we can replace the component C_i of G by $C_i[C_j \leftarrow C_k]$:

$$\frac{C_1 \mid \cdots \mid C_i \mid \cdots \mid C_m}{C_1 \mid \cdots \mid C_i[C_j \leftarrow C_k] \mid \cdots \mid C_m}, G_1 \sqsubseteq G_2$$

Now that BGL is fully presented, we hope the reader has a basis for reading and understanding the proof presented in Example 3.

6   Final Remarks

We believe that BGL offers a very suggestive step-by-step guidance when formal proofs are at stake. Once one gets familiar with the rules and graphs manipulation, it seems that proofs go on the flow smoother than other formal systems related to relation algebras. We have experienced this in an ongoing work that contains an axiomatization of a fragment of set theory in which some non-trivial results as, for example, Cantor's Theorem can be proved by strictly BGL diagrams.

In terms of expressive and proof power, BGL is equivalent to First Order Logic with equality having just binary relation symbols (nor constant neither function symbols) with negation restricted to formulas having at most two free variables. The proof of this fact will be provided in a future work.

We understand that the main contribution of this work is found in the step by step presentation of the BGL-system through examples that locally show the need for each rule and how it is applied. Thus we hope that this work will make BGL more accessible to a general audience.

References

1. Andréka, H., Bredikhin, D.A.: The equational theory of union-free algebras of relations. Algebra Universalis **33**, 516–532 (1995)
2. Curtis, S., Lowe, G.: Proofs with graphs. Sci. Comput. Program. **26**, 197–216 (1996)
3. de Freitas, R., Veloso, P.A., Veloso, S.R., Viana, P.: On graph reasoning. Inf. Comput. **207**, 1000–1014 (2009)
4. de Freitas, R., Veloso, P.A.S., Veloso, S.R.M., Viana, P.: A calculus for graphs with complement. In: Goel, A.K., Jamnik, M., Narayanan, N.H. (eds.) Diagrams 2010. LNCS (LNAI), vol. 6170, pp. 84–98. Springer, Heidelberg (2010). https://doi.org/10.1007/978-3-642-14600-8_11
5. Rensink, A.: Representing first-order logic using graphs. In: Ehrig, H., Engels, G., Parisi-Presicce, F., Rozenberg, G. (eds.) ICGT 2004. LNCS, vol. 3256, pp. 319–335. Springer, Heidelberg (2004). https://doi.org/10.1007/978-3-540-30203-2_23
6. Habel, A., Pennemann, K.-H.: Correctness of high-level transformation systems relative to nested conditions. Math. Struct. Comput. Sci. **19**, 245–296 (2009)

Schopenhauer's Partition Diagrams and Logical Geometry

Jens Lemanski[1]([✉])[iD] and Lorenz Demey[2][iD]

[1] Institute of Philosophy, FernUniversität in Hagen, Hagen, Germany
jens.lemanski@fernuni-hagen.de
[2] Center for Logic and Philosophy of Science, KU Leuven, Leuven, Belgium
lorenz.demey@kuleuven.be

Abstract. The paper examines Schopenhauer's complex diagrams from the *Berlin Lectures* of the 1820s, which show certain partitions of classes. Drawing upon ideas and techniques from logical geometry, we show that Schopenhauer's partition diagrams systematically give rise to a special type of Aristotelian diagrams, viz. (strong) α-structures.

Keywords: Arthur Schopenhauer · Logic diagrams · Partition diagram · Aristotelian diagram · Bitstring semantics · Logical geometry

1 Introduction

For almost 200 years the name of the philosopher Arthur Schopenhauer, who was born in Gdansk in 1788 and died in Frankfurt in 1860, was not associated with logic. It was not until the middle of the 2010s that it became known that for his *Berlin Lectures* in the 1820s, Schopenhauer composed a treatise on logic that covered the scope of an entire book. Since the discovery of this 'logica maior', philosophers, linguists and logicians have made numerous discoveries in Schopenhauer's work, of which only a few are mentioned here as examples: Schopenhauer's logic anticipates several important linguistic principles that later became prominent through the Vienna Circle [12], the Lvov-Warsaw School [11], and generative grammar [14]. Long before John Venn, Schopenhauer drew complex logic diagrams for n terms [23] and, at the same time as Joseph Gergonne, he extended Euler diagrams to the so-called Gergonne relations [26]. Furthermore, Schopenhauer already used logical notations which could have paved the way to mathematical logic towards the end of the 19th century, had they been known at that time [18].

However, Schopenhauer's logic is not only interesting from a historical point of view, but also offers numerous systematic points of departure for taking ideas further, rethinking old ones or developing new ones. In recent years, for example, Schopenhauer's approach has been modernized with the help of transition rules of elementary cellular automata [24]. A formalism called 'Schopenhauer diagrams' was developed to analyze processes of abstraction and reification in ontology and conceptual engineering [13,22]. For the present paper it is particularly noteworthy that a number of these Schopenhauer diagrams can provide

© Springer Nature Switzerland AG 2021
A. Basu et al. (Eds.): Diagrams 2021, LNAI 12909, pp. 149–165, 2021.
https://doi.org/10.1007/978-3-030-86062-2_13

general insights and theorems about Aristotelian diagrams in the contemporary framework of logical geometry [6].

In the present paper, we will further develop this latter approach, by underpinning it with new material from Schopenhauer's original manuscripts. We will show how Schopenhauer came to the modern idea of investigating not only Boolean algebras consisting of *propositions*, but also of *sets*. In particular, we will present a series of logic diagrams for complex partitions that Schopenhauer drew in his *Berlin Lectures*, and argue that these partition diagrams closely correspond to what are nowadays sometimes called (strong) α-structures [6,27]. Typical partition diagrams can be found below in Figs. 3 and 5.

The paper is organized as follows. We start in Sect. 2 by briefly discussing some key notions from logical geometry that are useful for studying Schopenhauer's logic. In Sect. 3 we then introduce Schopenhauer's logica maior, and describe in particular the logical context of those passages that are important for our argument. In Sect. 4 we show how Schopenhauer uses logic diagrams to visualize set partitions. Finally, in Sect. 5 we focus on a particularly interesting partition diagram from Schopenhauer's manuscripts, and argue that it gives rise to several (strong) α-structures, including a strong α_7-structure. Roughly speaking, Sects. 3 and 4 are primarily historically oriented, whereas Sects. 2 and 5 are of more systematic interest.

2   Aristotelian Relations and α-Structures

This section introduces some key notions from logical geometry that have turned out to be very fruitful for studying Schopenhauer's logic [6], and that will also take center stage in Sects. 4 and 5 of the present paper. We begin by discussing the *Aristotelian relations*, which can be characterized with various degrees of abstractness and generality [5,8]. For the purposes of this paper, it will be useful to consider a very general definition, in the setting of Boolean algebra [17].

Definition 1. *Let* $\mathbb{B} = \langle B, \wedge, \vee, \neg, \top, \bot \rangle$ *be an arbitrary Boolean algebra. Two elements* $x, y \in B$ *are said to be*

$\mathbb{B}$-contradictory	*iff*	$x \wedge y = \bot$ *and* $x \vee y = \top$,
$\mathbb{B}$-contrary	*iff*	$x \wedge y = \bot$ *and* $x \vee y \neq \top$,
$\mathbb{B}$-subcontrary	*iff*	$x \wedge y \neq \bot$ *and* $x \vee y = \top$,
in $\mathbb{B}$-subalternation	*iff*	$\neg x \vee y = \top$ *and* $x \vee \neg y \neq \top$.

More informal and familiar characterizations of the Aristotelian relations can be obtained from this definition by plugging in concrete Boolean algebras for $\mathbb{B}$. For example, we can take $\mathbb{B}$ to be a Boolean algebra of *propositions*. In this case, two propositions P and Q being contrary means that $P \wedge Q$ is contradictory while $P \vee Q$ is not tautological, i.e. P and Q cannot be true together, but can be false together. Similarly, there is a subalternation from P to Q iff P entails Q but not vice versa. For a second example, we can take $\mathbb{B}$ to be a Boolean algebra of *sets*, e.g. the powerset $\wp(D)$ of some domain of discourse D. In this second case, two sets X and Y being contrary means that $X \cap Y = \emptyset$ while $X \cup Y \neq D$, i.e. X and Y are disjoint but do not exhaust D. Similarly, there is a subalternation from X to Y iff $X \subseteq Y$ but not $X \supseteq Y$. The fact that Definition 1 allows us to deal not only with Aristotelian relations between propositions (as is usually done), but also between sets, will be absolutely crucial when we turn to Schopenhauer's diagrams. After all, the latter also represent relations between sets, viz. spheres/extensions of concepts (cf. Section 3). Concrete examples, which we will also discuss further, are e.g. the contrariety between fish and bird, the subalternation from bird to vertebrate, the contradiction between fish and ¬fish, and the subcontrariety between ¬fish and ¬bird.

Now that the Aristotelian relations have been defined relative to arbitrary Boolean algebras, we can likewise define the class of Aristotelian diagrams and one of its important subclasses: the so-called α-structures or α-diagrams [27].[1]

Definition 2. *Let $\mathbb{B}$ be as before, and consider a fragment $\mathcal{F} \subseteq B\backslash\{\top, \bot\}$. Suppose that $\mathcal{F}$ is closed under $\mathbb{B}$-complementation, i.e. if $x \in \mathcal{F}$ then $\neg x \in \mathcal{F}$. An Aristotelian diagram for $\mathcal{F}$ in $\mathbb{B}$ is a diagram that visualizes an edge-labeled graph $\mathcal{G}$. The vertices of $\mathcal{G}$ are the elements of $\mathcal{F}$, and the edges of $\mathcal{G}$ are labeled by the Aristotelian relations between those elements, i.e. if $x, y \in \mathcal{F}$ stand in some Aristotelian relation, then this is visualized according to the code in Fig. 1(a).*

Definition 3. *Let $\mathbb{B}$ be as before, and consider a natural number $n \geq 1$. An α_n-structure in $\mathbb{B}$ is an edge-labeled graph $\mathcal{G}$. The vertices of $\mathcal{G}$ form a fragment $\{x_1, \ldots, x_n, \neg x_1, \ldots, \neg x_n\} \subseteq B\backslash\{\top, \bot\}$, where all distinct x_i, x_j are pairwise $\mathbb{B}$-contrary, i.e. x_i and x_j are $\mathbb{B}$-contrary for all $1 \leq i \neq j \leq n$. The edges of $\mathcal{G}$ are labeled by the Aristotelian relations between those elements. An α_n-diagram in $\mathbb{B}$ is an Aristotelian diagram that visualizes such an α_n-structure in $\mathbb{B}$.*

Note that by Definition 2, Aristotelian diagrams are closed under complementation and only contain non-trivial elements (i.e. neither $\top$ nor $\bot$). The historical and systematic reasons for these restrictions are discussed in more detail in [33, Subsection 2.1]. Furthermore, the condition in Definition 3 regarding pairwise $\mathbb{B}$-contrariety between all distinct x_i, x_j immediately implies that there are several other Aristotelian relations in an α_n-structure as well. In particular, it follows that $\neg x_i$ and $\neg x_j$ are $\mathbb{B}$-subcontrary and that there are $\mathbb{B}$-subalternations from x_i to $\neg x_j$, for all $1 \leq i \neq j \leq n$. And of course, as in any Aristotelian diagram, it holds that x_i and $\neg x_i$ are $\mathbb{B}$-contradictory, for all $1 \leq i \leq n$. Several of the most well-known Aristotelian diagrams are indeed α-structures:

– The α_1-structure is simply a *pair of contradictory elements* (PCD); cf. Fig. 1(b). PCDs do not frequently appear in the literature, but they have considerable theoretical importance, since they can be thought of as the fundamental 'building blocks' for all other, larger Aristotelian diagrams [7,9].

---

[1] Strictly speaking, the term 'α-structure' refers to the (abstract) underlying graph, while the term 'α-diagram' refers to the (concrete) diagram. However, this distinction will not matter much in this paper, so we will usually not distinguish between these two terms, and follow Moretti [27] in simply talking about 'α-structures'.

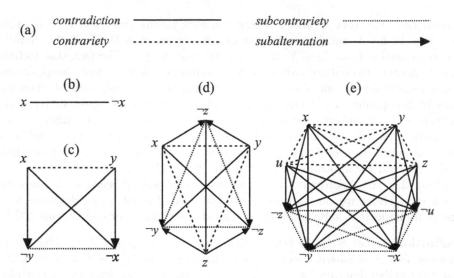

Fig. 1. (a) code for visually representing the Aristotelian relations; examples of (b) PCD, (c) classical square of opposition, (d) JSB hexagon, (e) Moretti octagon.

- The α_2-structure is a *classical square of opposition*; cf. Fig. 1(c). Without a doubt, this is the oldest and most well-known type of Aristotelian diagram.
- The α_3-structure is a so-called *Jacoby-Sesmat-Blanché (JSB) hexagon*, which is named after Jacoby [19], Sesmat [32] and Blanché [2]; cf. Fig. 1(d). After the classical square, this is the most well-known type of Aristotelian diagram.
- The α_4-structure is a so-called *Moretti octagon*, which is named after Moretti [27] (who drew it as a cube, rather than an octagon); cf. Fig. 1(e).

One of the key insights of logical geometry is that a given family of Aristotelian diagrams can have multiple *Boolean subtypes*, i.e. it is possible for two Aristotelian diagrams to exhibit exactly the same configuration of Aristotelian relations among their respective sets of elements, and yet have different Boolean properties [10]. The first concrete example of this phenomenon was pointed out by Pellissier [29], who showed that there are two Boolean subtypes of JSB hexagons: in a *strong* JSB hexagon, the join of the three pairwise contrary elements equals ⊤, whereas in a *weak* JSB hexagon, this join is not equal to ⊤. Such Boolean differences are nowadays usually characterized in terms of *bitstrings*, which are formally introduced below (for more details, see [10,34]).

Definition 4. *Let the Boolean algebra* $\mathbb{B}$ *and the fragment* $\mathcal{F}$ *be as before. The partition of* $\mathbb{B}$ *induced by* $\mathcal{F}$ *is defined as* $\Pi_{\mathbb{B}}(\mathcal{F}) := \{\pm x_1 \wedge \cdots \wedge \pm x_n \mid x_1, \ldots, x_n \in \mathcal{F}\} \setminus \{\bot\}$, *where* $+x_i = x_i$ *and* $-x_i = \neg x_i$. *For every* $y \in \mathcal{F}$ *we have* $y = \bigvee \{a_i \in \Pi_{\mathbb{B}}(\mathcal{F}) \mid a_i \leq y\}$. *The bitstring representation of* $y \in \mathcal{F}$ *keeps track of which* $a_i \in \Pi_{\mathbb{B}}(\mathcal{F})$ *enter into this join; for example, if* $\Pi_{\mathbb{B}}(\mathcal{F}) = \{a_1, a_2, a_3, a_4\}$ *and* $y = a_1 \vee a_3 \vee a_4$, *then* y *will be represented as the bitstring* 1011.

Using this technique, one can show, for example, that representing a strong JSB hexagon requires bitstrings of length 3 (so the join of its three pairwise contrary elements is $100 \lor 010 \lor 001 = 111$), whereas a weak JSB hexagon requires bitstrings of length 4 (so the join of its three pairwise contrary elements is $1000 \lor 0100 \lor 0010 = 1110 \neq 1111$). (Again, see [10,34] for more details.)

In general, determining the Boolean subtypes of a given type of Aristotelian diagrams is highly non-trivial [4]. However, for the specific subclass of α-structures, the situation is relatively straightforward, as is summarized by Theorem 1.

Theorem 1. – All α_1-structures (i.e., PCDs) require bitstrings of length 2,
- all α_2-structures (i.e., classical squares) require bitstrings of length 3,
- for $n \geq 3$, there are two Boolean subtypes of α_n-structures: (i) a strong subtype, which requires bitstrings of length n, and (ii) a weak subtype, which requires bitstrings of length $n + 1$.

Note that the important cutoff happens at $n = 3$. This is not a coincidence: because of their *binary* nature, the Aristotelian relations cannot capture the full Boolean complexity that may arise in larger sets [4]. Furthermore, note that the case $n = 3$ says that the family of JSB hexagons has two Boolean subtypes, viz. the strong JSB hexagons (requiring bitstrings of length 3) and the weak JSB hexagons (requiring bitstrings of length 4). In other words, Pellissier's original result on JSB hexagons [29] is thus subsumed as a special case of Theorem 1.

In a Boolean algebra $\mathbb{B} = \langle B, \land, \lor, \neg, \top, \bot \rangle$, a finite set $\Pi = \{x_1, \ldots, x_n\} \subseteq B \backslash \{\top, \bot\}$ (with $n \geq 2$) is said to be an *n-partition* of $\mathbb{B}$ iff (i) $x_i \land x_j = \bot$ for all distinct $x_i, x_j \in \Pi$ and (ii) $\bigvee \Pi = \top$. There is a clear correspondence between partitions and (strong) α-structures.[2] This is made fully precise in Theorem 2 below. Note that there is again a cutoff at $n = 3$, and that α_2-structures (i.e. classical squares of opposition) do *not* correspond to any partitions.

Theorem 2

1. Each 2-partition $\{x, \neg x\}$ gives rise to an α_1-structure with elements $\{x, \neg x\}$.
2. For $n \geq 3$, each n-partition $\{x_1, \ldots, x_n\}$ gives rise to a strong α_n-structure with elements $\{x_1, \ldots, x_n, \neg x_1, \ldots, \neg x_n\}$.

Concretely, each 2-partition corresponds to a PCD, each 3-partition to a strong JSB hexagon, and each 4-partition to a strong Moretti octagon.

3   The Context of Schopenhauer's Partition Diagrams

Schopenhauer's partition diagrams can be found in his logica maior, that is, in the *Berlin Lectures* that he held in the course of the 1820s. The logic in the *Berlin Lectures* [31, 234–368] can roughly be divided into four parts. The first part contains a doctrine of concepts [31, 242–260] and enriches classical positions

---

[2] For another perspective on the correspondence between partitions and Aristotelian diagrams, cf. [35].

of Aristotelian logic especially with diagrams and discussions on philosophy of language. The second part concerns the doctrine of judgement [31, 260–293]. It contains, among other things, treatises on the laws of thought, on truth, conceptual relations and contraposition. The third part concerns the theory of inferences [31, 293–356], in which Schopenhauer uses Euler diagrams to argue for the validity and naturalness of the original Aristotelian syllogistics. Furthermore, Stoic (propositional) logic and Aristotelian modal logic can also be found in this section. The short last part [31, 356–368] contains some additions and remarks about history and philosophy of logic, and about eristic dialectics.[3]

Schopenhauer's partition diagrams are given in the second part, more precisely in the treatise on relations between concepts. This treatise deals with many themes that had become popular through Kantian philosophy, and enriches them with logic diagrams. First, he explains the distinction between analytic and synthetic judgements by means of diagrams [31, 269–272]. The main part of this treatise, however, goes on to deal with Kant's theory of the four properties of judgements: quantity, quality, relation and modality. In contrast to Kant [20, III: 86ff.], Schopenhauer argues that this division should not simply be taken from textbooks of logic, since this leads to numerous problems [25], but that the properties of judgements only become apparent by analyzing the various ways in which two concepts can relate to one another. In order to find these relations, Schopenhauer makes use of geometric figures based on conceptual spheres. He therefore speaks of a 'clue of diagrams' ([31, 272], "Leitfaden sind die Schemata"). As a result, Schopenhauer uses six basic types of relational diagrams (RD) which can be described as follows.

Definition 5. *Let a circle or part of a circle in a given diagram represent a conceptual sphere. The diagrams RD1–6 in Fig. 2 depict the possible spatial positions of at least two conceptual spheres:*

> *RD1 Two conceptual spheres exactly overlap, so that only one sphere can be seen.*
> *RD2 One conceptual sphere completely contains another conceptual sphere.*
> *RD3 Two conceptual spheres are completely disjoint.*
> *RD4 A conceptual sphere includes two or more further spheres, such that the included spheres are mutually disjoint but do not exhaust the first sphere.*
> *RD5 Two conceptual spheres partly intersect each other.*
> *RD6 A conceptual sphere includes two or more further spheres, such that the included spheres do not intersect each other but do exhaust the first sphere.*

From a contemporary perspective, Schopenhauer's relational diagrams $RD1$–6 are clearly related to other diagrammatic systems.[4] In particular, $RD2$, $RD3$

---

[3] Schopenhauer's logica maior is thus structured in a way that had become standard in the history of logic (compare, for example, with the structure of William Ockham's *Summa Logicae* [28] or that of the *Port-Royal Logic* [1]), and that ultimately finds its roots in the division of Aristotle's logical works: (i) *Categories* on concepts/terms, (ii) *On interpretation* on judgements/propositions, and (iii) the remaining four works of the *Organon* on inferences/syllogisms.

[4] A terminological remark: we name a diagrammatic system after an author A to indicate that A drew or described at least some diagrams belonging to this system.

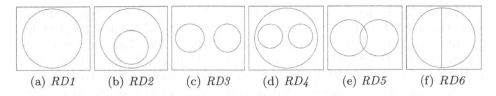

(a) *RD1*   (b) *RD2*   (c) *RD3*   (d) *RD4*   (e) *RD5*   (f) *RD6*

Fig. 2. Schopenhauer's six basic types of relational diagrams.

and *RD5* are Euler diagrams [15], *RD1*, *RD2*, *RD3* and *RD5* are Gergonne diagrams [16]. Note that *RD4* and *RD6*, which are more or less proper to Schopenhauer, depict strictly more than two conceptual spheres [31, 281].

In the *Berlin Lectures*, *RD1-6* are used to determine what the basic properties of judgements (*PJ*) are. For Schopenhauer, a proposition is called a 'judgement' when it expresses the relationship between at least two concepts. The first concept in the judgement is called the 'subject' (*S*), the second is called the 'predicate' (*P*). Using his six relational diagrams, Schopenhauer argues that there are six possible relations between two concepts *S* and *P*, which are listed below as *PJ$_S$* (the *S*-subscript stands for 'Schopenhauer'):

PJ_S1	universal		PJ_S4	negative
PJ_S2	particular		PJ_S5	disjunctive
PJ_S3	affirmative		PJ_S6	hypothetical

This treatise (and also further ones) shows that Schopenhauer on the one hand clearly orients his doctrine of judgements to Kant, but on the other hand also acts as a strong critic of him. After all, in his *Critique of Pure Reason*, Kant had also dealt with the *PJs*, which were to structure his entire system in many ways. Kant's so-called 'table of judgements' contains a total number of 12 *PJs*, grouped into 4 'titles' consisting of 3 *PJs* each. The titles are (*T1*) Quantity, (*T2*) Quality, (*T3*) Relation and (*T4*) Modality. Kant's 12 properties of judgements are listed below as *PJ$_K$* (the *K*-subscript stands for 'Kant'):

T1: Quantity	
PJ_K1	universal
PJ_K2	particular
PJ_K3	singular

T2: Quality			T3: Relation	
PJ_K4	affirmative		PJ_K7	categorical
PJ_K5	negative		PJ_K8	hypothetical
PJ_K6	infinite		PJ_K9	disjunctive

T4: Modality	
PJ_K10	problematic
PJ_K11	assertoric
PJ_K12	apodictic

Every PJ_S corresponds to some PJ_K, and just like Kant, Schopenhauer also classifies PJ_S1 and PJ_S2 under the title Quantity, PJ_S3 and PJ_S4 under Quality, and PJ_S5 and PJ_S6 under Relation. But unlike Kant, Schopenhauer rejects PJ_K10, PJ_K11, PJ_K12 from the title of Modality, PJ_K7 from Relation, PJ_K6 from Quality and PJ_K3 from Quantity, for various reasons. For example, Schopenhauer does not believe that modality is a property of the judgement, but rather of the one who judges, since modality only indicates the degree of certainty of the judge.

Schopenhauer argues with Aristotle that categoricity, i.e. PJ_K7, is not an independent property of judgements, but rather results from the cross-combination of $PJ_{K/S}1$ and $PJ_{K/S}2$ with PJ_K4/PJ_S3 and PJ_K5/PJ_S4. He thus holds that all judgements which represent a relationship between S and P are categorical. Drawing an analogy to Wittgenstein (TLP, 4.442), Schopenhauer's argument can be reformulated as follows: Kant's notion of 'categorical judgement' is logically quite meaningless; it simply indicates that the uttered proposition concerns a relationship between S and P. Schopenhauer reads from the set of Gergonne diagrams $\{RD1, RD2, RD3, RD5\}$ various instantiations of PJ_S1–4. These four properties of judgements traditionally originate from Aristotelian assertoric syllogistics; cross-combining them yields the following four categorical judgements:

	PJ_S3: affirmative	PJ_S4: negative
PJ_S1: universal	All S is P.	No S is P.
PJ_S2: particular	Some S is P.	Some S is not P.

Schopenhauer excludes the categorical judgements (PJ_K7) from Kant's title of relation $(T3)$, but retains the disjunctive (PJ_K8) and hypothetical (PJ_K9) judgements, i.e. those judgements which traditionally do not originate from Aristotelian syllogistics but rather from Stoic logic, and thus correspond in certain aspects to contemporary propositional logic [3]. Schopenhauer thus argues that PJ_K8 and PJ_K9 are not properties of *judgements* in the sense described above; after all, these properties do not concern any relationship between concepts S and P, but rather a relationship between two or more propositions (expressed by means of connectives such as "or" or "if ... then ...").

Nevertheless, Schopenhauer does integrate PJ_K8 and PJ_K9 into his own list (as PJ_S6 and PJ_S5, respectively), because the spatial combinations in the diagrams $RD2$, $RD4$ and $RD6$ provide him with an astonishing insight: connectives can not only be applied to combine *propositions* in order to obtain new, more complex propositions, but also to *concepts* in order to obtain new, more complex concepts. By applying logical connectives to concepts, Schopenhauer is thus able in his *Berlin Lectures* to develop complex partition diagrams.

4   Schopenhauer's Partition Diagrams

As we have just seen, Schopenhauer set up different properties of judgements (PJs) through a guideline of schemes, called $RD1-6$. At first he noticed that many PJ_Ks do not have a diagrammatic equivalent, but rather seem to be arranged arbitrarily. Moreover, Schopenhauer realized that Kant's title of relation $(T3)$ was problematic in several respects: categorical judgements (PJ_K7) do not have a unique function, but result from the cross-combination of the meaningful PJ under the title of quality $(T1)$ and quantity $(T2)$, i.e. PJ_K1, PJ_K2, PJ_K4, PJ_K5. Schopenhauer also noticed that hypothetical and disjunctive judgements, i.e. PJ_K8 and PJ_K9, are actually not regarded by Kant as relations between concepts. In the Kantian sense they are relations between judgements, which means that Kant inserts different criteria in his table of judgements.

Schopenhauer, however, argues that this problem can be solved with the help of the 'schematism of spheres'. Assuming that each concept in a judgement is represented by a circle in Euclidean space, in the case of two concepts there are several possible combinations of the two circles, which correspond by Definition 5 to the Gergonne relations. In contrast to $RD1$, in $RD2$ one circle contains a second one, in such a way that a third circle (disjoint from the second one) can be inserted, thus resulting in either $RD4$ or $RD6$, depending on whether the second and third circle jointly exhaust the first one (as in $RD6$) or not (as in $RD4$). As the insertion of further circles in a diagram such as $RD1$ can be repeated indefinitely, it is possible to create increasingly complex shapes in which a circle contains n other, mutually disjoint circles which are either jointly exhaustive of the first one (thus generalizing $RD6$) or they are not (thus generalizing $RD4$).

Schopenhauer is aware that diagrams such as $RD4$ open up many possibilities of interpretation, but he uses such diagrams mainly to prove the PJs under the titles of Quality and Quantity. For example, if the small contained circles in a diagram such as $RD4$ are designated with S_i (e.g. in the case of two small circles: S_1 and S_2) and the large containing circle with P, the following judgements can be read from $RD4$-type diagrams: (J1) All S_i is P; (J2) All that is not P is not S_i; (J3) Some P is S_i; (J4) Some P is not S_i. Next (and continuing the case of two small circles), he realizes that S_i in (J1–4) can actually be interpreted as S_1 or S_2. But to express the complex concept S_1 or S_2, the diagram must be drawn in such a way that the two small contained circles together completely exhaust the large containing circle (i.e. the complex concept S_1 or S_2 coincides exactly with the concept P, so that J4 no longer holds). Schopenhauer visualizes this by means of a large circle that is bisected by a line; cf. $RD6$. The large containing circle represents the concept P, while the two semicircles represent the disjoint and exhaustive subconcepts S_1 and S_2. In general, if a concept has n mutually disjoint and jointly exhaustive subconcepts, the large circle must be divided by $n - 1$ (non-intersecting) lines. We will call such diagrams *partition diagrams*, because the subconcepts are mutually disjoint and jointly exhaustive, and thus constitute a *partition* of the large concept.

Schopenhauer gives several examples, of which we briefly discuss two. First of all, consider the concept 'body', which can be divided into two disjoint and exhaustive subconcepts, viz. 'organic' and 'inorganic'. Schopenhauer thus divides the circle that represents the concept 'body' with a line into two halves that represent the subconcepts 'organic' and 'inorganic'; cf. the partition diagram in

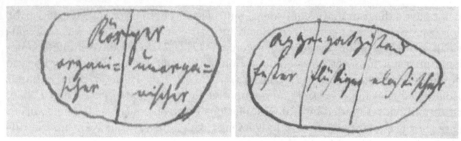

(a) Körper = body, organischer = organic, unorganischer = inorganic.

(b) Aggregatzustand = state of matter, fester = solid, flüßiger = liquid, elastischer = elastic.

Fig. 3. Schopenhauer's partition diagrams for (a) body and (b) state of matter.

Fig. 3a. Secondly, consider the concept 'state of matter', which can be divided into three mutually disjoint and jointly exhaustive subconcepts, viz. 'solid', 'liquid' and 'elastic'.[5] Schopenhauer thus divides the circle that represents the concept 'state of matter' with two lines into three equal parts that represent these three subconcepts; cf. the partition diagram in Fig. 3b.

Schopenhauer equates exclusive disjunction with contradiction, as is often done in contemporary logic as well [3,30,33]. This is based on the law of excluded middle. Using Fig. 3a as an example, Schopenhauer explains this as follows:

> "Here two judgements are connected in such a way that the affirmation of the one is the negation of the other; both can neither be negated nor affirmed at the same time: according to the law of thought of the excluded third." [31, 280]

For Schopenhauer, the partition diagram in Fig. 3a illustrates the contradiction of the two sub-concepts of 'body', whereas the following proposition describes this diagram by means of an exclusive disjunction:

$$\text{All bodies are either organic or inorganic.} \tag{1}$$

But Schopenhauer goes further and shows that the partition diagram not only facilitates knowledge *representation*, but also visual *reasoning*. If one adopts (1) as a premise, and adds an instance such as 'sea sponge' that belongs to the generic concept 'body', one can draw a conclusion including the subordinate concepts 'organic' and 'inorganic'. Schopenhauer takes the following example:

$$\text{A sea sponge is a body.} \tag{2}$$

---

[5] Nowadays we would probably make a different classification, for example we would certainly add 'plasma' to the states of matter. Schopenhauer represents these classifications according to the state of knowledge of the early 19th century. However, since he knows from the history of science that (structures of) concepts can change, he advocates an ontological relativism even for analytic judgements [21, Chap. 2.2.5f.].

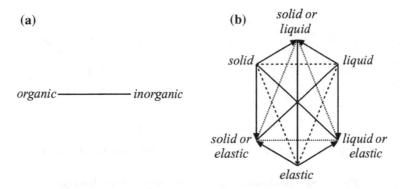

Fig. 4. (a) PCD and (b) strong JSB hexagon corresponding to Schopenhauer's 2- and 3-partitions for body and state of matter, respectively.

Thus, a sea sponge is either an organic or an inorganic body.   (3)

If we now have further information about the instance of the generic concept, e.g. to which subconcept the sea sponge must be assigned, we can infer (in the sense of Stoic logic) by means of proposition (3) and *modus tollendo ponens* to which subconcept it is not assigned, i.e. we can reason from (4) to (5). Vice versa, we can also use (3) and *modus ponendo tollens* to reason from (5) to (4).

A sea sponge is an organic body.   (4)

A sea sponge is not an inorganic body.   (5)

These ways of reasoning do not seem particularly spectacular, although it can be assumed that the partition diagram provides observational advantages, as it is easier and quicker to read than propositions (1–5). This becomes particularly evident when more complex diagrams are used. Schopenhauer initially drew the diagrams only to represent knowledge. Nevertheless, in several places in the text the added note "Illustrate!" can be found. We can assume that Schopenhauer used the frequently mentioned gesture of indication ('hindeuten') in his lectures to refer to specific regions of a given diagram.[6]

5   From Partition Diagrams to α-Structures

In this section we bring everything together, and show how to apply the insights from logical geometry (cf. Sect. 2) to Schopenhauer's partition diagrams

---

[6] Here Schopenhauer still folllows Kant very closely, who uses a similar square diagram in §29 of the Jäsche logic [20, IX: 108] in order to depict disjunctive judgements. In contrast to Schopenhauer, Kant describes in the text that one should use an x to mark the corresponding region of a disjunctive judgement. Unlike Kant, Schopenhauer's diagrams not only illustrate judgements, but also classes.

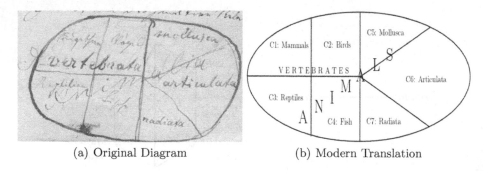

(a) Original Diagram
(b) Modern Translation

Fig. 5. Schopenhauer's most complex partition diagram

(cf. Sect. 3–4). In particular, Theorem 2 tells us that these partition diagrams correspond directly to certain Aristotelian diagrams, viz. α-structures. For example, the 2-partition in Fig. 3a gives rise to an α_1-structure, viz. the PCD in Fig. 4a, while the 3-partition in Fig. 3b gives rise to a strong α_3-structure, viz. the strong JSB hexagon in Fig. 4b.

The most complex partition diagram in Schopenhauer's logica maior is found in [31, 280], and is shown here as Fig. 5a from the original manuscripts. As can also be seen in the translation in Fig. 5b, it shows a large circle for the concept 'animals', which is then subdivided into seven subconcepts (C):

C	Original	Translation	C	Original	Translation
C1	Säugethier	mammals	C5	mollusca	mollusca
C2	Vögel	birds	C6	articulata	articulata
C3	Reptilien	reptiles	C7	radiata	radiata
C4	Fische	fish			

These subconcepts are mutually exclusive and jointly exhaustive, and thus constitute a 7-partition of the concept 'animals'. Appealing once again to Theorem 2, we find that this 7-partition gives rise to a strong α_7-structure, as shown in Fig. 6. This structure consists of the subconcepts C1–C7, which are pairwise contrary to each other, together with their complements (relative to 'animals'), which are pairwise subcontrary to each other. In order to make this more precise, note that C1–C7 can be viewed as the atoms of a Boolean algebra $\mathbb{B}_7$, which can be represented with bitstrings of length 7 [10], i.e. $\mathbb{B}_7$ is isomorphic to $\{0, 1\}^7$:

Subconcept	Bitstring	Complementary subconcept	Bitstring
C1: mammals	1000000	¬mammals = animals \ mammals	0111111
C2: birds	0100000	¬birds = animals \ birds	1011111
C3: reptiles	0010000	¬reptiles = animals \ reptiles	1101111
C4: fish	0001000	¬fish = animals \ fish	1110111
C5: mollusca	0000100	¬mollusca = animals \ mollusca	1111011
C6: articulata	0000010	¬articulata = animals \ articulata	1111101
C7: radiata	0000001	¬radiata = animals \ radiata	1111110

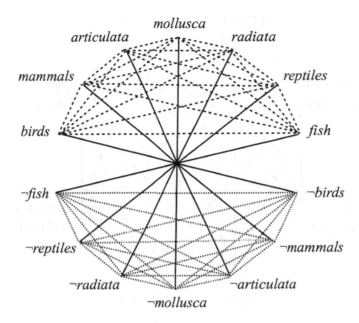

Fig. 6. Strong α_7-structure corresponding to Schopenhauer's 7-partition of 'animals'. For reasons of visual simplicity, the subalternation arrows are not drawn; they go from each 'positive' concept in the upper part of the diagram to each 'negative' concept in the lower part of the diagram (except for its contradictory, of course).

Interestingly, this 7-partition is itself hierarchically organized. Schopenhauer indicates that the subconcepts of 'mammals' (C1), 'birds' (C2), 'reptiles' (C3) and 'fish' (C4) together constitute the intermediate concept 'vertebrates', which is itself a subconcept of 'animals'. The bitstring representation of 'vertebrates' can easily be calculated in terms of the bitstrings of its four subconcepts:

$$\begin{array}{cccccccc}
\text{vertebrates} & = & \text{mammals} & \text{or} & \text{birds} & \text{or} & \text{reptiles} & \text{or} & \text{fish} \\
1111000 & = & 1000000 & \vee & 0100000 & \vee & 0010000 & \vee & 0001000
\end{array}$$

For certain reasoning purposes, it might not be required to subdivide the vertebrates into mammals, birds, reptiles and fish. In those circumstances, such a further subdivision would only yield unnecessary complexity, and should thus be dispensed with. Formally, this means that the original 7-partition of 'animals' reduces to a 4-partition, consisting of the concepts of 'vertebrates', 'mollusca', 'articulata' and 'radiata'. In terms of bitstring representations, this amounts to focusing exclusively on those bitstrings that have identical values in their first four positions, such as 1111000 and 0000100. Equivalently, one could say that we have moved from bitstrings of length 7 (corresponding to the original, fine-grained 7-partition) to bitstrings of length 4 (corresponding to the new, coarser 4-partition), by systematically collapsing the first four bits into a single bit, e.g. 1111000 and 0000100 reduce to 1000 and 0100, respectively. Appealing one final time to Theorem 2, we find that the coarsened 4-partition gives rise to a strong α_4-structure, i.e. a strong Moretti octagon, as shown in Fig. 7.

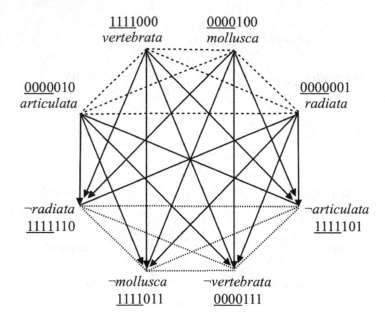

Fig. 7. Strong Moretti octagon corresponding to Schopenhauer's coarsened 4-partition of 'animals', incl. bitstring representations (with respect to the original 7-partition).

It bears emphasizing that 'vertebrates' is a primitive/atomic concept with respect to the coarsened 4-partition (where its bitstring representation 1000 contains just a single 1-bit), but it is a complex concept with respect to the original, more fine-grained 7-partition (where its bitstring representation 1111000 is itself the join of four other bitstrings). By contrast, 'mollusca' is an atomic concept with respect to the coarse 4-partition (where its bitstring representation 0100 contains just a single 1-bit) as well as with respect to the original, more fine-grained 7-partition (where its bitstring representation 0000100 again contains just a single 1-bit). This asymmetry between 'vertebrates' on the one hand and 'mollusca' (and 'articulata' and 'radiata') on the other hand captures the hierarchical nature of Schopenhauer's 7-partition of 'animals'.

6 Conclusion

In this paper we have shown how Schopenhauer's criticism of Kantian philosophy led him to the idea of representing Aristotelian relations between sets/concepts. To this end, he developed partition diagrams that went far beyond the diagrammatic techniques known at the time. Drawing upon ideas and techniques from logical geometry, we have shown that Schopenhauer's partition diagrams systematically give rise to a special type of Aristotelian diagrams, viz. (strong) α-structures. These incude a PCD (α_2), a strong JSB hexagon (α_3), a strong Moretti octagon (α_4) and a strong α_7-structure.

As systematic research on Schopenhauer's logic has only just begun, there are still many questions that require further investigation. For example, can one find similar ideas in published logic textbooks from the 19th century? Do partition diagrams play an important role in the interpretation of Schopenhauer's system, which has a structure based on the method of partition or *divisio* in Bacon's *De dignitate et augmentis scientiarum*? To what extent did Schopenhauer use these diagrams to make new discoveries for what was then called Stoic logic, which is considered the precursor of modern propositional logic?

Acknowledgements. We would like to thank Michał Dobrzański, Ingolf Max, Hans Smessaert and Margaux Smets for their valuable feedback on an earlier version of this paper. The research of the second author was supported by a Research Professorship (BOFZAP) from KU Leuven and the ID-N project 'Bitstring Semantics for Human and Artificial Reasoning' (BITSHARE).

References

1. Arnauld, A., Nicole, P.: Logic or the Art of Thinking. Cambridge University Press, Cambridge (1996)
2. Blanché, R.: Structures Intellectuelles. Vrin, Paris (1966)
3. Bocheński, I.: A Précis of Mathematical Logic. Reidel, Dordrecht (1959)
4. Demey, L.: Computing the maximal Boolean complexity of families of Aristotelian diagrams. J. Logic Comput. **28**, 1323–1339 (2018)
5. Demey, L.: Metalogic, metalanguage and logical geometry. Logique et Analyse **248**, 453–478 (2019)
6. Demey, L.: From Euler diagrams in Schopenhauer to Aristotelian diagrams in logical geometry. In: Lemanski, J. (ed.) Language, Logic, and Mathematics in Schopenhauer. SUL, pp. 181–205. Springer, Cham (2020). https://doi.org/10.1007/978-3-030-33090-3_12
7. Demey, L., Smessaert, H.: The interaction between logic and geometry in Aristotelian diagrams. In: Jamnik, M., Uesaka, Y., Elzer Schwartz, S. (eds.) Diagrams 2016. LNCS (LNAI), vol. 9781, pp. 67–82. Springer, Cham (2016). https://doi.org/10.1007/978-3-319-42333-3_6
8. Demey, L., Smessaert, H.: Metalogical decorations of logical diagrams. Logica Univ. **10**, 233–292 (2016)
9. Demey, L., Smessaert, H.: Aristotelian and duality relations beyond the square of opposition. In: Chapman, P., Stapleton, G., Moktefi, A., Perez-Kriz, S., Bellucci, F. (eds.) Diagrams 2018. LNCS (LNAI), vol. 10871, pp. 640–656. Springer, Cham (2018). https://doi.org/10.1007/978-3-319-91376-6_57
10. Demey, L., Smessaert, H.: Combinatorial bitstring semantics for arbitrary logical fragments. J. Philos. Logic **47**, 325–363 (2018)
11. Dobrzański, M.: Begriff und Methode bei Arthur Schopenhauer. Königshausen & Neumann, Würzburg (2017)
12. Dobrzański, M.: Problems in reconstructing Schopenhauer's theory of meaning: with reference to his influence on Wittgenstein. In: Lemanski, J. (ed.) Language, Logic, and Mathematics in Schopenhauer. SUL, pp. 25–45. Springer, Cham (2020). https://doi.org/10.1007/978-3-030-33090-3_3

13. Dobrzański, M., Lemanski, J.: Schopenhauer diagrams for conceptual analysis. In: Pietarinen, A.-V., Chapman, P., Bosveld-de Smet, L., Giardino, V., Corter, J., Linker, S. (eds.) Diagrams 2020. LNCS (LNAI), vol. 12169, pp. 281–288. Springer, Cham (2020). https://doi.org/10.1007/978-3-030-54249-8_22
14. Dümig, S.: The world as will and I-language: Schopenhauer's philosophy as precursor of cognitive sciences. In: Lemanski, J. (ed.) Language, Logic, and Mathematics in Schopenhauer. SUL, pp. 85–94. Springer, Cham (2020). https://doi.org/10.1007/978-3-030-33090-3_6
15. Euler, L.: Lettres à une Princesse d'Allemagne. Académie Impériale (1768)
16. Gergonne, J.D.: Essai de dialectique rationelle. Annales de Mathématiques Appliquées **7**, 189–228 (1817)
17. Givant, S., Halmos, P.: Introduction to Boolean Algebras. Springer, New York (2009) https://doi.org/10.1007/978-0-387-68436-9
18. Heinemann, A.-S.: Schopenhauer and the equational form of predication. In: Lemanski, J. (ed.) Language, Logic, and Mathematics in Schopenhauer. SUL, pp. 165–179. Springer, Cham (2020). https://doi.org/10.1007/978-3-030-33090-3_11
19. Jacoby, P.: A triangle of opposites for types of propositions in Aristotelian logic. New Scholasticism **24**, 32–56 (1950)
20. Kant, I.: Gesammelte Schriften (Akademie-Ausgabe). Ed. by Preußische/Göttinger/Berlin-Brandenburgischen Akademie der Wissenschaften. de Gruyter, Reimer (1900ff)
21. Lemanski, J.: World and Logic. College Publications, London (2021)
22. Lemanski, J., Dobrzański, M.: Reism, concretism and Schopenhauer diagrams. Studia Humana **9**, 104–119 (2020)
23. Lemanski, J., Moktefi, A.: Making sense of Schopenhauer's diagram of good and evil. In: Chapman, P., Stapleton, G., Moktefi, A., Perez-Kriz, S., Bellucci, F. (eds.) Diagrams 2018. LNCS (LNAI), vol. 10871, pp. 721–724. Springer, Cham (2018). https://doi.org/10.1007/978-3-319-91376-6_67
24. Matsuda, K.: Spinoza's redundancy and Schopenhauer's concision. An attempt to compare their metaphysical systems using diagrams. Schopenhauer-Jahrbuch **97**, 117–131 (2016)
25. Menne, A.: Die Kantische Urteilstafel im Lichte der Logikgeschichte und der modernen Logik. J. Gen. Philos. Sci. **20**, 317–324 (1989)
26. Moktefi, A.: Schopenhauer's Eulerian diagrams. In: Lemanski, J. (ed.) Language, Logic, and Mathematics in Schopenhauer. SUL, pp. 111–127. Springer, Cham (2020). https://doi.org/10.1007/978-3-030-33090-3_8
27. Moretti, A.: The Geometry of Logical Opposition. Ph.D. thesis, Neuchâtel (2009)
28. Ockham, W.: Summa Logicae (Op. Phil. I). Fransciscan Institute, St. Bonaventure, NY (1974)
29. Pellissier, R.: Setting n-opposition. Logica Univ. **2**(2), 235–263 (2008)
30. Prior, A.: Formal Logic. Oxford University Press, Oxford (1955)
31. Schopenhauer, A.: Philosophische Vorlesungen (Sämtliche Werke IX. Ed. by P. Deussen and F. Mockrauer). Piper & Co., Munich (1913)
32. Sesmat, A.: Logique II. Les Raisonnements. La syllogistique. Hermann, Paris (1951)
33. Smessaert, H., Demey, L.: Logical geometries and information in the square of opposition. J. Logic Lang. Inf. **23**, 527–565 (2014)

34. Smessaert, H., Demey, L.: The unreasonable effectiveness of bitstrings in logical geometry. In: Béziau, J.-Y., Basti, G. (eds.) The Square of Opposition: A Cornerstone of Thought. SUL, pp. 197–214. Springer, Cham (2017). https://doi.org/10.1007/978-3-319-45062-9_12
35. Smessaert, H., Shimojima, A., Demey, L.: Free rides in logical space diagrams versus Aristotelian diagrams. In: Pietarinen, A.-V., Chapman, P., Bosveld-de Smet, L., Giardino, V., Corter, J., Linker, S. (eds.) Diagrams 2020. LNCS (LNAI), vol. 12169, pp. 419–435. Springer, Cham (2020). https://doi.org/10.1007/978-3-030-54249-8_33

Revisiting Peirce's Rules
of Transformation for Euler-Venn Diagrams

Reetu Bhattacharjee[1,2(✉)] and Amirouche Moktefi[3]

[1] School of Cognitive Science, Jadavpur University, Kolkata, India
[2] Department of Mathematics, Mandsaur University, Mandsaur, India
reetu.bhattacharjee@meu.edu.in
[3] Ragnar Nurkse Department of Innovation and Governance,
Tallinn University of Technology, Tallinn, Estonia
amirouche.moktefi@taltech.ee

Abstract. Charles S. Peirce introduced in 1903 a set a transformation rules for Euler-Venn diagrams. This innovation contrasted with earlier practices where logicians rather extracted the desired information by a simple 'glance' at their diagrams. Also, Peirce's set of rules was the starting point of Sun-Joo Shin's more recent systems which, in turn, inspired most subsequent modern diagrammatic systems. Despite their significance, these rules got little attention from both diagram and Peirce scholars. In this paper, we revisit Peirce's rules of transformation and discuss the extent to which they 'survived' in modern diagrammatic systems. We will specifically consider their clarity and completeness to assess Peirce's assumption that some of his rules may be simplified while others may have been overlooked.

Keywords: Peirce · Euler-venn diagram · Rules of transformation

1 Introduction

Charles S. Peirce made significant contributions to logic diagrams. In addition to his work on the theory of diagrams, it is known that he improved the Euler-Venn's scheme and that he designed a fascinating system of Existential Graphs. His manuscripts continue to reveal remarkable advances, such as his recently rediscovered inclusion diagrams [5,15]. In this paper, we discuss one of Peirce's major, yet seldom noticed, innovations: his rules of transformation for Eulerian diagrams found in his manuscript 'On logical graphs' (1903), generally known as MS 479 [12][1]. This set of rules is historically significant for at least two reasons.

---

[1] Unfortunately, manuscript MS 479 has still not been properly published. It has only been partially reproduced and poorly edited in Peirce's *Collected Papers* [13]. This transcription, on which was based Shin's account, should be used with extreme caution. The manuscript is also not reproduced in Ahti-Veikko Pietarinen's edition of Peirce's existential graphs [15], but additional text and variants are included [14]. Apparently, Peirce intended to include his manuscript as a chapter in a volume of *Logical Tracts* [14, p. 72]. The original manuscript MS 479 is freely accessible on the Peirce Archive repository (https://rs.cms.hu-berlin.de/peircearchive/pages/search. php). The page numbers we indicate for MS 479 are the file titles in the Peirce Archive.

© Springer Nature Switzerland AG 2021
A. Basu et al. (Eds.): Diagrams 2021, LNAI 12909, pp. 166–182, 2021.
https://doi.org/10.1007/978-3-030-86062-2_14

First, it contrasts with earlier work on logic diagrams where no such formal rules were provided. Sun-Joo Shin argued that "Peirce was probably the first person that discussed the rules of transformation in a diagrammatic system" [18, p. 24]. Indeed, Peirce's predecessors generally invited their readers to detect the conclusion of an argument by a simple "glance" at the diagram that represents its premises [20, p. 15][2]. Peirce rather provided rules in accordance with which the diagram of the premises is to be transformed to produce the diagram of the conclusion from which the conclusion can be read off[3]. By introducing his rules, Peirce was "attempting to massage Euler diagrams into something that would possess more of the character of a logical *language* than a *diagram* or a *picture*" [14, p. 95]. As such, Peirce opened the way to a formal view of diagrams [9].

Second, Peirce's rules played a crucial role in the shaping of modern diagrammatic systems. Indeed, Shin's work, which is commonly regarded as the primary inspiration for subsequent systems [19], was itself based on the rules that Peirce has enumerated almost a century earlier [18, p. 28]. This legacy of Peirce is almost ironical when it is reminded that Peirce himself did not think highly of his Eulerian diagrams, sketched his rules rather loosely and did not believe the system to have the potential for significant growth [12] (see also [14, p. 84]). The formidable development of diagrammatic logic in recent decades does not support Peirce's scepticism, but it demonstrates the importance of his pioneering work on transformation rules in diagrammatic reasoning.

Despite their significance, Peirce's transformation rules attracted little attention, except for Shin's account [18, pp. 28–40]. Peirce scholars are justifiably more interested in Peirce's true *chef d'oeuvre*, his Existential graphs [2][4], while modern diagram scholars understandably discover those rules mainly through Shin's account (which does not reproduce Peirce's original formulations). In this paper, we revisit Peirce's rules of transformation and discuss the extent to which they 'survived' in modern diagrammatic systems. We will specifically consider their clarity and completeness to assess Peirce's assumption that some of his rules may be simplified while some others may have been overlooked [12]. For the purpose, we first review and discuss each of Peirce's six rules. To ease the reading of the paper, we discuss rules-1 to 3 in Sect. 2, rule-4 and its variations in Sect. 3 and rules-5 to 6 in Sect. 4. Finally we compare in Sect. 5 Peirce's set of rules with some modern diagrammatic systems, namely Shin's systems Venn-I and Venn-II [18] and the more recent system $Venn_{i_n}$ [4].

---

[2] Lewis Carroll is a remarkable exception here. See [7,8]. A comparison of Carroll's rules with those of Peirce is found in [10].

[3] Peirce explained that he used rules "in the sense in which we speak of the "rules" of algebra; that is, as a permission under strictly defined condition" [12]. In his entry on 'Symbolic Logic', published a year earlier, Peirce defined a rule as "a permission under certain circumstances to make a certain transformation" [11, p. 450].

[4] Peirce's mature Eulerian diagrams and Existential graphs were developed at the same time and share several features, including the formulation of transformation rules. But they differ significantly in their purpose: Eulerian diagrams served mainly for logical calculus while Existential graphs were designed for logical analysis. Roughly, calculus aims at carrying reasonings while analysis investigates them. On the opposition between calculus and analysis, see [3,11, p. 450].

2 Rules 1 to 3

Rule 1: "Any entire sign of assertion (i.e. a cross, zero or connected body of crosses and zeros) can be erased" [12, p. 042][5].

This particular rule basically helps us to erase certain information from a given conjunction of information. Using rule-1 we can get the diagram in Fig. 2 from the diagram in Fig. 1 by removing zero, cross and the connected body of crosses from the regions $(M - S - P)$, $((M \cap S) - P)$ and $((S - M - P) \cup ((S \cap P) - M))$ respectively.

Fig. 1. .   Fig. 2. .

Rule 2: "Any sign of assertion can receive any assertion" [12, p. 043].

Using rule-2 we can introduce new pieces of information in the form of disjunction. For example, the cross in the region $(S - P)$ and the zero in the region $(P - S)$ both receives cross to get the diagram in Fig. 4 from the diagram in Fig. 3.

Fig. 3. .   Fig. 4. .

Rule 3: "Any assertion which could permissively be written if there were no other assertion can be written at any time, detachedly" [12, p. 043].

Although there is no doubt that both rule-1 and rule-2 are quite intuitive, rule-3 does not turn out as such. How do we know which assertion is 'permissible' to be written in a diagram? Peirce never explained this particular rule nor gave any kind of example right after stating this rule[6]. But later, while showing how one can obtain conclusion in syllogistic reasoning using these six rules, Peirce mentioned that using rule-3 we can unify two diagrams [12, p. 046]. Now suppose we have the following two diagrams, Fig. 5 and Fig. 6.

The regions $((M \cap P) - S)$ and $((P \cap S) - M)$ in Fig. 5 are both blank. We don't know whether these two regions are empty or non empty and thus neither zero nor cross is permissible to be written in these regions. But if we consider the diagram in Fig. 6 together with the diagram in Fig. 5 then we have the new information 'Something are both M and P but not S and anything that

---

[5] Here cross and zero represents non-emptiness and emptiness of a region respectively. The connected lines between any of these symbols represent their disjunction

[6] This rule was written by Peirce on the margins of his manuscript, without further explanation. It seems to have been added later, as shown by the renumbering of the following rule.

Fig. 5. .   Fig. 6. .

is both P and S is also M' in our premise from the Fig. 6. Taking these two diagrams together we know which assertions are permissible in the regions ((M ∩ P) − S) and ((P ∩ S) − M) of the diagram in Fig. 5. Thus by using rule-3 we can introduce cross and zero in the regions ((M ∩ P) − S) and ((P ∩ S) − M) respectively and get the diagram in Fig. 7.

Fig. 7. .

Shin understood rule-3 in a similar manner and later used it as the basis of her 'unification rule'. However, rule-3 might not be merely a unification rule for diagrams. This rule let us introduce any new piece of information in a diagram if we have prior knowledge about it. For example, throughout MS 479 Peirce has mentioned that "nothing exists" is an absurd assertion [12, p. 034]. So we can take the universe to be always non-empty. Having this prior knowledge about the universe, take any diagram, say the diagram in Fig. 8. Now by using rule-3, we can introduce a connected body of crosses in Fig. 8 such that each region of the diagram has a cross of this connected body. The resulting diagram is in Fig. 9.

Fig. 8. .   Fig. 9. .

3   Rule 4

In this section, in a manner similar to Shin's exposition, Peirce's original rule-4 is divided for convenience into several sub-rules that are discusses here separately.

Rule 4: (i) "In the same compartment repetitions of the same sign, whether mutually attached or detached, are equivalent to one writing of it" [12, p. 043].
So both the diagrams in Fig. 10 and Fig. 11 are equivalent to the diagram in Fig. 12.
(ii) "Two different signs in the same compartment"
(a) "if attached to one another are equivalent to no sign at all and may be erased or inserted" [12, p. 043].

Fig. 10. .   Fig. 11. .   Fig. 12. .

The diagrams in Fig. 13 and Fig. 14 are equivalent.

Fig. 13. .   Fig. 14. .

(b) "But if they are detached from one another, they constitute an absurdity" [12, p. 043] (see Fig. 15).

Fig. 15. .

Point to be noted that all the above conditions are based on the presupposition that signs are not connected with any other signs in other compartments.

For the case where the same signs exists separately in the same compartment, the clause (i) of rule-4 is superfluous. Because, using rule-1 we can erase the extra sign and get the diagram in Fig. 12 from the diagram in Fig. 11. Again, if we have the diagram in Fig. 12, then we have the information that 'P is non-empty'. So by using rule-3 we can have the diagram in Fig. 11 from the diagram in Fig. 12. When the same signs exists mutually attached in the same compartment, we also do not need clause (i) of rule-4 to get the diagram in Fig. 10 from the diagram in Fig. 12. It can be done using rule-2. But we need rule-4(i) to get the diagram in Fig. 12 from the diagram in Fig. 10. A question might arise here – we have the information that 'P is non-empty or P is non-empty' (Fig. 10) and we know that 'P is non-empty' (Fig. 12) is always derivable from this information then why not use rule-3 to get the diagram in Fig. 12 from the diagram in Fig. 10? It is because, rule-3 alone lets us introduce certain information about which we have prior knowledge. It does not let us deduce anything from that information. So even if we know that 'P is non-empty or P is non-empty' (Fig. 10), we can not use rule-3 to derive 'P is non-empty' and get the diagram in Fig. 12.

Even if we need rule-4(i), it still needs modification. The condition that 'signs are not connected with any other signs in the other compartment' makes rule-4(i) incapable to get certain syntactically different looking diagrams which represents the same information. For example, the diagrams in Fig. 16 and Fig. 17 represents the same information but we cannot get the diagram in Fig. 17 from the diagram in Fig. 16 by using rule-4(i) unless we drop the condition 'signs are not connected with any other signs in the other compartment'.

Fig. 16. .

Fig. 17. .

In plain sight rule-4(ii-a) is also not needed here as we have the information that any region in diagram is 'either empty or non-empty' and thus by using rule-3 we get the diagram in Fig. 13 from the diagram in Fig. 14. The converse can be done by using rule-1[7]. But we need rule-4(ii-a) to get the diagram in Fig. 19 from the diagram in Fig. 18 as we cannot 'derive' the information '(S − P) is non-empty' from the information 'either (P ∩ S) is empty or (P ∩ S) is non-empty or (S − P) is non-empty' using rule-3. But again, similar to rule-4(i), the condition 'signs are not connected with any other signs in the other compartment' prevents us from doing so. The converse can be done using rule-2.

Fig. 18. .

Fig. 19. .

The rule-4(ii-b) seems more of a definition than a rule as it ended abruptly saying that "If they are detached from one another, they constitute an absurdity". In Peirce's Existential graphs, we know that empty oval is considered as "constantly false proposition or absurdity" [2, p. 219] i.e. it is considered to be a 'contradiction'(see [2,14,16]). Also, in existential graphs within a cut anything can be inserted [17, p. 647], in other words anything follows from contradictions. So, classical explosion rule was always present in Peirce's diagrammatic systems. In rule-4(ii-b), Peirce meant classical explosion by saying that "they constitute an absurdity" and everything follows from it. But, although presented in practice in Peirce's Existential graphs [17, p. 647] it is not mentioned explicitly here. So a modification regarding this rule is needed.

Rules-4(i) and (ii) were criticized by Shin due the usage of the words 'equivalence' and 'absurdity'. Shin claimed that "By analyzing clause (i) and (ii) of this rule, I will show that Peirce does not make a clear distinction between syntax and semantics either. This confusion leads him to several problematic treatments of diagrams" [18, p. 30] and Shin believed that "this reveals Peirce's lack of a distinction between syntax and semantics" [18, p. 35].

According to Shin, in rule-4(i) and (ii-a), Peirce used the word 'equivalent' to actually represent 'semantically equivalent' diagrams, not 'syntactically equivalent' diagrams. The main base for this argument, as shown by Shin [18, p. 31], is that the following diagrams in Fig. 20 will be considered 'equivalent' by rule-4(i) and (ii-a). But, although these diagrams represent the same facts, they are syntactically different looking.

---

[7] Shin also proposed to use rule 1 to get Fig. 12 and Fig. 14 from Fig. 11 and Fig. 13 respectively. She also proposed to use rule 2 to get Fig. 10 from Fig. 12.

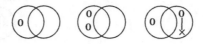

Fig. 20. .

Shin also pointed out the following three problems that occur when 'semantic equivalence' is taken into consideration.

(1) The possibility of semantically equivalent diagrams having a different syntactic form was not considered when Peirce criticized Lambert for having two different looking diagrams for the two equivalent proposition "Some A are B" (Fig. 21)[1] and "Some B are A" (Fig. 22)[12,18, p. 38].

Fig. 21. .   Fig. 22. .

(2) A deductive system, which is both sound and complete, lets us deduce a formula 'α' from a set of formula Γ if and only if α is a semantic consequence of Γ. So if α is semantically equivalent to some formula β then we can deduce α from β and vice-versa. But taking this assumption as a rule will make the existence of a deductive system unnecessary [18, p. 31].

(3) Peirce "did not have an accurate semantics to support his use of "equivalence" in a proper way" [18, p. 31]

Shin's criticisms can be disputed if we are reminded that 'Equivalence' is not always semantic since it can also be syntactic. Two different looking diagrams, say D and D', can be 'syntactically equivalent' if there is a rule that lets us get D' from D and vice-versa. The diagrams in Fig. 10, Fig. 11 and Fig. 12 all represent the same information i.e. 'P is non-empty' and rule-4(i) lets us get the diagrams from each other. Similar situation happens for the diagrams in Fig. 13 and Fig. 14. In [12, p. 041], Peirce already mentioned that two opposite signs, which are connected together in the same region, should annul each other and be equivalent to no sign at all. Then why did he need to construct rule-4(ii-a) that's says the same thing? The reason is that Peirce was trying to construct a rule that will let us get two syntactically different looking diagrams, which represents the same fact, from each other. That's why words like "equivalent to one writing of it" and "may be erased or inserted" was used respectively in rule-4(i) and (ii-a). It is true that we cannot just introduce a rule that would say that 'if formula α is a semantic consequence of a set of formulas Γ, then we can deduce α from Γ', but it is permissible, and even desirable, to have a rule that lets us get two equivalent but syntactically different looking diagrams from each other? In Shin's own system, she had the rule of splitting sequence where the diagram D_2 could be deduced from D_1 and both diagrams represent the same fact (see Fig. 23 [18, p. 123]).

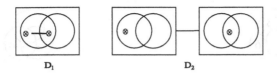

D_1 D_2

Fig. 23. .

One may reasonably argue that Peirce's opposition to Lambert's was motivated by the absence of rules in the latter to go from one diagram to it's semantically equivalent diagram. Shin's next objection that Peirce did not have accurate semantics didn't stand here as the notion 'equivalence' is used here 'syntactically'.

For rule-4(ii-b) Shin's objection was mainly regarding the notion of 'absurdity'. According to Shin there are two possible interpretations of 'absurdity'.

(1) "One is to take this phrase to mean that we should not be allowed to draw a diagram with two different kinds of signs in the same compartment. If so, this system has no way to represent a contradiction" [18, p. 32]. We argued earlier that this was not Peirce's interpretation of 'absurdity'.

(2) "The other interpretation of clause (ii-2), which seems to be more plausible, is that a diagram with "o" and "x" in the same compartment means absurdity. According to this interpretation, clause (ii-2) does not tell us how to transform a given diagram, but explains what assertion is made if a diagram has more than one character in a certain way. When we recall that these rules are stipulated to tell us what we are permitted to do in manipulating diagrams, it is rather puzzling why Peirce had to explain what a diagram means under these rules. What assertion is made in a diagram belongs to semantics, whereas the transformation rules belong to syntax. This clearly reveals Peirce's lack of a distinction between syntax and semantics" [18, p. 32]. This is again a same problem as we have dealt for the notion 'equivalence'. Peirce mentioned that whenever cross and zero exists detachedly in the same compartment it leads to absurdity way before introducing his rule of transformation. There was no need for him to again write it as a rule here. Also, in existential graphs, transformation rules are presented in a very much syntactic point of view, a fact that rather suggests an understanding of semantics and syntax. Yet, it is true that rule-4(ii-b) is not properly written and needs modification.

Rule 4: (iii) "If two contrary signs are written in the same compartments, the one being attached to certain others, P, and the other to certain others, Q, it is permitted to attach P to Q and to erase the contrary signs" [12, p. 043].

Using rule-4(iii) we can remove the contrary signs when it is in a disjunctive form. Peirce has given the following example where, by using the rule-4(iii), the diagram in Fig. 25 is obtained from the diagram in Fig. 24 [12, p. 044].

Fig. 24. .   **Fig. 25. .**

For rule-4(iii), Shin pointed out that if we have the following argument

"All S is P or some S is P.
No S is P.
Therefore, there is no S" [18, p. 33].

Then we have the diagrams in Fig. 26 where "first diagram in the following represents what the two premises convey. One of the o's in the first diagram is not attached to any other sign. Accordingly, the antecedent of clause (iii) is not satisfied. However, if we allow P (in clause (iii)) to be an empty sign, then we get the second diagram from the first one. After that, we need to add the second premise, "No S is P," to the second diagram. This is how we get the third diagram, which represents the conclusion of the previous syllogism. In order to get the rightmost diagram (which we want to get), we need to represent the second premise twice" [18, p. 33].

Fig. 26. .

If we follow the examples given by Peirce in MS 479 in [12, pp. 046–047] we will find that in the above case where two contradictory signs, '0' and '×', are in the same region 'S ∩ P' but only '×' is connected with another sign in 'S − P', we only erase the '×' and get the third diagram instead of the second one by applying rule-4(iii) to the first diagram. But again this condition, where one of the contrary signs is not connected with some other sign in some other region, is not mentioned precisely in rule-4(iii) and this rule needs to be modified.

4   Rules 5 to 6

Rule 5: "Any Area-boundary, representing a term can be erased, provided that if, in doing so, two compartments are thrown together containing independent zeros, those zeros be connected, while if there be a zero on one side of the boundary to be erased which is thrown into a compartment containing no independent zero, the zero and its whole connex be erased" [12, p. 044].

Rule-5 is a similar to rule-1, where we can erase information. By eliminating the curve M from the diagram in Fig. 27, we obtain the diagram in Fig. 28. Since, by the rule-4(i) the diagram in Fig. 28 is equivalent to the diagram in Fig. 29, the final diagram obtained after eliminating curve M is in Fig. 29.

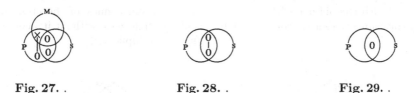

Fig. 27. .   **Fig. 28. .**   **Fig. 29. .**

Rule 6: "Any new Term-boundary can be inserted; and if it cuts every compartment already present, any interpretation desired may be assigned to it. Only where the new boundary passes through a compartment containing a cross the new boundary must pass through the cross, or what is the same thing a second cross connected with that already there must be drawn and the new boundary must pass between them, regardless of what else is connected with the cross. If the new boundary passes through a compartment containing a zero, it will be permissible to insert a detached duplicate of the whole connex of that zero, so that one zero shall be on one side and the other on the other side of the new boundary" [12, p. 044].

By Introducing the curve M, we get the diagram in Fig. 31 from the diagram in Fig. 30.

Fig. 30. .   **Fig. 31. .**

Rule-6 let us introduce a closed curve without changing the given information. In this particular rule Peirce mentioned what is to be happen to a cross or a zero after introducing a closed curve and there were several examples also. But in all of them there were no connected body of cross and zero together (see Fig. 32).

Fig. 32. .

So if we introduce the curve M in Fig. 32 which one of the following figures (Fig. 33 to Fig. 36) will we get as a result?

Fig. 33. A new cross was introduced and connected with the old one without taking the connected zero in consideration

Fig. 34. A new cross has been added with the old connex of the cross and only a single zero, without it's connex, has been duplicated

Fig. 35. Whole connex of the zero has been duplicated

Fig. 36. Connex of zero has been duplicated

Now both Fig. 34 and Fig. 35 are not a valid transformation from Fig. 32. By valid transformation we meant the validity notion used in Shin [18] or Venn$_{i_n}$ [4]. So we cannot get this two diagrams by using rule-6. Both Fig. 33 and Fig. 36 are valid. But the final figure that we will get from Fig. 32, by using rule-6, is Fig. 36. Figure 33 has been discarded since it doesn't precisely follow the conditions mentioned in the rule. While we introduce a new curve, a region having zero, is divided into two parts and each part should have a single detached zero. But this has not been done in this figure. Now in Fig. 36, everything mentioned in rule-6 has been followed. A new cross has been introduced and connected with the old connex of cross. Connex of zero has been duplicated— in this case we see two crosses for the connex of zero in the region ((P − S − M) ∪ ((P ∩ M) − S) ∪ (P ∩ S ∩ M)). This is because, while duplicating connex of zero, we find that the region containing cross has been divided in two parts. So two crosses have been introduced instead of one (like we did in Fig. 35).

Shin criticizes rule-5 and rule-6 saying that these rules do not exhaust all the possible cases as nothing about the existing '×' or the connected bodies of ×'s has been mentioned in them. Although this is true for rule-5 it is not so for rule-6. In rule-6 all the possible cases have been discussed. For rule-5, Peirce gave several examples in [12, pp. 046–047]. By examining these examples, we can say that after eliminating a curve, a cross or a connected body of crosses remains in the same region. But again it is not mentioned explicitly in the rule[8].

---

[8] Additional difficulties may appear when the number of closed curves increases, if the diagrams are not simple or reducible. Peirce occasionally used Venn diagrams for more than 3 curves. Some examples are found in [15]. On the construction of diagrams for n number of curves, see [6].

5   Comparison with Modern Diagrammatic Systems

We previously alluded to modern systems Venn I, Venn II and Venn$_{i_n}$ which are all based on Peirce's extended version of Venn diagrams [4,18]. Before proceeding, we need to mention that there are two more diagrammatic objects in Venn$_{i_n}$, 'names of individuals' and 'absence of individuals' (see [4]). If we exclude these objects then Venn$_{i_n}$ is similar to Shin's Venn-II system. From here onward, to ease the comparison with Peirce's rules, whenever we refer to Venn$_{i_n}$ we exclude the diagrammatic objects 'names of individuals' and 'absence of individuals' and anything regarding them. The main differences between these systems and Peirce's system are given in Table 1.

Table 1. Differences between the four systems

Primitive symbols	Peirce	Venn-I	Venn-II	Venn$_{i_n}$
Universe	Sheet of drawing	Rectangle ▭	Rectangle ▭	Rectangle ▭
Predicate	Closed curve ◯	Closed curve ◯	Closed curve ◯	Closed curve ◯
Emptiness	0	Shading	Shading	Shading
Non-emptiness	×	⊗	⊗	X
Disjunction	connecting 0's or ×'s or 0's and ×'s	connecting only ⊗'s	connecting ⊗'s or connecting two diagrams	connecting x's or connecting two diagrams

For simplicity, here onward we are going to use 'x' to represent 'non-emptiness' in Venn-I and Venn-II system also. Since the connecting line of Peirce (———) also connects diagrams in Venn-II and Venn$_{i_n}$, we have a new type of diagrams called compound diagrams (type-III diagrams for Venn$_{i_n}$ system [4]) where each of its components are called atomic diagrams (type-I or type-II diagrams for Venn$_{i_n}$ system. If a diagram consists of a single curve in a rectangle it is called a type-I diagram. If there are more than one curve then it is called a type-II diagram [4]). For example, the diagram in Fig. 37 is a compound diagram which represents the information 'Either All A are B and Some B are not A or Some A are not B and No A is B'. For Peirce's system we can represent this type of information of 'disjunctions of conjunctions' form by just converting the form into 'conjunction of disjunction'. So the corresponding diagram for Fig. 37 in Peirce's system is shown in Fig. 38. Generally, Peirce did not used any such compound diagrams but he did proposed an alternative way of representing a diagram when we deal with a complex form of information (see [12, p. 052]). There are no compound diagrams in the system Venn-I.

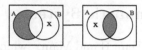

Fig. 37. .   Fig. 38. .

In the previous section we mentioned that rule-4 and rule-5 need modifications. Now suppose, we modify these rules accordingly, i.e.

(1) we remove the condition 'signs are not connected with any other signs in the other compartment' from rule-4(i) and (ii-*a*).
(2) For rule-4(ii-*b*), we add the condition that if 'it constitutes absurdity then anything follows'.
(3) For rule-4(iii), we add the condition that if two contrary signs are in the same compartment and only one of them is attached to some other sign, say R, in another region, then it is permitted to erase only the attached contrary sign and to keep the sign R as it is.
(4) For rule-5, after erasure of curve, the cross or connected body of cross will remain in the same position.

After this kind of modification it can be shown that Peirce's rules are adequate to perform any kind of transformations that are permitted in the other three systems. Table 2 shows which of Peirce's rules are analogous to the rules of the three systems.

For example, using types I-II diagrams, suppose we have the following two diagrams D_1 and D_2 (see Fig. 39). Now we get the diagram D_2 from D_1 by eliminating the x-node of the x-sequence in the region $(((A \cap B) - C) \cup ((A \cap C) - B) \cup (C - A - B) \cup ((B \cap C) - A))$ that falls in the shaded region $((A \cap C) - B)$ of the diagram D_1.

Fig. 39. .

In Peirce's system we get a similar transformation by using rule-4(iii) (see Fig. 40).

Table 2. Comparison of transformation rules among four systems

Types of Diagrams	Venn-I	Venn-II	Venn$_{i_n}$	Peirce
Type-I/II	The rule of erasure for Closed Curves	The rule of erasure for Closed Curves	Elimination Rules for Closed Curves	Rule-5
	The rule of erasure for Shading	The rule of erasure for Shading	Elimination Rules for Shading	Rule-1
	The rule of erasure for ⊗ or sequence of ⊗'s	The rule of erasure for ⊗ or sequence of ⊗'s	Elimination Rules for x or sequence of x's	Rule-1
	The rule of erasure of part of ⊗-sequence	The rule of erasure of part of ⊗-sequence	Elimination Rules for part of x-sequence	Rule-4(iii)
	The rule of spreading ⊗'s	The rule of spreading ⊗'s	Extension Rules for x-sequence	Rule-2
	The rule of introduction of basic regions (Closed Curves)	The rule of introduction of basic regions (Closed Curves)	Introduction Rules for Closed Curves	Rule-6
	The rule of conflicting information (Classical Explosion)	The rule of conflicting information (Classical Explosion)	Inconsistency Rules (Classical Explosion)	Rule-4(ii-*b*)
	The rule of unification of diagrams	The rule of unification of diagrams	Unification Rules	Rule-3
	N.A. As the universe can be either empty or non-empty	N.A. As the universe can be either empty or non-empty	Introduction Rules for x's	Rule-3
Type-III[a]	N.A. As there is no type-III diagrams	The rule of connecting diagram	Extension Rules for Diagrams	Rule-2
		The rule of splitting ⊗'s	Rules of Splitting Sequences	Not required here[b]
		The rule of the excluded middle	Rule of Excluded Middle	Rule-3
		The rule of conflicting information (Classical Explosion)for type-III diagram	Inconsistency Rules (Classical Explosion) for type-III diagram	Rule-4(iii)

[a] By rules for type-III diagrams, we only mean the rules using which type-I/II diagrams produce a type-III diagram.

[b] The rule of splitting sequences basically gives an equivalent type-III diagram of a type-I/II diagram. There is no change of information while using this rule. Thus it is not needed in Peirce's system, where we have only type-I or type-II diagrams.

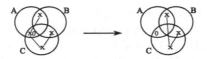

Fig. 40. .

Thus rule-4(iii) is analogous to the rule of erasure of part of ⊗-sequence in Venn-I and Venn-II or the elimination rule for part of x-sequence for the system Venn$_{i_n}$.

When we use type-III diagrams, we need an additional rule 'the rule of construction'. When dealing with such transformations in Peirce's case we need all together rule-4(i), rule-4(ii-*a*) and rule-2. For example, consider the following type-III diagrams, $D_1 - D_2$ and $D_3 - D_4$, in Fig. 41 and Fig. 42 respectively.

We get the diagram $D_3 - D_4$ from the diagram $D_1 - D_2$ through the following transformations (Fig. 43 to Fig. 45).

Fig. 41. .   Fig. 42. .

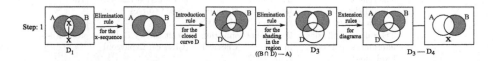

Fig. 43. .

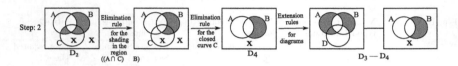

Fig. 44. .

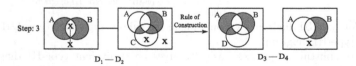

Fig. 45. .

In Peirce's system, by transforming the information from the form of 'disjunction of conjunction' to the form of 'conjunction of disjunction', we get corresponding diagrams of $D_1 - D_2$ and $D_3 - D_4$ in the Fig. 46 and Fig. 47 respectively. We get the diagram D_6 from the diagram D_5 by using the rules shown in Fig. 48.

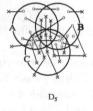

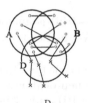

Fig. 46. .   Fig. 47. .

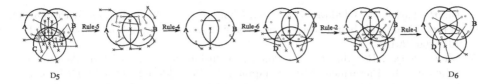

D_5 D_6

Fig. 48. .

6   Conclusion

In this paper, we exposed and discussed Peirce's set of rules for the transformation of Euler-Venn diagrams. We invoked its historical importance, then we identified the uses and shortcomings of each rule. We also considered the extent to which they 'survived' in modern diagrammatic systems.

Peirce himself conjectured that some of the rules may be simplified and some rules may have been overlooked. Such a task was more recently undertaken by Shin who argued that "(1) Some of the rules need to be clarified. (2) We need more rules to make this system complete. (3) Some semantic terminology (equivalence or absurdity) is used without clarification" [18, p. 35]. Our work partly corroborates Peirce's intuition and Shin's criticism. However, we demonstrate that only minor modifications are required. Moreover, such modifications were already implemented by Peirce in hi usage of the rules in the many examples that he provided. These examples were absent from Shin's account which was primarily based on the abridged transcription of manuscript MS 479 included in Peirce's *Collected Papers* [13].

A look at the original manuscript allowed us to return Peirce's original formulations and the modifications that his examples have suggested to him. Finally, we argued that slightly modified Peirce's rules are adequate to perform any kind of transformations that can be done to diagrams in the modern diagrammatic systems Venn-I, Venn-II and Venn$_{i_n}$.

Acknowledgment. We acknowledge with gratitude some fruitful suggestions given by Mihir Kumar Chakraborty and Ahti Pietarinen in the course of development of this research. The second author acknowledges support from TalTech internal grant SSGF21021.

References

1. Bellucci, F., Moktefi, A., Pietarinen, A.: Diagrammatic autarchy: linear diagrams in the 17th and 18th centuries. In: Burton, J., Choudhury, L. (eds.) First International Workshop on Diagrams, Diagrams, Logic and Cognition, CEUR Workshop Proceedings, vol. 1132, pp. 31–35 (2013). http://ceur-ws.org/Vol-1132/paper4.pdf
2. Bellucci, F., Pietarinen, A.V.: Existential graphs as an instrument of logical analysis: Part I. Alpha. Rev. Symb. Logic **9**(2), 209–237 (2016)
3. Bellucci, F., Moktefi, A., Pietarinen, A.V.: Simplex sigillum veri: Peano, Frege and Peirce on the primitives of logic. History Philos. Logic **39**(1), 80–95 (2018)

4. Bhattacharjee, R., Chakraborty, M.K., Choudhury, L.: Venn diagram with names of individuals and their absence: a non-classical diagram logic. Log. Univers. **12**(1), 141–206 (2018)
5. Bhattacharjee, R., Moktefi, A.: Peirce's inclusion diagrams, with application to syllogisms. In: Pietarinen, A.-V., Chapman, P., Bosveld-de Smet, L., Giardino, V., Corter, J., Linker, S. (eds.) Diagrams 2020. LNCS (LNAI), vol. 12169, pp. 530–533. Springer, Cham (2020). https://doi.org/10.1007/978-3-030-54249-8_50
6. Moktefi, A., Bellucci, F., Pietarinen, A.-V.: Continuity, connectivity and regularity in spatial diagrams for N terms. In: Burton, J., Choudhury, L. (eds.) DLAC 2013: Diagrams, Logic and Cognition, CEUR Workshop Proceedings, vol. 1132, pp. 31–35 (2014). http://ceur-ws.org/Vol-1132/
7. Moktefi, A.: Beyond syllogisms: carroll's (marked) quadriliteral diagram. In: Moktefi, A., Shin, S.-J. (eds.) Visual Reasoning with Diagrams, pp. 55–71. Birkhäuser, Basel (2013)
8. Moktefi, A.: Logic. In: Wilson, R.J., Moktefi, A. (eds.) The Mathematical World of Charles L. Dodgson (Lewis Carroll), pp. 87–119. Oxford University Press, Oxford (2019)
9. Moktefi, A.: Diagrammatic reasoning: the end of scepticism? In: Benedek, A., Nyiri, K. (eds.) Vision Fulfilled, pp. 177–186. Hungarian Academy of Sciences, Budapest (2019)
10. Moktefi, A., Bhattacharjee, R.: What are rules for? A Carroll-Peirce comparison. In: Diagrammatic Representation and Inference 12th International Conference, Diagrams 2021, LNAI. Springer (2021)
11. Peirce, C.S., Ladd-Franklin, C.: Symbolic logic. In: Baldwin, J.M. (ed.) Dictionary of Philosophy and Psychology, vol. 2, pp. 645–650. Macmillan, New York and London (1902)
12. Peirce, C.S.: On logical graphs. Houghton Library, Harvard University, MS 479. Accessible at: Peirce Archive (1903). https://rs.cms.hu-berlin.de/peircearchive/pages/search.php
13. Peirce, C.S.: Collected papers. In: Hartshorne, C., Weiss, P. (eds.) Harvard University Press, Cambridge, vol. 4. (1933)
14. Peirce, C.S.: Logic of the future. In: Pietarinen, A.-V. (eds.) 2, De Gruyter, Berlin (2020)
15. Pietarinen, A.-V.: Extensions of euler diagrams in peirce's four manuscripts on logical graphs. In: Jamnik, M., Uesaka, Y., Elzer Schwartz, S. (eds.) Diagrams 2016. LNCS (LNAI), vol. 9781, pp. 139–154. Springer, Cham (2016). https://doi.org/10.1007/978-3-319-42333-3_11
16. Pietarinen, A.-V., Bellucci, F., Bobrova, A., Haydon, N., Shafiei, M.: The blot. In: Pietarinen, A.-V., Chapman, P., Bosveld-de Smet, L., Giardino, V., Corter, J., Linker, S. (eds.) Diagrams 2020. LNCS (LNAI), vol. 12169, pp. 225–238. Springer, Cham (2020). https://doi.org/10.1007/978-3-030-54249-8_18
17. Roberts, D.D.: The existential graphs. Comput. Math. Appl. **23**(6–9), 639–663 (1992)
18. Shin, S.J.: The Logical Status of Diagrams. Cambridge University Press, Cambridge (1994)
19. Stapleton, G.: Delivering the potential of diagrammatic logics. In: Burton, J., Choudhury, L. (eds.) DLAC 2013: Diagrams, Logic and Cognition, CEUR Workshop Proceedings, vol. 1132, pp. 1–8 (2013). http://ceur-ws.org/Vol-1132/
20. Venn, J.: On the diagrammatic and mechanical representation of propositions and reasonings. Philos. Mag. **10**(59), 1–18 (1880)

Tractarian Notations

Francesco Bellucci$^{(\boxtimes)}$

University of Bologna, Via Azzo Gardino 23, 40122 Bologna, Italy
francesco.bellucci4@unibo.it

Abstract. In the *Tractatus* Wittgenstein presents two different notations for logic: the truth-tabular notation introduced at TLP 4.442, and the so-called N operator notation at TLP 5.502 (plus a third notation, the so called *ab*-notation, at TLP 6.1203). Gregory Landini (2007) has argued that both the truth-tabular notation and the N operator notation fulfill the Wittgensteinian ideal of having a language in which all and only logical equivalents have exactly one and the same expression. In this paper, I show that Landini's argument is mistaken, for it overlooks the crucial Tractarian distinction between truth-operation and truth-function.

Keywords: Ludwig Wittgenstein · Gregory Landini · Truth-tables · Notation · Logical equivalence

1 Tractarian Extensionality

At TLP 5.25–5.251 Wittgenstein distinguishes between a truth-operation and a truth-function. In the context of the sentential calculus, a proposition is a truth-function of elementary propositions (TLP 5) and is the result of truth-operations on elementary propositions (TLP 5.3). A truth-operation is the way in which a truth-function (a proposition) results from another truth-function; so '~ p' results from the application of the truth-operation '~' to the elementary proposition 'p'. Truth operations are iterative; thus '~~ p' is the result of two successive applications of the truth-operation '~' to the elementary proposition 'p'. One single truth-operation, the joint denial or Sheffer stroke, is capable of expressing all possible results of truth-operations on elementary propositions, i.e. the Sheffer stroke is a sole sufficient operator for the sentential calculus. A generalization of the Sheffer stroke to n elementary propositions is the so-called N operator, which Wittgenstein presents at TLP 5.5: 'every truth-function is a result of the successive application of the operation $(- - - - - \text{T})$ $(\xi, \ldots..)$ to elementary propositions'. The N operator generalizes the Sheffer stroke, which is a binary operation, to the simultaneous negation (represented by the left-hand parentheses) of n elementary propositions (represented by the right-hand parentheses). At TLP 5.502 Wittgenstein proposes to write 'N $\bar{\xi}$', where '$\bar{\xi}$' is a 'variable whose values are the terms of the expression in brackets, and the line over the variable indicates that it stands for all its values in the bracket'.

It is a central claim of the *Tractatus* that an operation does not characterize the *sense* of a proposition (TLP 5.25). The sense of a proposition is its agreement and disagreement with the possibilities of truth and falsity (or truth-possibilities) of elementary propositions (TLP 4.2); a proposition is the expression of agreement and disagreement with

© Springer Nature Switzerland AG 2021
A. Basu et al. (Eds.): Diagrams 2021, LNAI 12909, pp. 183–187, 2021.
https://doi.org/10.1007/978-3-030-86062-2_15

the truth-possibilities of elementary propositions (TLP 4.4); the expression of agreement and disagreement with the truth-possibilities of elementary propositions is the expression of its truth-conditions (TLP 4.431); therefore, the sense of a proposition is its truth-conditions, and the proposition expresses (i.e. shows, TLP 4.022) it, and expresses nothing else. An operation does not characterize the sense of a proposition because one and the same sense (agreement and disagreement with the truth-possibilities of elementary propositions) may be obtained by application of distinct truth-operations on elementary propositions. For example, '~ $(p \& \sim q)$' and '$p \supset q$' agree and disagree with the truth-possibilities of the elementary propositions 'p' and 'q' in precisely the same cases, i.e. they have the same truth-conditions, and thus the same sense. If '~' really characterized the sense of '~ $(p \& \sim q)$', it should equally characterize the sense of '$p \supset q$', because by the definition of sense '~ $(p \& \sim q)$' and '$p \supset q$' have the same sense. But no operation can be said to characterize the sense of a proposition in which it does not occur. Thus, '~' does not characterize the sense of '$p \supset q$', and therefore it neither characterizes the sense of '~ $(p \& \sim q)$', because the sense of these two propositions is the same (TLP 5.43).

2 Propositional Signs

Since the sense of a proposition is given by its truth-conditions, and since the expression of the truth-conditions of a proposition is the expression of its agreement and disagreement with the truth-possibilities of elementary propositions, the sign that expresses agreement and disagreement with the truth-possibilities of elementary propositions is a propositional sign, i.e. it expresses the truth-conditions of that proposition, or its sense. Thus a truth-table like that in Fig. 1, since it expresses, in the right-hand column, the proposition's agreement and disagreement with the truth-possibilities of the elementary propositions 'p' and 'q' given in the left-hand column, is a propositional sign (TLP 4.442). The propositional sign in Fig. 1 does not represent the truth-operation on elementary propositions from which the truth-function expressed in the right-hand column has been obtained. It only represents the *result* of some such truth-operation on elementary propositions. Truth-operations on elementary propositions which have one and the same truth-function as result are not distinguished in a truth-table. The reason is that since a truth-operation does not characterize the sense of a proposition, but only the result of truth-operations does, and this is the sense of the proposition, a propositional sign that should express the truth-operation by which a truth-function is obtained from elementary propositions would express something *foreign to the sense of the proposition*.

Landini (2007) takes Wittgenstein's goal to have been a notation in which all and only logical equivalent propositions have the same representation: "He hoped to demonstrate that a deductive calculus for logic can be supplanted by a representational system in which all and only logical equivalents have exactly one and the same expression. The representation of quantifier-free sentences in terms of their truth-conditions (or, alternatively, Venn's representation) offers just such a notation. As Wittgenstein sees matters, systems that employ different logical particles '&,' '∨,' ' ⊃,' '~,' etc., hide their formal ('internal') nature. Wittgenstein attempted to exploit the truth-table representation of propositions as evidence for his view that a proper representation would reveal that

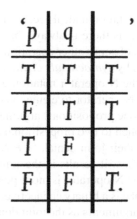

Fig. 1. A truth-table.

tautologies and contradictions are scaffolding." (Landini 2007, 124). In a notation for the sentential calculus in which the truth-table is the propositional sign, all and only logically equivalent propositions will be represented by the same propositional sign. In this notation, '$P \supset Q$', '$\sim (P \,\&\, \sim Q)$', '$(P \supset Q) \,\&\, (Q \lor \sim Q)$', '$\sim \sim (P \supset Q)$', etc. will have the same representation, namely the truth-table in Fig. 1.

Landini suggests, correctly in my opinion, that Wittgenstein hoped to extend this result to quantification theory with identity, and that he hoped to do it by means of the N operator. In fact, the truth-tabular method constitutes a decision procedure for sentential logic. Landini notes that Wittgenstain's project was doomed to failure in light of Church's later result that quantificational logic is undecidable. Here I want to focus on the sentential fragment, in order to understand whether Landini's idea is correct that the N operator notation is able to do what the truth-tabular notation does, i.e. to represent all and only propositional logical equivalents by the same propositional sign.

3   Tabular and Operational Notations

If the truth-tabular notation presented by Wittgenstein at TLP 4.442 is admitted as a legitimate notation for the sentential fragment of logic, we are at once provided with a means of distinguishing two different kinds of notations. I label them *tabular* and *operational*. Tabular notations are those in which only the *result* of a truth-operation on elementary propositions is represented. Operational notations are those in which truth-operations themselves are represented. Whenever an operation is represented, there is a possibility of representing the result of that operation by means of other operations, while when only the result of an operation is represented, this possibility is excluded.

The distinction between tabular and operational notations shows that, while it is correct that the truth-tabular notation that Wittgenstein presents at TLP 4.442 conforms to the notational ideal of having all and only logically equivalent propositions represented by the same sign, it is not the case that the N operator notation that Wittgenstein introduces at *TLP* 5.502 conforms to that ideal. In a nutshell, I want to argue that the N operator notation does not conform to the notational ideal because that notation is operational, not tabular, and in operational notations there is always the possibility of representing the result of one operation by means of some other operation, and therefore of representing logically equivalent propositions by *syntactically distinct* propositional signs.

Here is the explanation. In the N operator notation every truth-function is a result of successive application of the N operator to elementary propositions (TLP 5.5). The application of the N operator to one propositional argument 'P', represented as 'N(P)', results in its negation, '~ P'; applied to two propositional arguments 'P' and 'Q', represented as 'N(P, Q)', it results in their joint denial, '~ P & ~ Q'; applied to an arbitrary number of propositional arguments, it results in the joint denial of all the arguments (TLP 5.502). As I mentioned, the N operator is more powerful than the Sheffer stroke because whereas the Sheffer stroke is a binary connective, the N operator can be applied to any number of propositional arguments as the joint denial of all of them.

Now, Landini takes Wittgenstein to have thought that the N operator notation has a truth-tabular nature akin to the truth-tabular representation of propositional logic which Wittgenstein had presented at TLP 4.442, because in truth-tabular notation all and only logically equivalents are represented by the same sign. This cannot be quite right, however. In the N operator notation, 'NNN(P)' and 'N(P)' are logically equivalent but syntactically distinct formulas. The same is true, for example, of 'NN(P,Q)' and 'NN(Q,P)', which are logically equivalent but syntactically distinct. In order to make justice to Wittgenstein's claim concerning the N operator notation, Landini proposes – attributing the origins of these proposals to Wittgenstein himself – five equational 'rules of operation' (Landini 2007, 129–130) by means of which the equivalence of the representation of logically equivalent propositions is meant to be achieved:

(L1) $N(\xi_1, \dots \xi_n) = N(\xi_i, \dots, \xi_j), 1 \leq i \leq n,$ and $1 \leq j \leq n.$
(L2) $N(\dots\xi, \dots, \xi \dots) = N(\dots\xi, \dots).$
(L3) $N(\dots NN(\xi_1, \dots, \xi_n)\dots) = N(\dots\xi_1, \dots, \xi_n,\dots).$
(L4) $N(\dots N(\dots\xi, \dots, N\xi,\dots)\dots) = N(\dots).$
(L5) $NN(\gamma, N(\xi_1, \dots, \xi_n)) = N(N(\gamma, N\xi_1), \dots, N(\gamma, N\xi_n)).$

Clause (L1) corresponds to a generalization of the commutation rule '$\xi\gamma = \gamma\xi$'. Clause (L2) is a rule of elimination of equivalents. Clause (L3) corresponds to the rule of insertion and omission of double negation in whatever context it occurs ('$\xi = \sim \sim \xi$'). Clause (L4) allows us to delete or insert any argument of the form 'N($\dots\xi$, $\dots$, N$\xi,\dots$)', which is a tautology, in whatever context it occurs. Clause (L5) is a distribution rule. Landini takes Wittgenstein to have thought that (L1–5) make the N operator notation tabular: 'by application of (1)–(5) we can see how Wittgenstein thought that the N-operator recovers the features of truth-table representations' (2007, 130). Yet, Landini argues, (L1–5) are not to be taken as rules of logical equivalence in the proper sense: '[t]hese rules assert the sameness of certain *practices* of operation. They are not, therefore, identity statements' (*ibid.*).

It is far from clear in what sense (L1–5) are not rules of logical equivalence. A notation in which some rule concerning double negation is applicable is a notation in which in contexts like '$\xi = \sim\sim \xi$', syntactically distinct sentences appear on both sides of the '='. The '=' in fact states that syntactically distinct sentences are logically equivalent. On both sides of Landini's '=' in (L3), two syntactically distinct sentences in the N operator notation must appear. He says: 'N(NN(*p, q, r*) is to be regarded, *in some sense*, as the same as N(*p, q, r*)' (2007, 129). Now, the sense in which the former sentence is to be regarded as the same as the latter is that they are *logically* equivalent, not that they are *syntactically* equivalent. For were they syntactically equivalent, (L3) could not be applied, because (L3) states the logical equivalence of syntactically distinct sentences. One cannot apply it to sentences which are not syntactically distinct. By the same token, a notation in which a rule of commutation is applicable is a notation in which on the sides of '=' in '$\xi\gamma = \gamma\xi$' syntactically distinct sentences appear which the rule declares to be logically equivalent. And therefore on both sides of '=' in (L1) two distinct sentences in the N operator notation must appear.

Whenever a rule of logical equivalence of syntactically distinct sentences applies, it cannot be true that logical equivalent propositions have the same representation. It is not sufficient to do what Landini thinks Wittgenstein should do, namely to declare that these clauses are not rules of logical equivalence but rules of syntactical equivalence. Such a declaration is merely nominal and *cannot transform a rule of logical equivalence into a rule of syntactical equivalence.* Labeling them 'rules of practices of operation' rather than 'rules of logical equivalence' does not change the fact that they must succeed in capturing exactly what syntactically distinct sentences are logically equivalent. Landini is correct in saying that Wittgenstein's tabular notation fulfills the ideal of having all and only logical equivalents represented by the same sign. But he is wrong that the same is true of the N operator notation. In our terms, though they are equivalently expressive of the same fragment of classical logic, the truth-tabular notation of TLP 4.442 is tabular, whereas the N operator notation of TLP 5.502 is operational. Landini's (L1–5), being rules of logical equivalence, cannot turn an operational notation into a tabular one.

References

Landini, G.: Wittgenstein's Apprenticeship with Russell. Cambridge University Press, Cambridge (2007)

Wittgenstein, L.: Tractatus Logico-Philosophicus. Routledge & Kegan Paul, London (1922). Translated by C. K. Ogden

Equivalence Proof for Intuitionistic Existential Alpha Graphs

Arnold Oostra(⊠)📵

Universidad del Tolima, Ibagué, Colombia
noostra@ut.edu.co

Abstract. We give a formal proof of the mathematical equivalence between a proposed system of existential graphs and intuitionistic propositional calculus. Along the way, we obtain a new set of algebraic rules axiomatizing intuitionistic propositional logic.

Keywords: Existential graphs · Intuitionistic proposicional calculus · Strings

1 Introduction

Existential graphs were invented at the end of the 19th century by C.S. Peirce [10,11] and constitute a fully graphical version of classical logic. The system of Alpha graphs corresponds to the classical propositional calculus; Beta graphs are a graphical version of first-order logic with equality; Gamma graphs include, but are not limited to, modal logics and second-order logic. For a detailed account of existential graphs, see [12,15,16].

At the end of the 20th century, F. Zalamea strongly suggested the possibility of developing existential graphs for intuitionistic logic [14]. This is a formal system of symbolic logic that reflects the constructive principles of intuitionism [1,3,4]. Tarski pointed out that this logic is closely related to topology [13] and, on the other hand, the system of existential graphs is certainly a topological model for logic. Following Zalamea's anticipation, A. Oostra proposed a full system of intuitionistic existential graphs [5–7], followed by various other systems [8]. However, a complete proof of the equivalence of this graphical system and intuitionistic logic was lacking. Over the years, he developed with his students a proof method for this result [2,9].

In this paper, we outline a proof of the equivalence between the previously mentioned system of existential graphs and intuitionistic propositional calculus. In Sect. 2, we present the system of intuitionistic existential graphs at the Alpha, or propositional, level. The direction of the proof from the calculus to the graphs is straightforward, this is the content of the next section, where we also present intuitionistic logic. In Sect. 4 we introduce a linear presentation of intuitionistic Alpha graphs as strings and give an algebraic version of the rules of transformation. In the following section, we translate these strings back into traditional formulas and also attain the full equivalence. We finish in Sect. 6 with some conclusions.

© Springer Nature Switzerland AG 2021
A. Basu et al. (Eds.): Diagrams 2021, LNAI 12909, pp. 188–195, 2021.
https://doi.org/10.1007/978-3-030-86062-2_16

2   The System of Intuitionistic Alpha Graphs

This system uses the same Alpha graphs that Peirce used occasionally, but with a subtly different interpretation.

The components from which the intuitionistic Alpha graphs are built are:

- The plane surface without border upon which we draw all graphs, called the *sheet of assertion*;
- Propositions, symbolized by capital letters;
- Simple closed curves, called *cuts*;
- Curves called *scrolls*, and composed of two simple closed curves, one of them inside the other and the two intersecting at only one point;
- Curves called *double scrolls* and composed of three simple closed curves, two of them inside the other one – the inner curves do not touch each other at any point and each one intersects the outer curve at only one point.

We could also develop the system of intuitionistic existential graphs with multiple scrolls, containing any finite number of loops. Figure 1 shows the basic curves used in intuitionistic Alpha graphs.

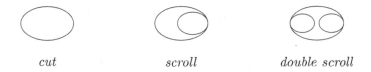

| cut | scroll | double scroll |

Fig. 1. The curves that can appear in intuitionistic Alpha graphs.

In both kinds of scrolls, the outer curve is called cut and the inner curves loops. The region limited by the cut and the loops is the outer area of the scroll and the interior of each loop is an inner area. An intuitionistic Alpha graph is defined as a diagram composed of a finite combination of letters, cuts, and scrolls drawn upon the sheet of assertion. There may be repeated letters but they all occupy different places. The cuts and scrolls do not touch the letters nor do they touch each other. Two graphs that can be continuously deformed into each other are equal.

At this point we adopt two additional conventions. Firstly: A simple cut enclosing a graph is an abbreviation of a scroll whose loop contains only an empty cut and whose outer area contains only the graph enclosed by the cut. Secondly: A double scroll with graphs in its areas is an abbreviation of a single scroll whose outer area contains only the same graph of the outer area of the double scroll, and whose loop contains only a double scroll with an empty outer area and the same loops as the original graph.

We denote $\mathcal{G}_L$ the set of all intuitionistic Alpha graphs composed with the letters of a given set L. We derive the interpretation of these graphs from the following clauses, which determine the difference with the usual graphs.

- The sheet of assertion is the universe of possibilities of truth.
- To draw a graph on the sheet means to assert its interpretation. To write a letter means to assert it. To write an empty cut means a contradiction.
- To draw two graphs means to assert both.
- To draw a scroll means to assert the implication whose antecedent is the graph in the outer area and whose consequent is the graph into the loop. Hence, by the first of our former agreements, to draw a graph enclosed in a cut without loops means to negate it.
- To draw a double scroll with an empty outer area means to assert the disjunction of the graphs enclosed in the two loops. Hence, to draw an arbitrary double scroll means to assert the implication whose antecedent is the graph in the outer area and whose consequent is the disjunction of the graphs into the loops.

From this we obtain the graphs for the primary propositional connectives, shown in Fig. 2.

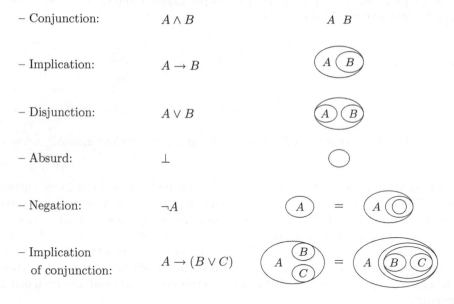

– Conjunction:	$A \wedge B$	$A\ B$
– Implication:	$A \rightarrow B$	
– Disjunction:	$A \vee B$	
– Absurd:	$\perp$	
– Negation:	$\neg A$	
– Implication of conjunction:	$A \rightarrow (B \vee C)$	

Fig. 2. Intuitionistic Alpha graphs for the basic connectives.

In general, an area is defined as a region of the sheet of assertion limited by curves, both cuts and loops. Topologically, this is a connected component of the complement of the curves. An area is even or odd if there is an even or odd number of curves around it, here we count both cuts and loops alike.

These are the rules of transformation allowed for intuitionistic Alpha graphs:

1 *Erasure.* In an even area, any graph may be erased. Any loop within an even area may be eliminated with its contents.
2 *Insertion.* In an odd area, any graph may be scribed. In an odd area limited externally by a cut, a loop containing any graph may be added to this cut.
3 *Iteration.* Any graph may be iterated in its own area, or in any area contained in this, that is not part of the graph to be repeated. Any loop may be iterated, with its contents, on its own cut.
4 *Deiteration.* Any graph may be erased if a copy of it persists in the same area or in any area around this. A loop with its contents may be erased if another loop with the same contents persists on its cut.
5 *Scrolling.* A scroll with empty outer area may be drawn around or removed from any graph on any area.

The reference to loops in rules **2** and **3** means that, in appropriate cases, we may transform a cut into a scroll, and a scroll into a double scroll. This is not applicable to double scrolls since we do not include multiple scrolls.

An intuitionistic Alpha graph G entails a graph H, which we denote $G \Rightarrow H$, if there exists a finite sequence of intuitionistic Alpha graphs $A_1, \ldots, A_n$ with $A_1 = G$, $A_n = H$ and such that each graph A_{i+1} follows from the previous graph A_i by applying any of the above rules, say **r**, which we denote $A_i \overset{\text{r}}{\Rightarrow} A_{i+1}$. Two graphs G, H are equivalent, and we write $G \Leftrightarrow H$, if $G \Rightarrow H$ and $H \Rightarrow G$.

3   From Formulas to Graphs

The basic connectives for intuitionistic propositional calculus (IPC) are implication $\rightarrow$, conjunction $\wedge$, disjunction $\vee$, and absurd $\bot$. Negation $\neg A$ is defined as $A \rightarrow \bot$. The set $\mathcal{F}_L$ of formulas in a set of letters L is defined inductively. The following formulas are taken as axioms for IPC [1,3]:

1. $A \rightarrow (B \rightarrow A)$
2. $(A \rightarrow (B \rightarrow C)) \rightarrow ((A \rightarrow B) \rightarrow (A \rightarrow C))$
3. $\bot \rightarrow A$
4. $(A \wedge B) \rightarrow A$
5. $(A \wedge B) \rightarrow B$
6. $(C \rightarrow A) \rightarrow ((C \rightarrow B) \rightarrow (C \rightarrow (A \wedge B)))$
7. $A \rightarrow (A \vee B)$
8. $B \rightarrow (A \vee B)$
9. $(A \rightarrow C) \rightarrow ((B \rightarrow C) \rightarrow ((A \vee B) \rightarrow C))$

The only inference rule is *modus ponens*: from formulas $A \rightarrow B$ and A we may proceed to B. A set of formulas Φ entails a formula F, which we denote $\Phi \vdash F$, if there exists a finite sequence of formulas $A_1, \ldots, A_n$ with $A_n = F$ and such that each formula A_i has the form of an axiom, or belongs to Φ, or follows from two previous formulas by *modus ponens*. A formula T entailed by the empty set, that is $\vdash T$, is called a theorem of IPC. Two formulas E, F are equivalent, and we write $E \approx F$, if $E \vdash F$ and $F \vdash E$.

Following the directions given in the previous section, we assign each formula F a unique graph $g(F)$. Transfer of the deduction relation is now derived from two facts. On the one hand, we can deduce the translation of each axiom from the empty sheet of assertion, Fig. 3 shows the proof of Axiom 1 as an example. Notice that in the first three steps we graphically prove theorem $A \to A$.

Fig. 3. Proof of Axiom 1 with intuitionistic Alpha graphs.

On the other hand, *modus ponens* corresponds to a graphical entailment, formally to $g(A \land (A \to B)) \Rightarrow g(B)$. Figure 4 shows a feasible proof.

Fig. 4. Proof of *modus ponens* with intuitionistic Alpha graphs.

Now we can remake every step of a deduction $E \vdash F$ graphically, and thus we obtain the following result.

Theorem 1. *If $E \vdash F$ then $g(E) \Rightarrow g(F)$, for any E, F formulas of IPC.*

4   A System of Strings

For the other direction of the sought equivalence, we introduce a formal theory for intuitionistic propositional logic. It has fewer symbols than the usual presentations, and its rules are inspired by the transformation rules of intuitionistic Alpha graphs. The alphabet comprises the chosen set L of propositional letters plus a constant $\top$ not in L, and two kinds of grouping marks: parentheses and square brackets. In the set of all finite sequences of this alphabet, we select the set $\mathcal{S}_L$ of strings by the following inductive clauses:

- Each letter of L is a string;
- the constant $\top$ is a string;
- $[\top]$ is a string;
- if S, T are strings then its juxtaposition ST is a string;
- if S, T are strings then $[S(T)]$ is a string;
- if S, T are strings then $[(S)(T)]$ is a string.

The interpretation of the strings as Alpha graphs is immediate: $[S(T)]$ is a scroll and $[(S)(T)]$ is a double scroll, while $\top$ can fill an empty area. In fact, every Alpha graph G translates into a string $s(G)$ in an (almost) unique way.

Next, we introduce a relation $\rhd$ in $\mathcal{S}_L$, here A, B, and C may be letters or arbitrary strings.

1. $AB \rhd BA$
2. $[(A)(B)] \rhd [(B)(A)]$
3. $A \rhd \top$
4. $AB \rhd A$
5. $A \rhd AA$
6. $A \rhd [(A)(B)]$
7. $[\top] \rhd A$
8. $[AB(C)] \rhd [AB(BC)]$
9. $A[(B)(C)] \rhd A[(AB)(C)]$
10. $A[AB(C)] \rhd A[B(C)]$
11. $[(A)(A)] \rhd A$
12. $[\top(A)] \rhd A$
13. $A \rhd [\top(A)]$

Now, a string S entails string T, which we denote $S \blacktriangleright T$, if there exists a finite sequence of strings $A_1, \ldots, A_n$ with $A_1 = S$, $A_n = T$ and such that $A_i \rhd A_{i+1}$ for each i. In fact, this is the smallest transitive relation on $\mathcal{S}_L$ that contains the relation $\rhd$. It is also reflexive by conditions 5 and 4, or 13 and 12. We require the following additional conditions of the relation $\blacktriangleright$.

1. If $A \blacktriangleright B$ then $AC \blacktriangleright BC$;
2. if $A \blacktriangleright B$ then $[B(C)] \blacktriangleright [A(C)]$;
3. if $A \blacktriangleright B$ then $[C(A)] \blacktriangleright [C(B)]$;
4. if $A \blacktriangleright B$ then $[(A)(C)] \blacktriangleright [(B)(C)]$.

In short, $\blacktriangleright$ is the smallest transitive relation on $\mathcal{S}_L$ that satisfies all seventeen clauses listed. The strings S, T are equivalent, and we write $S \blacklozenge T$, if $S \blacktriangleright T$ and $T \blacktriangleright S$.

With this relation $\blacktriangleright$ we can perform all Alpha transformation rules on strings. The meaning of the last four clauses is that if $S \blacktriangleright T$, then in any "even" position, we may substitute the string S by T, and T by S in any "odd" position. Therefore, for example, rules 3 and 4 allow us to erasure any graph in an even position and, simultaneously, to insert any graph in an odd position. A subtle inductive argument shows that all transformation rules are completely valid on strings [9]. Thus, we arrive at the following result:

Theorem 2. *If $G \Rightarrow H$ then $s(G) \blacktriangleright s(H)$, for any intuitionistic Alpha graphs G, H.*

5   From Graphs to Formulas

The way back from graphs to formulas now goes through the system of strings, which is in fact equivalent to the system of two-dimensional intuitionistic Alpha graphs. We achieve the translation of strings into formulas by means of a function $f : \mathcal{S}_L \to \mathcal{F}_L$ defined inductively as follows:

- $f(A) = A$ for every letter $A \in L$;
- $f(\top) = \bot \to \bot$;
- $f([\top]) = \bot$;
- $f(AB) = f(A) \wedge f(B)$,
- $f([A(B)]) = f(A) \to f(B)$;
- $f([(A)(B)]) = f(A) \vee f(B)$.

The proof in IPC of the seventeen rules stated in the previous section is possible, in fact most of them are straightforward. With this we arrive at the following result:

Theorem 3. *If $S \blacktriangleright T$ then $f(S) \vdash f(T)$, for any strings S, T.*

In general, the established correspondence is not bijective, since the composite function fs is not exactly the inverse of g and the same for the other cases. However, they are bijective "modulo equivalence", precisely, we have this result:

Theorem 4. $- fsg(F) \approx F$ for any formula F of IPC;
- $gfs(G) \Longleftrightarrow G$ for any intuitionistic Alpha graph G;
- $sgf(S) \blacklozenge S$ for any string S.

From this we conclude that the three ordered structures $\mathcal{F}_L/\approx$, $\mathcal{G}_L/\Longleftrightarrow$, and $\mathcal{S}_L/\blacklozenge$ are isomorphic, in fact all are isomorphic to the free Heyting algebra generated by L. Therefore, the three formal systems are logically equivalent and, in particular, the system of existential Alpha graphs is equivalent to intuitionistic propositional calculus.

6   Concluding Remarks

The rules listed in Sect. 4 constitute a novel presentation of intuitionistic propositional logic, given by thirteen rules and four quasi-rules. Although it is much longer than most of the usual versions, this axiomatization could be useful in some algebraic contexts.

On the other hand, the formulas–graphs–strings cycle that we have presented is also applicable to classical propositional logic and classical Alpha graphs. Furthermore, this method works for various sublogics of intuitionistic propositional calculus, such as implicative logic with conjunction, logics for which there thus are also systems of existential graphs in the style of C.S. Peirce.

References

1. Chagrov, A., Zakharyaschev, M.: Modal Logic. No. 35 in Oxford Logic Guides. Clarendon Press, Oxford (1997)
2. Fuentes, C.: Cálculo de secuentes y gráficos existenciales Alfa: Dos estructuras equivalentes para la lógica proposicional. Undergraduate thesis, Universidad del Tolima, Ibagué (Colombia) (2014)
3. Goldblatt, R.: Topoi. The Categorial Analysis of Logic. Elsevier, Amsterdam (1984)
4. Heyting, A.: Intuitionism. An Introduction. North-Holland, Amsterdam (1971)
5. Oostra, A.: Los gráficos Alfa de Peirce aplicados a la lógica intuicionista. Cuadernos de Sistemática Peirceana 2, 25–60 (2010)
6. Oostra, A.: Gráficos existenciales Beta intuicionistas. Cuadernos de Sistemática Peirceana 3, 53–78 (2011)
7. Oostra, A.: Los gráficos existenciales Gama aplicados a algunas lógicas modales intuicionistas. Cuadernos de Sistemática Peirceana 4, 27–50 (2012)
8. Oostra, A.: Representación compleja de los gráficos Alfa para la lógica implicativa con conjunción. Boletín de Matemáticas 26(1), 31–50 (2019)
9. Ortiz, J., Segura, J.: Gráficos Alfa intuicionistas. Undergraduate thesis, Universidad del Tolima, Ibagué (Colombia) (2018)
10. Peirce, C.S.: Collected Papers of Charles Sanders Peirce. Harvard University Press, Cambridge (1931–1958). 8 volumes
11. Peirce, C.S.: Logic of the Future. De Gruyter, Berlin (2019–2021). 3 volumes
12. Roberts, D.D.: The Existential Graphs of Charles S. Peirce. Mouton, The Hague (1973)
13. Tarski, A.: Der Aussagenkalkül und die Topologie. Fundamenta Mathematicae 31, 103–134 (1938)
14. Zalamea, F.: Pragmaticismo, gráficos y continuidad: hacia el lugar de C. S. Peirce en la historia de la lógica. Mathesis 13, 147–156 (1997)
15. Zalamea, F.: Peirce's Logic of Continuity. A Conceptual and Mathematical Approach. Docent Press, Boston (2012)
16. Zeman, J.J.: The Graphical Logic of C. S. Peirce. Ph.D. dissertation, University of Chicago, Chicago (1964)

Aaron Schuyler: The Missing Link Between Euler and Venn Diagrams?

Marcos Bautista López Aznar[1](✉), Walter Federico Gadea[1](✉),
and Guillermo Címbora Acosta[2](✉)

[1] University of Huelva, Huelva, Spain
[2] University of Sevilla, Sevilla, Spain

Abstract. The elementary idea of intersecting circles to express combinations of concepts was already present in Ramón Llull and Leibniz, while Venn is credited with the breakthrough that only with the intersection of two circles can we express all the required classes that arise by combining S and P in Boolean algebra. However, Aaron Schuyler published his popular manual *Principles of Logic for High Schools and Colleges* in 1869, ten years before Venn's work. Reading Venn's *Symbolic Logic,* we notice some similarities to this textbook, probably justified in the spirit of the age, as Venn never cited Schuyler. This American author made the same mistake as Venn in stating that the representation of the universal negative (E) was the only unambiguous diagram and shows how to represent the classes SP, $S\neg P$, $P\neg S$ and $\neg SP$ within the regions of the intersection of two circles. He also drew an original and novel diagram by eliminating the region between S and P from the Euler diagram that represents the universal negative proposition. Furthermore, their diagrams can be of great help in understanding the differences between Venn and Hamilton's proposals on the fundamental structure of propositions. Finally, Schuyler was one of the few 19th century authors who took the step of expressing propositions with a negative subject, something that did not make sense for Aristotelian logic, but it did for symbolic logic. For all these reasons, we consider Aaron Schuyler to be a good candidate for the missing link between Euler and Venn diagrams.

Keywords: Venn diagrams · Logic diagrams · History of diagrammatic notations

1 Schuyler's Role in the Venn-Hamilton Dispute

As we already know, during the 19th century there were many attempts to broaden the scope of the Aristotelian syllogism. In this sense, one of the most important tries was carry out by the doctrine of predicate quantification in the work of Hamilton, whose discussions with De Morgan motivated Boole to develop his theories [12]. With them, mathematical logic came into being, which would not have been possible without conceiving the proposition as the formal expression of an equality in which the functions of subject and predicate are interchangeable. According to Hamilton's doctrine of predicate quantification, a proposition expresses whether a part or the whole of the subject is the

© Springer Nature Switzerland AG 2021
A. Basu et al. (Eds.): Diagrams 2021, LNAI 12909, pp. 196–203, 2021.
https://doi.org/10.1007/978-3-030-86062-2_17

same or different from a part or the whole of the predicate, resulting in eight elementary propositions: X in toto = Y in toto; X in toto = Y ex parte; X ex parte = Y in toto; X ex parte = Y ex parte; X in toto ‖ Y in toto; X in toto ‖ Y ex parte; X ex parte ‖ Y in toto; X es parte ‖ Y ex parte (see Fig. 1.b). This version of the quantification of predicate became very popular, but Venn repeatedly criticized it in Symbolic Logic because Hamilton was undoubtedly still, in 1881, the biggest rival to Venn's own proposal. What we would like to highlight now is that some of the arguments used by Venn to criticize Hamilton were already present in Schuyler's work, who has defined himself as a debtor to Hamilton [17]. Aaron Schuyler was born in New York in 1828 and in 1875 became president of Baldwin University [11]. He wrote texts on Geometry, Algebra, Arithmetic and Logic, trying to establish the irreducible fundamental elements of each subject. For example, as Loomis said [14], he logically deduced twenty axioms of Geometry from only three fundamentals: Similarity, Equivalence, and Equality. The Schuyler diagrams that we will review in this paper first appeared in 1869 in *Principles of Logic for High Schools and Colleges,* a logic manual for students, accompanying a good part of the theoretical explanations. This manual was very popular in many colleges and universities in the United States [14] and perhaps it played a crucial role in the dispute over the fundamental structure of propositions that underlies inferences. Let us remember that in order to promote the development of new logics defined by formalism and mathematization, it was necessary to take the syllogism beyond the classical propositions: *All A is B* (A), *No A is B* (E), *Some A is B* (I), *Some A is not B* (O). In many cases, this discussion was carried out with diagrams.

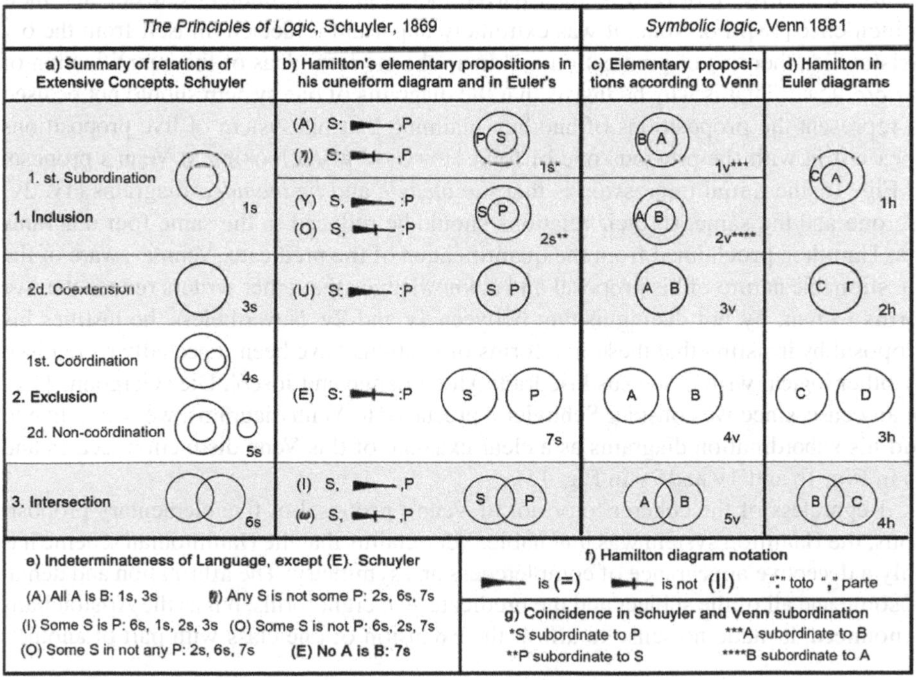

The Principles of Logic, Schuyler, 1869		Symbolic logic, Venn 1881	
a) Summary of relations of Extensive Concepts. Schuyler	b) Hamilton's elementary propositions in his cuneiform diagram and in Euler's	c) Elementary propositions according to Venn	d) Hamilton in Euler diagrams
1. st. Subordination 1. Inclusion — 1s	(A) S: ■▶— ,P / (ŋ) S: ■▶—+— ,P 1s* / (Y) S, ■▶— :P / (O) S, ■▶—+— :P 2s**	1v*** / 2v****	1h
2d. Coextension — 3s	(U) S: ■▶— :P S P	A B 3v	C Γ 2h
1st. Coördination — 4s / 2. Exclusion / 2d. Non-coördination — 5s	(E) S: ■▶—+— :P S P 7s	A B 4v	C D 3h
3. Intersection — 6s	(I) S, ■▶— ,P / (ω) S, ■▶—+— ,P S P	A B 5v	B C 4h

e) Indeterminateness of Language, except (E). Schuyler	f) Hamilton diagram notation
(A) All A is B: 1s, 3s ŋ) Any S is not some P: 2s, 6s, 7s	■▶— is (=) ■▶—+— is not (II) ":" toto "," parte
(I) Some S is P: 6s, 1s, 2s, 3s (O) Some S is not P: 6s, 2s, 7s	g) Coincidence in Schuyler and Venn subordination
(O) Some S in not any P: 2s, 6s, 7s (E) No A is B: 7s	*S subordinate to P ***A subordinate to B / **P subordinate to S ****B subordinate to A

Fig. 1. Elementary propositions of Schuyler, Venn and Hamilton [8, 17, 18].

Schuyler, having examined the work of Aristotle, Hamilton, and other important logicians of his time, reduces the elementary relations expressed in propositions to five [17], in a proposal clearly halfway between Aristotelian and class logic. The representation of these relations in diagrams without letters seems to us a great step in the attempt to formalize logic, which he defined as the science that deals with the formal laws of human thought [17]. The fundamental relations are those of inclusion (subordination and coextension), exclusion (coordinated and non-coordinated) and intersection (see Fig. 1a). He explains these relations using the concepts of species and genus, in what seems reminiscent of the Aristotelian paradigm. For example, the concepts of *Arabian horse* and *sheep* are not coordinated, because the *Arabian horse* is a species of the genus *horse*, while the *sheep* is not [17]. However, the concepts *horse* and *sheep* are coordinate, as species, to the genus *quadruped*.

Schuyler showed in his manual of logic the correspondence between the meaning of the cuneiform representations of the Scottish philosopher and the Leibniz-Euler diagrams. Now, Schuyler distinguishes two types of subordination, but only one type of universal exclusion (see Fig. 1b). Schuyler's comparison between the Hamilton and Euler diagrams clearly showed that there were three redundant figures in Hamilton (see Fig. 1b). In this way, Schuyler anticipated Venn's criticism of Hamilton by stating that his system had three superfluous forms [18].

Moreover, Hamilton had also argued that the fundamental relations expressed in any proposition could be represented using only four Euler diagrams [8] (See Fig. 1c). For his part, Venn reduces to five and only five the fundamental relations between classes that can be expressed by a proposition symbolically or diagrammatically: that of the inclusion of *A* by *B*, of *B* by *A*, their partial inclusion, their mutual exclusion and their coincidence [18]. For Venn, it was extremely important to detach himself from the old Aristotelian theories of predication and from the modern ideas of the quantification of the predicate. That is why he insists that the diagrams of one system should not be used to represent the propositions of another, claiming that his system of five propositions does not fit with the previous one of four. However, when looking at Venn's proposal in Fig. 1c, the initial impression is that the *include* and *be included* diagrams (1v, 2v) are one and the same, so their relations should be reduced to the same four diagrams that Hamilton proclaimed from the quantification of the predicate. Venn is aware of the questionable nature of his proposal and acknowledges that other writers reduce the five forms to four, by not distinguishing between 1v and 2v. Nevertheless, he justifies his proposal by insisting that these five forms of relations have been repeatedly recognised by other logical writers such as Friedrich Albert Lange and Joseph Diez Gergonne [18]. In any case, since we consider Schuyler a precursor to Venn diagrams, we would like to add his subordination diagrams as a clear example of this Venn distinction (see 1s and 2s in Fig. 1b and 1v and 2v in Fig. 1c).

Regardless of the coherence or not of Venn's proposal of five elementary propositions, the Hamilton system was untenable. Venn claims that the Hamiltonian scheme has only a deceptive appearance of completeness and symmetry; The affirmation and denial of some and all of the subject and the predicate give eight forms, but as the Aristotelians denounced, it made no sense to affirm the exclusion of one class with part of another, since in negative propositions the predicate is always taken universally. The quantification of the predicate postulated by Hamilton was also criticized by others authors and was finally defeated by class logic, which was represented by Venn diagrams in a simple and intuitive way. However, there was a second version of the quantification of the predicate, very similar to that maintained by Boole in 1854 [2], in which exclusion was not a primitive component of propositions themselves, but the basis of inferences in which the middle terms of two propositions are of different quality. For example, when formalizing the Aristotelian syllogism Camestres by quantifying the predicate we have: *toto A = part of B; toto C = part of non-B*; therefore *toto A = part of non-C* [1, 2]. Now, we must observe that the middle term *B* has a different quality in the premises (*B* versus *non-B*). This perspective allowed Charles Stanhope (1753–1816) and William Stanley Jevons (1835–1882) to build logic machines as early as the 19th century [9, 10, 15]. They considered that all propositions establish the equality of two terms, regardless of whether these terms are affirmative or negative (*No A is B = All A is part of non-B*). We must bear in mind that this way of proceeding avoids the inconsistency of affirming toto-partial exclusion. Finally, although Jevons reproached John Stuart Mill for the profound error into which he had fallen in underestimating the logical discovery of the quantification of predicate [10], class logic prevailed as the true paradigm of mathematical logic. Perhaps because exclusion seemed to be the most obvious relation between two terms in Euler diagrams.

2   A Shared Mistake: Does the Exclusion Really Have an Unambiguous Representation in the Euler Diagram?

Schuyler identifies ambiguity in language with the fact that a proposition can be represented in different diagrams (see Fig. 2e). According to him, diagrams are the only way to really communicate the meaning of a proposition precisely, that is, by pointing out what kind of extensional relation is established between two terms. The classical sentences A, E, I, O, are valid insofar as they conform to the structure of elementary relations. Due to the indefinite character of language, there are different ways in which each of these propositions can be represented, and this is a great obstacle in the development of the science of logic. Now, Schuyler states that *No S is P* (E) is the only representation that does not present ambiguity [17]. Schuyler's claim that the Euler diagram [6] representing the universal negative was the only unequivocal structure was in keeping with the deeply held 19th century belief that exclusion was a primitive component of propositions. In 1881, Venn will repeat the same mistake. For him, this proposition can only be diagrammed as exclusion in non-intersecting circles, in the same way that two separate circles can only mean the universal negative (E). Due to this supposed evidence, exclusion was taken as a constitutive element of logical propositions, contributing to the decline of predicate quantification. But none of them analysed the representation of E in the same way as the rest of the Euler diagrams. *No A is B* can be transformed into *All A is non-B*, so we have the same ambiguity problem that we had with *All A is B*. That is, *non-B* is undetermined in the universal negative (E) in the same way that *B* is undetermined in universal affirmative (A): *All non-B is A?* Euler diagrams do not give a precise answer to either A or E. But there is an important difference between *All A is some non-B* ($\neg A \lor$

B) and *All A is all non-B* (A $\veebar$ B). And although Venn diagrams do allow us to represent both propositions, Venn himself failed to understand the difference between inclusive and exclusive disjunctions [7]. On the other hand, the astonishing diagram in the next section suggests that Schuyler was aware of the ambiguity of exclusion diagram with respect to *non-P* and *non-S*.

3   No *Non-s* is *Non-p*: A Surprising Diagram

Schuyler analyses the universal negative (E) in the sentences *No S is P* and *No non-S is non-P*. In the second diagram of Fig. 2, Schuyler eliminates the circle that limits the region of *P* to express *All non-S is P* and *All non-P is S*.

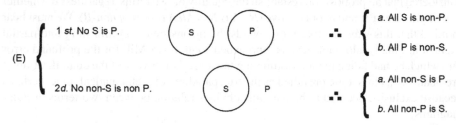

Fig. 2.   Inferences by means of privatives (E). Schuyler [17].

This is a very novel diagram for us, but it also seems difficult to interpret. In it we see that everything outside *S* (that is, *non-S*) is *P* because since *P* is not limited, now there is no region for *non-S_non-P*. Thus, it is correct that *No non-S is non-P*. On the other hand, we see that everything that is not *P* has to be *S*, because having eliminated the circle of *P* there is no area for *non-P_non-S*. Later, if we consider that within *S* only *non-P* is possible, then the representation also claims that the *PS* class has been removed. In this case, the diagram would amount to an exclusive disjunction. It seems likely that Schuyler took the idea of eliminating the *P* circle from W.D. Wilson, who had already represented the particular propositions I and O using cut circles, although without making them disappear entirely [20] (see Fig. 3). Therefore, we believe that Schuyler anticipated Venn's idea of representing propositions by removing regions from Euler diagrams. In addition, to carry out this operation, he had to consider negative terms such as *non-P* on a purely formal level. As we know, one of the limits of Aristotelian logic in the 19th century was its refusal to operate with negative terms such as *non-human* or *non-horse*, claiming that such terms lacked material truth. For the same reason, starting from two negative premises, a materially true conclusion could not be reached. In fact, some of the logicians who call themselves Aristotelians still today refuse to use infinite or indefinite terms under the same conditions as affirmative terms [16]. This topic was vehemently discussed in the Middle Ages, when some writers in the twelfth and thirteenth centuries adopted a principle called "conversion by contraposition": It states that *Every S is P* is equivalent to *Every non-P is non-S* and *Some S is not P* is equivalent to *Some non-P is not non-S* [4]. Later, in the 18th century, Ploucquet, anticipating the quantification of the predicate, admitted that two negative premises can generate a valid conclusion, which was an important advance in the development of symbolic logic [13]. But the definitive step towards the new mentality in the use of negative terms was taken by De Morgan and Boole. They were the first to utilise the concept of the universe of discourse to legitimise the so-called infinite terms.

Schuyler's treatment of negative terms seems to us halfway between Aristotelian conceptions and new forms of logical analysis. On the one hand, Schuyler converts propositions like Richard Whately (1787–1863), who supported the principles of opposition and obversion in his work: *Every S is P = Every non-P is non-S*, and *Some A is not B* is equivalent to *Some A is non-B*, and thus it converts to *Some non-B is A* [19]. On the other hand, Schuyler is less modern than authors such as Stanhope [9] or Lewis Carroll [3], who did admit syllogisms with two negative premises. Schuyler, like Venn, conceives, converts, and represents the classical propositions A, E, I, O using negative terms as subjects and predicates. However, none of them took the step of labelling the diagrams themselves with negative terms. Perhaps, both identify the class with the concept that affirms the presence of the quality or attribute that distinguishes it. But prioritizing the term *A* over *non-A* seems to us a psychological vestige from Aristotelian metaphysics [1, 5]. Despite this, the diagram Schuyler made to illustrate the transposition of E (see Fig. 2) deserve to be considered an advance that promoted reflection on how to exhaustively draw the mutually exclusive classes that make up Boole's universe of discourse.

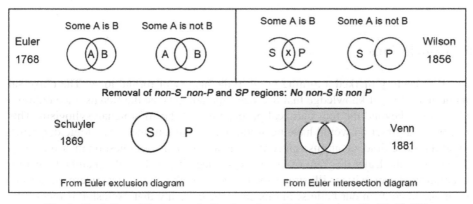

Fig. 3. Evolution of the removal of regions from Euler diagrams [6, 17, 18, 20].

4 How to Represent Four Classes Using Only Two Intersecting Circles

A diagram builder wondering how to represent exhaustively and exclusively all the possibilities of the universe of discourse could easily find the solution if he carefully scrutinised the diagrams of the particulars I, and O from the perspective of the privatives that Schuyler presents in Fig. 4. At this point, we would be closer to becoming aware that it is also possible to express the universal propositions A and E by eliminating regions of two intersecting circles in the way Venn did (see Fig. 3). It is obvious that Schuyler does not specify which region indicates each possibility, because it is obvious to anyone. For example, *non-S_non-P* can only be represented in the area outside both circles. Schuyler's exposition of the I and O diagrams from the perspective of the privative ones shows that the Euler diagram for particular propositions is enough to express all the possible combinations of two given classes *x* and *y*, and their contraries. However, Schuyler did not take the step of representing universal premises using two intersecting circles. The step was finally taken by John Venn, who was more familiar with Boole's work and the understanding of propositions as eliminating possibilities from the universe of discourse.

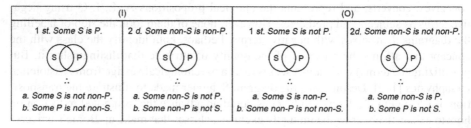

(I)		(O)	
1 *st. Some S is P.*	2 *d. Some non-S is non-P.*	1 *st. Some S is not P.*	2*d. Some non-S is not non-P.*
∴	∴	∴	∴
a. Some S is not non-P.	*a. Some non-S is not P.*	*a. Some S is non-P.*	*a. Some non-S is P.*
b. Some P is not non-S.	*b. Some non-P is not S.*	*b. Some non-P is not non-S.*	*b. Some P is not S.*

Fig. 4. Inferences by means of privatives (I), (O). Schuyler [17].

5 Conclusions

The 19th century marked a turning point in the history of formal logic. The universal mathematisation of knowledge that had been spreading since the Renaissance required taking logic beyond the four classical propositions of the Aristotelian syllogism. This process of abstraction would have been impossible without the use of logic diagrams, but at the same time, diagrams can confuse us when we do not question their meaning. Considering the Euler diagram of the universal negative as the only unambiguous representation of propositions illustrates the danger of making uncritical interpretations of an image to confirm our beliefs; in this case, the certainty that exclusion is one of the bases of the proposition. The Venn logic paradigm, which established class relations of exclusion, inclusion and intersection as the fundamentals of propositions, prevailed over the paradigm of quantification of the predicate, although the latter had allowed to create the first logical machines in history. Aaron Schuyler, in his book *Principles of Logic for High School and Colleges,* gives us some of the keys to understanding the criticisms that John Venn launched in *Symbolic Logic* against Hamilton and the defenders of quantification of the predicate. At the same time, his diagrams seem to us to be an intermediate link between the Leibniz-Euler and Venn diagrams. First, because they show the step of expressing propositions with a negative subject, which do not make sense for Aristotelian logic and do for symbolic logic. Second, because all classes that result from the combination of two variables are clearly expressed within two intersecting circles. Third, because the surprising and original diagram of *No non-P is non-S* puts into practice, perhaps for the first time, the suggestive idea that Leibniz-Euler diagrams can be manipulated in new ways to eliminate certain classes from the representation. In any case, the truth is that in the 19th century the use of diagrams and the idea of the universe of discourse were very widespread. It is quite possible that Schuyler, Venn, and other logicians were influenced by the same ideas of their time that ultimately served as the basis for both their mistakes and their achievements. There is probably no single link between the Euler and Venn diagrams. However, Schuyler diagrams seem to us a representative species of the historical context in which the turbulent passage from Aristotelian logic to formal logic was taking place. We believe that Aaron Schuyler deserves a place in the history of logic diagrams. His effort to graphically illustrate all the explanations in his logic manual is noteworthy. Furthermore, some of his diagrams seem absolutely original and deeply suggestive to us, and perhaps they were truly inspiring to others in their time.

References

1. Aznar, M.B.L.: Diagramas lógicos de Marlo para el razonamiento visual y heterogéneo: válidos en lógica matemática y aristotélica. University of Huelva, Spain (2020). http://rabida. uhu.es/dspace/handle/10272/19769
2. Boole, G.: An Investigation of the Laws of Thought, on which are Founded the Mathematical Theories of Logic and Probabilities. Dover, New York (1854)
3. Carroll, L.: The Game of Logic. Macmillan, London (1886)
4. De Rijk, L. M.: Logica Modernorum: the origin and early development of the theory of supposition, vol. 2. Van Gorcum, Assen (1967)
5. Descartes, R., Tweyman, S.: Meditations on First Philosophy: In Focus. Routledge, London (1993)
6. Euler, L.: Lettres à une Princesse d'Allemagne. Emile Saisset, Charpentier, Paris (1843)
7. Gardner, M.: Logic Machines and Diagrams. McGraw-Hill, New York (1958)
8. Hamilton, W.: Lectures on Metaphysics and Logic, vol. IV. William Blackwood and Sons, Edinburgh, London (1860)
9. Harley, R.: The Stanhope demonstrator. Mind 4(14), 192–210 (1879). https://doi.org/10.1093/ mind/os-4.14.192
10. Jevons, W.: The Substitution of Similars, The True Principle of Reasoning, Derived from a Modification of Aristotle's Dictum. Macmillan and Co., London (1869)
11. Kullman, D.E.: Aaron Schuyler 1828–1913. http://sections.maa.org/ohio/ohio_masters/sch uyler.html. Accessed 10 Apr 2021
12. Laita, L.M.: Influences on Boole's logic: the controversy between William Hamilton and Augustus De Morgan. Ann. Sci. 36(1), 45–65 (1979). https://doi.org/10.1080/000337979002 00121
13. Lombraña, J.V.: Historia de la Lógica. Universidad de Oviedo, Gijón (1989)
14. Loomis, E.: Life and appreciation of Dr. Aaron Schuyler. Cornell University, USA (1936)
15. Mays, W., Henry, D.P.: Jevons and logic. Mind 62(248), 484–505 (1953). https://doi.org/10. 2307/2964507
16. Oriol, M., Gambra, M.: Lógica Aristotélica. Dykinson, Madrid (2015)
17. Schuyler, M.A.: Principles of Logic for High School and Colleges. Wilson, Hinkle & Co., Cincinnati (1869)
18. Venn, J.: Symbolic Logic. MacMillan, London (1881)
19. Whately, R.: Elements of Logic. Harper and Brothers Publishers, New York (1853)
20. Wilson, W.D.: An Elementary Treatise on Logic: Designed for Use of Schools and Colleges as Well as for Private Study and Use. D. Appleton, New York (1856)

Validity as Choiceless Unification

Frank Thomas Sautter[1]($\boxtimes$) (iD) and Bruno Ramos Mendonça[2]($\boxtimes$) (iD)

[1] Universidade Federal de Santa Maria, Santa Maria, Brazil
`ftsautter@ufsm.br`
[2] Universidade Federal da Fronteira Sul, Erechim, Brazil
`bruno.ramos@uffs.edu.br`

Abstract. We propose a variant of Euler's Diagrammatic Method for Categorical Syllogistic. According to this variant, a categorical syllogism is valid without existential import if, and only if, there is one, and only one, way to unify the representations of the premises, and the representation of the conclusion follows from this unification.

Keywords: Euler · Venn · Peirce · Diagrammatic method · Syllogism

1 Introduction

Eulerian diagrams [2] constitute a paradigmatic example of graphic systems that depict logical relations between terms through relations of intersection and separation of closed figures in a plane. At least since Venn [10], it is well-known that Euler diagrams have three major representational disadvantages. First, these diagrams fail to meet a *maxim of quantity*: in general, they express more information than what is carried by a categorical proposition.[1] For instance, to express 'All As are Bs' with Euler diagrams we need to draw a closed figure denoting A inside a closed figure denoting B, but we cannot do this without overspecifying the relation between these terms. As a consequence of this fact, Euler diagrams do not accept *information integration*: to add the representation of a new proposition to an already set up diagram, we need to modify the previous construction to make it compatible with the new information [10, pp. 113-114]. Due to this lack of information integration, to demonstrate the validity of an argument through Euler diagrams, we need to carry out a proof by cases considering several alternative diagrammatic constructions of the same set of premises. In sum, as a final disadvantage, Euler's system does not satisfy *information unification*: in general, there are multiple ways of jointly depicting a set of premises with these diagrams and, consequently, the assessment of the validity of an argument often requires scanning each one of these possibilities.

To solve these problems, Venn proposed to distinguish the representation of terms from the representation of propositions [10, pp. 103 ff.]. In Venn diagrams,

---

[1] We are here exploring an analogy with Grice's *maxim of quantity*: 'Make your contribution as informative as is required (for the current purposes of the exchange) [and] do not make your contribution more informative than is required' [3, p. 45].

© Springer Nature Switzerland AG 2021
A. Basu et al. (Eds.): Diagrams 2021, LNAI 12909, pp. 204–211, 2021.
https://doi.org/10.1007/978-3-030-86062-2_18

a structure of partially overlapping closed figures represents the set of considered terms. Then, we represent propositions through the insertion of special marks in specific parts of the diagram. In this way, Venn offered a diagrammatic system that guarantees the possibility to integrate and unify distinct pieces of information. Notwithstanding, Venn and Euler's diagrams are different kinds of graphic representations. Therefore, we should ask: is it possible to provide an *Eulerian* solution for Euler diagrams' representational problems? Perhaps we can offer at least a partially positive answer to this inquiry. Further exploring a proposal outlined by C. S. Peirce, we will see that we can offer an enhancement of Euler diagrams that, although still does not meet the maxim of quantity and information integration, at least enjoys information unification.

In some recently rediscovered manuscripts, Peirce sketched a modification of Euler diagrams that promises an Eulerian partial solution for their representational issues. Figure 1(a) shows the usual graphical representation of a general term, while Fig. 1(b) shows Peirce's novelty (See [6, p. 282] and [8, p. 934]). Peirce adopts the following convention: The enclosure inwardness in Fig. 1(a) represents a positive term, while the enclosure inwardness in Fig. 1(b) represents a negative term[2]. Peirce uses both types of representation at the same time, although one of them suffices[3].

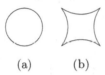

(a)   (b)

Fig. 1. Types of enclosure inwardness: (a) concave and (b) convex.

Peirce distinguishes universal syllogisms [6, p. 283] (i.e., syllogisms with universal propositions only), particular syllogisms (i.e., syllogisms with universal and particular propositions) [6, p. 283], and spurious syllogisms [6, p. 284] (syllogisms with particular propositions only). Universal categorical propositions are represented as total inclusions in his treatment of the syllogistic, although he also briefly discusses their representation as total exclusions [8, p. 934]. The present paper is organized as follows. Section 2 reconstructs Peirce's treatment of universal syllogisms as trios of total inclusions, and Sect. 3 treats them as trios of total exclusions. Section 4 treats particular syllogisms differently from Peirce's approach to them[4]; it is an original, although a natural, approach to

---

[2] Peirce's novelty was recently examined by Bhattacharjee and Moktefi [1], Moktefi and Pietarinen [5], and Pietarinen [9].

[3] Peirce [4, p. 1410] distinguishes two purposes of a system of logical symbols: if a system is devised for the investigation of logic, it requires economy of kinds of sign [7, p. 222-223]; and, if a system is devised as a calculus, no such a restriction is required. Thus, Peirce's novelty is not aimed at foundational studies.

[4] His treatment is akin to Euler's treatment.

particular syllogisms. A notion of validity[5] as choiceless unification underlies the treatment of universal syllogisms as trios of total inclusion and our novel treatment of particular syllogisms.

2　Universal Syllogisms as Trios of Total Inclusion

If negative terms are allowed, there are four types of universal categorical propositions[6]. Figure 2(a) represents 'All X is Y', Fig. 2(b) 'All non-X is Y' ('Each thing is X or Y')[7], Fig. 2(c) 'All X is non-Y' ('No X is Y'), and Fig. 2(d) 'All non-X is non-Y' ('All Y is X')[8].

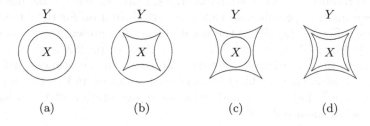

Fig. 2. Universal categorical propositions as total inclusions.

　　In the context of negative terms, there are eight types of valid universal syllogisms expressed as relations of total inclusion[9]. Figure 3(a) displays BARBARA, Fig. 3(c) CELARENT and Figs. 3(e) and 3(g) 'disjunctive syllogisms'.

　　The other four types of valid universal syllogisms are reducible[10] to the above ones by applying the rules of immediate inference given in Fig. 4.

　　In what follows, by unification we mean the obtainment of a figure that contains the representations of the three terms and preserves the mutual relations among them depicted in the premises. To obtain a unification, the middle term must have similar representations, in a geometric sense, in both premises. If this is not the case, we need to apply rules of immediate inference to produce such a situation[11]. Figure 5(a) shows the choiceless unification of the representations of

---

[5] We deal exclusively with terms without existential import. The validity dependent on existential import can also be treated by our approach, but some cases, such as DARAPTI, FELAPTON and FESAPO, require special rules.

[6] Peirce refers to them as 'assertions of non-existence' [6, p. 282].

[7] Thus, the introduction of negative terms expands the expressive power of Categorical Syllogistic.

[8] Peirce [6] does not mention this last assertion of non-existence.

[9] It is not difficult to prove that these eight types are the only valid types of universal syllogisms.

[10] As the rules in Fig. 4 express relations of logical equivalence, no priority is claimed to the first four syllogisms, at least from a logical point of view.

[11] This procedure will be better explained and exemplified in the case of particular syllogisms.

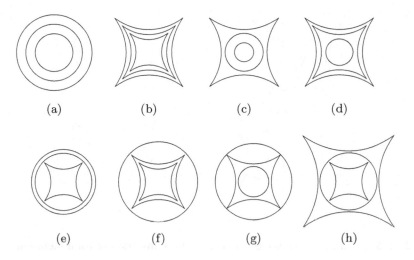

Fig. 3. Valid universal syllogisms as trios of total inclusion.

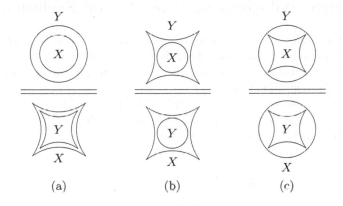

Fig. 4. Rules of immediate inference for total inclusion.

'All X is Y' and 'All Y is non-Z'. In this example, there is only one way in which we can unify the diagrams of the premises, and, consequently, there is no choice to be made. On the other hand, Fig. 5(b) illustrates the choiceful unification of the representations of 'All X is non-Y' and 'All Y is Z'. At the top, we have the representations of the two unifications compatible with the premises.

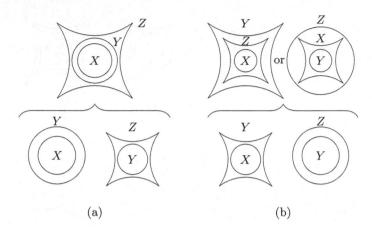

Fig. 5. Examples of choiceless and choiceful unification of total inclusions.

3   Universal Syllogisms as Trios of Total Exclusion

Universal syllogisms, represented as trios of total exclusion, do not allow the testing of validity by unification. Therefore, this section will be brief, aiming only to show the existence of a second possibility of homogeneous representation and testing of universal syllogisms.

Figure 6(a) represents 'All X is Y', Fig. 6(b) depicts 'All non-X is Y' ('Each thing is X or Y'), Fig. 6(c) exhibits 'All X is non-Y' ('No X is Y'), and Fig. 6(d) portrays 'All non-X is non-Y' ('All Y is X').

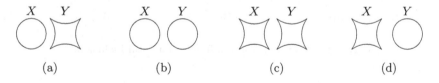

Fig. 6. Universal categorical propositions as total exclusions.

The Fig. 7(i) represents the valid universal syllogism represented in the Fig. 3(i)[12], for $i =$ a, b, c, d, e, f, g, h, in which the cutted enclosures represent the middle term[13].

---

[12] Each row in an item represents a premise.

[13] In a sense, universal syllogisms, represented as trios of total exclusion, also admit a type of unification, in which the middle term needs to be represented dissimilarly, in a geometric sense, in the premises, that is, it needs to be represented with a concave enclosure in one premise and with a convex enclosure in the other.

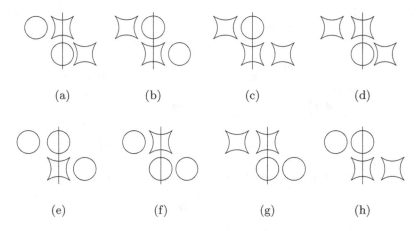

(a)   (b)   (c)   (d)

(e)   (f)   (g)   (h)

Fig. 7. Valid universal syllogisms as trios of total exclusion.

4   Particular Syllogisms

Figure 8(a) and Fig. 8(b) display Euler's solution [2, p.340] for the diagrammatic representation of the particular affirmative proposition 'Some A is B' and the particular negative proposition 'Some A is not B', respectively. If we preserve the interpretation of the closed figures given in the representation of universal propositions, then both diagrams have an excess of information, namely, both inform 'Some A is B' (or, equivalently, 'Some B is A'), 'Some A is not B', and 'Some B is not A'[14]. Figure 8(c) to Fig. 8(f) introduce an economical solution for the diagrammatic depiction of particular propositions. This proposal is also uniform with the representation of universal propositions. Figure 8(c) and Fig. 8(f) exhibit two ways of picturing the particular affirmative proposition 'Some A is B', and Fig. 8(d) and Fig. 8(e) furnish two ways of representing the particular negative proposition 'Some A is not B'[15].

In the unification process, in addition to the equivalences between Fig. 8(c) and Fig. 8(f) and between Fig. 8(d) and Fig. 8(e), we can use the *conversio simplex* of the particular proposition, showed in Fig. 9(a), and the following special rule, showed in Fig. 9(b): if there is an instance of a term X, depicted by a point at the top of Fig. 9(b), and all X is Y, represented by the concentric circles at the middle of Fig. 9(b), then *this instance* is also an instance of the term Y, as pictured at the bottom of Fig. 9(b).

---

[14] Euler is right in using indexes, a second type of representation for particular propositions, but he fails in the implementation.

[15] In accordance with the representation of universal propositions by total inclusion, we adopt Fig. 8(c) and Fig. 8(e) as primary and, only in very special situations in the unification process, we need to resort to Fig. 8(d) and Fig. 8(f).

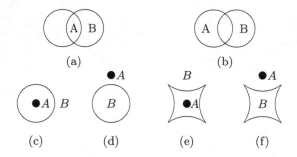

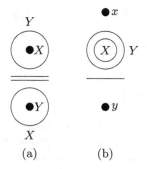

Fig. 8. Particular categorical propositions.

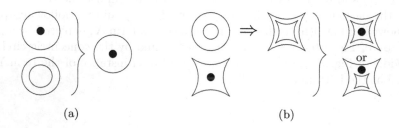

Fig. 9. Rules of immediate inference for partial inclusion.

Figure 10(a) exhibits the successful unification of the valid mode IA of the Third Figure of Categorical Syllogistic. Figure 10(b) shows the unsuccessful unification of the invalid mode AO of the First Figure of Categorical Syllogistic. The arrow indicate the application of the immediate rule of inference. It is unsuccessful because there are two unifications compatible with the premises, that is, it is a choiceful unification.

Fig. 10. Examples of choiceless and choiceful unification of particular syllogisms.

5   Conclusive Remarks

In this paper we argued that it is possible to offer an answer to Venn's criticism of Eulerian diagrams by exploring and further modifying an improvement, originally proposed by Peirce, of these representations. Against Venn, we have shown that, even though these new Eulerian diagrams are still hyper-informative and do not accept information integration, they now satisfy information unification, at least when it comes to the depiction of valid arguments. Some questions remain unanswered, though. For example, in opposition to what happens in the other cases, the use of rule expressed by Fig. 9(b) promotes a special kind of unification that cannot be reduced to a simply amalgamation of the representations of the premises. Is this really a heterogeneous case of unification? We still don't know it. In any case, this and other questions are a theme for future investigation.

References

1. Bhattacharjee, R., Moktefi, A.: Peirce's inclusion diagrams, with application to syllogisms. In: Pietarinen, A.-V., Chapman, P., Bosveld-de Smet, L., Giardino, V., Corter, J., Linker, S. (eds.) Diagrams 2020. LNCS (LNAI), vol. 12169, pp. 530–533. Springer, Cham (2020). https://doi.org/10.1007/978-3-030-54249-8_50
2. Euler, L.: Letters of Euler, Vol. I: On Different Subjects in Natural Philosophy Addressed to a German Princess. J. & J. Harper, New York (1833)
3. Grice, H.P.: Logic and conversation. In: Cole, P., Morgan, J.L. (eds.) Syntax and Semantics, Volume 3: Speech Acts, pp. 41–58. Academic Press, New York (1975)
4. Hartshorne, C., Weiss, P., Burks, A. (eds.): The Collected Papers of Charles Sanders Peirce (Electronic Edition). InteLex Corporation, Charlottesville (1994)
5. Moktefi, A., Pietarinen, A.-V.: Negative terms in Euler diagrams: Peirce's solution. In: Jamnik, M., Uesaka, Y., Elzer Schwartz, S. (eds.) Diagrams 2016. LNCS (LNAI), vol. 9781, pp. 286–288. Springer, Cham (2016). https://doi.org/10.1007/978-3-319-42333-3_25
6. Peirce, C.S.: On logical graphs [Euler and EGs]. In: Pietarinen, A.-V. (ed.) Logic of The Future: Writings on Existential Graphs. Volume 1: History and Applications, pp. 282–291. De Gruyter, Berlim (2020). https://doi.org/10.1515/978-3-110-6514-09_012
7. Peirce, C.S.: On logical graphs. In: Pietarinen, A.-V. (ed.) Logic of The Future: Writings on Existential Graphs. Volume 1: History and Applications, pp. 211–261. De Gruyter, Berlim (2020). https://doi.org/10.1515/978-3-110-6514-09_009
8. Peirce, C.S.: Definitions for Baldwins's dictionary. In: Pietarinen, A.-V. (ed.) Logic of The Future: Writings on Existential Graphs. Volume 3: Pragmaticism and Correspondence, pp. 922–949. De Gruyter, Berlim (forthcoming)
9. Pietarinen, A.-V.: Extensions of Euler diagrams in Peirce's four manuscripts on logical graphs. In: Jamnik, M., Uesaka, Y., Elzer Schwartz, S. (eds.) Diagrams 2016. LNCS (LNAI), vol. 9781, pp. 139–154. Springer, Cham (2016). https://doi.org/10.1007/978-3-319-42333-3_11
10. Venn, J.: Symbolic Logic. Macmillan, London (1881)

Truth Tables Without Truth Values: On 4.27 and 4.42 of Wittgenstein's *Tractatus*

Tabea Rohr[✉]

Archives Henri-Poincaré, Université de Lorraine, Nancy, France

Abstract. In 4.27 and 4.42 of his *Tractatus* Wittgenstein introduces quite complicated formulas, which are equivalent to 2^n and 2^{2^n}. This paper shows, however, that the formulas Wittgenstein presents fit particularly well with the way he thinks about truth values, logical connectives, tautologies, and contradictions. Furthermore, it will be shown how Wittgenstein could have avoided truth values even more radically. In this way it is demonstrated that the reference to truth values can indeed be substituted by talking of existing and non-existing facts.

1 Introduction

In his *Tractatus*, Wittgenstein presents a very useful diagrammatic device for logic: truth tables. Unlike Frege, however, Wittgenstein does not assume the independent objects "truth" and "falsehood." This paper aims to show how this philosophical position is reflected in Wittgenstein's formulas, which calculate the size and number of such tables, and in his truth tables themselves.

In 4.27, Wittgenstein presents a formula whose purpose is to calculate the possible combinations "with regard to the existence of n atomic facts"[1] (*"Bezüglich des Bestehens und Nichtbestehens von n Sachverhalten."* Note, however, that the word "existence," which occurs in both English translations[2] of the *Tractatus*, is not used in the German version):

$$K_n = \sum_{\nu=0}^{n} \binom{n}{\nu} \tag{1}$$

This formula is much more complicated than 2^n, which is equivalent to Wittgenstein's formula K_n.[3]

---

[1] Here, and henceforward, I quote the translation by Odgen [11], if not otherwise specified.

[2] Pears/McGuinness [9] translate the explanation of 4.27 as the "possibilities of existence and non-existence" for "*n* state of affairs.".

[3] This equivalence can easily be checked with the binomial theorem: $(x + y)^n = \sum_{k=0}^{k=n} \binom{n}{k} x^{n-k} y^k$. In the special case we now have $x = y = 1$, which yields: $2^n = (1+1)^n = \sum_{k=0}^{k=n} \binom{n}{k} 1^{n-k} 1^k = \sum_{k=0}^{k=n} \binom{n}{k} = K_n$.

© The Author(s) 2021
A. Basu et al. (Eds.): Diagrams 2021, LNAI 12909, pp. 212–220, 2021.
https://doi.org/10.1007/978-3-030-86062-2_19

As a reminder:

$$\sum_{\nu=0}^{n} \binom{n}{\nu} = \frac{n!}{0!(n-0)!} + \frac{n!}{1!(n-1)!} + \cdots + \frac{n!}{n!(n-n)!} \qquad (2)$$

Similarly, Wittgenstein calculates the number of possibilities "with regard to the agreement and disagreement of a proposition with the truth possibilities of n elementary propositions," which he later also calls "truth-conditions" [11, 4.431], in 4.42 with:

$$L_n = \sum_{\kappa=0}^{K_n} \binom{K_n}{\kappa} \qquad (3)$$

L_n is equivalent to 2^{2^n}.

In his *Companion to Wittgenstein's Tractatus*, Max Black only points out the equivalence of the formulas K_n and L_n with 2^n and 2^{2^n}, respectively ([1, p. 215 and p. 222]; see also: [2, 198] and [3, 116]). However, I will argue that Wittgenstein actually has good reasons to present K_n instead of 2^n, and L_n instead of 2^{2^n}: Firstly, in his formulas, Wittgenstein does not assume truth values, which he argues are not independent objects, but he only assumes atomic facts. Furthermore, in L_n, Wittgenstein distinguishes between the different numbers of possible truth conditions, thereby separating out the cases in which no or all possible truth conditions are chosen: contradiction and tautology.

In what follows, I will firstly explain the combinatorial approach which underlies the different formulas and thereby show that truth values are not considered in Wittgenstein's formulas. Secondly, I will show how L_n fits with Wittgenstein's attitude towards tautology and contradiction. Thirdly, I will argue that Wittgenstein also presents the truth table in 4.31 and 4.442 in the way he does because of his rejection of truth values as independent objects. Finally, I will follow Wittgenstein's approach to its logical end by introducing pure tables of atomic facts ("*Sachverhaltstabellen*"), or truth tables without truth values.

2   The Formulas Explained from a Combinatorial Point of View

$\binom{n}{\nu}$, a fragment of K_n, and k^n, the generalized version of 2^n, are the standard formulas for calculating two of the four standard tasks in combinatorics:[4]

permutation with repetition	permutation without repetition
combination with repetition	combination without repetition

k^n is the formula used to calculate the number of possible permutations with repetition. $\binom{n}{\nu}$ is the formula for calculating the number of possible combinations without repetition. Hence, with K_n and 2^n one calculates the number of possible

---

[4] In a combination, in contrast to a permutation, the order of selection does not matter.

combinations of existing atomic facts or, as it is usually expressed, truth values, in quite different ways.

With 2^n one calculates the number of possibilities to choose arbitrarily n times one of two truth values. This can be illustrated as an urn problem. Consider an urn of (black and red) truth-values balls. One now asks how many ways there are to take out (and put back immediately) a truth-value ball from the urn n times. In this situation, the order is important, because it matters which truth value is assigned to which elementary proposition.

Thus, in our urn problem, one asks for the number of possibilities for putting truth-value balls at places denoted by elementary propositions. In the following picture, for example, p_1 is true and all other elementary propositions are false.

$\binom{n}{\nu}$, which occurs in K_n, instead indicates the number of possibilities for choosing ν of n elementary propositions. Thus, the balls in our urn do not represent truth values but rather elementary propositions. From this urn, ν balls are chosen without repetition. When a ball representing an elementary proposition is chosen, this means that the elementary proposition is true or, as Wittgenstein puts it, that it expresses an atomic fact. When a ball representing an elementary proposition is not chosen, this means that the elementary proposition is false or, to put it in other words, that the atomic fact does not exist.

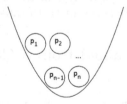

For $\nu = 1$, for example, there are n different possibilities. Choosing the ball labeled with p_1, for example, means that only p_1 is true and any other proposition is false. For $\nu = n$, however, there is only one possibility, namely to choose every ball. The balls taken from the urn do not have to be placed in a specific order. It is only significant which balls are chosen.

With K_n one calculates the sum of the possibilities for all ν from 0 to n.[5] Thus, truth values are not considered in K_n at all, but only the existence and non-existence of atomic facts.

That an atomic fact exists is, according to Wittgenstein, the meaning of the expression "a proposition is true":

> If the elementary proposition is true, the atomic fact exists; if it is false the atomic fact does not exist. [11, 4.25]

In 4.28, the paragraph just after 4.27, in which he presents his formula, Wittgenstein explains:

> To these combinations [of atomic facts to exist, and the others not to exist] correspond the same number of possibilities of the truth and falsehood of n elementary propositions. [11, 4.28]

However, Wittgenstein does not use the words "true" and "false" in order to denote independent objects:

> It is clear that to the complex of the signs "F" and "T" no object (or complex of objects) correspond. [11, 4.441]

Thus, instead of "the fact exists" one can say "the proposition is true," but not that "the proposition denotes the truth."[6]

2^{2^n} underlies the same combinatorial approach as 2^n, and L_n the same as K_n. 2^{2^n} can be illustrated as an urn problem as follows: take a ball representing a truth value and put it in a specific place, which is now not simply denoted by an elementary proposition anymore but by one of the 2^n truth value assignments. The picture on the next page, for example, illustrates that only one truth-value assignment is true, namely, that all elementary propositions are true. Thus, it represents an n-ary conjunction.

$\binom{K_n}{\kappa}$, which is part of L_n, indicates the number of possibilities for choosing κ of K_n possible combinations of elementary propositions. Thus, the balls in the

---

[5] A similar explanation of K_n is made by Morris [4], endnote 6 to chapter 5. Morris, however, does not contrast this to 2^n. Zalabardo [12, 187–188] points out that 2^n is the number of subsets of a set of n state of affairs. This is correct, but in this explanation it is hard to see why Wittgenstein did not choose 2^n as a formula in the first place.

[6] It is central to Wittgenstein's philosophy of logic that there are no logical objects or concepts. This sits in sharp contrast especially to Frege and also to Russell. (See [5, 59–60], [4, 205–206], [6, 52–62 and 86–93] and [3, 97–100]).

Frege first introduced the idea that propositions denote truth values, which are objects, and names of logical connectives denote functions, just like names of other concepts do. This idea is fundamental to Frege's logicism. Since for Frege mathematical propositions do have a content, and parts of mathematics are logical in nature, logic also must have a content (see [8], chapter 1). Wittgenstein in contrast already points out in his Notebook [10] in an entry dated 25 December 1914 that his "fundamental idea" is "that the logical constants are not proxies". This is also expressed in 4.0312 of the *Tractatus*.

urn do not represent elementary propositions, as in $\binom{n}{\nu}$, but rather combinations of the elementary propositions. The choice of a ball representing a combination of elementary propositions represents the agreement with the existence of the facts described by the elementary propositions occurring in the combination, and with the non-existence of the facts described by the elementary propositions not occurring in the combination. That a ball is not chosen represents the respective disagreement. Since, according to sentence 1 and 2 of the *Tractatus*, the world is defined as all facts that exist, if there were n elementary propositions, one of the balls would represent the world.

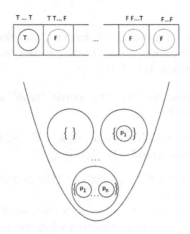

3   Tautology and Contradiction

Besides the different combination approaches, there is another important distinction between 2^{2^n} and L_n: while one calculates all possibilities at once with 2^{2^n}, with L_n, one calculates *separately* the possibilities for every number ("κ") of combinations of elementary propositions chosen. In this way, Wittgenstein already separates the "two extreme cases" [11, 4.46]: to choose no or all elementary propositions. He calls these cases "tautology" and "contradiction." They are considered in L_n by the last and the first summand, $\binom{K_n}{K_n}$ and $\binom{K_n}{0}$. Both summands are 1 for any result of K_n.

Hence, Wittgenstein's perspective of considering "the range which is left to the facts by the proposition" [11, 4.463] is already opened up by L_n. This angle leads him to the conclusion that tautology and contradiction cannot "in any way determine reality" [11, 4.463], because "[t]autology leaves to reality the whole infinite logical space; contradiction fills the whole logical space and leaves no point to reality" [11, 4.463]. In other words, to choose no or any combination of elementary propositions does not specify our world, which is the totality of existing facts.

4   Wittgenstein's Truth Tables

With the explanation of of formula L_n in mind it is possible to understand why the truth conditions are ordered the way they are in Wittgenstein's truth tables. In particular, the truth table for three elementary propositions in 4.31 seems quite chaotic or at least less reasonable than the standard order:

p	q	r
T	T	T
F	T	T
T	F	T
T	T	F
F	F	T
F	T	F
F	F	T
F	F	F

However, if one counts the occurrences of "T" in the table, one understands the idea behind Wittgenstein's order: Wittgenstein first lists the case that three elementary propositions are chosen, then the cases that two of three are chosen, then that one of three is chosen and, at the end, that no proposition is chosen. In the normal form this is mixed up.

The way Wittgenstein presents the truth table in 4.442 also fits with the way Wittgenstein presents his formulas. In this truth table, Wittgenstein only marks the possible combinations of atomic facts with "T" and leaves a blank for the impossible combination.

p	q	
T	T	T
F	T	T
T	F	
F	F	T

This seems quite strange. Taking into account that Wittgenstein eschews the assumption of truth values as independent objects, this notation becomes less puzzling: Wittgenstein does not use "T" and "F" to assign truth values. In the left columns, he uses "T" to mark the elementary propositions that describe a fact that exists, and "F" to mark the elementary propositions that describe a fact that does not exist. In the last column he uses "T" and "F" to mark the agreement and disagreement with the existence and non-existence of the atomic facts described by the elementary propositions. Since there are no other options besides agreement and disagreement, one can only use "T" in the last column to mark the agreement and stipulate that a blank in the last column expresses disagreement [11, 4.43]. This is just what Wittgenstein does when he only uses "T"s in the last column of the table in 4.422.

5   Truth Tables Without Truth Values

Wittgenstein could even have avoided using "T" and "F," which are usually considered to denote truth values, more consistently. In 4.31 he presents truth possibilities, which he had defined just before in 4.3 as "the possibilities of the existence and non-existence of the atomic facts."

Wittgenstein could have also just listed the different combinations of atomic facts. Thus, instead of the truth table in 4.31, he could have simply presented the following table:[7]

$$
\begin{array}{c|c}
p & q \\
\hline
 & q \\
\hline
p & \\
\hline
\end{array}
$$

This is a pure table of atomic facts ("*Sachverhaltstabelle*") or, to put it in other words, a truth table without truth values.

Avoiding signs for truth and falsehood in a truth table for a particular logical connective, such as the one introduced in 4.442, is more complicated. However, Wittgenstein introduces an alternative notation "if the sequence of the truth-possibilities is once for all determined" [11, 4.442], namely:

$$(TT–T) \ (p,q) \ \text{or} \ (TTFT) \ (p,q)$$

Such a linear notation can be easily transformed into one without truth values. One could only give a list of the possible combinations of atomic facts:

$$((p,q), \ (q), \ ())$$

In this notation, the sign for the conditional is completely substituted by brackets and commas. This illustrates Wittgenstein's remark that "[l]ogical operation signs are punctuations" [11, 5.4611].

---

[7] Black [1, 217] suggests that Wittgenstein's table in 4.31 could be read as follows:

$$
\begin{array}{ll}
p \ \text{and} & q \\
\text{not-}p \ \text{and} & q \\
p \ \text{and not-}q \\
\text{not-}p \ \text{and not-}q
\end{array}
$$

This would be in accordance with the "same-level interpretation," according to which "T and F serve merely as indicators of the positive and negative quality of the proposition signs to which they are attached" [1, 217]. Black assumes that this interpretation is correct [1, 218]. He contrasts it with the "new level" interpretation according to which "the T's and F's in the truth-table must [...] be understood to stand for truth and falsehood respectively" [1, 216]. Ricketts [7] also stresses that truth tables are "object language expressions" and not "metalinguistic devices."

To propose that p does not hold with "not-p," as Black suggests, is nevertheless kind of circular: One uses logical connectives within the truth table, and truth tables themselves, to express a logical connection.

If we now want to transform this into a truth table, we could just insert this possible combination into a third column and – just as in 4.442 – leave a blank for the impossible combinations. The truth table for the conditional would then look like this:

$$
\begin{array}{c|c|c}
p & q & (p, q) \\
\hline
p & & (p, \ ) \\
\hline
& q & \\
\hline
& & (\ , \ )
\end{array}
$$

This looks a bit artificial. It shows, however, that Wittgenstein's attempt to substitute talk of "true propositions" with talk of "existing facts" (*"bestehenden Sachverhalten"*) could be realized even more consistently. In combination with the reading of his formulas K_n and L_n presented in this paper this shows that "T" and "F" can in fact be perceived as mere abbreviations and not as names for objects *sui generis*. Thus, Wittgenstein managed to introduce truth tables without presupposing truth values.

Acknowledgments. I would like to thank the audience of the *UNILOG* in Vichy, 2018, the audience of the *Tractatus* conference in Vienna, 2018, Göran Sundholm, Wolgang Kienzler, and three anonymous referees for helpful comments.

References

1. Black, M.: A Companion to Wittgenstein's Tractatus, 5th edn. Cornell University Press, Ithaca (1992)
2. McGinn, M.: Elucidating the Tractatus. Wittgenstein's Early Philosophy of Logic and Language. Clarendon Press, Oxford (2006)
3. Milne, P.: Tractatus 5.4611: 'Signs for logical operations are punctuation marks'. In: Sullivan, P., Potter, M. (eds.) Wittgenstein's Tractatus. History and Interpretation, pp. 97–124. Oxford University Press, Oxford (2013)
4. Morris, M.: Wittgenstein and the Tractatus. Routledge, New York (2008)
5. Peterson, D.: Wittgenstein's Early Philosophy. Three Sides of the Mirror. University of Toronto Press, Toronto (1990)
6. Potter, M.: Wittgenstein's Notes on Logic. Oxford University Press, Oxford (2009)
7. Ricketts, T.: Pictures, logic and the limits of sense in Wittgenstein's Tractatus. In: Sluga, H., Stern, D. (eds.) The Cambridge Companion to Wittgenstein, pp. 59–99. Cambridge University Press, Cambridge (1996)
8. Rohr, T.: Freges Begriff der Logik. Mentis, Paderborn (2020)
9. Wittgenstein, L.: Tractatus logico-philosophicus. Routledge, London (1995). Translated by D. F. Pears and B. F. McGuinness
10. Wittgenstein, L.: Notebooks 1914–1916. Ed. by G. H. Wright with an English translation by G. E. M. Anscombe. Blackwell, Oxford (2004)
11. Wittgenstein, L.: Tractatus logico-philosophicus. Routledge, London (2005). Translated by C. K. Odgen
12. Zalabardo, J.: Representation and Reality in Wittgenstein's Tractatus. Oxford University Press, Oxford (2019)

Open Access This chapter is licensed under the terms of the Creative Commons Attribution 4.0 International License (http://creativecommons.org/licenses/by/4.0/), which permits use, sharing, adaptation, distribution and reproduction in any medium or format, as long as you give appropriate credit to the original author(s) and the source, provide a link to the Creative Commons license and indicate if changes were made.

The images or other third party material in this chapter are included in the chapter's Creative Commons license, unless indicated otherwise in a credit line to the material. If material is not included in the chapter's Creative Commons license and your intended use is not permitted by statutory regulation or exceeds the permitted use, you will need to obtain permission directly from the copyright holder.

Combining and Relating Aristotelian Diagrams

Leander Vignero[✉]

Centre for Logic and Philosophy of Science, KU Leuven, Leuven, Belgium
`leander.vignero@kuleuven.be`

Abstract. Combining and relating logical diagrams is a relatively new area of study in the community of people who work on Aristotelian diagrams. Most attempts now have been relatively *ad hoc*. In this paper I outline a more systematic research program inspired by a category-theoretic perspective. As concerns the logical diagrams, I will mainly focus on the Demey–Smessaert tradition.

Keywords: Aristotelian diagrams · Logical diagrams · Logical geometry · Category theory · Categorification · Square of opposition

1 Introduction

Even though the research interest in Aristotelian diagrams might be at a 500-year high, there is relatively little work on relating and combining Aristotelian diagrams. Two notable exceptions here are Pizzi [5] and Demey and Steinkrüger [3], but their work is not systematic.

In this paper, I propose and explore an approach to relating and combining Aristotelian diagrams based on category theory. Two central authors in the modern literature on Aristotelian diagrams are Demey and Smessaert who call their research program *logical geometry*. I will mainly elaborate on their research tradition, since it is both influential and mathematically sophisticated.

The remainder of this paper is organised as follows. In Sect. 2, we recap some basic definitions and discuss what structure-preserving maps between diagrams should look like. In Sect. 3, we use our morphisms to see what the right notions of coproduct and product of two diagrams comprehend. In Sect. 4, we consider a couple of examples. Finally, I propose some future topics within this research line in Sect. 5. Throughout the paper, we will use the diagrams as *visual proofs* that given functions are indeed well-defined morphisms.

I would like to thank Lorenz Demey for his valuable help and support. I would also like to thank Sylvia Wenmackers, Wouter Termont, and the BITSHARE-team. I acknowledge my doctoral fellowship of the Research Foundation Flanders (Fonds Wetenschappelijk Onderzoek, FWO) through Grant Number 1139420N.

© Springer Nature Switzerland AG 2021
A. Basu et al. (Eds.): Diagrams 2021, LNAI 12909, pp. 221–228, 2021.
https://doi.org/10.1007/978-3-030-86062-2_20

2 Relating Diagrams

Before we can continue, we should set the stage. We quickly restate a couple of definitions that can be found in Smessaert and Demey [6]. Let us start out with the logical relations.

Definition 1. *Logical diagrams and logical relations*
A logical diagram is a couple (F, B) where F is a fragment of the Boolean algebra B.[1] We say that $x, y \in B$ are in:

1. *B-bi-implication (BI_B) iff $x = y$,*
2. *B-left-implication (LI_B) iff $x < y$,*
3. *B-right-implication (RI_B) iff $y < x$,*
4. *B-contradictory (CD_B) iff $(x \wedge y) = 0_B$ and $(x \vee y) = 1_B$,*
5. *B-contrary (C_B) iff $(x \wedge y) = 0_B$ and $(x \vee y) < 1_B$,*
6. *B-subcontrary (SC_B) iff $(x \wedge y) > 0_B$ and $(x \vee y) = 1_B$,*
7. *B-unconnectedness (Un_B) iff none of the above holds.*

The relations 1–3 are called implication relations and relations 4–6 are called opposition relations. We denote the set of logical relations by $\mathfrak{R}$.

As mentioned above, Demey and Smessaert [2] have a formal mechanism for relating logical diagrams, their notion of Aristotelian isomorphism:

Definition 2. *Aristotelian isomorphism*
An Aristotelian isomorphism $f : (F_1, B_1) \to (F_2, B_2)$ is a bijection $f : F_1 \to F_2$ such that for all logical relations $R \in \mathfrak{R}$ and all $x, y \in F_1$ we have that $x R_{B_1} y$ iff $f(x) R_{B_2} f(y)$.

We can weaken this definition somewhat to get a notion of morphism. To do this, we consider that Smessaert and Demey impose an order relation $\leq_i$ on the set of logical relations [6]:

Definition 3. *The $\leq_i$ relation*
There is an informativity order $\leq_i$ on $\mathfrak{R}$ which is given by: $Un \leq_i LI$, $Un \leq_i RI$, $Un \leq_i C$, $Un \leq_i SC$, $LI \leq_i BI$, $RI \leq_i BI$, $C \leq_i CD$, and $SC \leq_i CD$.

We are now ready to formulate our new notion of morphism, which we will call infomorphisms:

Definition 4. *Infomorphisms*
An infomorphism $f : (F_1, B_1) \to (F_2, B_2)$ consists of a function (also denoted by) $f : F_1 \to F_2$ that satisfies the following condition: for all $x, y \in F$ it holds that if $x R_{B_1} y$, then $f(x) R'_{B_2} f(y)$ with $R \leq_i R'$.

---

[1] A good introduction to Boolean algebras can be found in Givant and Halmos [4].

It is easy to see that the logical diagrams with infomorphisms constitute a category, which we will denote by $\mathbb{A}_\mathbb{D}$.[2] Perhaps it is time for a couple of philosophical musings. We should care about infomorphisms because they relate logical diagrams in the following way: an infomorphism $f : (F_1, B_1) \to (F_2, B_2)$ is a way of identifying a subfragment of F_2 within the ambient space B_2 that has opposition-implication structure that is at least as informative as the structure induced by B_1 on F_1.

Example 1. Syllogistics and First-order logic
Consider the following example from the literature on logical diagrams. One can consider the fragment $F := \{\forall x(Sx \to Px), \forall x(Sx \to \neg Px), \exists x(Sx \wedge Px), \exists x(Sx \wedge \neg Px)\}$ in (the Lindenbaum–Tarski algebras of) both syllogistics (SYL) and first-order logic (FOL). We have an infomorphism

$$f : (F, FOL) \to (F, SYL) : x \mapsto x. \tag{1}$$

The visualizations in Fig. 1 speak for themselves: f is obviously well-defined, there is no need for algebraic manipulation. This works because syllogistics is a deductively stronger system than first-order logic. Indeed, syllogistics has the existential import axiom: $\forall x(Sx \to Px) \to \exists x(Sx \to Px)$.

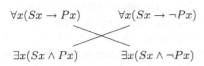

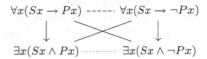

(a) Diagram of the fragment with first-order logic as the ambient system.

(b) Diagram of the same fragment with syllogistics as the ambient system.

Fig. 1. Arrows denote implications, dashed lines denote contrarities, dotted lines denote subcontrarities, full lines denote contradictions.

Example 2. S5 + modal collapse
Consider the following fragment $F := \{\Box p, \Box \neg p, \Diamond p, \Diamond \neg p\}$. We can now consider F within the following two ambient systems: the Lindenbaum–Tarski algebras of the modal language induced by $S5$ and $S5$ with modal collapse.[3] The diagrams are drawn in Fig. 2. By adding the modal collapse axiom, the diagram degenerates into a pair of contradictory propositions (PCD). Indeed, $\Box p$ is equivalent to $\Diamond p$, and $\Box \neg p$ is equivalent to $\Diamond \neg p$ under $S5$ + modal collapse. A glance at Fig. 2 teaches us that the following is an infomorphism:

$$g : (F, S5) \to (F, S5 + \text{modal collapse}) : x \mapsto x. \tag{2}$$

---

[2] Given that the use of category theory is relatively novel to most philosophers, I would like to point the interested reader to Awodey's superb primer [1].

[3] In philosophical logic, one calls the following axiom 'modal collapse': $\Diamond \varphi \to \Box \varphi$.

(a) F with ambient system S5.

(b) F with ambient system S5 + modal collapse.

Fig. 2. Arrows denote implications, dashed lines denote contrarities, dotted lines denote subcontrarities, full lines denote contradictions. We do not draw the equivalencies $\Box p \leftrightarrow \Diamond p$ and $\Box \neg p \leftrightarrow \Diamond \neg p$.

By focusing on structure-preserving maps, we can better appreciate the structure itself. This gets to the philosophical core of this paper: logical diagrams have their own mathematical structure, which we will refer to here as opposition-implication structure. At the highest level of abstraction, this is what philosophers (and particularly those who work in the Demey-Smessaert tradition) are studying. Furthermore, I claim that the infomorphisms are the straightforward way to study this opposition-implication structure. The examples and constructions throughout this paper serve to back up these claims.

It is interesting to note that the opposition-implication structure that is studied in logical geometry is in fact distinct from Boolean-algebra structure. That is, the infomorphisms between diagrams (F_1, B_1) and (F_2, B_2) do not correspond one-to-one with the Boolean-algebra homomorphisms between B_1 and B_2. In fact, this was already emphasized by Demey and Smessaert [2]. They provide an example of an Aristotelian isomorphism that does not give rise to a Boolean isomorphism. The precise relations between opposition-implication structure and Boolean-algebra structure are beyond the scope of this paper and will be explored in forthcoming work.

3   Combining Diagrams

Now that we have proposed a way to relate diagrams, category theory allows us to consider limit and colimit constructions. In particular, we will consider the universal mapping properties (UMPs) associated to the coproduct and product. This means that we will relate diagrams through a third diagram satisfying such a UMP. As a refresher, the UMPs are summarized in Fig. 3. We will now show the existence of coproducts and products in the category, $\mathbb{A}_{\mathbb{D}}$, of logical diagrams with infomorphisms.

Theorem 1. $\mathbb{A}_{\mathbb{D}}$ *has binary coproducts. Let* (F_1, B_1) *and* (F_2, B_2) *be Aristotelian diagrams. Then* $(F_1, B_1) + (F_2, B_2) := (F_1 + F_2, B_1 + B_2)$ *with*

$$F_1 + F_2 := \{(x, 1_{B_2}), (1_{B_1}, y) \mid x \in F_1, y \in F_2\}, \tag{3}$$

$B_1 + B_2$ *is the coproduct of* B_1 *and* B_2 *in the category of Boolean algebras,*

$$i_1 : (F_1, B_1) \to (F_1 + F_2, B_1 + B_2) : x \mapsto (x, 1_{B_2}), \tag{4}$$

and

$$i_2 : (F_2, B_2) \to (F_1 + F_2, B_1 + B_2) : y \mapsto (1_{B_1}, y) \tag{5}$$

as injections.

Proof. It is clear that i_1 and i_2 are infomorphisms. Moreover, $i_1(x)R_{B_1+B_2}i_1(y)$ iff $xR_{B_1}y$ for $x, y \in F_1$ by construction. Similarly, $i_2(x)R_{B_1+B_2}i_2(y)$ iff $xR_{B_2}y$ for $x, y \in F_2$. Furthermore, $(x, 1_{B_2})Un_{B_1+B_2}(1_{B_1}, y)$ for all $x \in F_1$ and $y \in F_2$. This also follows by construction.[4]

Now let $f : (F_1, B_1) \to (F, B)$ and $g : (F_2, B_2) \to (F, B)$ be infomorphisms. There is a unique candidate φ that might satisfy the UMP:

$$\varphi : (F_1, B_1) + (F_2, B_2) \to (F, B) : z \mapsto \begin{cases} f(x) \text{ if } z = (x, 1_{B_2}), \\ g(y) \text{ if } z = (1_{B_1}, y). \end{cases} \tag{6}$$

We only need to check that φ really is an infomorphism.

- $(a, 1_{B_2})$ and $(1_{B_1}, b)$:
  Given that $(a, 1_{B_2})Un_{B_1+B_2}(1_{B_1}, b)$ for all $a \in F_1$ and $b \in F_2$, it is clear that $Un \leq_i R$ where $\varphi(a, 1_{B_2})R_B\varphi(1_{B_1}, b)$.
- $(x, 1_{B_2})$ and $(y, 1_{B_2})$:
  Now, consider $(x, 1_{B_2})R_{B_1+B_2}(y, 1_{B_2})$, we need to show that $R \leq_i R'$ where $\varphi(x, 1_{B_2})R'_B\varphi(y, 1_{B_2})$. We have already remarked that $(x, 1_{B_2})R_{B_1+B_2}(y, 1_{B_2})$ iff $xR_{B_1}y$. Given that f is an infomorphism, we know that $R \leq R''$ where $f(x)R''_Bf(y)$. By definition of φ, we obtain that $R' = R''$ and therefore $R \leq_i R'$
- $(1_{B_1}, w)$ and $(1_{B_1}, z)$:
  Analogous to the previous case.

This concludes the proof.

The coproduct turns out to be a fragment that consists of the disjoint union of the original fragment and its ambient structure is such that it is strong enough to maintain the logical relations that were present within the original fragment, but weak enough not to impose logical relations between the 'elements' from the different fragments. One of the niceties is that we have also given an explicit construction of both the fragment and its ambient Boolean algebra.

Before we can work on the products, we first need the following lemma:

Lemma 1. *Suppose that (F_1, B_1) and (F_2, B_2) are Aristotelian diagrams. Let $x_1, y_1 \in F_1$ and $x_1 R_{B_1} y_1$, and $x_2, y_2 \in F_2$ and $x_2 R'_{B_2} y_2$. Then the relation R induced by $B_1 \times B_2$ between (x_1, x_2) and (y_1, y_2) is the meet of R and R', taken in $(\mathfrak{R}, \leq_i)$.*

---

[4] In this paper, we only consider cases where the fragments (as per usual) exclude the top and bottom elements of the algebra. If one wishes to include them, however, only the obvious and desired (!) relations are induced by the constructions. *Mutatis mutandis*, all the proofs yield the same well-behaved results —with a proviso in Lemma 1. The tweaks just involve the consideration of the special cases involving the top and bottom elements.

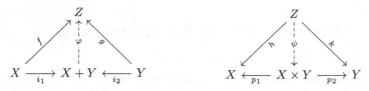

(a) UMP for the coproduct of X and Y.

(b) UMP for the product of X and Y.

Fig. 3. Let me quickly reiterate the meaning of the UMPs. (a) The fact that $X+Y$ is the coproduct of X and Y means the following: for any f and g there exists a unique map φ such that the diagram in subfigure (a) commutes. (b) The fact that $X \times Y$ is the product of X and Y means the following: for any h and k there exists a unique map ψ such that the diagram in subfigure (b) commutes.

Proof. This follows from a simple yet tedious case split. Let us consider an exemplary case: $aCD_{B_1}b$ and $cC_{B_2}b$. We need to show that $(a,c)C_{B_1 \times B_2}(c,d)$. Indeed, $C = CD \wedge C$. Clearly $(a,c) \wedge (b,d) = (0_{B_1}, 0_{B_2})$, since $a \wedge b = 0_{B_1}$ and $c \wedge d = 0_{B_2}$ *ex hypothesi*. Yet $c \vee d \neq 1_{B_2}$, since $cC_{B_2}d$. This entails that $(a,c) \vee (b,d) \neq (1_{B_1}, 1_{B_2})$. Hence $(a,c)C_{B_1 \times B_2}(b,d)$.

Theorem 2. $\mathbb{A}_\mathbb{D}$ *has binary products. Let* (F_1, B_1) *and* (F_2, B_2) *be Aristotelian diagrams. Then* $(F_1, B_1) \times (F_2, B_2) := (F_1 \times F_2, B_1 \times B_2)$. *Here,* $F_1 \times F_2$ *is the Cartesian product of* F_1 *and* F_2 *and* $B_1 \times B_2$ *is the product of* B_1 *and* B_2 *in the category of Boolean algebras. We also have*

$$p_1 : (F_1 \times F_2, B_1 \times B_2) \to (F_1, B_1) : (x,y) \mapsto x \tag{7}$$

and

$$p_2 : (F_1 \times F_2, B_1 \times B_2) \to (F_2, B_2) : (x,y) \mapsto y \tag{8}$$

as projections.

Proof. First we need to prove that p_1 and p_2 are in fact well-defined infomorphisms. Let us prove that p_1 is well-defined, the other case (p_2) is analogous. Proving that a p_1 is an infomorphism amounts to showing that the following holds:

if $(x_1, y_1)R_{B_1 \times B_2}(x_2, y_2)$, then $x_1 R'_{B_1} x_2$ with $R \leq_i R'$.

But this is exactly what follows from Lemma 1. Indeed, consider any $x_1, x_2 \in F_1$ and $y_1, y_2 \in F_2$ and $x_1 R' x_2$, then $R = R' \wedge R'' \leq R'$. It is now established that $((F_1 \times F_2, B_1 \times B_2), p_1, p_2)$ is a good candidate for a product.

Now let $f : (F, B) \to (F_1, B_1)$ and $g : (F, B) \to (F_2, B_2)$, we need to show that there is a unique function $\varphi : (F, B) \to (F_1 \times F_2, B_1 \times B_2)$ such that $f = p_1 \circ \varphi$ and $g = p_2 \circ \varphi$. There is only one candidate $\varphi := (f, g)$. We need to check that this is indeed an infomorphism. By hypothesis, f and g are infomorphisms, so for any $z_1, z_2 \in F$ with $z_1 R_B z_2$ we have that $f(z_1)R'_{B_1}f(z_2)$, $g(z_1)R''_{B_2}g(z_2)$ with $R \leq R'$ and $R \leq R''$. Let us investigate the relation between $(f,g)(z_1)$

and $(f, g)(z_2)$. Lemma 1 teaches us that $(f, g)(z_1)(R' \wedge R'')_{B_1 \times B_2}(f, g)(z_2)$. But given that $R \leq R'$ and $R \leq R''$, we obtain that $R \leq (R' \wedge R'')$. This shows that (f, g) is indeed an infomorphism. This concludes the proof.

4   Examples

Now that we have developed all these ways to relate diagrams, we should consider some examples. As mentioned above, the most well-known example is the square of opposition. So let us have a look at what happens if we take the coproduct and the product of two squares of opposition. To do this, we need only follow the constructions outlined in Theorem 1 and Theorem 2. The diagrams are outlined in Fig. 4 and Fig. 5 respectively.

Being able to draw the diagrams allows to analyze them easily. For instance, if one considers Fig. 4a, one sees that it consists of two disjoint squares of opposition. What is interesting is that the coproduct of $(\{\Box p, \Box \neg p, \Diamond p, \Diamond \neg p\}, S5)$ with itself is isomorphic to $(\{\Box p, \Box \neg p, \Diamond p, \Diamond \neg p, \Box q, \Box \neg q, \Diamond q, \Diamond \neg q\}, S5)$, as witnessed by Fig. 4b. Figure 4 clearly shows the Aristotelian isomorphism of two diagrams (F_1, B_1) and (F_1, B_2), does not entail the isomorphism of their ambient Boolean algebras B_1 and B_2. Similarly, it would require a lot of messy algebra to see that the diagram in Fig. 5 consists of two disjoint Buridan octagons [3]. Now, this can be spotted in a single glance!

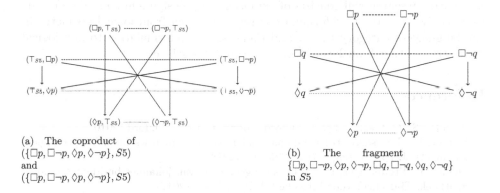

(a)   The   coproduct   of
$(\{\Box p, \Box \neg p, \Diamond p, \Diamond \neg p\}, S5)$
and
$(\{\Box p, \Box \neg p, \Diamond p, \Diamond \neg p\}, S5)$

(b)   The   fragment
$\{\Box p, \Box \neg p, \Diamond p, \Diamond \neg p, \Box q, \Box \neg q, \Diamond q, \Diamond \neg q\}$
in $S5$

Fig. 4. Two isomorphic Aristotelian diagrams that consist of two disjoint squares of opposition.

228   L. Vignero

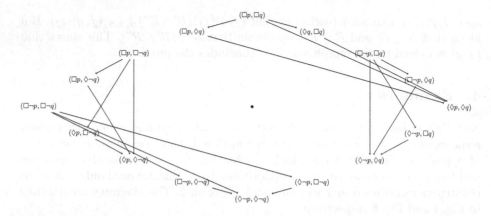

Fig. 5. The product of $(\{\Box p, \Box \neg p, \Diamond p, \Diamond \neg p\}, S5)$ and $(\{\Box q, \Box \neg q, \Diamond q, \Diamond \neg q\}, S5)$, two squares of opposition. Contradictory statements are obtained by reflection across the dot. The arrows are implications. Note that this product is isomorphic to two disjoint Buridan octagons [3].

5   Conclusion and Outlook

In this paper, we have proposed a type of morphism, the infomorphisms, that turns the class of logical diagrams into an apparently well-behaved category, $\mathbb{A}_\mathbb{D}$. Future research will consists of exploring these insights in more detail. Particularly, we will try to reformulate existing research from logical geometry in the language of category theory and then use this reformulation to link logical geometry to other branches of logic and computer science.

References

1. Awodey, S.: Category Theory. Oxford University Press, Oxford (2010)
2. Demey, L., Smessaert, H.: Combinatorial bitstring semantics for arbitrary logical fragments. J. Philos. Log. **47**(2), 325–363 (2018)
3. Demey, L., Steinkrüger, P.: De logische geometrie van johannes buridanus' modale achthoek. Tijdschrift voor Filosofie **79**(2), 217–238 (2017)
4. Givant, S., Halmos, P.: Introduction to Boolean Algebras. Springer, New York (2009). https://doi.org/10.1007/978-0-387-68436-9
5. Pizzi, C.: Generalization and composition of modal squares of oppositions. Log. Univers. **10**(2), 313–325 (2016)
6. Smessaert, H., Demey, L.: Logical geometries and information in the square of oppositions. J. Logic Lang. Inform. **23**(4), 527–565 (2014)

Residuation in Existential Graphs

Nathan Haydon[1(✉)] and Ahti-Veikko Pietarinen[1,2]

[1] Tallinn University of Technology, Tallinn, Estonia
{nathan.haydon,ahti.pietarinen}@taltech.ee
[2] Research University Higher School of Economics, Moscow, Russia

Abstract. Residuation has become an important concept in the study of algebraic structures and algebraic logic. Relation algebras, for example, are residuated Boolean algebras and residuation is now recognized as a key feature of substructural logics. Early work on residuation can be traced back to studies in the logic of relations by De Morgan, Peirce and Schröder. We know now that Peirce studied residuation enough to have listed equivalent forms that residuals may take and to have given a method for arriving at the different permutations. Here, we present for the first time a graphical treatment of residuation in Peirce's Beta part of Existential Graphs (EGs). Residuation is captured by pairing the ordinary transformations of rules of EGs—in particular those concerning the cuts—with simple topological deformations of lines of identity. We demonstrate the effectiveness and elegance of the graphical presentation with several examples. While there might have been speculation as to whether Peirce recognized the importance of residuation in his later work, or whether residuation in fact appears in his work on EGs, we can now put the matter to rest. We cite passages where Peirce emphasizes the importance of residuation and give examples of graphs Peirce drew of residuals. We conclude that EGs are an effective means of enlightening this concept.

Keywords: Residuation · Existential graphs · Charles Peirce · Cuts · Lines of identity

1 Introduction

As discussed by Pratt [19] and Maddux [13], De Morgan described the residuation laws in the form of Theorem K in 1860 [3,6]. The resulting equivalences state that given any three relations—a, b, and c—and well known relation operations—relational composition (;), complement (͞), converse (˘) and relational containment/inclusion ($\sqsubseteq$)—the following are equivalently defined:

$$a;b \sqsubseteq c \iff \breve{a};\bar{c} \sqsubseteq \bar{b} \iff \bar{c};\breve{b} \sqsubseteq \bar{a} \,. \tag{1}$$

Residuation can broadly be thought of as an inverse operation, much like how division is inverse to multiplication and subtraction to addition. In the context

Supported by (Haydon) the ESF funded Estonian IT Academy research measure (2014-2020.4.05.19-0001) and (Pietarinen) the Basic Research Program of the HSE University and the TalTech grant SSGF21021.

© Springer Nature Switzerland AG 2021
A. Basu et al. (Eds.): Diagrams 2021, LNAI 12909, pp. 229–237, 2021.
https://doi.org/10.1007/978-3-030-86062-2_21

of relations discussed here, residuation gives a remainder when relational composition is denied or converted (as in the equivalences above).

Another example is found in what is called the residuation property (RES), which shows how residuation also acts like implication:

$$(\text{RES}) \quad p \wedge q \sqsubseteq r \iff q \sqsubseteq p \to r.$$

This is related to the deduction theorem (and to currying), and the property plays an important role in characterizations of a range of implications from classical to intuitionistic (Heyting algebra) implication. Residuation is now recognized as a key property of substructural logics [10, 14] and, following Lambek, of categorical grammar [7, 8].

Given the significance that has been placed on residuation in the study of relations since, it is perhaps curious that Charles S. Peirce, who studied De Morgan's work closely and who went on to make significant contributions to the algebra of relations (e.g. his 'dual' and 'general' algebras of relatives), seems to have placed little emphasis on it in his later work. Maddux even describes the omission of residuation (De Morgan's Theorem K) in Peirce's later work as "puzzling" [13, p. 435].

We now know that Peirce did emphasize the residuation property above in his characterization of propositional logic as occurs in his 1880 algebra of the copula paper, §4, which presents a calculus of the consequence relation [11]. In it, two meanings of the copula $\prec$ (or $\overline{\prec}$) are delineated by using two signs: (1) the consequence relation (Peirce's sign of illation) $\Rightarrow$; and (2) the material implication $\to$. An expression of the form $x \Rightarrow y$ is called a *sequent*, according to the proof-theoretic terminology. Then the calculus of the copula (a Boolean algebra) consists of the following axiom and rules:

1. Identity: (Id) $x \Rightarrow x$
2. Peirce's Rule:
$$\frac{x \wedge y \Rightarrow z}{x \Rightarrow y \to z} \,(\text{PR})$$

3. Rule of Transitivity:
$$\frac{x \Rightarrow y \quad y \Rightarrow z}{x \Rightarrow z}(\text{Tr})$$

The double line in (PR) means that the lower sequent can be derived from the upper sequent and vice versa. The second rule, here renamed as *Peirce's Rule*, is probably the first formulation of the *law of residuation*: that the material implication is a right residual of conjunction.

We also now know that Peirce studied residuation enough to have listed equivalent forms residuals may take and to have given a method for arriving at the different permutations [16, CP 4.343] [19]. While this helps confirm Peirce's awareness of residuation and to assuage some doubts about the scope of his insights, it does not help explain why Peirce seems to have placed much less emphasis in his later work on a concept whose importance he had—and when looking back on it perhaps should have—so emphasized.

Pursuing Peirce's potential connection to residuation from another direction, it is equally curious that Peirce makes no direct mention of his earlier algebraic studies of residuation in his later presentation of Existential Graphs (EGs). Given that Peirce often cites EGs as the culmination of his earlier work on relations (with his algebraic studies of residuation, no doubt, as one), along with his insistence that EGs should be the "logic of the future" [18], it would be problematic if such a concept was left without representation.

Of course this is not the case. We remedy the seeming omission here by showing how residuation is naturally presented in EGs. Given the relatively sparse syntax of the graphs and that residuation can easily be represented without any changes to the syntax or transformation (i.e., inference) rules, it would seem rather that Peirce's supposed omission might be due to a belief that the other rules of EGs suffice to enlighten the concept. The presentation of residuation in EGs given here is the first of its kind—in particular, the first for its quantificational Beta extension that includes lines of identity.

This paper presents residuation in the context of relations and relational operations and sets aside for the time the functional characterization in terms of Galois connections.[1] The aim is to help situate Peirce's work in the development of residuation (following, in particular, the work of Maddux and Pratt cited above), to present the beginnings of a graphical presentation of residuation, and to address the connection between residuation and Peirce's work on Existential Graphs.

2   Beta Graphs and Relational Operations

We begin with a short introduction to EGs and the diagrammatic presentation of the operations needed to represent residuation in a logic of relations. We assume basic familiarity with the interpretation and transformation rules of EGs. Helpful introductions to Peirce's EGs can be found in [18,20]. The richer algebraic/categorical framework upon which this work relies can be found in [5]. We save an extended treatment of residuation in the latter context for subsequent work. Here, we stick rather to the perspective from relation algebra, leveraging the more traditional notation for relation algebras found for example in [1,4,21]. Though it predates Peirce's EGs, a helpful introduction along these lines is given by Peirce in his "Note B" [15]. What follows can be seen as a graphical treatment á la the later EGs of the algebraic work given in this note.

A general binary relation is scribed on the sheet with an ingoing 'wire' serving as a placeholder for the domain of the relation and an outgoing wire signalling the codomain. These wires represent the collection of individuals who might satisfy/stand in the relation presented.

$$-R-\quad\text{(a)}\qquad\qquad -R\!\!-\!\!S-\quad\text{(b)}\qquad\qquad \ominus\!\!R\!\!\ominus\quad\text{(c)}\qquad\qquad (2)$$

---

[1] On the history of residuation (adjunctions) and its relation to Galois connections, see [2].

Operations from relation algebra and their corresponding EGs are given in (2), where (a) is a general relation R, (b) is relational composition $R; S$, and (c) is complement $\bar{R}$. Given the ⊸loves⊸ and ⊸benefits⊸ relations we can for example express 'lovers of benefactors' as ⊸loves⊸benefits⊸ and 'lovers of non-benefactors' as ⊸loves⊸benefits⊸. .

Relations have a definite order such that reversing the domain and codomain gives a different relation. Changing the domain and codomain of the "x loves y" relation, for example, forms the converse relation "y is loved by x".[2] In [5] it is shown that lines of identity in fact obey the equations of a special Frobenius algebra. Graphically, this involves the addition of 'cups' and 'caps' (‹ , ›) that serve as markers for keeping explicit track of the bending of wires and the respective domain and codomain for each relation. For example, EGs of (a) a relation R, (b) its converse $\check{R}$, and (c) relation inclusion/containment $R \sqsubseteq S$ are given below in (3).

$$\text{⊸R⊸} \quad \text{(a)} \qquad\qquad \text{‹R›} \quad \text{(b)} \qquad\qquad \left(\!\!\begin{array}{c}\text{R}\\\text{S}\end{array}\!\!\right) \quad \text{(c)} \qquad (3)$$

The addition of cups and caps are important since the initial presentations of residuation in De Morgan's and Peirce's works depend on tracking the converse (and other) relations. Importantly, whereas the single cut represents complement/negation, a nested cut represents inclusion/containment relation (Peirce's "scroll"). The use of 'cups' and 'caps' as endcaps in this context is to show that the domain of R is preserved in the domain of S and that the codomain of R is preserved in the codomain of S.

More discussion on relation algebras can be found in [4]. For a detailed translation of the EG syntax into first-order logic, relation algebras, and a discussion of the transformation rules in this context, see [5]. With relational composition, complement, converse, and inclusion expressible graphically in the syntax, we have the relational operations needed to present residuation in EGs.

3   Residuation in Existential Graphs

Given relational composition (;), left and right residuals take the form of division.

$$a \sqsubseteq c/b \quad \Longleftrightarrow \quad a;b \sqsubseteq c \quad \Longleftrightarrow \quad b \sqsubseteq a\backslash c. \qquad (4)$$

Peirce enumerated several equivalent forms residuals may take (Schröder lists many more in [22]). The list depends on which operations are taken as primitive.

---

[2] While Peirce emphasizes this ordering in his algebraic work, he says little about reading such an ordering off the graphs. The few such places on the ordering of the 'hooks' around the relation terms appear in the early drafts of EGs from late 1886, in which Peirce notices how the connections of the lines to the relations should be read "clockwise" or "counterclockwise" (their converses) "beginning at the left/right" of the relation term; see [18, pp.220,263,295,302,303]. A further advantage of the notation in [5] is that the ordering is always explicit.

We begin by adding to (4) the residuals in terms of complement $(\bar{\ })$ and converse $(\breve{\ })$ relations. This allows us to fairly directly convert the residuation laws into a form amenable to the syntax of the EGs. We also use $(\dagger)$ to represent what Peirce calls *relative sum*, which is the dual to relative composition.

$$a \sqsubseteq c/b \qquad \Longleftrightarrow \qquad a;b \sqsubseteq c \qquad \Longleftrightarrow \qquad b \sqsubseteq a\backslash c \qquad (4)$$

$$a \sqsubseteq (\bar{c};\breve{b})^- \qquad \Longleftrightarrow \qquad a;b \sqsubseteq c \qquad \Longleftrightarrow \qquad b \sqsubseteq (\breve{a};\bar{c})^- \qquad (5)$$

$$a \sqsubseteq c\dagger\breve{b} \qquad \Longleftrightarrow \qquad a;b \sqsubseteq c \qquad \Longleftrightarrow \qquad b \sqsubseteq \breve{a}\dagger c \qquad (6)$$

$$\cdots \qquad\qquad\qquad \cdots \qquad\qquad\qquad \cdots$$

Example 1 (Residuation laws in EGs). Let us begin with EGs of (4) that correspond to the row of equations in (5).

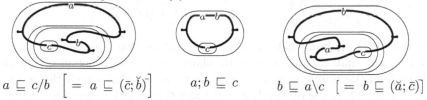

$$a \sqsubseteq c/b \; \Big[= a \sqsubseteq (\bar{c};\breve{b})^- \Big] \qquad a;b \sqsubseteq c \qquad b \sqsubseteq a\backslash c \; \big[= b \sqsubseteq (\breve{a};\bar{c})^- \big]$$

Relational composition (;) is represented in the graphs by connecting, via a line of identity, the respective outgoing and ingoing wire for the relations a and b. Relational inclusion/containment is captured by nested cuts, i.e. Peirce's scroll, and complement and converse are likewise represented as discussed in Sect. 2.

Only two graphical transformations are needed to represent the residuals in the side columns of (5), depicted by the left and right graphs above. One transformation is to add an S- or Z-shaped bend to a line of identity (cf. 'cups' and 'caps' producing converses). The other is to add a double cut around subgraphs. This has the effect of changing the consequent in the newly directed implication (see Remark 2). Both are straightforward transformation rules in EGs.

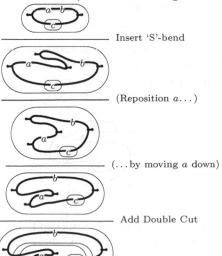

Insert 'S'-bend

(Reposition a...)

(...by moving a down)

Add Double Cut

The right residual is derived as follows. The transformation to the final graph begins by inserting an S-shaped bend in the line of identity (employing 'cups' and 'caps') between the composition of a and b. We then grab a and pull it down to the left. Finally, add a double cut around the bottom subgraph to form the consequent of the new implication. Notice that in the process we switch the endcap that preserves the codomain of the relation. This topographical deformation yields the converse on a. Performing similar operations (now with a Z-shaped bend) on b yields the graph on the left in Example 1. The simple transformations of bending wires to reposition subgraph relations and adding/removing a double cut are sufficient to capture the long list of equivalences given by Peirce in [16, CP 3.341].

Remark 1. Peirce's Rule of 1880 (residuation) is of particular importance, as the full distributivity laws can be deduced from it and the standard lattice rules [9]. The rules of Modus Ponens, LEM and *Ex Falso* are deducible similarly.

Remark 2. Peirce's Rule of 1880 has a particularly clear representation in EGs. Graphical representation of logical constants brings out the relation between logical connectives as *adjunctions*. The following rules are provable in EGs (Alpha):

$$\frac{A\ B\ \textcircled{C}}{A\ (\!(B\ \textcircled{C})\!)}\ \text{(RG1)} \qquad \frac{A\ (\!(B\ \textcircled{C})\!)}{A\ B\ \textcircled{C}}\ \text{(RG2)}$$

The rule (RG1) is immediate from the observation that it is permissible to add a double cut, namely the scroll with blank areas. The other direction (RG2) follows from the observation that permits removing that scroll. Hence Peirce's Rule is justified by the observational element that entitles the addition/removal of scrolls with blank areas. Indeed according to Peirce, "every copula is so closely connected with a conjunction that the notation should show the connection" [18, p. 428], concluding that "copulas are nothing but conjunctions" [18, p. 426].

Remark 3. On one loose manuscript leaf (RS 104, c.1903) Peirce formulated residuation as the pair of $A\,B \prec C$ and $A \prec C \curlyvee \overline{B}$. Here his algebra of logic notations $\curlyvee$ (aggregate) and $\overline{}$ (vinculum) correspond to logical disjunction and negation, respectively. Above and below these two consequence relations he wrote in the language of EGs: "$AB\ \ C$" and "$A\ \textcircled{C}\ B$", respectively. Taking the blank to mean the derivation along the consequence relation, one can move between these two graphs solely by an addition (top-down) or erasure (bottom up) of a blank scroll.

Example 2. We give Peirce's rule for the cyclic permutation of terms [19] in graphs. In Peirce's words the rule is: "the three letters may be cyclically advanced one place in the order of writing, those which are carried from one side of the copula to the other being both negatived and converted" [16, CP 3.341].

$$a;b \sqsubseteq c \qquad\qquad\qquad \breve{c};a \sqsubseteq \breve{b}$$

Again the equivalent expression is captured by the transformation rules in EGs—in this case rotating the subgraphs clockwise by pulling b down and around now also raising ⓒ around and up.

It is worth comparing the topological transformations used in the derivations above with an equivalent derivation using first-order logic.[3] When these moves are put into graphical notation many steps are found to be either roundabout or to have little content, such as to introduce, label, re-label, and eliminate excess variables.

4   Two Further Examples of Residuation in EGs

Two further examples highlight the efficiencies gained by the graphical treatment of residuation for derivations and for further thought.

It is known that in axiomizations of relation algebras, such as by [23], Theorem K can replace the axioms governing the rules between involution and distribution over Boolean join [12, p. 25]. One version of the key axiom is $\ddot{x}; \overline{x;y} \sqsubseteq \bar{y}$.

In "Note B" [15] Peirce also presents the following important equations: $\mathbb{I} \sqsubseteq \overline{\check{x};\bar{y}}$ and $x; \check{\bar{x}} \sqsubseteq \mathbb{I}$, where $\mathbb{I}$ is the identity relation for relational composition. These equations correspond to the *linear negation operation* in linear logic [19].

Example 3. We show that these equations follow from simple identities on x and the topographical moves described in Sect. 3.

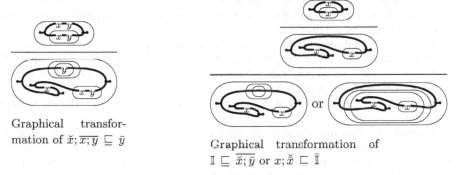

Graphical transformation of $\check{x}; \overline{x;y} \sqsubseteq \bar{y}$

Graphical transformation of $\mathbb{I} \sqsubseteq \overline{\check{x};\bar{y}}$ or $x;\check{\bar{x}} \sqsubseteq \mathbb{I}$

The equations for the right residual, $a\backslash b$, are given by the following [19]:

$$a\backslash b \quad = \quad \bar{\check{a}} \dagger b \quad = \quad (\check{a};\bar{b})^{\smallsmile} \tag{7}$$

Example 4. The rightmost equation is expressed by the EG below left. We clearly see that the residual has the form of an implication.

$(\check{a};\bar{b})^{\smallsmile}$

---

[3] One such derivation is [21, p. 42] which we forego here due to space limitations. Schmidt speculates that equational reasoning using predicate logic results in derivations that are six times longer than the corresponding algebraic handling of relations [p. xi].

5 Concluding Remarks

This last example brings us back to a common place where Peirce draws specific attention to residuation. He singles out the graph on the right, equivalent to the one on the left and which has a very nice vertical symmetry, to represent the key feature of necessary reasoning. In this sense, residuation is a general logical principle that has a *maximum level of abstractness*. Being maximally abstract means that such principles add nothing to the premises of the inference which they govern [17, NEM 4:175, 1898].

References

1. Brink, C., Kahl, W., Schmidt, G.: Relational methods in computer science. Springer, Wien (1997). https://doi.org/10.1007/978-3-7091-6510-2. ISBN 3-211-82971-7
2. Erné, M. Adjunctions and Galois connections: origins, history and development. In: Denecke, K., Erné, M., Wismath, S. (eds.) Galois Connections and Applications, pp. 1–138. Springer, Dordrecht (2004). https://doi.org/10.1007/978-1-4020-1898-5_1
3. De Morgan, A.: On the Syllogism, no. IV, and on the Logic of Relations. Trans. Cambridge Phil. Soc. **10**, 331–358 (2016)
4. Givant, S.: Introduction to Relation Algebras. Springer, Cham (2017). https://doi.org/10.1007/978-3-319-65235-1
5. Haydon, N., Sobociński, P.: Compositional diagrammatic first-order logic. In: Pietarinen, A.-V., Chapman, P., Bosveld-de Smet, L., Giardino, V., Corter, J., Linker, S. (eds.) Diagrams 2020. LNCS (LNAI), vol. 12169, pp. 402–418. Springer, Cham (2020). https://doi.org/10.1007/978-3-030-54249-8_32
6. Heath, P.: On the Syllogism, and Other Logical Writings. Routledge, London (1966)
7. Lambek, J.: The mathematics of sentence structure. American Math. Monthly **65**(3), 154–170 (1958)
8. Lambek, J.: Pregroups and natural language processing. Math. Intell. **28**(2), 41–48 (2006). https://doi.org/10.1007/BF02987155
9. Ma, M., Pietarinen, A.-V.: Peirce's sequent proofs of distributivity. In: Ghosh, S., Prasad, S. (eds.) ICLA 2017. LNCS, vol. 10119, pp. 168–182. Springer, Heidelberg (2017). https://doi.org/10.1007/978-3-662-54069-5_13
10. Ma, M., Pietarinen, A.-V.: A graphical deep inference system for intuitionistic logic. Logique & Analyse **245**, 73–114 (2018)
11. Ma, M., Pietarinen, A.-V.: Peirce's Calculi for classical propositional logic. Rev. Symbolic Logic **13**(3), 509–540 (2020)
12. Maddux, R.D.: Relation algebras. In: Brink, C., Kahl, W., Schmidt, G. (eds.) Relational Methods in Computer Science. Springer, New York (1997). ISBN 3-211-82971-7
13. Maddux, R.D.: The origin of relation algebras in the development and axiomatization of the calculus of relations. Studia Logica **50**, 421–455 (1991)
14. Ono, H.: Substructural logics and residuated lattices: an introduction. In: Hendriks, V.F., Malinowski (eds.) Trends in Logic, vol. 20, pp. 177–212 (2003)
15. Peirce, C.S.: Note B. In: Peirce, C.S. (ed.) Studies in Logic by Members of Johns Hopkins University. Little, Brown, pp. 187–203 (1883)

16. Peirce, C.S.: The collected papers of Charles S. Peirce (CP). Hartshorne, C., Weiss, P., Burks, A.W. (eds.) Harvard University Press, Cambridge (1931–66)
17. Peirce, C.S.: The New Elements of Mathematics by Charles S. Peirce (NEM). In: Eisele, C. (ed.) The Hague: Mouton (1976)
18. Peirce, C.S.: Logic of the future: writings on existential graphs. In: Pietarinen, A.-V. (ed.) Volume 1: History and Applications, volume 2/1: The Logical Tracts; 2/2: The 1903 Lowell Lectures; volume 3: Pragmaticism and Correspondence. De Gruyter (2019)
19. Pratt, V.: Origins of the Calculus of Binary Relations. In: Proceedings Seventh Annual IEEE Symposium on Logic in Computer Science, vol. 1, pp. 248–254 (1992)
20. Roberts, D.D.: The Existential Graphs of C.S. Peirce. Mouton, The Hague (1973)
21. Schmidt, G.: Relational Mathematics. Cambridge University Press, Cambridge (2010)
22. Schröder, E.: Vorlesungen über die Algebra der Logik (Exakte Logik). Algebra und Logik der Relative. B.G. Teubner, Leipzig, Dritter Band (1895)
23. Tarski, A.: On the calculus of relations. J. Symbolic Logic 6, 73–89 (1941)

On Identity in Peirce's Beta Graphs

Javier Legris(✉) iD

Interdisciplinary Institute of Political Economy of Buenos Aires, CONICET-University of Buenos Aires, Av. Córdoba, 2122, C1120 AAQ Buenos Aires, Argentina
javier.legris@fce.uba.ar

Abstract. Charles S. Peirce achieved, by the line of identity, a rich and useful *analysis* of quantification in the Beta Graphs that can be easily translated into the standard existential quantifier of First Order Logic. In this paper I claim that the way the line of identity expresses *identity relation* does not correspond to the usual understanding in the standard classical First Order Language with identity (FOL=). It will be argued that the line of identity cannot be used to express equations as in FOL=, but it expresses *individual identity* (in an ontological sense).

Keywords: Diagrammatic logic · Existential Graphs · Logical identity · History of logic

1 Introduction: Peirce's Beta System for Existential Graphs

In the last decade of the 19th century, Charles S. Peirce started developing a proof system for mathematical logic by means of *diagrammatic* notation: his *Existential Graphs* (EGs). Peirce regarded EGs as his masterpiece in logic and worked on it until his death in 1914, formulating and testing different presentations and applying his ideas on deduction and proof to them. Peirce did not achieve an ultimate formulation that he could regard as satisfactory. Peirce developed them with *formation rules* for their signs (*graphs*) representing the logical structures of sentences and with *rules of transformation* ("illative rules") to prove the conclusion of an argument from given premises (see, v. g., [1] 4.423). Most of the formulations were not published, but they remained in manuscripts and they have been intensively studied and expanded in recent times on the basis of examining unpublished manuscripts kept in his *Nachlass* at Harvard Library.

Following the general framework previously achieved by his algebraic approach in 1885, Peirce formulated two systems: the *Alpha System*, for propositional logic; and the *Beta System*, for predicate logic. Besides, he sketched the *Gamma System*, aimed at developing modal and higher-order Logic. Peirce achieved a full account of the first two systems, while the third remained in a fragmentary form. Beta received a systematic formulation by John Zeman in 1964 and by Don Roberts in 1973. In both cases, the adequacy of the Beta system with respect to classical First-Order Logic (FOL) was proven (see [2] and [3], appendix 4). Roberts' book turned out to be the standard presentation of EGs.

The Beta System is an extension of the Alpha System. Thus, Beta, like Alpha, is based on the *sheet of assertion*, a blank surface. Every graph written ("scribed") on it

© Springer Nature Switzerland AG 2021
A. Basu et al. (Eds.): Diagrams 2021, LNAI 12909, pp. 238–245, 2021.
https://doi.org/10.1007/978-3-030-86062-2_22

is understood as the assertion of a true sentence. Beta includes the *cut*: a closed line drawn on the assertion sheet as the sign for negation. The conjunction of two sentences is implicit in their spatial arrangement of sentences in the sheet of assertion.

The new sign in Beta is the *line of identity*. Peirce describes it as a "heavy line":

"The line of identity is a Graph [...], is a heavy line with two ends and without other topical singularity [...], not in contact with any other sign except at its extremities." ([1] 4.416).

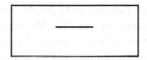

Fig. 1. The line of Identity.

Figure 4 represents the assertion that some individual object exists: "Something exists" (as Peirce wrote in a manuscript for the third Lowell Lecture in 1903, [4] p. 21). Other signs, which are called *spots*, corresponding to predicates, can be attached to the extremities of the line of identity. The following graph represents the sentence "Something is on the matt" (where "is on the matt" functions as a *spot*) (Fig. 2).

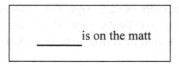

Fig. 2. A predicate attached to the line of identity. Existential quantification.

Besides, the following graph expresses the fact that "Something is P and Q", that is, the predicates P and Q are predicates of at least one individual.

Fig. 3. Two predicates attached to a line of identity.

With the aid of the cut, Peirce formulates the universal quantification, so that a sentence "Everything is P" is rendered into the Beta system as

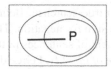

Fig. 4. Universal quantification.

In short, the Beta system includes two kinds of lines as basic logical signs: (i) "lines of separation" ("seps" or cut, closed lines) and (ii) "lines of identity" ("heavy" lines). Due to my interest in the analytical features of the line of identity, I will not go into detail about very important logical aspects of Peirce's Beta graphs, such as multiple quantification and the dependence of quantifiers (but see [5]). Moreover, the well-known problems concerning the connections between lines of identity will be overlooked. However, an important feature of the Beta system should be highlighted: *individual variables* are not necessary, since the line of identity fulfils their function.

2   The Logic of Identity in the Beta System

Identity statements were included in the first systems for predicate logic developed at the origins of mathematical logic (a clear example is given in Gottlob Frege's *Begriff-sschrift*). Putting aside logicists' attempts in foundations of mathematics, identity was indispensable in order to express arithmetical (and also set-theoretical) principles in the form of equations. In his paper of 1885, Peirce added identity to his algebraic notation (with the craw foot and quantifiers as basic signs) in a "second-intentional logic". The relation of Identity was designated by the sign '1' (different from the usual equation sign) as "a special token of second intention" (see [1] 3.398) and was characterized by the definition:

$$1[i\,j] = \pi[x](x[i]x[j] + \sim x[i] \sim x[j]),$$

Corresponding to the principle "to say that things are identical is to say that every predicate is true of both or false of both". It seems to have an ontological import, but Peirce also formulated a more general principle "if i and j are identical, whatever is true of i is true of j", symbolised in the current standard notation as

$$i = j \text{ if } \forall i\,\forall j(i = j \rightarrow (x[i] \rightarrow x[j])).$$

Hence his conception at that time followed not only the well-known Leibniz's principles of identity of *indiscernibles* but also the principle of substitution for variables. Two functions of identity were implied in this characterization: (a) the expression of the sameness of content, giving rise to substitution *salva veritate*, and (b) an ontological function related to indiscernibility.

Now, Roberts argued that the Beta System "takes account of individual identity and individual existence" ([3] p. 47). The previous Fig. 1, interpreted as **"Something exists"**, could also be understood as the statement in the standard notation for First-Order Logic with Identity (FOL=):

$$\exists x \exists y\, x = y,$$

which includes not only quantifiers but also the identity sign and indicates the identity of the individuals at the extremity of the line. Therefore, the line of identity would also designate *identity*. In Beta there would be *only one sign* for both notions.

According to these ideas, the graph represented in Fig. 3 should be interpreted in the standard notation for FOL= as

$$\exists x\, \exists y\, ((Px \wedge Qy) \wedge x = y).$$

This would render the full meaning of a line of identity when spots are attached to it.

On this basis, Beta graphs can be used to represent statements concerning identity alone, such as the assertion of differences (or "inequalities"), like in the following graph (Fig. 5):

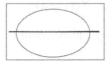

Fig. 5. Inequalities.

where a line of identity is crossing through an empty cut. It should be interpreted, then, as "There are at least two different individuals", that is, in the standard notation:

$$\exists x \exists y(\neg x = y)$$

If the two extremities of a line of identity are joined, then we would have the assertion of the existence of an identical individual:

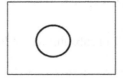

Fig. 6. Existence.

represented by '$\exists x\, x = x$' in the standard notation for FOL=. Note that it is not the case of an empty cut; in this graph a *thick* curved closed line is drawn. If this line is crossed by a cut, the following graph is obtained (Fig. 7):

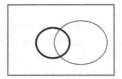

Fig. 7. Inexistence.

that should correspond to the negation of what Fig. 6 is asserting.

A more striking case is represented by the following graph:

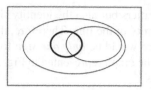

Fig. 8. Law of identity.

According to the conventions for constructions of Beta graphs, this is a *universal statement concerning identity:* the "law of identity", expressed in the standard notation as

$$\forall x(x = x).$$

Therefore, Fig. 8 represents a logic principle of FOL= and a proof using Beta rules can be found in [3] p. 62.

Finally, the following graph also represents a universal statement:

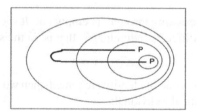

Fig. 9. Principle of substitution.

being read off as the "principle of substitution" of FOL=:

$$\forall x \, (x = x \to Px \to Px).$$

As it is well known, both principles, identity and substitution, accurately capture the logic notion of identity, which is indispensable for the construction of first-order theories.

3 Existential Graphs as an Analytical Tool

EGs are the result of an iconic idea of logical notation. For Peirce, mathematical proof and deduction in general were a process accomplished by means of *icons* (see, *inter alia*, [1] 3.363). An icon is not essentially tied to visual aspects of the represented structures. As Peirce wrote in his MS 293, a diagram is an icon of a set of rationally related objects. In a diagram the *analytical* role of icons turns out to be essential. (Peirce writes explicitly, "icon [or analytic picture]", [1] 1.275.) The purpose of EGs was "to dissect the operations of inference into as many steps as possible" ([1] 4.424). Hence, "exact" or *mathematical* logic had mainly an analytical role. In 1893 he wrote that the aim of logic was "to analyse reasoning and see what it consists in" ([1] 2.532 and 4.134).

According to Peirce, *analysis* means the decomposition of something into its basic elements. It applied above all to concepts in general (see [1] 1.294) The problem is to determine why EGs would provide the *right* analysis of logical concepts. Bellucci and Pietarinen tried to provide an answer to the aforementioned question about this special role of Existential Graphs which implies a particular idea of analysis. The key idea can be understood as *uniqueness of decomposition*:

"A system, constructed so as to employ the least amount of logical machinery and the least number of logical objects, forces us to the correct analysis of propositions. To say that an analysis is correct can in the first place mean nothing more than this." ([6] p. 210).

According to Peirce, the system of EGs provides "the only method by which all connections of relatives can be expressed by a single sign", as "the System of Existential Graphs recognizes but *one mode* of combination of ideas" (MS 482, 1897 and MS 490, 1906, see [6] p. 211) (my emphasis).

4 The Function of the Line of Identity

It can be assumed that the line of identity has an analytical function in the sense of the *uniqueness of decomposition* described above. The *line of identity* is a *sole* sign that *cannot be decomposed* and this is shown by the actions performed in constructing the sign. Notwithstanding, the line of identity deserves an accurate clarification.

Peirce adopted the idea of using lines to express links between different "units" (or "spots", see [11], p. 3). "I note that lines may be treated as monads", wrote Peirce at that time (see Peirce Ms 714, p. 1, 1889, quoted in [3], p. 22). Hence, a line can represent a (sole) individual. Later, by developing his EGs, Peirce elaborated on this idea. Peirce wrote:

"A heavy line shall be considered as a continuum of contiguous dots; and since contiguous dots denote a single individual, such a line without any point of branching will signify the identity of the individuals denoted by its extremities, and the type of such unbranching line shall be the Graph of Identity, any instance of which (on one area, as every Graph-instance must be) shall be called a Line of Identity." ([1] 4.561).

Hence, lines of identity consist of a series of dots. Furthermore, "a dot merely asserts that some individual object exists" ([1] 4.567). A line of identity can be understood as a "stretched" dot or a (continuous) series of dots, expressing the existence of some individual object. The *monadic* nature of the line of identity is clear. Thus, it can be interpreted as an existential quantifier when it is attached to a spot. In addition, according to Roberts, when the line of identity is not attached to a spot, it could be understood as expressing an *identity sentence* (see [3] pp. 53 ff.).

Anyway, some questions remain unsolved. Since Beta lacks individual variables, the dependence of quantifiers cannot be determined by the use of them. Furthermore, the graph for the principle of substitution (Fig. 9) can be transformed (according to the standard rules of transformation for Beta graphs—see [3], p. 56) into the graph (Fig. 10):

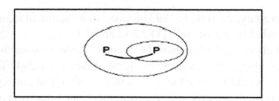

Fig. 10. Identity of Predicates.

corresponding to $\forall x\ (Px \to Px)$. In the standard notation for classical first-order logic with identity, the sentences $\forall x\ (Px \to Px)$ and $\forall x\ (x = x \to (Px \to Px))$ are trivially equivalent, but this last sentence does not mean the same as the principle of substitution. This difficulty can be extended to individual terms in general. In the Beta System, signs for individual constants are not specified; the non-logical signs are the spots, corresponding to predicates.

The graph represented by Fig. 1 is also problematic. Peirce understands it as saying "Something exists" (in the universe of discourse), but this sentence is generally regarded as a semantic presupposition, not expressible in the notation. Sometimes, it has been represented by a sentence such as '$\exists x\ x = a$', but in the Beta System the expression of individual constants is not taken into account.

5   Conclusions

From the preceding discussion, some consequences can be drawn and some issues can be raised. The line of identity primarily expresses the existential quantifier in an accurate way and, with the aid of the cut, the universal is obtained. Now, the previous discussion suggests that the notion of identity represented by it corresponds to *individual identity* and *not to identity as used in equations* (that is, the usual idea of identity as an equivalence relation). Surely, this is related to the difficulties concerning the representation of individual terms in the Beta System and the expression of identity sentences independently from quantification (in fact, how to express a = b in the Beta System?). Additionally, this fact makes clear that the EG in no way served as a logic basis for mathematical theories. Individual identity is an ontological issue connected, for example, with the principle of the identity of indiscernibles and it was debated among logicians and philosophers at the turn of the 20th century. It should not be overlooked that the very role of identity in logic was far from clear at the beginnings of mathematical logic. For example, it has been shown before that, in his algebra of logic, Peirce defined identity as a second order predicate. All these issues deserve further investigation.

Hence, it seems reasonable to confine the line of identity to the expression of (a) the existence of an individual (individual identity) and (b) the existential quantifier. It must be observed that the completeness proofs for the Beta System carried out by Zeman and Roberts are limited to classical FOL without identity.

Acknowledgements. The author wishes to thank the anonymous referees for their valuable criticism to a previous draft of this paper. This work was supported by the research projects

PIP 11220170100463CO (CONICET, Argentina), PICT 2017 0506 (ANPCyT, Argentina) and UBACYT 20020170100684BA (University of Buenos Aires).

1. References

1. Peirce, C.S.: Collected Papers. 8 volumes, vols. 1–6 ed. by Charles Hartshorne & Paul Weiss, vols. 7–8 ed. by Arthur W. Burks, pp. 1931–1958. Harvard University Press, Cambridge
2. Zeman, J.J.: The graphical logic of C. S. Peirce. Ph.D. dissertation, Department of Philosophy, University of Chicago (1964)
3. Roberts, D.: The Existential Graphs of Charles S. Peirce. Mouton, The Hague (1973)
4. Peirce, M.S.: (R) 462 (1903). https://www.unav.es/gep/Port/ms462/ms462.html
5. Pietarinen, A.-V.: Exploring the beta quadrant. Synthese **192**(4), 941–970 (2015). https://doi.org/10.1007/s11229-015-0677-5
6. Bellucci, F., Pietarinen, A.-V.: Existential graphs as an instrument for logical analysis. Part 1: alpha. Rev. Symbolic Logic **9**(2), 209–237 (2016). https://doi.org/10.1017/S1755020315000362. ISSN 1755-0203

Peirce's Diagrammatic Solutions
to 'Peirce's Puzzle'

Ahti-Veikko Pietarinen[1,2](✉) (iD)

[1] Tallinn University of Technology, Tallinn, Estonia
ahti.pietarinen@taltech.ee
[2] Research University Higher School of Economics, Moscow, Russia

Abstract. We present Peirce's own solution to what is known as 'Peirce's Puzzle' in formal semantics and pragmatics. In his mostly unpublished writings, Peirce analyses some sentences in the modal extension of his Beta Existential Graphs (that is, in a diagrammatic system of quantified first-order logic with tinctures) and in algebraic logic. These diagrams represent the pragmatic idea of information states that support or fail to support sentences with (non-existential) indefinites and modalities. The interpretation of such sentences presupposes a graphic-pragmatic criterion of cross-identification.

Keywords: Existential graphs · Peirce's puzzle · Modality · Indefinites · Tinctures · Line of identity

1 Introduction

Peirce's Puzzle states that in first-order predicate logic (and in its origins in Peirce's general algebra of logic), the sentence A is semantically equivalent to B [1,3,4,6]:

A: There is a married pair and if the husband fails the wife suicides.
B: If every married man fails some married woman suicides.

However, the common understanding of these sentences rather is that by uttering *A* we mean something stronger and more specific than what we mean by uttering *B*, namely that "... if her husband fails then *she* suicides" (*A*).

Peirce re-examined the puzzle, which he first presented in 1906 [3], in the Logic Notebook (LN) notes on Sept 6–7, 1908 [319r-320r], and in letters to P. Carus in 1908 and to J.H. Woods in 1913. He uses quantification over "states of information". He takes the meaning of *A* to be that there is some married couple of which, *under all "conceivable states of the universe,"*[1] the wife will suicide or else her husband would not have failed [4].

---

[1] Epistemic phrases such as "states of information", "circumstances", "states of knowledge", "states of affairs", "certain recognized states" were all used in Peirce's late writings, in the order of prevalence.

Supported by the Basic Research Program of the HSE University and the TalTech grant SSGF21021.

© Springer Nature Switzerland AG 2021

A. Basu et al. (Eds.): Diagrams 2021, LNAI 12909, pp. 246–250, 2021.
https://doi.org/10.1007/978-3-030-86062-2_23

I consider the graph   ⟨wife of—fails—suicides⟩ $\Sigma_i \Sigma_j \, w_{ji} \cdot (\overline{f}_i \, \maltese \, s_i)$.

A = There is a married pair and if the husband fails the wife suicides.

B = If every married man fails some married woman suicides.

$$\Sigma_m \Sigma_n \Sigma_p \Sigma_q \, w_{nm} \cdot \overline{f}_n \, \maltese \, w_{qp} \cdot s_p.$$

Let w be a state of things and f_{wu} means u fails in state of things w.

$$A' = \Sigma_i \Sigma_j \Pi_w \, \overline{f}_{wj} \, \maltese \, \overline{w}_{wji} \, \maltese \, s_{wi}$$

There is a man and a woman and under all circumstances if the man is married to the woman and he fails she will suicide.

$$A = \Sigma_i \Sigma_j \Pi_w \, w_{ji} \cdot (\overline{f}_{wj} \, \maltese \, s_{wi})$$

There is a married couple and under all circumstances if the husband fails the wife will suicide.

$$B = \Pi_w \Sigma_i \Sigma_j \, w_{ji} \cdot (\overline{f}_{wj} \, \maltese \, s_{wi})$$

Under all circumstances, there is a married couple and if all husbands fail some married woman will suicide.

$$C = \Sigma_j \Pi_w \Sigma_i \, w_{ji} \cdot (\overline{f}_{wj} \, \maltese \, s_{wi})$$

There is a married man and under all circumstances if he fails some wife of his will suicide.

Next day he adds shading (tincture) and considers their four variants:

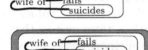

There is a married woman and should her husband fail she will commit suicide (under the actual circumstances). But it is not said that his failure will have any connexion with her suicide.

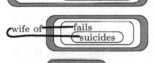

Under all circumstances there would be a married woman who, should her husband fail, would commit suicide.

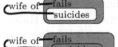

There is a married woman and under all circumstances, *the fact would be* (Qu[estion]: What precisely does this mean?) that if her husband fails she will commit suicide.

⟨wife of—fails—suicides⟩   There is a married woman; and if her husband *might* fail she will suicide.

⟨wife of—fails—suicides⟩   There is a married woman and should her husband fail she might commit suicide.

In contemporary terms, Peirce's thought is that we need to take into account the utterer's *information state* that the tinctures expose to view. Such information states either *support* or *fail to support* the sentences (or *neither*, in which case a gap obtains). An information state that supports A is for the utterer to have in mind an individual who, *in all states of things he considers conceivable*, commits suicide in case her husband fails. Peirce thus draws a connection between the antecedent and consequent not present in the first, indicative conditional ("will") or in its corresponding tincture-free graphs.

Since one needs reference to individuals as mental constructs (discourse referents) that have to persist across a range of conceivable states in the mind of the speaker, Peirce anticipates the problem of *quantification into modal contexts*. He presupposes that a pragmatic criterion for *cross-identification* of individuals is given and well understood between the utterer and interpreter [5].

Interestingly, in LN these sentences are immediately followed by another example of the same sort [4]. Here, instead of the circumstances or information states, what is quantified over are the *moments of time t*, analysed in his other diagrammatic (i.e., algebraic) notation:

Let h_{uv} mean u and v live in the same house. c_{tu} means that u comes home on say t. e_{tv} means v eats dinner on say t.

$\Sigma_u\Sigma_v\Pi_t\, h_{uv} \cdot (\bar{c}_{tu} \,\psi\, e_{tv})$. There is somebody living in a house who has one or other of two things, either he eats dinner every day he comes home or is on such terms with another person in the house that every day on which he comes home that other eats dinner. [That is,] Some person is on such terms with some person (himself or another) living in the same house that if the former comes home the latter eats dinner.

$\Pi_t\Sigma_u\Sigma_v\, h_{uv} \cdot (\bar{c}_{tu} \,\psi\, e_{tv})$. On every day somebody or other eats dinner unless there be somebody living in the same house who does not come home, or what is the same thing, on every day on which if everybody that lives in a house at all comes home somebody in the same house eats dinner.

$\Sigma_u\Pi_t\Sigma_v\, h_{uv} \cdot (\bar{c}_{tu} \,\psi\, e_{tv})$. Every day there is somebody that does not come home unless there be somebody living in the same house who eats dinner. [That is,] There is some person of whom it is true that on every day on which he comes home somebody living in the house with him eats dinner.

Peirce's notes in LN continue with a sketch of logical languages that have values such as "true sometimes", "under certain circumstances", the "limit" values, and so on. In general, Peirce's goal both in the information-states and in the moments-of-times versions of the puzzle is to develop a method for logically analysing natural-language modalities interspersed with indefinite expressions.

Peirce's insight was that indefinites not only impart contingent existential information but also contribute to which information states of the speaker support or fail to support such sentences. Tinctures indicate that the *entire range of the states of things* or *of the moments of times* that the speaker considers *conceivably possible* is to be taken into account in the evaluation of these graphs. The method suggest an involvement of a pragmatic principle of cross-identification [5]:

(**Pragmatic Cross-Identification**). When a line protrudes into tinctured enclosures, the individual actually represented in the utterer's mind must be identical to the individuals in all those information states that the tinctured enclosure opens to the view during utterer's evaluation of that sentence (that is, comes to be represented on the *verso* of the sheet of assertion).

Peirce took the resulting scrolls whose outloop enclosures are tinctured to signify not conditionals *de inesse* (material conditionals) but *hypotheticals*. In the latter, the sentence need not be necessarily true in case the antecedent turns out to be false. Those are known as *variably strict conditionals* [2,9]. In all of Peirce's examples, thus, a modality appears that influences the way in which conditional sentences are to be interpreted. This influence equally holds for disjunctive, non-conditional sentences, as seen in many of Peirce's own examples here.

2   Discussion and Conclusions

Peirce had shown in his examples that the puzzle arises also for disjunctive sentences, and so the proposals in terms of strict implications [8] will not suffice. in general, Peirce's Puzzle raises three interrelated questions:

1. What are the scope relations of quantifiers in these modal sentences?
2. How are indefinite noun phrases used in conditionals and disjunctions?
3. Do semantic and pragmatic considerations go hand in hand when logical analyses are performed on natural language assertions?

His own diagrammatic (that is, graphical and algebraic) analyses of the puzzle are noteworthy as they are the first to give rise to quantification into modal contexts adjoined with a pragmatic interpretation of information states. On January 6, 1909 letter draft to Paul Carus Peirce states the proposal to be "a small improvement upon existential graphs" [4]. This small improvement nonetheless has some wider repercussions, including the notion of the line of identity (which is not a ligature) that can cross the boundaries between tinctured and non-tinctures areas (in Peirce's terms a "reference" [4,6]).

The wider philosophical conclusion from Peirce is that one must insist on the *reality of some possibilities*. His analysis of the puzzle, which contemporary research still needs to address in full, makes an important case for this conclusion. "Admitting no reality but existence" would be, as Peirce in the light of these examples is confident enough to conclude, an "absurd result" [3].

Further, the Gricean principles of quantity and quality of information (which derive from Peirce's theory of signs [7]), including the sincerity conditions, contribute to the meaning of sentences exemplified in the puzzle. When interpreting relevant sentences the speaker is assumed to possess information about the wife in question, or at least to posit her as the topic (*ens rationis*) of the discourse. Sentences are asserted in those states, and in those states only, which the utterer considers *conceivably possible*.

References

1. Dekker, P.: Dynamics and pragmatics of 'Peirce's Puzzle'. J. Semantics **18**, 1–31 (2001)
2. Ma, M., Pietarinen, A.-V.: Peirce's Dragon Logic of 1901. Preprint (2019)
3. Peirce, C.S.: Prolegomena to an apology for pragmaticism. Monist **16**, 492–546 (1906)
4. Peirce, C.S.: Logic of the Future: Writings on Existential Graphs, Three Volumes. Gruyter, Berlin (2019–2021). A.-V. Pietarinen (ed.)
5. Pietarinen, A.-V.: Peirce's Pragmatic theory of proper names. Trans. Charles S. Peirce Soc. **46**, 341–63 (2010)
6. Pietarinen, A.-V.: Two papers on existential graphs by Peirce. Synthese **192**, 881–922 (2015)

7. Pietarinen, A.-V., Bellucci, F.: H. Paul Grice's Manuscript on 'Peirce's General Theory of Signs'. Int. Rev. Pragmatics **7**, 128–75 (2015)
8. Read, S.: Conditionals are not truth-functional: an argument from Peirce. Analysis **52**, 5–12 (1992)
9. Stalnaker, R.: A theory of conditionals. Stud. Logical Theory **2**, 98–112 (1968)

What Are Rules for? A Carroll-Peirce Comparison

Amirouche Moktefi[1(✉)] and Reetu Bhattacharjee[2,3]

[1] Ragnar Nurkse Department of Innovation and Governance, Tallinn University of Technology, Tallinn, Estonia
amirouche.moktefi@taltech.ee
[2] School of Cognitive Science, Jadavpur University, Kolkata, India
[3] Department of Mathematics, Mandsaur University, Mandsaur, India
reetu.bhattacharjee@meu.edu.in

Abstract. Unlike their predecessors, Lewis Carroll and Charles S. Peirce introduced rules to manipulate diagrams. As such, they opened the way to a formal view of diagrams. However, their motivations seem different, and even opposite: Pierce's rules are present for the purposes of analysis while Carroll's rules enable an epistemological strategy to ease calculus.

Keywords: Carroll diagram · Venn diagram · Peirce · Rules

1 Introduction

It is well known that Lewis Carroll and Charles S. Peirce made important contributions to logic diagrams. The former designed mature diagrams with a closed universe and an effective representation of categorical propositions [5]. The latter introduced an influential theory of diagrams, significant improvements in Euler-Venn diagrams and a fascinating system of existential graphs [11]. However, it is less known that, in contrast to their predecessors, both introduced sets of rules for diagrammatic reasoning. This shift anticipated the development of modern diagrammatic systems [13]. Carroll's rules are found in his treatise on symbolic logic [2], while Peirce's are exposed in a manuscript on logical graphs [10]. This paper discusses the role of rules in the diagrammatic systems of Carroll and Peirce. For the purpose, we first compare their methods with those of their predecessors to understand what difference those rules make in the manipulation of the diagrams. Then we investigate the independent motivations that led Carroll and Peirce to introduce their rules.

2 An Instructive Development

Let us consider the diagrammatic solution of a simple problem of elimination, as they were commonly addressed by logicians in Carroll and Peirce's time [7]. Suppose we are offered two premises: "All x are m" and "No m is y" and are asked what conclusion follows from them. Figure 1 shows how this problem is solved using the methods of Leonhard Euler, John Venn, Carroll and Peirce respectively (to ease comparison, we dropped Carroll's existential import for the first premise).

© Springer Nature Switzerland AG 2021
A. Basu et al. (Eds.): Diagrams 2021, LNAI 12909, pp. 251–254, 2021.
https://doi.org/10.1007/978-3-030-86062-2_24

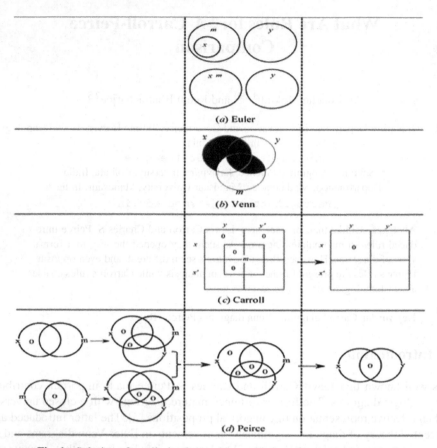

Fig. 1. Solution of an elimination problem with four diagrammatic methods

Euler combines the premises into a diagram where x is strictly inside m which is strictly outside y. However, Euler's followers observed that it is unknown what the exact relation between x and m is. Hence, we need two diagrams to depict all options, as in Fig. 1(a). Then, we search the relation between x and y that is common to the two diagrams in order to obtain the conclusion: "No x is y". To overcome this difficulty, Venn introduced a scheme where only one diagram would be needed. We simply draw a 3-term diagram and shade the empty regions: x *not-m* and m y, as in Fig. 1(b). A glance shows that the region x y is empty. Thus, the conclusion is: "No x is y".

Carroll's method resembles Venn's since he represents the premises on his 3-term diagram by marking the empty compartments with a '0', and obtains the diagram on the left in Fig. 1(c). However, unlike Venn, Carroll transfers information from this diagram to a 2-term diagram that will deliver the conclusion. This transfer is made in accordance with strict rules. Here, we use the rule that if a quarter of the 3-term diagram has two '0's, one in each cell, then one has to mark the corresponding quarter of the 2-term diagram with '0'. We obtain the diagram on the right in Fig. 1(c) which shows the conclusion: "No x is *y*". Peirce exhibits all the steps of the solution. He represents the premises on separate 2-term diagrams and adds a curve to each to introduce the missing term. Then, he merges the resulting diagrams to obtain a 3-term diagram that exhibits all the information contained in the premises. By applying a transformation rule, similar to Carroll's, Peirce eventually gets a 2-term diagram that exhibits the conclusion: "No *x* is *y*".

We observe that Venn reduced the number of diagrams required for the solution of such problems, but Carroll and Peirce did the opposite: they added diagrams to reach the conclusion. This is particularly clear in the elimination step where Venn extracts the conclusion by a simple "glance" at the 3-term diagram that represents the premises [14, p. 15]. Instead, Carroll and Peirce used rules to exhibit that conclusion on a 2-term diagram. Both made what Peirce called "an instructive development" [10].

3   What Are Rules for?

Carroll and Peirce developed independently their sets of rules. Although Carroll owned a copy of [9], the two logicians apparently never interacted [4]. Despite their apparent resemblance, their approaches differ on some important aspects. First, Carroll did not explicitly define what rules are. He loosely used the term 'rules' in the sense of instructions or procedures, as he did in the numerous games that he invented [8]. Peirce had a more formal conception of rules which he used "in the sense in which we speak of the "rules" of algebra; that is, as a permission under strictly defined condition" [10]. Also, Carroll referred to rules of transfer, a wording that suggests distinct unrelated diagrams, while Peirce rather speaks of rules of transformation, a terminology that suggests that diagrams are being reworked.

Finally, Carroll's rules were formulated specifically for syllogistic problems and would require serious modifications for complex problems [3]. Peirce's do not refer to a specific context of use. Yet, their function seems straightforward. Indeed, Peirce was not enthusiastic about calculus problems such as the one we discussed above. He rather worked on the analysis of reasoning by breaking it into elementary steps:

> [T]he purpose of a system of logical symbols [...] is simply and solely the investigation of the theory of logic, and not at all the construction of a calculus to aid the drawing of inferences. These two purposes are incompatible, for the reason that the system devised for the investigation of logic should be as analytical as possible, breaking up inferences into the greatest possible number of steps, and exhibiting them under the most general categories possible; while a calculus would aim, on the contrary, to reduce the number of processes as much as possible, and to specialize the symbols so as to adapt them to special kinds of inference. [12, p. 450]

Interestingly, Peirce's rules for Euler-Venn diagrams were developed at the same time as his Existential graphs, which also made a thorough use of transformation rules, and were specifically designed for the purposes of analysis [1].

If analytical purposes justify Peirce's 'instructive development', they hardly do for Carroll who was rather concerned with calculus. A close look shows that Carroll did not actually oppose to the reading of the conclusion directly from the diagram of the premises. However, he still recommends transferring information to another diagram to ease the reading and reduce risks of error:

> The best plan, for a *beginner*, is to draw a *Biliteral* Diagram alongside [the Triliteral Diagram], and to transfer, from the one to the other, all the information he can. He can then read off, from the Biliteral Diagram, the required Propositions. After a little practice, he will be able to dispense with the Biliteral Diagram, and to read off the result from the Triliteral Diagram itself. [2, p. 53]

This passage shows that Carroll considered the usage of rules as an epistemological strategy to increase the confidence of the user in the outcome of his manipulation [6]. Hence, Carroll makes a beautiful transition between Venn and Peirce in that he approves of both procedures, depending on the context of use and the ease of the user.

Acknowledgement. The first author acknowledges support from TalTech internal grant SSGF21021.

References

1. Bellucci, F., Moktefi, A., Pietarinen, A.-V.: Simplex sigillum veri: Peano, Frege and Peirce on the primitives of logic. Hist. Philos. Logic **39**(1), 80–95 (2018)
2. Carroll, L.: Symbolic Logic. Macmillan, London (1897)
3. Moktefi, A.: Beyond syllogisms: Carroll's (marked) quadriliteral diagram. In: Moktefi, A., Shin, S.-J. (eds.) Visual Reasoning with Diagrams. SUL, pp. 55–71. Birkhäuser, Basel (2013). https://doi.org/10.1007/978-3-0348-0600-8_4
4. Moktefi, A.: Are other people's books difficult to read? The logic books in Lewis Carroll's private library. Acta Baltica Historiae et Philosophiae Scientiarum **5**(1), 28–49 (2017)
5. Moktefi, A.: Logic. In: Wilson, R.J., Moktefi, A. (eds.) The Mathematical World of Charles L. Dodgson (Lewis Carroll), pp. 87–119. Oxford University Press, Oxford (2019)
6. Moktefi, A.: Diagrammatic reasoning: the end of scepticism? In: Benedek, A., Nyíri, K. (eds.) Vision Fulfilled, pp. 177–186. Hungarian Academy of Sciences, Budapest (2019)
7. Moktefi, A.: The social shaping of modern logic. In: Gabbay, D., et al. (eds.) Natural Arguments: A Tribute to John Woods, pp. 503–520. College Publications, London (2019)
8. Morgan, C. (ed.): The Pamphlets of Lewis Carroll: Games, Puzzles & Related Pieces. LCSNA, New York (2015)
9. Peirce, C.S. (ed.): Studies in Logic. Little, Brown, and Company, Boston (1883)
10. Peirce, C.S.: On logical graphs. Houghton Library, Harvard University, MS 479 (1903). Peirce Archive (https://rs.cms.hu-berlin.de/peircearchive/)
11. Peirce, C.S.: Logic of the Future, edited by A.-V. Pietarinen. De Gruyter, Berlin (2020)
12. Peirce, C.S., Ladd-Franklin, C.: Symbolic logic. In: Baldwin, J.M. (ed.) Dictionary of Philosophy and Psychology, vol. 2, pp. 645–650. Macmillan, New York (1902)
13. Stapleton, G.: Delivering the potential of diagrammatic logics. In: Burton, J., Choudhury, L. (eds.) DLAC 2013: Diagrams, Logic and Cognition, CEUR Workshop Proceedings, vol. 1132, pp. 1–8. http://ceur-ws.org/Vol-1132/
14. Venn, J.: On the diagrammatic and mechanical representation of propositions and reasonings. Philos. Mag. **10**, 1–18 (1880)

A Diagrammatic Representation
of Hegel's *Science of Logic*

Valentin Pluder[1]($\boxtimes$) and Jens Lemanski[2]

[1] Philosophisches Seminar, Universität Siegen, Siegen, Germany
valentin.pluder@uni-siegen.de
[2] Institut für Philosophie, FernUniversität in Hagen, Hagen, Germany
jens.lemanski@fernuni-hagen.de

Abstract. In this paper, we interpret a 19th century diagram, which is meant to visualise G.W.F. Hegel's entire method of the *Science of Logic* on the basis of bitwise operations. For the interpretation of the diagram we use a binary numeral system, and discuss whether the anti-Hegelian argument associated with it is valid or not. The reinterpretation is intended to make more precise rules of construction, a stricter binary code and a review of strengths and weaknesses of the critique.

Keywords: Diagrammmatic representation · G.W.F. Hegel · Bitstring semantics · Boolean algebra · Logic · Contradictions · Dialectic

1 Introduction

Diagrams have experienced periods of boom and bust in the history of logic and mathematics [6]. In the 19th century, especially Hegelians took up the fight against diagrams [8], although some diagrams can be found in Hegel's manuscript remains [4]. Today, there are moderate Hegelians who are either open-minded about the use of diagrams or even promote them intensively [7]. In the 19th century, it were mainly Hegel's opponents who provided diagrams to prove that the core ideas of Hegel's entire system could be represented with a few diagrams.

In the following, we draw on the first edition of Friedrich Heinrich Allihn's *Antibarbarus Logicus* published in 1850 [1]. Allihn, a critic of Hegel, attempts to brutally simplify Hegel's entire dialectical method in a diagram about the size of half a text page. From a logical perspective, it is interesting to note that Allihn's diagram uses a binary code and bitwise operations to represent the logical relations within Hegel's *Science of Logic*, and rhombuses to represent the unfolding and rejoining of contradictory opposites. In this paper, we will first introduce Allihn's book and his diagram (Sect. 2), then provide a modern logical interpretation of the Hegel diagram (Sect. 3), and finally discuss whether it appropriately describes Hegel's system or not (Sect. 4).

© Springer Nature Switzerland AG 2021
A. Basu et al. (Eds.): Diagrams 2021, LNAI 12909, pp. 255–259, 2021.
https://doi.org/10.1007/978-3-030-86062-2_25

2   Allihn's *Antibarbarus Logicus* and the Hegel Diagram

Allihn is an influential orthodox representative of J. F. Herbart's philosophy in the 19th c.: he represents a normative and purely formal logic that is completely separate from psychology, empiricism and metaphysics. At its core are the law of identity [1, i, p. 4] and the associated law of noncontradiction [1, i, p. 15]. Allihn rejects grasping concepts via negative properties, which does not allow to form contradictory concepts [1, i, p. 8]. Moreover, he denies the idealistic and constructive character of thinking [1, i, p. 12]. These points alone make him a passionate opponent of Hegel's philosophy and the Hegel school [3, chap. 15].

However, *Antibarbarus Logicus* – contrary to what the title suggests – is not only a polemic against Hegel. It is also a supposedly conservative and slender presentation of what Allihn believes to be the only true logic. In doing so, he opposes the logical pluralism in the 19th century. Book i contains chapters on logic, book ii on fallacies. In the latter, Hegel serves him as a cautionary example of formal fallacies and invalid reasoning that go hand in hand with a fusion of logic and metaphysics.

Allihn ends the book on fallacies with an attempt to formalise the structure of Hegel's *Science of Logic* by a diagram that is widely unknown today. His aim is to reveal the actual simple and uniformly repetitive character of Hegel's logic:

> Hegel's logic is the tragic product of one and the same dialectical hurdy-gurdy, which plays the same boring melody including the same dissonances for all texts. [1, ii, p. 34]

The supposed diversity of Hegelian logic is deduced from a zero point and, for all its supposed differentiation, never gets beyond this zero point [1, ii, p. 35]. To illustrate this, Allihn uses a diagram (Fig. 1a) including a bit alphabet of + and − and some quasi-Boolean operations to represent negation, conjunction, etc. Unfortunately, the explanation of the formalisation used is very poor:

> The methodical formula is: thesis, antithesis and synthesis, or opposition, unity, opposition, unity and so on. If we now designate pure thinking with +− which is = 0, then + would be pure being, the position, and − pure nothingness, the negation. [1, ii, p. 54]

Both the notation and the assignment to the Hegelian terms raise several problems of interpretation. Instead of discussing these in detail, we propose a reinterpretation (Fig. 1b) in the following section.

3   Interpretation of Allihn's Hegel Diagram

Let $\mathfrak{A}$ be a system which consists of five relations or bit manipulations $\mathbb{R} = \{$DIV, CONT, CONJ, ADD-1, ADD-0$\}$, a binary alphabet $\Sigma = \{0, 1\}$ by which relata or concepts can be described in terms of bitstrings, and a starting point 10.

Two types of $\mathbb{R}$ in $\mathfrak{A}$ can be distinguished: $(\mathbb{R}^-)$ negative relations and $(\mathbb{R}^+)$ affirmative relations. $\mathbb{R}^-$ indicate relations between relata that are either in contradiction with each other (CONT) or result from the division of another relatum (DIV). $\mathbb{R}^+$ indicate relations that either conjugate two relata into a third (CONJ) or add a 1 (ADD-1) or a 0 (ADD-0) to a bitstring of a relatum.

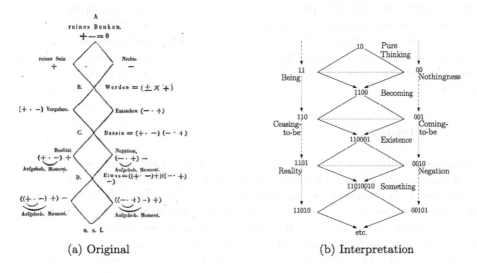

(a) Original   (b) Interpretation

Fig. 1. Allihn's Hegel diagram

Figure 1b is a diagram of $\mathfrak{A}$, in which relations are represented by lines or arrows and relata by bitstrings, which correspond to concepts such as *pure thinking, being, nothingness* etc. Arrows represent $\mathbb{R}^+$ such that a number of bits is added to an already existing relatum (arrowshaft) and thus forms a new relatum (arrowhead). $\mathbb{R}^-$ is represented either by a normal line to represent division or by a dotted line to show contradictions of relata.

Type	Relation	Diagram	Example
$\mathbb{R}^-$	CONT	Dotted line	1101 CONT 0010
	DIV	Straight line	110001 DIV 110(1) & 001(0)
$\mathbb{R}^+$	CONJ	Straight arrow	110 & 001 CONJ 110001
	ADD-1	Dashed arrow	110 ADD-1 1101
	ADD-0	Dotted arrow	001 ADD-0 0010

In the first step, the bitstring 11 and its contradiction 00 are the result from DIV of 10 into 1 and 0 as well as ADD-1 to 1 and ADD-0 to 0. To these bitstring and to their successors all subsequent bit manipulations are then applied according to the diagram. If $k \geq 2$ is the length of a particular bitstring indicating a relatum in CONT relation, then $2k$ is the length of the bitstring indicating the relatum at the arrowhead of a CONJ relation.

For the relata, however, only the two bits on the right of a string are important, as they indicate the process of thesis (e.g. 11), antithesis (e.g. 00) and synthesis (e.g. 1100). Thus one can read from Fig. 1b a simple directed graph that begins at 11 (thesis) and breaks off e.g. at 11010010 (synthesis): 11 – 00 – (11)00 – (1)10 – (0)01 – (1100)01 – (11)01 – (00)10 – (110100)10. The last bit in a string indicates whether a concept has a positive or negative connotation, e.g. *reality* = 1, *negation* = 0. The bits in brackets can be seen as the sublated historical process of thought, which is always remembered or preserved less in thesis and antithesis than in the preceding synthesis, but nevertheless grows steadily from CONT relata to CONT relata.

4   Discussion

Although the diagram is the critique of a Hegel opponent, it actually illustrates well two aspects of Hegel's process of thought: the rhombuses strung together illustrate the continued representation or differentiation of concepts into contradictory opposites and their subsequent conjunction into a unity [3, chap. 13]. The extension of the bitstring illustrates the gain in determination in this history of thought: each contradictory relatum (which is why we have represented Ahlinn's + with 11 and − with 00) extends the string of opposites by one bit each. This gain in determination is preserved in the subsequent conjunction.

On the other hand, Allihn also misses Hegel in essential points. For example, Hegel does not deduce from *pure thinking*, but tries to think *pure being*. The earlier simple concepts do not remain in isolation [2, p. 362]. They are sublated in favour of later concepts which can only be seen at the bitstrings. The later concept – once *becoming* has been passed through – stand on the side of *being* and are not = 0. Allihn's conception here seems influenced by Schelling's philosophy of identity [5]. The opposition of *reality* and *negation* is different from that of *pure being* and *nothingness* [4, chap. VI.2]. Despite these and several points of criticism not mentioned here, however, the originality of presenting Hegel's thinking in bitstrings and bitwise operationen still remains to be emphasised.

References

1. Allihn, F.H.T.: Antibarbarus Logicus, enthaltend (i) einen kurzen Abriss der allgemeinen Logik (...) und (ii) die Lehre von den Trugschlüssen (...), 1st edn. Plötz, Halle (1850)
2. de Boer, K.: Hegel's account of contradiction in the science of logic reconsidered. J. Hist. Philos. **48**(3), 345–373 (2010)
3. Ficara, E.: The Form of Truth: Hegel's Philosophical Logic. De Gruyter, Berlin; Boston (2020)
4. Harris, F.: Hegel's Development: Night Thoughts (Jena 1801–1806), 2nd edn. Clarendon, Oxford (1983)
5. Houlgate, S.: Schelling's critique of Hegel's science of logic. Rev. Metaphysics **53**(1), 99–128 (1999)

6. Lemanski, J.: Periods in the use of Euler-type diagrams. Acta Baltica **5**(1), 50–60 (2017)
7. Maybee, J.E.: Picturing Hegel: An Illustrated Guide to Hegel's Encyclopaedia Logic. Lexington, Plymouth (2009)
8. Pluder, V.: The limits of the square: Hegel's opposition to diagrams in its historical context. In: Vandoulakis, I., Béziau, J.-Y. (eds.) The Exoteric Square of Opposition. Birkhäuser, Cham (2021)

Jin Yuelin's Simplification of Venn Diagrams

Xinwen Liu[1] and Ahti-Veikko Pietarinen[2,3(✉)]

[1] Institute of Philosophy, Chinese Academy of Social Sciences, Beijing, China
liuxw-zxs@cass.org.cn
[2] Tallinn University of Technology, Tallinn, Estonia
ahti-veikko.pietarinen@ttu.ee
[3] Research University Higher School of Economics, Moscow, Russia

Abstract. Chinese logician and philosopher Jin Yuelin published in 1935 a textbook *Logic* (in Chinese) in which he proposed proving the distributive laws by a slightly non-standard version of Venn diagrams. In Jin Yuelin's modification some segments of the circles are marked with dashed instead of continuous lines, namely those that following the meet and join operations encircle the regions outside of the meet regions, as well as those that encircle the regions inside of the join regions. Hence the validity of distribution of meets over joins in the first distributive law and joins over meets in the second is observed by the sameness of the diagrams with exactly the same dashed and continuous line segments. This slight modification removes the need for shading empty regions and liberates one from using the cross mark "x" or some other 'non-visuals' for existence while freeing shading for some other uses (as e.g. opined by Venn and Peirce) and reducing clutter. In addition to such implications to the theory of Venn diagrams, we expose this little-known detail from the history of Venn diagrams and assess the factors that contributed to its discovery, such as whether Jin Yuelin decided to apply the dashed line following his reading of Peirce's broken-cut notations.

Keywords: Venn diagrams · Jin Yuelin · Dashed circles · Distributivity · Peirce

1 Introduction

Expressiveness and visual design features of Venn diagrams (VD) have a certain give-and-take relation. Historically, Charles S. Peirce's development upon VD in 1898–1903 increased their expressiveness with diagrammatical apparatus, including lines, circles, cross marks, shadings [9, 10], as well as changes in the shapes of the circles [5, 8]. Motivated by the low expressive power of standard VD (and the apparent decrease of their iconicity when increasing their expressiveness), Peirce established his own logical systems known as Existential Graphs (EG; [8]). Although both VD and EG use the circle or oval as a primitive notion, they interpret it very differently. A circle in a VD represents a class or a set, in EG it represents (among others) the operation of negation (the cut) that negates the propositions enclosed by the oval.

Supported by the Basic Research Program of the HSE University and the National Social Science Fund China 20&ZD046.

© Springer Nature Switzerland AG 2021
A. Basu et al. (Eds.): Diagrams 2021, LNAI 12909, pp. 260–263, 2021.
https://doi.org/10.1007/978-3-030-86062-2_26

2 Venn Diagrams and Existential Graphs

In manuscript R 481 of c.1900 (unpublished until [8]), Peirce compares the fundamental notions of the two theories of VD and EG. He notices that the circle in Euler diagrams represents the *logical breadth* of the sheet while in EG it represents the *logical depth*. One can distinguish between truth and assertions in the latter, and together with the cuts, hypotheticals become assertible. As modality naturally arises from the hypotheticals quite naturally, it was not long after, by 1903, that Peirce proposes to add the Gamma part to EG. Gamma concerns modal logic with a primitive diagrammatic notion of the "broken cut":

The broken cut, which encloses any other graph, represents a weakened negation of "possibly not". Together with the *continuous circle* that represents the *contradictory* negation, Peirce's innovation gave rise to a number of systems of modal logic [3].

In the 1906 "Prolegomena" [7] Peirce presented modalities in terms of *tinctures*. In some of the examples of that paper (and still more in the unpublished drafts, see [8]), Peirce now used also the broken circles. However, those were not cuts but *polarity markers* to distinguish between *oddly* and *evenly enclosed* nested circles, thus applied merely to contribute to the (visual) perspicuity of graphs.

Nevertheless, after paper's publication its ideas influenced many Cambridge philosophers and logicians, especially F.P. Ramsey, C.K. Ogden and G.E. Moore, and even that of L. Wittgenstein. Ogden had transcribed parts of the paper under V. Welby's tutelage in 1911. Ramsey had studies those notes by the time his critical review on Wittgenstein's *Tractatus* appeared in 1924.

3 Venn Diagrams in Jin Yuelin's 1935 Book *Logic*

What can be added to these storylines is that a Chinese logician and philosopher, Jin Yuelin (1895–1984) was also influenced by Peirce's 1906 paper. For one thing, he adopted the type-toke distinction in the philosophy of logic [2]. Jin Yuelin may have become acquainted with the paper during his stay in Cambridge and other places in Europe between 1922 and 1925. China's leading scholar in philosophy of logic with Wang Hao among his students, Jin Yuelin attended Wittgenstein's and Moore's classes and became much influenced by Ramsey. Then in 1935 he published his textbook *Logic* (in Chinese) [2], the first modern logic textbook in China to introduce first-order predicate calculus. From then on, modern logic entered in China as a science.

In *Logic* (pages 209–11) Jin Yuelin considered the following distributivity laws:

(1) $(A \cap B) \cup (A \cap C) = A \cap (B \cup C)$
(2) $(A \cup B) \cap (A \cup C) = A \cup (B \cap C)$

As to the proof of the first one, Jin Yuelin proposed a non-standard notation: diagrams with dashed segments of circles. Jin Yuelin's proof goes on as follows:

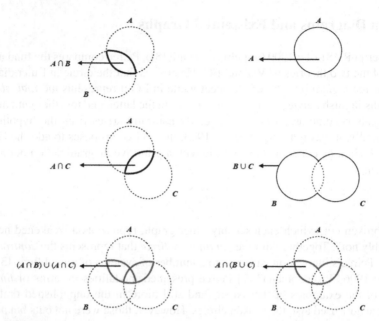

As the two columns show the same diagram as the conclusion (the inner red lines in the bottom left diagram are without significance), the original proposition is established. A similar argument can easily be produced for the second distributivity law.

4   Some Remarks on Dashed Circles in Venn Diagrams

We present a couple of comments on Jin Yuelin's proposal.

1. The arrows in the original figures of the diagrams in his 1935 book are not elements of the language of diagrams and can be deleted without loss of expressiveness.
2. Segments of circles with dashed instead of continuous lines are those that following the meet and join operations, encircle the regions outside of the meet regions as well as those that encircle the regions inside of the join regions, respectively.
3. The operation of combining two diagrams changes a continuous circle into a dashed circle, since newly introduced minimal regions represent empty classes. For example, on the right column above, the circle A becomes partly broken after its combination with the diagram representing $B \cup C$.
4. The dashed segments of circles correspond to the shading of VD that in late 19th century was proposed by Venn, Peirce and others in order to abolish the regions "which are made to vanish" by striking "out (by shading) those which are made to vanish by the data of the problem" [1, p. 73].
5. The dashing amounts to an alternative shortcut (a Jin Yuelin-style 'free ride') in which the validity of the distribution of meets over joins in the first distributive law and joins over meets in the second is observed by the sameness of the diagrams that have exactly the same dashed and continuous line segments.

6. A proof by such 'pictures' is of course not a fully rigorous proof as any of the classes may be empty. But the method has some virtues to recommend itself, such as removing the need for shading empty regions. This liberates one from using the cross mark "x" or some other 'non-visuals' for existence while permitting applications of shading for some other purposes (as e.g. opined by Venn and Peirce; [4, 6]). The need for alternative shadings and thus the amount of clutter in VD is reduced.

7. There is one more dashed VD that appears on page 211 of *Logic*, this time within the rectangular box that denotes the (finite) universe, with regions of VD that correspond to the basic Boolean laws:

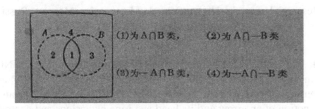

The rectangular box around the VD is presented to make sense of complementation. This exterior region (which did not appear in Venn's earlier proposal) is not labelled to denote some particular universe.

5   Conclusion

Jin Yuelin's incidental proposal appearing in his 1935 book *Logic* produces a notation to simplify VD with broken-line circles. This decreases neither expressiveness nor iconicity of VD while relinquishing shading for other logical uses in theories of VD.

References

1. Couturat, L.: The Algebra of Logic. Open Court, Chicago (1914)
2. Jin, Y.: Logic. Tsinghua University Press, Beijing (1935). (in Chinese)
3. Ma, M., Pietarinen, A.-V.: Gamma graph calculi for modal logics. Synthese **195**, 3621 (2017)
4. Moktefi, A., Pietarinen, A.-V.: On the diagrammatic representation of existential statements with Venn diagrams. J. Logic Lang. Inform. **24**(4), 361–374 (2015). https://doi.org/10.1007/s10849-015-9228-1
5. Moktefi, A., Pietarinen, A.-V.: Negative terms in Euler diagrams: Peirce's solution. In: Jamnik, M., Uesaka, Y., Elzer Schwartz, S. (eds.) Diagrams 2016. LNCS (LNAI), vol. 9781, pp. 286–288. Springer, Cham (2016). https://doi.org/10.1007/978-3-319-42333-3_25
6. Moktefi, A., Shin, S.-J.: A history of logic diagrams. In: Gabbay, D., et al. (eds.) Logic: A History of Its Central Concepts, pp. 611–682. North-Holland, Amsterdam (2012)
7. Peirce, C.S.: Prolegomena to an apology for pragmaticism. Monist **16**(4), 492–546 (1906)
8. Peirce, C.S.: Logic of the future: writings on existential graphs. In: Pietarinen, A.-V. (ed.), vol. 1–3. De Gruyter, Berlin & Boston (2019–2021)
9. Pietarinen, A.-V.: Extensions of Euler diagrams in Peirce's four manuscripts on logical graphs. In: Jamnik, M., Uesaka, Y., Elzer Schwartz, S. (eds.) Diagrams 2016. LNCS (LNAI), vol. 9781, pp. 139–154. Springer, Cham (2016). https://doi.org/10.1007/978-3-319-42333-3_11
10. Shin, S.-J.: The Logical Status of Diagrams. Cambridge University Press, Cambridge (1994)

Venn Diagrams with "Most": A Natural Logic Approach

Xinwen Liu[1] and Ahti-Veikko Pietarinen[2,3]($\boxtimes$)

[1] Institute of Philosophy, Chinese Academy of Social Sciences, Beijing, China
liuxw-zxs@cass.org.cn
[2] Tallinn University of Technology, Tallinn, Estonia
ahti-veikko.pietarinen@ttu.ee
[3] Research University Higher School of Economics, Moscow, Russia

Abstract. This note exposes a little-known fact originally proposed by Nicholas Rescher in 1965, that the generalized second-order quantifier "Most" and the rules governing its behavior can be incorporated into Euler-Venn diagrams with an iconic notion of an arrow and its head and vane extensions and contractions. The objective is then to analyse this work further and to link it with the related but independently developed recent work in the area of natural logic.

Keywords: Venn diagrams · Rescher quantifier · Most · Arrow · Natural logic

1 Introduction

Nicholas Rescher and Neil A. Gallagher [1] introduced in 1965 a new item of the diagrammatic apparatus, a kind of an arrow, into Venn Diagrams (VD), with the purpose of diagrammatizing what is known as the "Rescher-quantifier" ([2], p. 18), namely the proportional generalized determiner "Most" denoting a binary relation. This quantifier is of second-order and not definable in standard first-order predicate logic. Thus the extension of the traditional system of VD by the diagrammatic representation of "Most" increases the system's expressiveness. Rescher & Gallagher also developed some rules for VD that involved two types of propositions:

U: "Most S is P" and W: "Most S is not P".

The objective of this paper is to analyze their work and to link it to some related work that recently and independently has been developed in the area of *natural logic*.

2 Rescher's Diagram

Rescher's work is seldom mentioned in research on diagrammatic logic and its history (but see [10] for a recent study on cognitive effectiveness of reasoning with proportional

Basic Research Program (HSE University); National Social Science Fund China 20&ZD046

© Springer Nature Switzerland AG 2021
A. Basu et al. (Eds.): Diagrams 2021, LNAI 12909, pp. 264–268, 2021.
https://doi.org/10.1007/978-3-030-86062-2_27

Euler diagrams, including "Most"). In 1964 Rescher published a note "Plurality Quantification" [3] of two pages which presents a generalization of first-order quantification to what is known as the Rescher Quantifier. Next year he published with his student a paper "Venn Diagrams for Plurative Syllogisms" [1].[1] That paper introduced a new item of the diagrammatic apparatus into Venn diagrams, in order to diagrammatize the Rescher-quantifier, namely an arrow that connects two line segments:

In this arrow notation the two lines are called the *head* of the arrow (the line towards which the arrow points to) and the *vane* of the arrow (from which the arrow points away from). The arrow and its lines can cross the boundaries of regions. Their purpose is to show which *minimal regions* are combined into one region.

The plural phrase with the determiner "Most" can have many meanings associated to it. No uniquely agreed semantics exists in the literature on generalized quantifiers, though 'more than a half' is the baseline. In VD, the function of the arrow is to indicate that the region comprising all minimal regions on which the vane of the arrow falls is of greater cardinality than the region comprising all minimal regions on which the head of the same arrow falls. Thus the clause "Most *A*s are *B*s" is to be presented by one of the following two diagrams (redrawn here as they appear in [4], p. 127):

Fig. 1. Most As are Bs

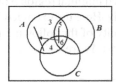

Fig. 2. Most As are Bs

In these two diagrams, Fig. 1 encodes the information that the region 2 is of greater cardinality than the region 1. The diagram of Fig. 2 encodes the information that the region comprising the minimal regions 5 and 6 is of greater cardinality than the region comprising the minimal regions 3 and 4. Both diagrams communicate the same proposition, namely "Most *A*s are *B*s", in which the determiner "Most" interpreted to mean *more than half of As are Bs.*

Aside from thus increasing expressiveness of standard VD, diagrammatic rules for the arrow notation were presented as follows ([4], pp. 127–128):

(R1) The vane of an arrow may always be *extended.*
(R2) The head of an arrow may always be *contracted.*
(R3) The vane of an arrow may always be *contracted out of a shaded region.*
(R4) The head of an arrow may always be *extended into a shaded region.*
(R5) An arrow may always be *drawn from a starred region into a shaded one.*

---

[1] Republished as Chapter VII of *Topics in Philosophical Logic* ([4], 126–33). Obviously, then, VD were a topic in philosophical logic long before their resurgence in the 21st century ([5], 395–22).

(R6)   In diagramming the premises of a plurative syllogism, *if* (1) both the heads of the two arrows overlap in one region, and (2) both the vanes overlap in one region, and (3) the head of each arrow overlaps in one region with the vane of the other, *then* a non-emptiness mark can be placed in the vane-overlap region.

We illustrate the workings of these rules by the following example. Beginning with the extended VD of Fig. 2, one can add the diagrammatic information of "All *B*s are *C*s" as another premise and obtain the diagram presented in Fig. 3 with two shaded regions. According to (R3) the vane of the arrow may then be contracted out of the shaded minimal region 5, to obtain VD in Fig. 4.

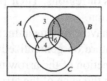   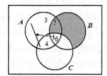

Fig. 3. Most As are Bs and all Bs are Cs   **Fig. 4.** Most As are Bs and all Bs are Cs (R3).

Three more steps, namely the head of the arrow contracted out of the minimal region 4 by (R2), the vane extended into the minimal region 4 by (R1), and the head extended into the shaded minimal region 5 by (R4), will then lead to the diagrams of Figs. 5, 6 and 7, respectively. Here the extended VD in Fig. 7 reads "Most *A*s are *C*s". This procedure tests the validity of the following plurative syllogism: "Most *A*s are *B*s, All *B*s are *C*s; therefore, Most *A*s are *C*s".

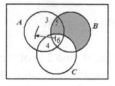   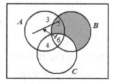   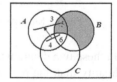

Fig. 5. (R2)   **Fig. 6.** (R1)   **Fig. 7.** (R4)

(This example is from [1] with a corrected complete arrow appearing in VD of Fig. 7)

3   Two Comments on Rescher's Rules

First, rules (R1)–(R6) are strictly speaking not rules for *logical inference* but pertain to the diagrammatic *testing procedure* applied to check for the validity of certain valid plurative syllogisms. They are rules for manipulating VD and as such are indecisive as to which minimal regions arrows should be extended to or contracted from, in contradistinction to inference rules that determine the permissibility of inferential steps. For example, (R4) says that the head of an arrow may always be extended into a shaded region. Then, given the diagram in Fig. 3, we have three (minimal) shaded regions, the region 5, the region $\overline{A}BC$, and the region $B\overline{C}$. The rule (R4) does not determine to which shaded region the head of the arrow is to be extended.

Second, Rescher's rules only deal with standard syllogistic two-premise arguments. That is, they only work with a simplified fragment of syllogistic logic. Can the idea be brought to bear on arguments that have more than two premises?

4 Arrow for "Most" in Natural Logic

Recent work by Lawrence S. Moss and others have led to the development of the theory of *natural logic*. For example, in [6] a sound and complete proof system is given for the logical system with sentences of the form.

All X are Y, Some X are Y, and *Most X are Y.*

These are interpreted on finite models with the meaning of "most" as "strictly more than half". The proof system is syllogistic without variables, and the problem of adding sentences of the form *No X are Y* to yield a larger syllogistic fragment with the determiner "*No*" is left as an open problem ([6], p. 125). In a follow-up study [7], a Rescher-like arrow notation was rediscovered as a *digraph* for the quantifier "most", defined as: For V any finite set and A_v is a finite set for $v \in V$, a digraph $G = (V, \rightarrow)$ is obtained in a natural way, defined as $u \rightarrow v$ if and only if *most A_u are A_v* ([7], 3701).

5 Conclusion

Proportional plurality quantifier has interesting formal features, such as "the duality implication that MOST $\Rightarrow$ not-MOST-not and also the failure (in contrast to SOME and ALL) to be self-commutable", which is the "sort of qualification, standing apart from all-or-something approach of the usual quantifiers, that is interestingly applicable and useful in the domain of human affairs" [8, pp. 1–2]. Rodgers ([9], p. 147) adds that "many statements expressible in higher order logic simply cannot be expressed in diagrammatic systems. Addressing this issue requires the addition of higher order concepts to Euler diagram reasoning systems". Adding the quantifier "most" it the first step in this direction in VD. New diagrammatic inference systems can be obtained from research on natural logics, adding precision to the semantics of "Most" leading to improvements upon Rescher's original rules.

References

1. Rescher, N., Gallagher, N.: Venn diagrams for plurative syll. Phil. Stud. **16**, 49–55 (1965)
2. Gabbay, D., Guenther, F. (eds.): Handbook of Philosophical Logic. D. Reidel (1989)
3. Rescher, N.: Plurality quantification. J. Symb. Log. **27**, 373–374 (1964)
4. Rescher, N.: Topics in Philosophical Logic. Springer, Heidelberg (1968). https://doi.org/10.1007/978-94-017-3546-9
5. Gabbay, D.M., Guenthner, F. (eds.): Handbook of Philosophical Logic, vol. 4. Springer, Heidelberg (2001). https://doi.org/10.1007/978-94-017-0456-4
6. Endrullis, J., Moss, L.S.: Syllogistic logic with "most." In: de Paiva, V., de Queiroz, R., Moss, L.S., Leivant, D., de Oliveira, A.G. (eds.) WoLLIC. LNCS, vol. 9160, pp. 124–139. Springer, Heidelberg (2015). https://doi.org/10.1007/978-3-662-47709-0_10

7. Lai, T., Endrullis, J., Moss, L.S.: Majority digraphs. Proc. Am. Math. Soc. **144**(9), 3701–3715 (2016)
8. Rescher, N.: Plurality quantification revisited. Phil. Inq. **26**, 1–6 (2004)
9. Rodgers, P.: A survey of Euler diagrams. J. Visual Lang. Comput. **25**, 134–155 (2014)
10. Sato, Y., Mineshima, K.: Human reasoning with proportional quantifiers and its support by diagrams. In: Jamnik, M., Uesaka, Y., Elzer, S. (eds.) Diagrammatic Representation and Inference. LNAI, vol. 9781. Springer, Cham (2016). https://doi.org/10.1007/978-3-319-42333-3_10

New Representation Systems

New Representations of Modal Functions

Pedro Falcão[(✉)]

São Paulo, Brazil
pedroalonsofalcao@gmail.com

Abstract. In this paper we show how to represent any modal function (i.e. a function expressed by a formula of the propositional modal logic S5) as a tuple of truth-functions, and we provide a nice graphical representation of the modal functions in terms of colorations of edges of certain complete bipartite graphs.

Keywords: Modal logic · Clone theory · Graph theory

1 Introduction

For the sake of clarity, in the first part of this paper we will deal with diagrams representing some simple operations on simple structures; in the second part we will deal with diagrams representing less simple operations on the same simple structures; and in the third part we will deal with different kinds of operations on some less simple structures. Some rather technical definitions will be necessary at some points, but we will try to avoid heavy formalism. The objective of this paper is to point to the train of thought that leads to the more complex diagrams at the end of Sect. 4, hoping that you will find them as nice as we do.

2 Boolean Operations

2.1 Unary Boolean Operations

The *unary Boolean operations* are: identity (id), negation ($\neg$), *verum* ($\top$), and *falsum* ($\bot$). We start by considering the action of these operations on the structures A1 and A2, shown in the Fig. 1.

I express my gratitude to Rodrigo Ramos for the fruitful discussions that we had on the topics presented here, and for the great help he provided while turning this material into the LaTeX format. I would also like to thank Roderick Batchelor, Melina Bertholdo, Luiza Ramos, Tomás Troster, and the reviewers for their comments on the preliminary versions of this paper, and Levi Magalhães for the enthusiasm we shared when considering the prototypes of these diagrams.

© Springer Nature Switzerland AG 2021
A. Basu et al. (Eds.): Diagrams 2021, LNAI 12909, pp. 271–278, 2021.
https://doi.org/10.1007/978-3-030-86062-2_28

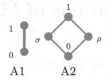

Fig. 1. Structures A1 and A2

(A1 and A2 are part of the family of n-dimensional cubes. A1 can be interpreted as the set of subsets of a set with one element, and A2 as the set of subsets of a set with two elements. Modal logicians might interpret the elements of these structures as intensional propositions, i.e. sets of possible worlds, on a model with n possible worlds.)

A nice and simple way to represent the unary Boolean operations on A1 is shown in Fig. 2, where the operations are represented by directed graphs denoting their actions over the structures.

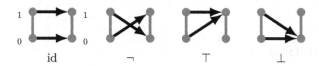

Fig. 2. Graph diagrams representing the unary Boolean operations on A1.

We can use the same idea to represent the unary Boolean operations on A2, as presented in Fig. 3.

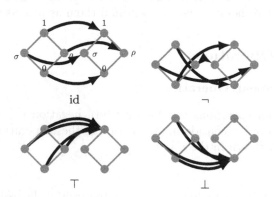

Fig. 3. Graph diagrams representing the unary Boolean operations on A2

Nice and simple as they are, these representations have serious limitations: they are only fit for dealing with unary operations. A method to represent binary operations using graphs is provided in the next section.

2.2 Binary Boolean Operations

The following definition can be easily generalized for (not only Boolean) operations of arbitrary arities.

The *graph representation of a binary operation on a finite set A* is a function *from* the edges of the complete bipartite graph whose parts are copies of A *to* the elements of A. An example considering a binary operation on A1 is given in Fig. 4.

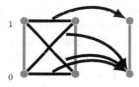

Fig. 4. Example of a graph representation of a binary operation (conjunction) on A1

A way to make this representation more compact is to give colors to the nodes and to the edges of the graph. If we use white for 1 and black for 0, the (first half of the) binary Boolean operations on A1 will look like the ones in Fig. 5.

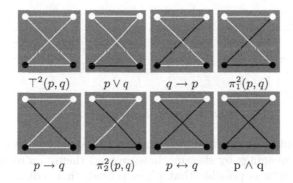

Fig. 5. Some binary Boolean operations on A1

In each of the diagrams in Fig. 5, the nodes on the left represent the range of values of the variable p, while the nodes on the right represent the range of values of the variable q. This means that we can associate the edges connecting these nodes with the rows of a truth-table. Colouring these edges is in a way equivalent to filling a truth-table with truth-values. This approach to the Boolean operations on A1 is in the spirit of Alharbi [1], where Boolean operations of higher arities are also considered.

We will consider the graphs representing binary (including *modal*) operations on A2 (and on A3), but we will first return to the unary operations.

3   Modal Operations

3.1   Unary Modal Functions

A nice way to represent the modal functions in general, and particularly the unary modal functions, is to use the method of subtables, developed in Massey [5] by restricting the semantics presented in Kripke [4] to the propositional case. In this context, a *model for the propositional modal logic S5* is defined as a pair $\langle W, a \rangle$, where W is a non-empty subset of the set of *all* models for *classical* propositional logic and a $\in$ W.

When we restrict ourselves to the case of a single variable, the set of all models of classical propositional logic is simply the usual truth-table for a single variable, and therefore the set of its non-empty subsets can be represented in the *unary modal table* in Fig. 6. This modal table consists of three *subtables*. In its leftmost side, the first subtable contains both the values T and F, the second subtable contains only the value T, and the third subtable contains only the value F. The first subtable represents the cases where p is contingent (a proposition is *contingent* if it is both possibly true and possibly false), while the second and third subtables represent the cases where p is *rigid*, i.e. not contingent.

It is not widely known that there are only 16 classes of non-equivalent formulas with a single variable in propositional S5. This simple fact can be seen by appreciating that every *unary modal function* corresponds to a distribution of T's and F's in the unary modal table (cf. Fig. 6):

p	$\top$	$\Diamond$	$\neg\Box$	∇	$\neg\nabla^-$	id	$\neg\Diamond\lor\nabla^+$	∇^+	$\neg\nabla^+$	$\Box\lor\nabla^-$	$\neg$	∇^-	Δ	$\Box$	$\neg\Diamond$	$\bot$
T	T	T	T	T	T	T	T	T	F	F	F	F	F	F	F	F
F	T	T	T	T	F	F	F	F	T	T	T	T	F	F	F	F
T	T	T	F	F	T	T	F	F	T	T	F	F	T	T	F	F
F	T	F	T	F	T	F	T	F	T	F	T	F	T	F	T	F

Fig. 6. Unary modal functions

Since many of these functions are fairly exotic for many readers, we will try to clarify their meanings by showing how we read them. $\top p$ reads 'a tautology with p'; $\Diamond p$ reads 'it is possible that p'; $\neg\Box p$ reads 'it is not necessary that p'; ∇p reads 'it is contingent that p'; $\neg\nabla^- p$ is read 'it is not contingently false that p'; id(p) is read p; $\neg\Diamond \lor \nabla^+ p$ is read 'it is either impossible or contingently true that p'; $\nabla^+ p$ is read 'it is contingently true that p'; $\neg\nabla^+ p$ is read 'it is not contingently true that p'; $\Box \lor \nabla^- p$ is read 'it is either necessary or contingently false that p'; $\neg p$ is read 'not p'; $\nabla^- p$ is read 'it is contingently false that p'; Δp is read 'it is rigid that p', $\Box p$ is read 'it is necessary that p'; $\neg\Diamond p$ is read 'it is impossible that p' and, finally, $\bot p$ is read 'a contradiction with p'. The notation used here for the more exotic unary modal functions was devised by Roderick Batchelor. See [2].

3.2   Unary Modal Operations on A2

We now consider the unary modal functions as operations on A2. A way to simplify the representation of unary Boolean operations on A2 (cf. Fig. 3) is to use colors to denote the nodes, and coloured circles to denote their images. This would give us something like what is shown in a diagonal of Fig. 7. Notice that there are $4^4 = 256$ unary operations on A2, and while we represent here only the 16 that happen to be modal operations, the diagrams could represent the other 240 as well. When we think of the unary modal functions as operations on A2 (Fig. 7), we associate the gray colors (white and black) with the rigid values (necessary and impossible) and the colorful colors (green and yellow) with the contingent values (contingently true and contingently false). The operations are presented here in the order in which they appear in Fig. 6; we label the operations using the *moody truth-functions*, presented in the next subsection.

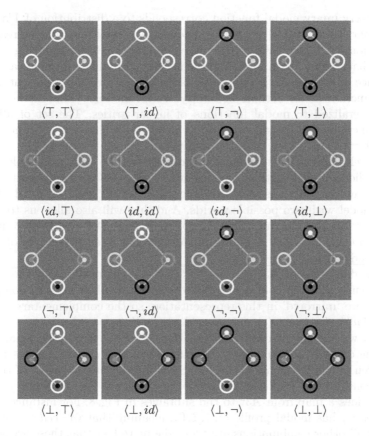

Fig. 7. Unary modal operations on A2 – labeled as moody truth-functions (Color figure online)

3.3   Modal Functions as Moody Truth-Functions

There is yet another way to think about the unary modal functions (or indeed the modal functions in general): we can think of them as *moody truth-functions* (cf. Falcão [3]). Notice that in Fig. 6 we have juxtaposed the complementary subtables, forming, as it were, a second (unary) truth-table. So we can think of a unary modal function as a pair $\langle f_1, f_2 \rangle$ where $f_i (i \in \{1, 2\})$ is a unary Boolean function and the rule for $\langle f_1, f_2 \rangle(p)$ is: if p is contingent, apply f_1; if p is not contingent, apply f_2. For instance, $\Diamond p$ can be represented as $\langle \top, id \rangle(p)$, and $\Box p$ can be represented as $\langle \bot, id \rangle(p)$. When we think about a moody truth-function $\langle f_1, f_2 \rangle$ as an operation on A2, the rule is: if p is colorful, apply f_1; if p is gray, apply f_2.

4   Binary Modal Functions

Just as every unary modal function corresponds to a distribution of T's and F's in the unary modal table (cf. Fig. 6), the binary modal functions correspond to distributions of T's and F's in the binary modal table, shown in Fig. 8. Notice that we have again juxtaposed complementary subtables, forming, as it were, a sequence of 8 classical (2-ary) truth-tables, and so we can represent a binary modal function as an octuple of binary truth-functions. This method can be easily generalized to modal functions of higher arities. The set of all binary modal functions will be called μ^2, and the set of all binary truth-functions will be called τ^2.

It is well known that a formula of propositional S5 with n variables is valid iff it is valid in models with 2^n possible worlds. This is what allowed us to treat the unary modal functions as certain operations on A2, since A2 can be thought of as a model with two possible worlds. And that will also allow us to consider the binary modal functions as certain operations on A4. But in fact, it is also fruitful to consider them as operations on a chain of substructures of A4, viz. A1, A2 and A3.

In the diagrams for A3 (cf. Fig. 10) the warm colors stand for the sets with two elements, and the cold colors for sets with one element. The complementary sets are 'mirrored' in this representation, so the complementary pairs are: $\langle white, black \rangle$, $\langle red, purple \rangle$, $\langle orange, blue \rangle$, $\langle yellow, green \rangle$.

When we use the moody truth-functions to think about the binary modal operations on (e.g.) A3, in order to determine which $g_i (i \in \{1, ..., 8\}) \in \tau^2$ "acts" for a given pair of elements of A3, we check for the *modal profile* of that pair, i.e. we consider the sequence $\langle \Diamond(\pm p_1 \wedge \pm p_2) \rangle$ where $\pm p_i$ is either p_i or $\neg p_i$. This provides a map from $A3^2$ to the subtables in Fig. 8. For instance, the pair $\langle red, orange \rangle$ has modal profile $\langle 1, 1, 1, 0 \rangle$, meaning that $\Diamond(\neg red \wedge \neg orange)$ is false but all other combinations are true (the mere fact that these are sets with two possible worlds is enough to conclude that). So the line F F is (the only line) absent from its corresponding subtable, and so $f(red, orange) = g_2(red, orange)$. g_2 will also "act" for the pair $\langle black, black \rangle$, as it has the modal profile $\langle 0, 0, 0, 1 \rangle$. It is interesting that no pair of elements of A3 has the modal profile $\langle 1, 1, 1, 1 \rangle$, i.e. in order to find a pair of *independent propositions* (represented in the first subtable in Fig. 8), and so use g_1, we need to resort to A4.

The representations of the binary operations presented on Figs. 9 and 10 might perhaps look a bit clumsy, but we will argue that they are pretty readable. Take, for instance, the representation of π_1^2. π_1^2 is read 'the projection of the first argument', and its representations on Figs. 9 and 10 do look like projections. Considering the representations of $p \to q$ we notice some interesting facts about the implication: it is not symmetric, it is always true when the antecedent is impossible, it is always true when the consequent is necessary, it is always true when the antecedent is equal to the consequent, and it is at least as big as its consequent. The representation of $p \leftrightarrow q$ is also very nice, as the black lines show exactly the relation of complementarity between the elements of A2 and A3.

So far, we have not considered any binary modal operation on A2 and A3. Perhaps the most famous binary modal function is the *strict implication* $\Box(p \to q)$. Its representation can be obtained by painting black each non-white edge in the graph for $p \to q$.

p q	$f \in \mu^2$	$g_i \in \tau^2$	p q	$f \in \mu^2$	$g_i \in \tau^2$	p q	$f \in \mu^2$	$g_i \in \tau^2$	p q	$f \in \mu^2$	$g_i \in \tau^2$
T T			T T			T F			T T		
T F		g_1	T F		g_3	F T		g_5	F T		
F T			F F			F F			T F		g_7
F F			F T			T T			F F		
T T			T T			T T			T T		
T F		g_2	F T		g_4	T F		g_6	F F		g_8
F T			F F			F T			T F		
F F			T F			F F			F T		

Fig. 8. Binary modal table

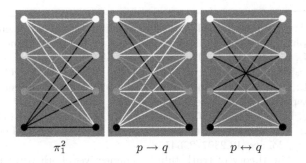

$\pi_1^2 \qquad\qquad p \to q \qquad\qquad p \leftrightarrow q$

Fig. 9. Binary operations on A2

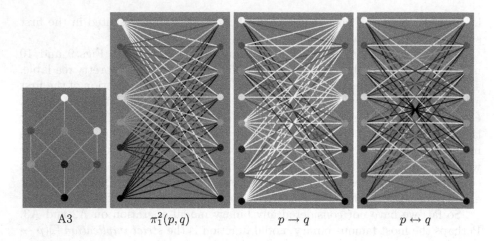

Fig. 10. Binary operations on A3 (Color figure online)

5   Conclusion

We've built a computer program that takes as input a list of binary modal functions (expressed either as a formula of propositional S5 or as an octuple of truth-functions) and returns a graphical representation of their actions over the structures A1, A2 and A3. Its full version is not available yet, but a sample with a given list can be seen at: https://editor.p5js.org/osnola2/present/XAlP8ohWv. To change the displayed function, the user should press the keys a or z. The functions are labeled in the moody truth-functions notation. All the diagrams in Figs. 5, 7, 9 and 10 were generated by this program.

References

1. Alharbi, E.: Truth graph: a novel method for minimizing boolean algebra expressions by using graphs. In: Pietarinen, A.-V., Chapman, P., Bosveld-de Smet, L., Giardino, V., Corter, J., Linker, S. (eds.) Diagrams 2020. LNCS (LNAI), vol. 12169, pp. 461–469. Springer, Cham (2020). https://doi.org/10.1007/978-3-030-54249-8_36
2. Batchelor, R.: Clone theory: modal functions (2020, unpublished manuscript)
3. Falcão, P.: Aspectos da teoria de funções modais. Master's thesis, University of São Paulo (2012). https://doi.org/10.11606/D.8.2012.tde-11042013-104549
4. Kripke, S.A.: A completeness theorem in modal logic. J. Symb. Log. **24**(1), 1–14 (1959). https://doi.org/10.2307/2964568
5. Massey, G.J.: The theory of truth tabular connectives, both truth functional and modal. J. Symb. Log. **31**(4), 593–608 (1966). https://doi.org/10.2307/2269695

Diagramming Imprecise and Incomplete Temporal Information

Hakob Barseghyan(✉) 📷

University of Toronto, Toronto, ON M5S 11, Canada
hakob.barseghyan@utoronto.ca

Abstract. Incomplete and imprecise temporal data is abundant in various branches of science and technology as well as everyday life (e.g., "*A* began after 1066 but before 1069 and ended after 1245", "*B* took place no later than 156 BC"). While point-circles and lines/bars have been traditionally used to depict *precise* temporal points and intervals, it is unclear how imprecise and incomplete temporal data can be effectively visualized or even represented. This paper suggests an intuitive diagrammatic notation for visualizing both imprecise and incomplete temporal information. It suggests using traditional whiskers with edges to depict temporal imprecision and whiskers without edges to depict incomplete temporal entities. This notation can be easily incorporated into linear temporal visualizations, such as historical timelines, Gantt charts, and timetables, to identify gaps in temporal information. The paper lays down the diagrammatic elements of the notation and illustrates their applicability to all standard relations between temporal entities. It also shows how these elements can be combined to produce complex timelines. Some possible future directions are also outlined.

Keywords: Imprecise temporal data · Incomplete temporal data · Temporal diagramming · Temporal representation

1 Introduction

Temporal data is often imprecise and/or incomplete. Such imprecision and incompleteness are widespread in all branches of academic history, in museum and library studies, in medicine, as well as everyday life. Temporal records often involve imprecise and/or incomplete *intervals* (e.g., "*A* began sometime in December 1999 and ended in January 2000", "*B* ended sometime before 1456") and *points* (e.g., "*P* happened between July 1 and July 25, 1543", "*Q* took place after 156 BC"). While visualizing precise and complete temporal data is relatively straightforward (point-circles and lines/bars have been traditionally used to depict respectively temporal points and temporal intervals), visualizing imprecise and incomplete temporal data is not a simple task.

Part of the problem has to do with the absence of a consensus on how imprecise and incomplete temporal information is to be *represented*. Most approaches to representing and reasoning about temporal data have traditionally been concerned with *precise* data [5, 6, 12, 16]. In the recent years, there has been an increased focus on representing and

© Springer Nature Switzerland AG 2021
A. Basu et al. (Eds.): Diagrams 2021, LNAI 12909, pp. 279–286, 2021.
https://doi.org/10.1007/978-3-030-86062-2_29

reasoning about imprecise temporal information [1, 11]. Yet, despite a growing body of research on imprecise temporal data, there is still work to be done before a standard accepted account of imprecise and incomplete data is reached.

In addition, the existing approaches to visualizing imprecise and incomplete temporal information are far from intuitive, not easily scalable, and/or difficult to incorporate into traditional historical timelines. For example, the PlanningLine approach [3] can be useful in project management and planning, but its scalability is questionable. Specifically, it is unclear how the black imprecision caps can stack on large historical timelines. The approach also assumes the use of color to convey meaning, which may not be ideal from accessibility standpoint. Another suggestion is to visualize temporal imprecision by paint strips, bricks, and weights [9]. While intuitive, this approach is not easily scalable as it risks introducing too much clutter, especially on large historical timelines. The common issues with linear timelines, such as scalability, are addressed by the triangular model where temporal entities are depicted in a two-dimensional space [7, 15]. The triangular model can handle large amounts of temporal information, yet it assumes a steep learning curve and cannot be easily incorporated into linear timelines used in historical fields. Thus, there is still a need in a relatively simple notation for portraying incomplete and imprecise temporal information on linear timelines.

This paper offers a simple and intuitive extension to customary linear temporal visualizations commonly found on historical timelines, Gantt charts, timetables, etc. Central to this notation is the use of traditional whiskers with edges to depict temporal imprecision and whiskers without edges to depict incomplete data. With this standardized approach to diagraming precise/imprecise and complete/incomplete temporal intervals and temporal points, this paper aims to provide a useful tool for depicting various temporal entities and relations.

The paper is structured as follows. Section 2 lays down the main diagrammatic elements of the notation. Section 3 demonstrates that the notation allows easy visualization of all standard relations between temporal intervals, between intervals and points, and between points. Section 4 shows how these basic visual elements can be combined to produce more complex temporal diagrams. Section 5 discusses some possible future directions in diagraming and representing temporal information.

2 Building Blocks

There is an ongoing debate in literature concerning the primitive temporal unit – temporal points (instants) or temporal intervals (periods). Some authors consider time points as a primitive unit and construct intervals as bounded by two points [10, 14], while others take intervals as primitive and construe time points as a very short interval or an interval with a coinciding beginning and end [4]. There are also accounts where points and intervals are ascribed the same level of importance [8]. Here, I will not take any sides on this issue, but will provide diagrammatic tools for visualizing both temporal points and temporal intervals. Such an impartial approach with respect to the underlying temporal ontology will help ensure that the notation can visualize any imprecise or incomplete temporal information and can be effectively used by all parties regardless of their position on which temporal unit(s) should be considered as primitive.

Achich et al. construe imprecise time *points* as disjunctive ascending sets [1]. For example, "*A* took place sometime in the 1980s" involves an imprecise temporal point, which they represent by the disjunctive ascending set {1980...1989}. Regardless of whether one agrees with this approach, it is uncontroversial that temporal uncertainty typically involves lower and upper bounds that indicate the beginning and the end of the imprecision interval. Thus, an imprecise temporal point P can be characterized by its beginning $P^{(B)}$ and end $P^{(E)}$. In traditional linear diagrams temporal points are usually depicted as little point-circles. Adding *whiskers* to such point-circles to visualize temporal imprecision seems to be an intuitive approach (Fig. 1).

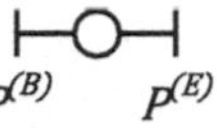

Fig. 1. A depiction of a temporal point that took place sometime between $P^{(B)}$ and $P^{(E)}$.

The same approach can be used to indicate the *incompleteness* of temporal data. In some cases, we only know that P happened sometime before $P^{(E)}$, or only that it happened sometime after $P^{(B)}$. Such instances of incomplete temporal data can be visualized by skipping the edge of the respective whisker (Fig. 2).

Fig. 2. Depictions of $\left(P^{(B)} < P\right)$ and $\left(P < P^{(E)}\right)$ respectively.

When the timing of an event is completely unknown, we can omit both edges to indicate that. To see why such a usage might be useful, we must appreciate that incompleteness is best understood in a *relative* rather than *absolute* sense. After all, for any recorded event, we can, at the very least, say that it happened sometime between the Big Bang and the present. But such an approach is not very useful in either everyday situations or academic settings. A historical record stating that a certain political event happened in the Middle East before the Late Bronze Age collapse can, of course, be amended to say that it happened *sometime after humans settled in the Middle East* and before the Late Bronze Age collapse, but that may not be helpful from the perspective of historical scholarship. For a historian, such an amended record is as good as incomplete, since it doesn't indicate a beginning boundary relative to the temporal period under study. Similarly, a medical record stating that a patient had a surgery sometime before 2020 can be easily amended to say that the patient had a surgery *sometime after their birth* and before 2020. Yet, such an amendment doesn't say much to a physician, who wishes to know the age of the patient at the time of the surgery. Thus, the incompleteness of temporal data is best understood as not absolute but relative incompleteness: arguably all incomplete temporal records are incomplete relative to the context that defines the specific temporal period within which they are expected to be dated. Therefore, it might be useful to indicate that the timing of a certain event is completely unknown within that context and invite future research.

The same approach can be easily extended to temporal *intervals*. Temporal interval A can be understood as a temporal entity with duration, i.e. as bounded by the beginning A^- and end A^+, such that $(A^- < A^+)$. As with temporal points, both the beginning and the end of an interval can be imprecise. Such an imprecision can be easily depicted by adding imprecision whiskers to bars traditionally used on timelines (Fig. 3).

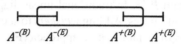

$$A^{-(B)} \qquad A^{-(E)} \qquad\qquad A^{+(B)} \qquad A^{+(E)}$$

Fig. 3. A depiction of an interval that began sometime between $A^{-(B)}$ and $A^{-(E)}$ and ended sometime between $A^{+(B)}$ and $A^{+(E)}$.

Temporal intervals can be not only imprecise but also *incomplete*; such situations are common in all fields dealing with temporal information. We can visualize such scenarios on timelines by omitting the respective edges of imprecision whiskers (Fig. 4).

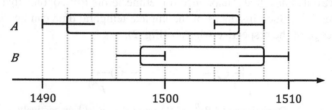

Fig. 4. Interval A started sometime after 1490 and ended between 1504 and 1508. Interval B started before 1500 and ended before 1510.

As in the case of temporal points, sometimes we have no knowledge about the beginning or the end of a temporal interval (once again, the incompleteness here is to be understood as *relative* to the given context). Such scenarios can be depicted by skipping both edges of the imprecision whisker.

When the temporal information is complete and known with precision, the whiskers are no longer needed and the notation collapses into the customary usage of points and bars on linear temporal diagrams. This adds to the intuitiveness of the notation and makes it easy to incorporate into traditional timelines.

3   Applicability to Temporal Relations

It is essential to ensure that the new notation can visualize the whole range of relations that can obtain between temporal entities. The latter include relations between temporal intervals, relations between temporal intervals and temporal points, and relations between temporal points. Let us consider these in turn.

The traditional taxonomy of relations between precise intervals by Allen [4] has been recently extended to also apply to imprecise temporal intervals [1]. The current notation allows to visualize all of these relations (Fig. 5).

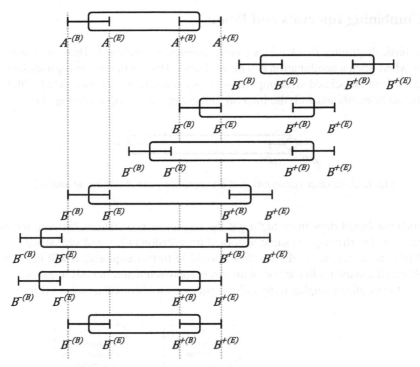

Fig. 5. Visualizations of relations between two intervals (from top to bottom): Before (A, B), Meets (A, B), Overlaps (A, B), Starts (A, B), During (A, B), Ends (A, B), and Equals (A, B).

The relations between imprecise time interval and imprecise time point, defined by Achich et al. [1], are also straightforwardly visualizable in this notation (Fig. 6).

It is obvious that the relations between imprecise time points (i.e. *before*, *after*, and *equals*) can also be easily visualized using this notation.

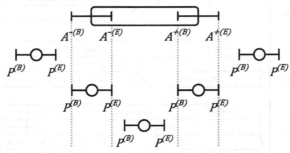

Fig. 6. Visualizations of relations between temporal point P and interval A (from left to right): Before (P, A), Starts (P, A), During (P, A), Ends (P, A), and After (P, A).

4   Combining Intervals and Points

These basic diagrammatic elements can be seamlessly combined. There are many scenarios where such a combination can be advisable. Thus, if the beginning and/or end of an interval are associated with important events, the time points associated with these events can be combined with the interval to highlight the association (Fig. 7).

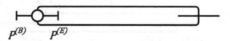

$$P^{(B)} \qquad P^{(E)}$$

Fig. 7. A succinct visualization of an imprecise point P starting an interval.

Such combined depictions highlight the respective associations between temporal entities, avoid portraying the same historical imprecision twice, and save space.

While the scalability of the notation should be further explored, it will likely be as scalable as the standard linear bar/point-circle approach that it intends to extend. It can be used to visualize complex temporal information on historical timelines (Fig. 8).

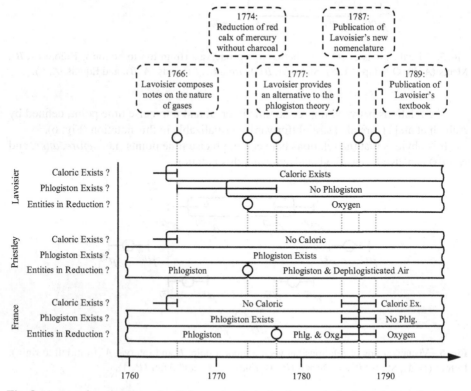

Fig. 8. A complex timeline visualizing some key developments in late eighteenth-century chemistry (adopted from [13]).

The timeline presents the evolution of answers provided by two individual epistemic agents – Lavoisier and Priestley – as well as the community of French chemists to three key questions: those concerning the existence of caloric and phlogiston and the question of which entities are involved in reduction reactions. It highlights the relations between some key events and transitions in the respective agents' belief systems. For instance, it shows that Priestley's 1774 experiment of the reduction of red calx of mercury resulted in changes both in Lavoisier's and Priestley's theories of reduction, but didn't immediately affect the beliefs of the wider community of French chemists.

In addition to *summarizing* our historical knowledge of the episode, the timeline also highlights many important *gaps* in our knowledge. The various whiskers on the timeline identify the imprecision and/or incompleteness of our historical records. For example, it shows that Lavoisier rejected the idea of phlogiston sometime between 1766 and 1777 and invites future research on locating a more precise timing of this transition.

5 Future Directions

Taking this notation as a starting point, we can identify a number of future research directions. First, it will be interesting to see if the notation can be extended to apply to:

- *Inclusive imprecision ranges*, e.g., "*A* took place on or before 1177 BCE" or "*B* took place on or after March 20, 2000".
- *Imprecision intervals with holes*, e.g., "*A* started sometime between 1989 and 1991, but not in 1990".
- *Known order with unknown timing*, e.g. "A happened before B, and B happened before C, but the timing of these events is unknown". Situations like these are abundant in academic history, where we often know that a sequence of events happened within, say, a lifetime of a person or within a certain historical period, but we lack any knowledge concerning their timing.
- *Uncertain temporal information* (where 'uncertainty' is understood as the degree of doubt that the respective epistemic agent has towards a given temporal information), e.g., "Maybe *P* happened in 1177 BCE", or "*Q* likely took place after *P*" (see [2] and references therein).

The notation also indicates that a new approach to *representing* incomplete temporal entities and relations between them is needed. While we can easily depict sequences of incomplete points and intervals, the definitions of temporal relations given by Achich et al., such as *before*, *meets*, *overlaps*, or *starts*, do not apply to *incomplete* temporal entities [1]. Their workaround is to reduce partial incompleteness to imprecision; for instance, they interpret "the journey begins by the June 5, 2019" as saying that it begins sometime between June 3, 2019 and June 7, 2019. Leaving aside the linguistic objection that in standard English, "by time *P*" means "before or at time *P*" (but not slightly *after P*), one can argue that the introduction of June 3, 2019 as a beginning bound seems rather arbitrary. A historian will likely object to such random amendments to historical records. Thus, it is important not to reduce incompleteness to imprecision, but to provide a set of definitions for temporal relations that will work with *incomplete* data.

References

1. Achich, N., Ghorbel, F., Hamdi, F., Metais, E., Gargouri, F.: Representing and reasoning about precise and imprecise time points and intervals in semantic web: dealing with dates and time clocks. In: Hartmann, S., Küng, J., Chakravarthy, S., Anderst-Kotsis, G., Tjoa, A.M., Khalil, I. (eds.) DEXA 2019. LNCS, vol. 11707, pp. 198–208. Springer, Cham (2019). https://doi.org/10.1007/978-3-030-27618-8_15
2. Achich, N., Ghorbel, F., Hamdi, F., Métais, E., Gargouri, F.: Approach to reasoning about uncertain temporal data in OWL 2. Procedia Comput. Sci. **176**, 1141–1150 (2020)
3. Aigner, W., Miksch, S., Thurnher, B., Biffl, S.: PlanningLines: novel glyphs for representing temporal uncertainties and their evaluation. In: Proceedings of the Ninth International Conference on Information Visualisation (IV 2005), pp. 457–463 (2005)
4. Allen, J.: Maintaining knowledge about temporal intervals. Commun. ACM **26**, 832–843 (1983)
5. Anagnostopoulos, E., Batsakis, S., Petrakis, E.G.M.: CHRONOS: a reasoning engine for qualitative temporal information in OWL. Procedia Comput. Sci. **22**, 70–77 (2013)
6. Artale, A., Franconi, E.: A survey of temporal extensions of description logics. Ann. Math. Artif. Intell. **30**, 171–210 (2000)
7. Billiet, C., Van de Weghe, N., Deploige, J., De Tré, G.: Visualizing and reasoning with imperfect time intervals in 2-D. IEEE Trans. Fuzzy Syst. **25**(6), 1698–1713 (2017)
8. Bochman, A.: Concerted instant-interval temporal semantics I: temporal ontologies. Notre Dame J. Formal Logic **31**(3), 403–414 (1990)
9. Chittaro, L., Combi, C.: Visualizing queries on databases of temporal histories: new metaphors and their evaluation. Data Knowl. Eng. **44**, 239–264 (2003)
10. Dean, T., McDermott, D.: Temporal data base management. Artif. Intell. **32**, 1–55 (1987)
11. Ghorbel, F., Hamdi, F., Métais, E.: Dealing with precise and imprecise temporal data in crisp ontology. Int. J. Inf. Techno Web Eng. (IJITWE) **15**(2), 30–49 (2020)
12. Gutierrez, C., Hurtado, C., Vaisman, A.: Temporal RDF. In: Gómez-Pérez, A., Euzenat, J. (eds.) ESWC 2005. LNCS, vol. 3532, pp. 93–107. Springer, Heidelberg (2005). https://doi.org/10.1007/11431053_7
13. Levesley, N., Barseghyan, H.: Diagraming late-eighteenth century chemistry. In Barseghyan, H., Patton, P., Shaw. J. (eds.) Visualizing Worldviews (Forthcoming)
14. McDermott, D.: A temporal logic for reasoning about process and plan. Cogn. Sci. **6**(2), 101–155 (1982)
15. Qiang, Y., et al.: Analysing imperfect temporal information in GIS using the triangular model. Cartogr. J. **49**(3), 265–280 (2012)
16. Zekri, A., Brahmia, Z., Grandi, F., Bouaziz, R.: τOWL: a systematic approach to temporal versioning of semantic web ontologies. J. Data Seman. **5**(3), 141–163 (2016)

Comics and Diagrams: An Introductory Overview

Andrea Tosti[✉] ⓘD

Lancaster University, Bailrigg, Lancaster L1 4YW, UK
a.tosti@lancaster.ac.uk

Abstract. Although both so-called 'data comics' and 'comics geographies' fields have been defined as emerging, there is a lack of structured and multidisciplinary studies that deal with comics' diagrammatic nature. The parallelism between comics and diagrams, dear to many comics makers and some scholars, is more than a mere graphic suggestion or similarity. Unlike the written word and other visual art forms, comics, through the multi-vectorial narrative skills typical of the page layout, overcomes alphabetic writing's linearity to open up to synchronic and parallel space-time narratives. Comics also favours ellipses, spatial dislocations, and micro-narrations in larger narratives more naturally than in other artistic and narrative forms. Moreover, comics make possible, differently than in literature, a strong involvement of the reader in constructing alternative paths. The grid is the element that brings comics back under the category of diagrams through its ability to temporalize space and create hierarchies and relationships between the parts (panels). Through some examples, this paper offers an introductive overview of the many possibilities offered by a diagrammatic reading of comics, demonstrating how even the most straightforward grid configurations can convey complex concepts.

Keywords: Comics · Diagrams · Maps

1 Introduction: The Impossible Definition(s)

Describing what comics and diagrams are is a challenging job. As for diagrams, a non-specialist but exhaustive definition could be: "Diagrams are schematic figures or patterns comprising lines, symbols, or words to which meanings are attached" [2, p. 397]. Eddy's definition of diagrams fit comics. We could describe comics as schematic figures AND patterns comprising lines AND/OR symbols AND/OR words to which the meanings are attached. Both diagrams and comics do not constitute a hybridization of the elements listed above but their own categories. Furthermore, as Cates notes [4], many comics use a pictogrammatic graphic style used in diagrammatic representation.

The definition of comics is also problematic. As Groensteen said: "searching for the essence of comics is to be assured of finding not a shortage but a profusion of responses" [8, p. 12]. The broad definition(s) of comics includes objects very different from each other due to the context they are created and used. Although this vast number of variants,

© Springer Nature Switzerland AG 2021
A. Basu et al. (Eds.): Diagrams 2021, LNAI 12909, pp. 287–294, 2021.
https://doi.org/10.1007/978-3-030-86062-2_30

every comics reader knows comics are self-evident. Paraphrasing what Wittgenstein said about games, if we look at comics, we won't see "something that is common to all, but similarities, relationships, and a whole series of them at that. To repeat: don't think, but look!" [1, p. 209].

This invocation is fascinated when it refers to comics, a medium that could confirm that seeing is itself a form of thinking able to override reading and overcome the logo-centric and linear approach to information. As Miodrag says, "All narrative forms can, analeptically or proleptically, override their diegetic sequencing, but […] only comics can potentially override textual progression" [13, p. 143], through reading the page that can "be seen and read in both linear and nonlinear, holistic fashion" [10, p. 48].

1.1 Comics, Maps and Diagrams

Holistic and nonlinear reading is also typical of cartography. The relationships between comics and maps have been widely explored [15, 16, 18]. As the eye can wander on the map, building imaginary routes so it can also do so on the comics page, violating the 'Z-Path' (left to right, top to bottom) [5, 11] in favour of erratic paths [17] as it happens for infographics [12]. In fact, comics, through the multi-vectorial (or 'multi-order') narrative skills typical of its page layout, overcomes the sequentiality to open up also to synchronic spatial and temporal narratives [6]. Furthermore, comics and maps can work together to create new paths within a space that is both narrative and geographic [16]. Therefore, multilinearity is an attribute of both maps and comics, but Hadler [9] notes that it also belongs to diagrams. Within certain limits, simultaneity and multilinearity allow free exploration of maps, diagrams and comics: one (the author) first builds and then (the reader) explores an ambient made up of space, discrete units or logical operators. Comics, therefore, live in a constant tension between a linear dimension, which involves the individual graphic discrete units (the panels), and a spatial dimension that concerns the page's layout. The plastic and semantic relationships between these images - adjacent and not - which their "coexistence in praesentia" [8, p. 23] makes possible are the ones that make comics into a text.

Beyond the purely narrative aspect, in this article, we are interested in whether and to what extent comics can be a device that the reader can reshape to transform the latter into a user capable of creating or discovering new meanings. As is the case with diagrams, considering that "the essential thing about the diagram is that it is made in order for something to emerge from it, and if nothing emerges from it, it fails" [3, p. 290]. There are comics in which the diagrammatic character is made explicit [e.g., 4], but 'diagrammaticity' is an element present - with varying intensity - even in those comics considered more 'traditional' by a large audience.

1.2 The Grid

The grid is both the framework, the foundations and the surface of the comics building. In common usage, the term 'grid' indicates a regular subdivision of the space into equal portions, but in the comics field, it is used colloquially to indicate any organizational structure of the comics page: the layout. In this article, we will deal with table grid, colloquially also called *waffle-irons*. More generally, the grid is also the element that creates time through the fragmentation of space, both geographic and typographical.

Fresnault-Deruelle [7] notes that the signifying function of spatialization for the page's narrative purposes or, better, of the cage ('grid'), emerges overwhelmingly. More than other elements, the grid brings comics back under diagrams, as it can introduce hierarchical principles and consequentiality between the parts in the comics' narrative, creating a discontinuity between the panels. This discontinuity is recomposed thanks to the wider system represented by the page layout. As Miodrag notes [14], "Narrative breakdown - the dispersal of content into discrete, interdependently interwoven units - has few parallels in other media". Groensteen suggests that spatial and topographical parameters ('spatiotopie', according to the author's neologism[1]) condition the ways in which the story is told. The page composition is a communication resource in its own right. Whatever the comics are, a grid itself warns us that narration is happening on the page. "But comics is not only an art of fragments, of scattering, of distributions; it is also an art of conjunction, of repetition, of linking together" [8, p. 22]. To summarize, the grid in the comics plays several vital functions: it connects separate moments but physically co-present on the page; assigns to the latter different hierarchy importance; transforms space into a narratively oriented temporality, that is, it allows the passage from a sequential narrative to a spatial narrative. Furthermore, the grid creates the possibility of spatial relationships not necessarily temporal or chronological, but that also express logical associations and correlations.

Naturally, more complex grids will correspond to greater complexity both of the hierarchy and connections between the panels and the mobility of the gaze, but this does not mean to affirm the existence of "'zero degree' of spatiotopical expression. On the contrary, "[the regular grids] express a vision of the world founded on the notion of order, on Cartesian logic, on rationality" [8, p. 49].

In this article, we will deal above all with the apparently more straightforward configurations, precisely to demonstrate both there is no zero degree of spatial expression and those apparently simple grids, comparable to lattice graphs, allow complex narratives. However, most of all, comics make possible, differently than in literature, a strong involvement of the reader in constructing alternative narrative paths.

Through the examples shown below, we try to demonstrate how comics can easily and naturally show non-chronological and intertwined narrative lines, different points of view, contemporary events, temporal dislocations and complex information networks proposed to the reader with great clarity.

---

[1] "The spatio-topia is the point of view that can be had on comics before thinking about any single comics, and starting from which it is possible to think about a new performance of the medium" [8, p. 23].

2   The Evolutionary Structure

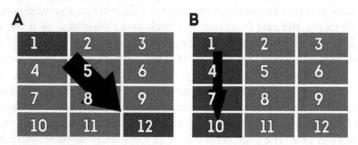

Fig. 1. Description of alternative routes in *A Short History of America*.

In Robert Crumb's one page comics *Short History of America* (1979)[2], the general context is always present to the reader's eye, who can jump from one panel to another. Through twelve panels arranged on four strips, Crumb describes the evolution of a single plot of land to show how anthropization and industrialization had profoundly and negatively affected that land. In this apparently linear grid, we can still identify the different functions performed by the grid.

- It is a linear representation system of time, based on the sequentiality of the alphabetic reading, which allows us to read it following the numerical order in Fig. 1 ('z-path'). The contiguity of the panels gives rise to a consequential and 'evolutionary' narrative, in which each panel acquires meaning in comparison with the one that precedes it and with what follows it, as in a timeline;
- The grid allows overwriting a linear reading thanks to the physical coexistence on the page and the perception of coexistence on a temporal level of the individual panels/moments. Thus, the reader's eye can create its own paths, comparing temporally more distant events, which simultaneously may (Fig. 1B) or may not (Fig. 1A) be spatially contiguous.
- The grid creates an effective system offering a specific type of orientation and an order of an aesthetic-organizational nature, which gives the user/reader an idea of organization and, therefore, of a world independent of each subsequent 'reading'.

We can undertake sinuous comparative paths searching for comparisons not only nonlinear but that find resonances and influences back and forth in time. The erratic wandering of the gaze on the page – experimentally confirmed [17] - is, to varying degrees, typical of comics use but is favoured by at least one row and one column. In the comic strip format, an alphabetical orientation prevails. Also, in the case of comics pages called multi-panel or polyptych, in which a continuous background is divided into

---

[2] A black and white version can be found at <https://klaustoon.wordpress.com/2020/06/19/an-analysis-of-a-short-history-of-america-by-robert-crumb/>, while a color version can be found at <http://www.fumettologica.it/wp-content/uploads/2014/05/A-Short-History-Of-America-By-Robert-Crumb-670x472.jpg>.

panels (like a geographical map divided into sections by parallels and meridians), the general image works both as a map and as an abstract [18, pp. 49–53].

It is not a question here of doubt there is a preferential order of reading or that cartoonists cannot guide the reading path of their readers [11]. It must be stressed that what we have often defined as a wandering gaze up to this point can instead be guided to infer new narratives and information without betraying the general meaning of the work. Comics is a device that can offer different and coherent interpretative possibilities, a generator of multiple meanings that can be used and not just read, just like a diagram.

3 Micro-narrations

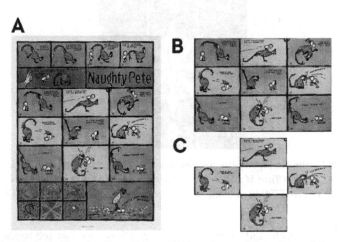

Fig. 2. Charles Forbell, *Naughty Pete,* 7 Sept 1913, and two elaborations of the same comics. (Color figure online)

We have seen how it is possible to select alternative patterns to the usual sequential ones to create different meaning paths or summaries and elaborate hypotheses, experiment with jumps and temporal dislocations and play with the *what if*. In Charles Forbell's Sunday Page proposed below (Fig. 2A), we find an apparently less regular grid. Actually, we can isolate some partially autonomous stories, even if inserted in the general flow. These stories are visually separated into blocks thanks to the skilful use of both the layout and colours. The section we are interested in analyzing is the central one, in which Pete first jumps on the back of the rooster and then fights with the latter (Fig. 2B). The different chromatic dominances (green and yellow) of the alternating panels creates two distinct narrative lines. The yellow, the brightest colour, highlights the three critical moments of the sequence: the jump on the back of the rooster, the unsaddling, the fight (two panels) and while the four green panels are those that tell the intermediate and less convulsive phases.

The micro-sequence in yellow (Fig. 2C) can be effectively read counterclockwise - starting from the first panel at the top. Therefore, the two sequences are integrated because they tell the same episode with different rhythms and intensity but maintain a certain autonomy. Both paths lead to the same conclusion but offering different nuances and syntheses. Both contain the topical information (the unsaddling and the conflict) necessary to move on to the following sections, but following the green path, the final confrontation is caused only by the threat of violent action while, following the yellow one, the struggle is caused by Pete's aggressive reaction.

4 Centripetal Narrations

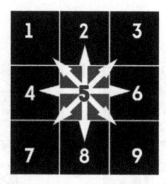

Fig. 3. Diagram of the centripetal structure in Bianca Bagnarelli's comics illustrating the article by Selin Davis *I Couldn't Turn My Abortion Into art* (The New York Times: <https://opinionator. blogs.nytimes.com/2014/07/02/i-couldnt-turn-my-abortion-into-art/>), 2 July 2014.

In this work[3] - a delicate and intense work on abortion - we can see much better how diagrammatic and sequential nature coexist peacefully. Bagnarelli introduces an even stronger idea of circularity of the gaze, using a grid of nine panels and concatenating them in a centripetal manner to the central one. These connections express the woman's relationship with the hypothetical future life she carries within. In explaining the drama, the author finds a synthesis that did not suffer from brutal simplification or incompleteness. Bagnarelli builds a sort of whirlwind tale. The central panel represents the fulcrum of this graphic vortex. The entry point through which the reader begins a comics is conventionally identified in the upper left panel, just as the lower right panel represents the exit point. Here, the most likely entry point for the reader's gaze will be the central panel that imposes the hierarchy of meaning by acting as the graphic centre of the story and the centre of the woman's thoughts. The image of the fetus - that is the focus of the woman's thoughts - also serves as the work's visual title. The comics proposed here is perceived as a whole. Therefore, the reader perceives the other images - that are connected to the central one and marginalized by it - simultaneously. In this struggle between looking and reading, between the whole (the page) and the particular (the panels and their sequential reading, the Z-path), lies the visual and perceptive explanation of another conflict: that

[3] The comics can be found at < https://images.vogue.it/gallery/23655/Big/0121f912-69eb-40fa-8a77-823ecbac1843.jpg>.

between story and vision. The user of the comics page perceives, especially given the absence of words, the grid as a unique structure, meaningful than building its own particular path. There are also symmetrical references (hands and feet to the right and left, the woman's head up and down) that offer emotional associations, not purely temporal but (also) conceptual. What counts here is the expression of the protagonist's emotions expressed by the whirlpool shape of the grid (Fig. 3).

5 Conclusions

This article analyzed the literature relating to a spatial and sequential use of comics to make them fall within the field of diagrams, albeit considering the specific differences. The similarities of the comics with the diagrams and, in particular with the maps, were then highlighted to demonstrate not only their affinities but also some possibilities of cooperation. It was chosen to analyze regular layouts to demonstrate that the diagrammatic nature of comics does not belong specifically to those explicitly diagrammatic comics but can be found in most incarnations of the medium. Although some obvious constraints and limitations, we have seen how the grid, the topographical gaze, and the sequential narrative - or narratives - cooperate to multiply the medium's narrative possibilities. We also saw how involving the reader in an operation inherent in reading a story and the active construction of other meanings, through operations of dislocation, recombination and partial rewriting of the narrative material provided. The examples provided have demonstrated how this proliferation of paths does not betray the general meaning of the works analyzed but is instead capable of both expanding it, synthesizing it, and even creating new narrative and meaningful paths. Through associations less linked to a sequential approach, new paths could lead to new and unprecedented meanings. Therefore, comics could be considered not only as a narrative tool but as a critical, ambiguous and polysemic instrument of analysis. Studies about the diagrammatic nature of comics are not yet very numerous, but the possibilities of a collaboration between these two machines generating multiple meanings could lead to exciting and fruitful developments.

References

1. Bambrough, R.: Universals and Family Resemblances. Proc. Aristotelian Soc. New Ser. **61**(1960), 207–222 (1961)
2. Eddy, M.D.: Diagrams. In: Grafton, A., Blair, A., Goeing, A.S. (eds.) A Companion to the History of Information, pp. 397–440. Princeton University Press, Princeton (2020)
3. Bazzul, J., Kayumova, S.: Toward a social ontology for science education: introducing Deleuze and Guattari's assemblages. Educ. Philos. Theory **48**(3), 284–299 (2016)
4. Cates, I.: Comics and the grammar of diagrams. In: Ball, D.M., Kuhlman, M.B. (eds.) The Comics of Chris Ware, pp. 90–102. Univ. Press of Mississippi, Jackson (2010)
5. Cohn, N.: The Visual Languages of Comics: Introduction to the Structure and Cognition of Sequential Images. The Univ. of Chicago Press, London (2013)
6. del Rey Cabero, E.: Beyond linearity: holistic, multidirectional, multilinear and translinear reading in comics. Comics Grid J. Comics Scholarsh. **9**(1), 5 (2019)
7. Fresnault-Deruelle, P.: Du linéaire au tabulaire. Communications **24**(1), 7–23 (1976)

8. Groensteen, T.: System of Comics. Univs. Press of Mississippi, Jackson (2007)
9. Hadler, F., Irrgang, D.: Nonlinearity, multilinearity, simultaneity: notes on epistemological structures. In: Moura, M., Sternberg, R., Cunha, R., Queiroz, C., Zeilinger, M. (eds.) Proceedings of the Interactive Narratives, New Media & Social Engagement International Conference, pp. 70–87 (2014)
10. Hatfield, C.: Alternative Comics: An Emerging Literature. Univ Press of Mississippi, Jackson (2009)
11. Jain, E., Sheikh, Y., Hodgins, J.: Inferring artistic intention in comic art through viewer gaze. In: SAP 2012, pp. 55–62. ACM Press, Los Angeles (2012)
12. Majooni, A., Masood, M., Akhavan, A.: An eye-tracking study on the effect of infographic structures on viewer's comprehension and cognitive load. Inf. Vis. 17(3), 257–326 (2017)
13. Miodrag, H.: Comics and Language: Reimagining Critical Discourse on the Form. Univ. Press of Mississippi, Jackson (2013)
14. Miodrag, H.: Narrative breakdown in The Long and Unlearned Life of Roland Gethers by Hannah Miodrag. Comics Forum (2013b). <https://comicsforum.org/2013/03/27/narrative-bre akdown-in-the-long-and-unlearned-life-of-roland-gethers-by-hannah-miodrag/>. Accessed 25 Apr 2021
15. Moore, A.: Cubes, shadows and comic strips - a.k.a. interfaces, metaphors and maps? In: Whigham, P.A., McLennan, B.R. (eds.) SIRC 2004: A Spatio-Temporal Workshop, Proceedings of the 16th Annual Colloquium of the Spatial Information Research Centre; University of Otago. Dunedin, New Zealand, pp. 97–102 (2004)
16. Moore, A.B., Nowostawski, M., Frantz, C., Hulbe, C.: Comic strip narratives in time geography. ISPRS Int. J. Geo-Inf. 7(7), 245 (2018)
17. Omori, T., Ishii, T., Kurata, K.: Eye catchers in comics: controlling eye movements in reading pictorial and textual media. In: Paper Presented at the 28th International Congress of Psychology, Beijing (2004). <http://www.cirm.keio.ac.jp/media/contents/2004ohmori.pdf>. Accessed 25 Apr 2021
18. Tosti, A.: Heterodox origin of comics: calligrams, chronologies and maps. Todas as Letras 21(1), 45–68 (2019)

Analysis of Diagrams

Image Schemas and Conceptual Blending in Diagrammatic Reasoning: The Case of Hasse Diagrams

Dimitra Bourou[1,2]([✉]), Marco Schorlemmer[1,2], and Enric Plaza[1]

[1] Artificial Intelligence Research Institute, IIIA-CSIC,
Bellaterra, Barcelona, Catalonia, Spain
dbourou@iiia.csic.es

[2] Dept. Ciències de la Computació, Universitat Autònoma de Barcelona,
Bellaterra, Barcelona, Catalonia, Spain

Abstract. In this work, we propose a formal, computational model of the sense-making of diagrams by using the theories of image schemas and conceptual blending, stemming from cognitive linguistics. We illustrate our model here for the case of a Hasse diagram, using typed first-order logic to formalise the image schemas and to represent the geometry of a diagram. The latter additionally requires the use of some qualitative spatial reasoning formalisms. We show that, by blending image schemas with the geometrical configuration of a diagram, we can formally describe the way our cognition structures the understanding of, and the reasoning with, diagrams. In addition to a theoretical interest for diagrammatic reasoning, we also briefly discuss the cognitive underpinnings of good practice in diagram design, which are important for fields such as human-computer interaction and data visualization.

Keywords: Diagrammatic reasoning · Image schema · Conceptual blending · Hasse diagram · Formal specification · Sense-making

1 Introduction

Diagrams are often advocated as an effective way to represent and reason with information—'a picture is worth a thousand words' goes the aphorism used in English and other languages—and their advantages over purely textual representations have been studied extensively [24,35,38,40].

An important formal contribution to better understand the relative advantage of one representation formalism over another has been the theory of observational advantage put forward by Stapleton et al. [38], which stems from Shimojima's early work on the efficacy of representations [35]. This theory characterizes the advantage that particular representations have for visualizing certain semantics, because their structure makes some information directly observable. In contrast, other representations require some transformation steps to provide

© The Author(s) 2021
A. Basu et al. (Eds.): Diagrams 2021, LNAI 12909, pp. 297–314, 2021.
https://doi.org/10.1007/978-3-030-86062-2_31

the same information. Therefore, this advantage of making some information directly observable can be a criterion for the suitability of a representational formalism for a particular reasoning task.

Nonetheless, the theory of Stapleton et al. is based on an abstract character-isation of 'observation,' defined in terms of translations to and from an abstract syntax for diagrams, not taking into account its actual geometry, and the active role of the observer in the interpretation [26,40]. Such approaches have been fruitful when applied to the study of reasoning with diagrammatic represen-tations, once the interpretation of their syntax is clearly defined. However, we believe they do not fully capture the way we make sense of geometrical configu-rations in the first place, and the reason they afford certain interpretations and reasoning tasks, and not others.

In this paper we propose one way to formally and computationally model this sense-making process, drawing from the theory of image schemas and concep-tual blending originating in cognitive linguistics [12,13,19,22]. We take 'sense-making' to be the process by which humans structure percepts into meaningful constructs [41]. We hereby model the sense-making of diagrams in particular as conceptual blends of image schemas with the geometric configuration that con-stitutes a diagram. To the best of our knowledge, modeling the sense-making of diagrams in this manner is novel, and we believe it could be of value for shedding further light into the efficacy of diagrammatic representations and their utility for fields pertaining to human-human or human-machine communication.

To illustrate our approach, we will use the particular example of a Hasse diagram (Fig. 1; left). Its geometry comprises a configuration of several points, some of which intersect pairwise with lines. The points are also positioned in specific locations relative to each other. Two of the possible ways for an observer to make sense of, for instance, points e, b and a, and the lines eb and ba that connect them, in Fig. 1 are that:

1. point e with b, and b with a, form two pairs of entities that are linked by lines eb and ba, respectively
2. points e, b, and a are increasing grades on a scale, with direction from e, to b and then towards a.

This understanding of the geometric configuration allows for the emergence of inferences such as the following: since a, b and e represent some quantities such that b is more than e and a is more than b, then a must be more than e. Accord-ing to Stapleton et al. [38], such interpretations are 'direct' in the sense that they require zero transformation steps on the geometric configuration. Moreover, dif-ferent conclusions can be drawn depending on whether the 'scale' or the 'link' conceptualisation is at play; the former imbues the sense of quantity, while the latter, the sense of symmetric association. In general, diagrams, taken as geo-metric configurations, do not bring up a unique way of making sense of them, that is, they do not have a one-to-one mapping with semantics. Therefore, both the geometric configuration and the semantics of a diagram are distinct from each other and from the diagram as we make sense of it.

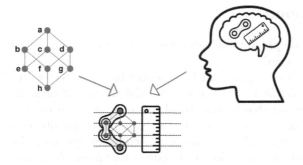

Fig. 1. Visual overview of our model. The geometry of the Hasse diagram (left), and the interpreted diagram (bottom), are distinct. The latter emerges only when schemas (right) are integrated with the geometry, giving rise to the interpreted diagram as a blend (bottom).

Our modeling view of the processes underlying the above scenario is that, although the inferences appear to arise directly from some geometric configurations, we consider that they emerge within a conceptual blend of certain image schemas with these configurations. Image schemas, like LINK and SCALE, are mental structures acquired by all humans at a very early age [19,22]. According to the homonymous theory, humans can make sense of stimuli in their environment by unconsciously integrating image schemas with them. The conceptual blending theory examines in detail the principles under which this integration takes place [13].

Hence, in our model, perceptual stimuli (i.e., the geometric configuration of a Hasse diagram) become meaningful because they prompt the conceptual blending of image schemas with them. More precisely, we describe this unconscious process as constituted by the activation of those image schemas that are useful for inference in the current context, and their subsequent integration with the stimuli, by way of establishing suitable correspondences with it. Given these correspondences, a conceptual blend can be constructed, whereby the geometric elements are structured into a coherent, integrated unit through image schemas, and give rise to the diagram as made sense of by the observer (Fig. 1; bottom). We implement our model by formalizing the geometric configuration, the internal structure of the image schemas, the correspondences between the two, and, ultimately, by computing their blend. We further show that our model can account for several direct inferences afforded by Hasse diagrams.

The remainder of this paper is organised as follows: Sect. 2 introduces the key ideas directly related to this work. Section 3 presents our blending model and the inferences resulting therein. Section 4 reviews existing frameworks in diagrammatic reasoning, and existing formalisations of image schemas and conceptual blending. Section 5 explains how our work complements the existing approaches to diagrammatic reasoning, and how it could be developed and applied in the future.

2   Background

In this section we present the theoretical background upon which our computational model is based.

The literature of diagrammatic reasoning has been very valuable for formally studying the informational content, and the efficacy of diagrams for inference. To that end, an one-to-one and total mapping between the syntax (geometric configuration) and the semantics of the diagram is typically assumed [26]. However, as our example in Sect. 1 shows, a certain configuration does not have one possible abstract interpretation. Relatedly, many researchers have suggested that the interpretation of diagrams entails a constructive and imaginative process on the part of the observer [7,26]. This is in agreement with the claims of enactive cognition.

The enactive cognition paradigm posits that cognition is the sense-making of self-sustaining agents who bring their own original meaning upon their environment [41]. Therefore, in our case study, an enactive cognition approach would posit that no geometric configuration is meaningful in itself, but it prompts the observer to unconsciously structure it into a meaningful diagram by activating suitable frames (in our case, image schemas), and integrating them appropriately with the configuration.

Image schemas fit the role of such frames because they are mental structures formed early in life, constituting structural contours of repeated sensorimotor contingencies, such as CONTAINER, SUPPORT, VERTICALITY and BALANCE [19,22]. These mental structures are acquired by experiencing (for instance) our bodies being balanced, trying to maintain our balance, supporting an object, etc. Repeated experiences of the same kind lead to to the formation of a mental structure reflecting what is invariant among them. This mental structure, called image schema, is a gestalt; it consists of components, in a specific relational structure, which can be systematically integrated with other domains, structure them, and enable conceptual meaning to arise in the mind of the observer. This is related to the phenomenon of mental visualization, i.e., seeing something in our 'mind's eye', such as visualizing a generic chair when hearing the word 'chair' [17,18]. Mental visualization is necessary for inference and prediction, and image schemas have been proposed to enable such visualization [25, pp. 513, 519–520].

Sense-making as integration of image schemas with other domains can be described though the theory of conceptual blending. The central claim of this theory is that a systematic process of building correspondences between different mental spaces underlies diverse instances of sense-making. Mental spaces are "small conceptual packets constructed as we think and talk, for purposes of local understanding and action." [13, p. 40] They comprise coherent and integrated chunks of information, containing entities, and relations or properties that characterise them. To construct a blend, some pairs of elements from two mental spaces (called input spaces) must be put in correspondence with each other, and merged into the same entity in a new mental space (called blended space). This process allows properties of both corresponding elements to come together in the blend, leading to the emergence of novel structure and thus novel meaning.

3 Approach

As explained above, image schemas can lead to inferences as a result of their internal structure. We capture the structure of each schema formally with a typed first-order logic (FOL) theory, following existing conceptual descriptions of image schemas, or experimental work, when available. The geometry of the diagrams is captured in the same way, additionally using some existing Qualitative Spatial Reasoning (QSR) formalisms to represent topological and geometrical aspects in a manner compatible with human cognition.

We hereby present a case study of sense-making of diagrams, modeling it as a conceptual blending of image schemas with the corresponding geometric configuration. The integration of the image-schematic space with the geometric space follows the principles of conceptual blending, i.e., establishing a cross-space correspondence between these two spaces. Formalising these correspondences allows us to compute the conceptual blend that characterises the diagram as the combination of image schemas with the geometric configuration based on category-theoretic colimits [33].

3.1 Diagrammatic Syntax and Its Formalisation

The geometric configuration of Fig. 1 follows the convention of Hasse diagrams, representing the transitive reduction of a partially ordered set (poset). Typically Hasse diagrams are two-dimensional but this is not a requirement. They consist of edges and vertices, drawn as points and lines. Each point represents one element of the poset. Assuming elements x, y and z of the poset, ordered by the '$<$' relation, then the lines between points are drawn according to the following syntactic rules:

- If $x < y$ then x is shown in a lower position than y in the configuration;
- x and y are connected by a line in the diagram iff $x < y$ or $y < x$, and there is no element z such that $x < z$ and $z < y$;
- lines may intersect with each other, but each one intersects with exactly two points

Therefore, the vertical position of the geometric elements in the configuration of a Hasse diagram has a proper syntactic role, representing the direction of ordering [8]. Consequently, the minimal and the maximal element are always visualised as the lowest and highest points respectively. A poset can be graded, or have ranks, when all maximal chains have the same finite length [37, p. 99]. This intuitively means that there exist groups of incomparable elements that are the same number of steps away from the minimum element. To emphasise this structure, the Hasse diagrams of these posets can be—optionally—drawn with the elements of the same rank as horizontal hyperplanes, as is the case in the Hasse geometric configuration of Fig. 1.

In order to describe the geometric configurations at hand, we draw from some formal systems developed in the QSR literature. In particular, we require a logical formalism that can capture topological relations and relative positions of the elements in a configuration. Some suitable formalisms are [9] and [16] respectively. Geometric entities can be characterised as being of type point, line or region, and we can describe and reason about their precise topological configuration [9]. The relative position of two-dimensional objects, of any shape, with respect to each other, can also be formalised [16]. This is done by denoting the position (right, right-front, front, left-front, etc.) of an object relative to another.

3.2   A Formal Model of Sense-Making

In this subsection, we present the formalization of the geometric configuration, of the image schemas to be integrated with it, the correspondences between the two, and, finally, their blends. We present here the theories for image schemas LINK, PATH, VERTICALITY, and SCALE [19,22].[1] In our model, we have described the structure of each image schema and of the geometric configuration of the diagram with a typed FOL theory. Our formalisations were guided by the existing literature, mainly [19,22]. For the formalisation of the geometric configuration, the aforementioned QSR formalisms are also needed. This way, we can declare instances of geometric types (points and lines) and describe the topological relations and relative position between all pairs of these instances. In Appendix A we show some details of our formalizations. In the remainder of this section, however, we describe our model in a more intuitive and informal manner.[2] The category theoretical colimit is an abstract operation that can be applied on any kind of mathematical object. In our case, it is applied to logical theories. Having specified some correspondences between elements of these theories, the computation of their colimit yields a new theory where all counterpart elements (types, predicates or functions) are merged into the same element, and the remaining elements and axioms of both theories are also included (see also Appendix A). This mathematical framework is apt to model conceptual blending [33].

LINK. The prototypical LINK schema consists of two distinct linked entities, and a link connecting them. Being linked constrains two entities with respect to each other, i.e., they are bound in some way, due to being in the same relation. More concretely, being linked is a symmetric and irreflexive property. Our formalisation reflects this structure.

PATH. The PATH schema consists of a source, a goal, and a path. The path consists of a series of adjacent locations that connect the source with the goal. By the structure of the schema, it is obvious that, if someone is on a certain

---

[1] The image schemas, and the correspondences selected, are those that lead to a blend consistent with the set theoretical semantics of the Hasse diagram. This blend serves as a proof of concept, but it is not the only possible blend to model the sense-making of this diagram.

[2] The complete executable specifications are available in https://drive.google.com/drive/folders/1jcQdJT0qbnAua3uXIgTEW8zV3kF_2R14?usp=sharing and the sense-making of more diagrams is modeled in [5].

location of the path, then they have already traversed all prior locations, and that contiguous locations serially lead from the source to the goal without branching. Therefore, the PATH schema is axiomatised as a total order; a collection of serially neighboring locations with the source and goal as the terminal locations in this series.

VERTICALITY. This schema reflects the structure of our experience of standing upright with our bodies resisting to gravity, or of perceiving upright objects like trees. Thus, VERTICALITY involves a simple distinction between up and down. The VERTICALITY schema comprises an axis and a base, or the ground, as a reference point [34]. Therefore, we model VERTICALITY as a unique vertical axis with its base.

SCALE. The SCALE schema comprises an ordered set of several grades. Unlike VERTICALITY however, it does not imply a particular geometric orientation. SCALE has a cumulative property (if someone has 15 euros, they also have 10); consequently, we formalise it as a total order on grades.

Hasse Configuration. The Hasse configuration of Fig. 1 has eight points (*a* to *h*) and twelve lines (*ba*, *ca*, etc.). Each line intersects with a pair of points. The logical theory modeling this configuration states the topology and orientation relations among all entities of the configuration with predicates such as *intersects* [9], and *right_back* [16], respectively.

Overall Blend Network. The sense-making of the Hasse configuration is modeled as the conceptual blending of image schemas with it (Fig. 2). Some image schemas form blends among them, and the elements of these blends are subsequently put in correspondence and blended with the geometric configuration.

Specifically, linked entities of LINK are put in correspondence with contiguous locations of the PATH, giving rise to the CHAIN image-schematic blend, comprising a path of linked entities/locations.[3] The *linked* and *contiguous* predicates are also put in correspondence (see Appendix A). Subsequently, CHAIN is put in correspondence with the geometric configuration in the following way: The blended entities/locations of the CHAIN are put in correspondence with points that intersect with the same line. The link of the LINK schema is put in correspondence with the line itself. This means that the sequence of points connected by lines in the Hasse configuration (e.g., points *h*, *e*, *b*, and *a* in Fig. 1), is in correspondence with an instance of the CHAIN image-schematic blend with contiguous entities/locations. Specifically, this image-schematic blend is put in correspondence with the geometry so that the source is the geometrically lowest point, and the goal is the geometrically highest one (*back*, and *front* of all other points, respectively, to use the terminology of [16]).

---

[3] Here, we extend the convention of typesetting image schemas in small caps, to include also our own proposed image-schematic blends, CHAIN and VERTICAL-SCALE. Note that the resulting, blended type 'entity/location' now models the structure afforded by CHAIN, as reflected in the union of the axioms of LINK and PATH in the blend. See also Appendix A and [4] for more details on the construction of the blend.

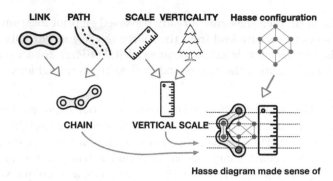

LINK   PATH   SCALE VERTICALITY   Hasse configuration

CHAIN   VERTICAL SCALE

Hasse diagram made sense of

Fig. 2. Image schema blends modeling the sense-making of the Hasse configuration.

Regarding the SCALE and VERTICALITY schemas, they are blended into the VERTICAL-SCALE image-schematic blend. The correspondences between SCALE and VERTICALITY allow the construction of a blend which integrates quantitatively ordered grades of SCALE with vertically ordered marks of VERTICALITY, giving rise to blended levels (dashed horizontal lines in Fig. 2). As for the correspondences of the VERTICAL-SCALE blend with the geometric configuration, the levels are put in correspondence with points, with respect to their geometric ordering. For instance, the level that is immediately above the base is put in correspondence with points e, f, and g, resulting in their integration into the same level in the final blend.

Guided by all the aforementioned correspondences, the VERTICAL-SCALE and the CHAIN image-schematic blends, as well as the geometric configuration, are all blended into a final blend (Fig. 2; bottom right) which has the structure of an ordering that is schematic and geometric at the same time. In other words, this complex network of cross-space correspondences enables the computation of one final blend, whereby the Hasse configuration is structured into a single, coherent gestalt.[4]

Blended Structure and Inferences. The resulting blended space integrates geometric and image-schematic aspects, providing more meaningful structure to the geometric configuration. Within this integrated structure, a variety of inferences emerge. Blending the LINK schema with the geometric configuration of two points intersecting with the same line, gives rise to the interpretation that these points participate in some relationship, and are contingent upon one another in exactly the same way. The two points, together with the line, comprise a single whole; the *LinkSchema*. Blending CHAIN with the geometric configuration structures any set of serially linked shapes into an unitary configuration, i.e., a chain.

---

[4] Viewing our model as a network of blends such as LINK-VERTICAL and SCALE-PATH is equally possible and mathematically equivalent to the blends presented, because the colimit operation is associative.

Ultimately, VERTICAL-SCALE and CHAIN, blended with the geometric configuration of the Hasse diagram, yield the Hasse diagram as an observer makes sense of it: as several chains of linked elements, arranged at several levels of generality along a down-up axis. It is important to clarify that the geometric configuration as a graded structure with directionality from point h to point a (reflected in the logical axioms of PATH, VERTICALITY, and SCALE, which jointly appear in the blend—see Appendix A), is neither geometric nor image-schematic. This graded structure emerges in a conceptual blend whose entities and properties are both geometric and image-schematic at the same time. This is only possible because the cognitive structure of the image schemas (reflected in their logical axioms) is blended with the geometric structure, and it yields a variety of inferences within the blend. First, points on the same horizontal plane (e.g., b, c, and d) are construed as being on the same level (dashed horizontal lines in Fig. 2). Second, some points are transitively above others, such as $above(a, e)$, $above(d, h)$ and so on. Notice that this predicate now models the structure afforded by VERTICAL-SCALE. Finally, through this blended $above$ predicate, and the CHAIN, the points that are serially and pairwise linked, form six maximal chains. All of these chains have points a and h, the geometrically uppermost and lowermost points, as their goal and source.

4   Related Work

In the literature of diagrammatic reasoning, it is often posited that the efficacy of diagrams lies in the sharing of structural properties between the geometric configuration and the semantics of a diagram [28,39]. These properties allow observers to make some inferences directly. Therefore, the more the properties of the geometry of a diagram match the properties of a given semantics, the more efficacious the diagram is to represent this semantics [35,38]. A similar framework, called Semiotic-Conceptual Analysis, is proposed by Priss [29]. This framework attempts to explain how meaning is represented with diagrams, language etc. and indeed accounts for various phenomena, such as polysemy, and whether a certain representational format is advantageous for some semantics.

Several research groups have worked on formalizing image schemas and relations among them. Rodriguez and Egenhofer provide a relational algebra based on the CONTAINER and SURFACE schema, used to model, and reason about, spatial relations of objects inside a room [31]. Kuhn formalised image schemas as ontology relations using functional programming, in a relatively abstract and general way [21]. Others concretise their formalisations more, using bigraphs [42], or QSR [15]. Such formalisms imbue topological and other properties into the schemas. The latter also formalises the interrelations of image schemas, as families of logical theories, constructed from combinations of primitive components. Embodied Construction Grammar formalises [2] and implements [6] language understanding by putting in correspondence the components of specific schemas (image schemas, and other kinds of schemas) with phonemes. The framework also incorporates an additional formalism (x-schema) allowing the modeling of inference.

Finally, regarding a formal view of conceptual blending, the blending process has been described though a general, mathematical theory [3,10,33]. This is done through amalgams, obtained by generalising the input spaces as much as necessary to find commonalities, and blending parts of them towards a consistent and novel output. This framework, together with image schemas, has been used to interpret an icon by blending a description of the schema with a QSR description of the icon [11]. This approach is a conceptual equivalent of the current computational model. In the same direction, other work formalises the blending of given mental spaces, in order to obtain inference as novelty. Related to our approach, Goguen [14] applied algebraic specifications and their category-theoretic operations for modeling the cognitive understanding of space and time when solving a riddle. Building on this work, Schorlemmer et al. [32] modeled the process of solving a riddle, using blending and typed FOL specifications of image schemas. The interrelations between amalgams, Goguen's framework, and our current model of blending are discussed in [33]. All aforementioned work contributes valuable, useful formalisations of blending as a creative process.

5   Discussion

The predominant logical approach to diagrammatic reasoning requires a level of abstraction which does not allow for fully taking into account the spatial structure of the geometry, the embodiment of the observer, and the interaction of the two. We believe embodied experiences—whose invariants are crystalised in the form of image schemas—can provide additional insight into the process of understanding and reasoning with a diagram. We present a computational model of this perceptual structuring process, through the integration of image schemas with the geometry of a diagram. To the best of our knowledge, this approach is a novel and valuable theoretical contribution to the diagrammatic reasoning literature. Our work is also directly relevant for human computer interaction and data visualization because, as we explain below, it has the potential to unravel guiding principles towards more intuitive visualizations.

5.1   Diagrammatic Inference with Image-Schematic Blends

Given our modeling of diagram understanding as emerging from conceptual blends of image schemas with geometric configurations, in our Hasse diagram, the facts that: (a) point a is above point h (b) points h, e, b and a form a CHAIN and (c) points b, c and d are on the same level, all can be quickly inferred from the geometric configuration. To make inference (a), for instance, an observer may mentally visualise a physical path of linked locations, starting at location h, extending towards higher locations e and b, up to a, which lies above h and the rest of the locations traversed in the path. This mental visualisation facilitates the inference that $h < a$ directly from the Hasse diagram. Mental visualization is indeed necessary for inference, and image schemas are the mental structures that enable it [25, pp. 513, 519].

Eye-tracking experiments have shown that subjects can make inference (a) for one transitive step without physically manipulating the diagram they were shown [36], and this is interpreted using Shimojima's theory of direct inference [35].

Other inferences made possible through mental visualisation, modeled as the final blended space that integrates the structure of all four described image schemas with the geometry, are (Fig. 2): the transitive ordering of points in terms of their grade on the VERTICAL-SCALE, the inference that the point on the source of the CHAIN is ordered before all others (corresponding semantically to the minimal element), that the point on the goal is after all others (maximal element), and the existence of distinct instances of CHAIN (including all maximal chains).

In the present work, our main goal was to create a cognitively-inspired model of the sense-making of diagrams, not to make claims about human cognition. Consequently, we have not undertaken any psychological experiments. However, our claims about the cognitive structure of Hasse diagrams are consistent with experiments showing that being upright, as opposed to slanted, and explicitly showing levels, makes Hasse diagrams more efficacious (i.e., interpreted faster) [20]. Moreover, other work on diagrammatic reasoning also claims that Hasse diagrams prioritize visualizing the structure of the order they represent, through a vertical organization, and explicit visualization of levels [8]. Levels corresponding to elements with the same rank, i.e., same number of steps away from the minimum element, are geometrically orthogonal to the vertical axis. In fact, this axis is the one intended to be interpreted, and elements of the same rank are indeed not comparable semantically with respect to the ordering.

5.2 Efficacy of Diagrammatic Representations

According to the view of efficacy that we have discussed, some geometric configurations are more efficacious for representing a given semantics, than others. This phenomenon is attributed to some geometric configurations having more similar properties with certain semantics, than others do [35,38]. A Hasse diagram would then be considered very efficacious to represent a partial order, because the geometric arrangement of shapes along a vertical axis has a transitive and asymmetric property, as does a partial order. A diagram whose geometric configuration did not have these properties, or worse, had contradicting ones (e.g., symmetry), would be less efficacious to represent poset semantics. Euler diagrams, for example, have different properties. Representing that $Q \subseteq P$ and $P \cap R = \emptyset$ with the Euler diagram of Fig. 3, makes the inference that $Q \cap R = \emptyset$ directly observable [38]. A Hasse diagram can also represent this scenario, as well any possible constellation of sets. Then why are Hasse diagrams predominantly used to represent posets, and Euler diagrams for set membership?

According to our framework, the higher efficacy of Euler diagrams for set membership and inclusion, and of Hasse diagrams for poset semantics, can be explained as follows: The geometry of the Hasse diagram, comprising shapes that are one above another and grouped in parallel horizontal lines, as well as

the semantics of a poset, are easy to put in correspondence with the VERTICAL-
SCALE schema. These correspondences enable constructing blends whereby the
aforementioned inferences (transitivity, existence of maximal elements, minimal
elements, and maximal chains) emerge. In contrast, the geometric configuration
of an Euler diagram, comprising closed curves that are inside one another, is
more compatible with the CONTAINER schema; the boundary, inside and outside
of the CONTAINER schema can be put into correspondence with the boundary,
interior and exterior of closed curves.

The above is also true for the semantics of set membership. Thus, having
mentally structured the diagram as comprising physical containers, we can mentally visualize the impossibility of Q being inside P and inside R at the same
time. Similarly, the CONTAINER schema also fits the semantics of set membership.
This is in agreement with Priss's suggestion that observers find set membership
and inclusion easier to read from Euler than Hasse diagrams, because the former
enable mentally visualising the impossibility of Q exiting P and approaching R
[30]. In fact, it has been proposed that our understanding of abstract set theoretical notions also rests on the same image schematic structures [23].

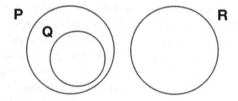

Fig. 3. A simple Euler diagram.

5.3 Conclusions and Future Work

In this paper we have provided a formal framework of sense-making of a diagram, as a creative, active process on the part of the observer, involving the
conceptual blending of image schemas with the geometry of the diagram. Most
previous computational work making reference to conceptual blending and image
schemas only considered creative and problem solving tasks. However, the more
fundamental process of sense-making is also a creative and active process that
can be explained with conceptual blending. In our work, this view serves as the
conceptual foundation.

In future investigations we would like to model a broader range of
image schemas, enabling us to examine alternative—including erroneous—
interpretations of diagrams. Moreover, we aim to characterise formally what it
means for a diagram to be efficacious, in the context of our framework. To achieve
both goals, we are currently expanding our framework with the role of the semantics, as well as by including some formal criteria for selecting well-integrated, consistent blends that are useful for reasoning with diagrams. This approach would
be a cognitively plausible way to model possible interpretations of diagrams.

The outputs of our model could guide designers of various visual representations, so they can design them with the aim to match the properties of the geometric syntax with the intended meaning. For example, following up with the comparison we made in the previous subsection, if a designer wants to visually represent some ordinal values, a tool based on our framework might recommend the use of a vertical geometric configuration and not a horizontal one. This is because a VERTICAL-SCALE is likely to map to such a configuration and lead to a blend with the intended semantics. In contrast, if a designer wants to represent the notion of belonging in a group, a tool would recommend a configuration with topological containment, because its semantics are similar. Various such recommendations can be made precise thanks to our model and could contribute to new tools directed at designers.

Acknowledgments. The present research was supported by CORPORIS (PID2019-109677RB-I00) funded by Spain's *Agencia Estatal de Investigación*; by DIVERSIS (201750E064) and DIAGRAFIS (202050E243) funded by CSIC; and by the *Ajuts a grups de recerca consolidats* (2017 SGR 172) funded by *AGAUR/Generalitat de Catalunya*.

A   FOL Specifications of Image Schemas and Blends

In this appendix we present some more technical detail of our formalizations and their implementation. The complete formalisation using CASL [1] and HETS [27] can be found in https://drive.google.com/drive/folders/1jcQdJT0qbnAua 3uXIgTEW8zV3kF_2R14?usp=sharing.

A.1   The Hasse Geometric Configuration

The specification of the Hasse geometric configuration defines sorts corresponding to geometric entities, and predicates that define their relations in terms of orientation and topology. Below, we present a fragment of this description, for points a, b, e, and h (here constants c_1, c_2, c_5, and c_8) and lines r_{21} and r_{52}, which we discuss as an example throughout the paper.

$rightBackPP(c_1, c_2) \wedge rightBackPP(c_1, c_5) \wedge leftBackPP(c_5, c_8) \wedge backPP(c_1, c_8) \wedge backPP(c_2, c_5)$

$intersectLP(r_{21}, c_1) \wedge intersectLP(r_{21}, c_2) \wedge intersectLP(r_{52}, c_2) \wedge intersectLP(r_{52}, c_5) \wedge \ \cdots$

A.2   Image Schemas

To model image schemas we followed the principles of algebraic specification, declaring for each schema several types, defining operations and predicates on types, and modeling their behavior by stating axioms over these operations and predicates.

The LINK schema specification

$\forall s \in LinkSchema : linked(anEnt(s), anotherEnt(s))$

$\forall l \in Link \; \exists!s \in LinkSchema : l = link(s)$

$\forall x, y \in Entity \; \forall s : LinkSchema : linked(x, y) \Leftrightarrow (anEnt(s) = x \wedge anotherEnt(s) = y)$
$\quad \vee \; (anEnt(s) = y \wedge anotherEnt(s) = x)$

$\forall x, y \in Entity : linked(x, y) \Leftrightarrow linked(y, x)$

$\forall x \in Entity : \neg linked(x, x)$

The PATH schema specification

$\forall p \in Path \; \exists!s \in SPGschema : path(s) = p$

$\forall s \in SPGschema : inPath(source(s), path(s))$

$\forall s \in SPGschema : inPath(goal(s), path(s))$

$\forall s \in SPGschema : source(s) \neq goal(s)$

$\forall s \in SPGschema \; \forall l \in Location : \neg(isFollowedBy(l, source(s), path(s))$
$\quad \vee \; isFollowedBy(goal(s), l, path(s)))$

$\forall s \in SPGschema \; \forall l \in Location : inPath(l, path(s)) \wedge l \neq source(s)$
$\quad \Rightarrow \exists!k \in Location : inPath(k, path(s)) \wedge isFollowedBy(k, l, path(s))$

$\forall s \in SPGschema \; \forall l \in Location : inPath(l, path(s)) \wedge l \neq goal(s)$
$\quad \Rightarrow \exists!m \in Location : inPath(m, path(s)) \wedge isFollowedBy(l, m, path(s))$

$\forall k, l \in Location \; \forall p \in Path : isFollowedBy(k, l, p) \vee isFollowedBy(l, k, p)$
$\quad \Rightarrow inPath(k, p) \wedge inPath(l, p)$

$\forall s \in SPGschema \; \exists!l \in Location : inPath(l, path(s)) \wedge placed(trajector(s), l)$

$\forall k, l \in Location; p \in Path : contiguous(k, l) \Leftrightarrow isFollowedBy(k, l, p)$
$\quad \vee \; isFollowedBy(l, k, p)$

The VERTICALITY schema specification

$\forall s \in VerticalitySchema : inAxis(base(s), axis(s))$

$\forall m \in Mark : \neg above(m, m)$

$\forall s \in VerticalitySchema; m \in Mark : inAxis(m, axis(s)) \wedge (m \neq base(s)) \Rightarrow above(m, base(s))$

The SCALE schema specification

$\forall c \in Scale \; \exists!s \in ScaleSchema : scale(s) = c$

$\forall c \in Scale; \; x, \; y \in Grade : inScale(x, c) \wedge inScale(y, c) \wedge (x \neq y) \Rightarrow more(x, y) \vee more(y, x)$

$\forall c \in Scale; \; x, \; y \in Grade : more(x, y) \Rightarrow \exists!c \in ScaleSchema : inScale(x, scale(c))$
$\quad \wedge \; inScale(y, scale(c))$

$\forall x, \; y \in Grade : less(x, y) \Leftrightarrow more(y, x)$

$\forall x, \; y \in Grade : more(x, y) \Leftrightarrow \neg more(y, x)$

$\forall x \in Grade : \neg more(x, x)$

$\forall x, \; y, \; z \in Grade : more(x, y) \wedge more(y, z) \Rightarrow more(x, z)$

A.3   The Hasse Blend

Blends are computed with HETS as category-theoretic colimits, once the correspondences between instances of image schemas and the Hasse configuration have been established by means of spans of morphisms. Below we show a fragment of the blend, namely the constants and axioms of the blend combining CHAIN with the geometric configuration, and some theorems deducible from the blend. As mentioned, this blend includes signatures and axioms of the LINK and PATH schemas, as well as those of the geometric configuration. We denote those coming from LINK in orange, those from PATH in purple, and those from the geometry in teal.

Declaration of constants:

$lsc_{21}, lsc_{31}, lsc_{41}, lsc_{52}, lsc_{53}, lsc_{62}, lsc_{64}, lsc_{73}, lsc_{74}, lsc_{85}, lsc_{86}, lsc_{87} \in LinkSchema$

$p_1, p_2, p_3, p_4, p_5, p_6 \in Path$ $s_1, s_2, s_3, s_4, s_5, s_6 \in SPGschema$

$c_1, c_2, c_3, c_4, c_5, c_6, c_7, c_8 \in Point$ $r_{21}, r_{31}, r_{41}, r_{52}, r_{53}, r_{62}, r_{64}, r_{73}, r_{74}, r_{85}, r_{86}, r_{87} \in Line$

Axioms:

$rightBackPP(c_1, c_2) \wedge rightBackPP(c_1, c_5) \wedge leftBackPP(c_5, c_8) \wedge backPP(c_1, c_8)$
$\wedge\ backPP(c_2, c_5)$
$intersectLP(r_{21}, c_1) \wedge intersectLP(r_{21}, c_2) \wedge intersectLP(r_{52}, c_2) \wedge intersectLP(r_{52}, c_5) \wedge \ \ldots$

$\forall x, y \in Point\ \forall s \in LinkSchema : linked(x, y)$
    $\Leftrightarrow (anEnt(s) = x \wedge anotherEnt(s) = y) \vee (anEnt(s) = y \wedge anotherEnt(s) = x)$
$\forall s \in LinkSchema : linked(anEnt(s), anotherEnt(s))$
$\forall l \in Line : exists!s \in LinkSchema : link(s) = l$
$\forall x, y \in Point : linked(x, y) \Leftrightarrow linked(y, x)$
$\forall x \in Point : \neg linked(x, x)$

$anEnt(lsc_{21}) = c_2 \wedge anotherEnt(lsc_{21}) = c_1 \wedge link(lsc_{21}) = r_{21}$
$anEnt(lsc_{52}) = c_5 \wedge anotherEnt(lsc_{52}) = c_2 \wedge link(lsc_{52}) = r_{52} \wedge \ \ldots$
$\forall s \in SPGschema : source(s) \neq goal(s)$
$\forall s \in SPGschema\ \forall l \in Point : \neg (isFollowedBy(l, source(s), path(s))$
    $\vee isFollowedBy(goal(s), l, path(s)))$
$\forall s \in SPGschema\ \forall l \in Point : inPath(l, path(s)) \wedge l \neq source(s)$
    $\Rightarrow \exists!k \in Point : inPath(k, path(s)) \wedge isFollowedBy(k, l, path(s))$
$\forall s \in SPGschema\ \forall l \in Point : inPath(l, path(s)) \wedge l \neq goal(s)$
    $\Rightarrow \exists!m \in Point : inPath(m, path(s)) \wedge isFollowedBy(l, m, path(s))$
$\forall k, l \in Point\ \forall p \in Path : isFollowedBy(k, l, p) \vee isFollowedBy(l, k, p)$
    $\Rightarrow inPath(k, p) \wedge inPath(l, p)$
$\forall p \in Path \exists!s \in SPGschema : path(s) = p$
$\forall s \in SPGschema : inPath(source(s), path(s))$
$\forall s \in SPGschema : inPath(goal(s), path(s))$
$\forall s \in SPGschema \exists!l \in Point : inPath(l, path(s)) \wedge placed(trajector(s), l)$
$\forall k, l \in Point\ \forall p \in Path : linked(k, l) \Leftrightarrow isFollowedBy(k, l, p) \vee isFollowedBy(l, k, p)$

$path(s_1) = p_1 \wedge source(s_1) = c_8 \wedge goal(s_1) = c_1$

$path(s_2) = p_2 \wedge source(s_2) = c_8 \wedge goal(s_2) = c_1 \wedge \ldots$

$isFollowedBy(c_2, c_1, p_1) \wedge isFollowedBy(c_5, c_2, p_1) \wedge isFollowedBy(c8, c_5, p_1)$

$isFollowedBy(c_3, c_1, p_2) \wedge isFollowedBy(c_5, c_3, p_2) \wedge isFollowedBy(c8, c_5, p_2) \wedge \ldots$

Theorems deducible from the blend (about points linked by lines and located in paths):

$linked(c_1, c_2) \wedge linked(c_2, c_5) \wedge linked(c_5, c_8) \wedge \ldots$

$inPath(c_1, p_1) \wedge inPath(c_2, p_1) \wedge inPath(c_5, p_1) \wedge inPath(c8, p_1)$

$inPath(c_1, p_2) \wedge inPath(c_3, p_2) \wedge inPath(c_5, p_2) \wedge inPath(c8, p_2) \wedge \ldots$

References

1. Astesiano, E., et al.: CASL: the common algebraic specification language. Theor. Comput. Sci. **286**(2), 153–196 (2002)
2. Bergen, B., Chang, N.: Embodied construction grammar in simulation-based language understanding. In: Construction Grammars: Cognitive Grounding and Theoretical Extensions, vol. 3, pp. 147–190. John Benjamins (2005)
3. Bou, F., Plaza, E., Schorlemmer, M.: Amalgams, colimits, and conceptual blending. In: Concept Invention. CSCS, pp. 3–29. Springer, Cham (2018). https://doi.org/10.1007/978-3-319-65602-1_1
4. Bourou, D., Schorlemmer, M., Plaza, E.: A cognitively-inspired model for making sense of Hasse diagrams. In: Proceedings of International Conference of the Catalan Association for Artificial Intelligence (2021)
5. Bourou, D., Schorlemmer, M., Plaza, E.: Modelling the sense-making of diagrams using image schemas. In: Proceedings of the Annual Meeting of the Cognitive Science Society (2021)
6. Bryant, J.E.: Best-fit constructional analysis. Ph.D. thesis, EECS Department, University of California, Berkeley, August 2008. http://www2.eecs.berkeley.edu/Pubs/TechRpts/2008/EECS-2008-100.html
7. Cheng, P.C.H., Lowe, R.K., Scaife, M.: Cognitive science approaches to understanding diagrammatic representations. In: Blackwell, A.F. (ed.) Thinking with Diagrams, pp. 79–94. Springer, Dordrecht (2001). https://doi.org/10.1007/978-94-017-3524-7_5
8. Demey, L., Smessaert, H.: The relationship between Aristotelian and Hasse diagrams. In: Dwyer, T., Purchase, H., Delaney, A. (eds.) Diagrams 2014. LNCS (LNAI), vol. 8578, pp. 213–227. Springer, Heidelberg (2014). https://doi.org/10.1007/978-3-662-44043-8_23
9. Egenhofer, M.J., Herring, J.R.: Categorizing binary topological relations between regions, lines, and points in geographic databases. Technical report, Department of Surveying Engineering, University of Maine (1991)
10. Eppe, M., Maclean, E., Confalonieri, R., Kutz, O., Schorlemmer, M., Plaza, E., Kühnberger, K.U.: A computational framework for conceptual blending. Artif. Intell. **256**, 105–129 (2018)
11. Falomir, Z., Plaza, E.: Towards a model of creative understanding: deconstructing and recreating conceptual blends using image schemas and qualitative spatial descriptors. Ann. Math. Artif. Intell. **88**, 457–477 (2019)
12. Fauconnier, G., Turner, M.: Conceptual integration networks. Cogn. Sci. **22**(2), 133–187 (1998)
13. Fauconnier, G., Turner, M.: The Way We Think. Basic Books, New York (2002)

14. Goguen, J.: Mathematical models of cognitive space and time. In: Reasoning and Cognition, pp. 125–128. Keio University Press (2006)
15. Hedblom, M.M.: Image Schemas and Concept Invention: Cognitive, Logical, and Linguistic Investigations. Springer, Cham (2020). https://doi.org/10.1007/978-3-030-47329-7
16. Hernández, D.: Relative representation of spatial knowledge: the 2-D case. In: Mark, D.M., Frank, A.U. (eds.) Cognitive and Linguistic Aspects of Geographic Space, pp. 373–385. Springer, Dordrecht (1991). https://doi.org/10.1007/978-94-011-2606-9_21
17. Jackendoff, R.: Semantics and Cognition, vol. 8. MIT Press, Cambridge (1983)
18. Jackendoff, R.: Consciousness and the Computational Mind. MIT Press, Cambridge (1987)
19. Johnson, M.: The Body in the Mind. University of Chicago Press, Chicago (1987)
20. Körner, C., Albert, D.: Comprehension efficiency of graphically presented ordered sets. In: Kallus, K., Posthumus, N., Jimenez, P. (eds.) Current Psychological Research in Austria. Proceedings of 4th Scientific Conference Austrian Psychological Society, pp. 179–182. Akademische Druck - u. Verla, Graz (2001)
21. Kuhn, W.: An image-schematic account of spatial categories. In: Winter, S., Duckham, M., Kulik, L., Kuipers, B. (eds.) COSIT 2007. LNCS, vol. 4736, pp. 152–168. Springer, Heidelberg (2007). https://doi.org/10.1007/978-3-540-74788-8_10
22. Lakoff, G.: Women, Fire, and Dangerous Things. University of Chicago Press, Chicago (1987)
23. Lakoff, G., Núñez, R.E.: Where mathematics comes from: how the embodied mind brings mathematics into being. AMC 10(12), 720–733 (2000)
24. Larkin, J.H., Simon, H.A.: Why a diagram is (sometimes) worth ten thousand words. Cogn. Sci. 11(1), 65–100 (1987)
25. Mandler, J.M., Cánovas, C.P.: On defining image schemas. Lang. Cogn. 6(4), 510–532 (2014)
26. May, M.: Diagrammatic reasoning and levels of schematization. In: Iconicity. A Fundamental Problem in Semiotics, pp. 175–194. NSU Press, Copenhagen (1999)
27. Mossakowski, T., Maeder, C., Lüttich, K.: The heterogeneous tool set, HETS. In: Grumberg, O., Huth, M. (eds.) TACAS 2007. LNCS, vol. 4424, pp. 519–522. Springer, Heidelberg (2007). https://doi.org/10.1007/978-3-540-71209-1_40
28. Palmer, S.E.: Fundamental aspects of cognitive representation. In: Cognition and Categorization, pp. 259–302. Erlbaum, Hillsdale (1978)
29. Priss, U.: Semiotic-conceptual analysis: a proposal. Int. J. Gen. Syst. 46(5), 569–585 (2017)
30. Priss, U.: A semiotic-conceptual analysis of Euler and Hasse diagrams. In: Pietarinen, A.-V., Chapman, P., Bosveld-de Smet, L., Giardino, V., Corter, J., Linker, S. (eds.) Diagrams 2020. LNCS (LNAI), vol. 12169, pp. 515–519. Springer, Cham (2020). https://doi.org/10.1007/978-3-030-54249-8_47
31. Rodríguez, M.A., Egenhofer, M.J.: A comparison of inferences about containers and surfaces in small-scale and large-scale spaces. J. Vis. Lang. Comput. 11(6), 639–662 (2000)
32. Schorlemmer, M., Confalonieri, R., Plaza, E.: The Yoneda path to the Buddhist monk blend. In: Proceedings of the Joint Ontology Workshops 2016. CEUR Workshop Proceedings, CEUR-WS.org (2016)
33. Schorlemmer, M., Plaza, E.: A uniform model of computational conceptual blending. Cogn. Syst. Res. 65, 118–137 (2021)

34. Serra Borneto, C.: Liegen and stehen in German: a study in horizontality and verticality. In: Cognitive Linguistics in the Redwoods, pp. 459–506. Cog Linguist, Mouton de Gruyter (1996)
35. Shimojima, A.: On the efficacy of representation. Ph.D. thesis, Indiana University (1996)
36. Shimojima, A., Katagiri, Y.: An eye-tracking study of exploitations of spatial constraints in diagrammatic reasoning. Cogn. Sci. **37**(2), 211–254 (2013)
37. Stanley, R.P.: Enumerative Combinatorics. Cambridge University Press, Cambridge (2011)
38. Stapleton, G., Jamnik, M., Shimojima, A.: What makes an effective representation of information: a formal account of observational advantages. J. Logic. Lang. Inf. **26**(2), 143–177 (2017)
39. Stenning, K.: Seeing Reason: Image and Language in Learning to Think. Oxford University Press, Oxford (2002)
40. Stenning, K., Lemon, O.: Aligning logical and psychological perspectives on diagrammatic reasoning. Artif. Intell. Rev. **15**(1–2), 29–62 (2001)
41. Varela, F.J.: Organism: a meshwork of selfless selves. In: Tauber, A.I. (ed.) Organism and the Origins of Self, pp. 79–107. Springer, Dordrecht (1991). https://doi.org/10.1007/978-94-011-3406-4_5
42. Walton, L., Worboys, M.: An algebraic approach to image schemas for geographic space. In: Hornsby, K.S., Claramunt, C., Denis, M., Ligozat, G. (eds.) COSIT 2009. LNCS, vol. 5756, pp. 357–370. Springer, Heidelberg (2009). https://doi.org/10.1007/978-3-642-03832-7_22

Open Access This chapter is licensed under the terms of the Creative Commons Attribution 4.0 International License (http://creativecommons.org/licenses/by/4.0/), which permits use, sharing, adaptation, distribution and reproduction in any medium or format, as long as you give appropriate credit to the original author(s) and the source, provide a link to the Creative Commons license and indicate if changes were made.

The images or other third party material in this chapter are included in the chapter's Creative Commons license, unless indicated otherwise in a credit line to the material. If material is not included in the chapter's Creative Commons license and your intended use is not permitted by statutory regulation or exceeds the permitted use, you will need to obtain permission directly from the copyright holder.

Through the Eyes of an Archeologist: Studying the Role of Prior Knowledge in Learning with Diagrams

Erica de Vries$^{(\boxtimes)}$ (iD)

Univ. Grenoble Alpes, LaRAC, 38000 Grenoble, France
Erica.deVries@univ-grenoble-alpes.fr

Abstract. The study of learning with diagrams predominantly embraces a cognitivist viewpoint in which learning and problem solving are conceptualized as interplay between internal mental and external physical representations. We start by questioning two of its underlying assumptions originating in objectivist epistemology: the directness of diagrams as iconic representations and the sameness, in terms of cultural and linguistic background, of the producers and users of diagrams. In the remaining of the contribution, a relativistic viewpoint is adopted according to which reality is socially and experientially based and representation crucially depends on an observer within a context. Consequently, we investigate a theoretical situation in which learners have knowledge of neither the content nor the type of diagram. As a method of study, we adopt the perspective of an archeologist who encounters inscriptions of unknown civilizations, and in doing so, we uncover the role of abductive reasoning in learning with diagrams. A formal framework is developed with recursive rules for further studying the distinctions between pictures and diagrams, literal (denotative), figurative (connotative), and metalinguistic meaning, as well as internal and external representations. The framework is illustrated with a set of content-free test patterns. Some directions for evaluating the framework are presented in the conclusion.

Keywords: Prior knowledge · Learning · Abduction · Multiple visual languages

1 Introduction

The study of learning with diagrams predominantly embraces a cognitivist viewpoint in which learning and problem solving are conceptualized as interplay between internal mental and external physical representations [1, 25, 30, 41]. Notwithstanding the success of this approach, recent research demonstrates that cognition also crucially depends on bodily experience and on material and sociocultural aspects of the environment [26]. In this contribution, we examine learning with diagrams from a relativist viewpoint in which the construction of reality is socially and experientially based, and in which representation crucially depends on an observer within a context. We start by examining two assumptions originating in objectivist epistemology: the *directness* of diagrams as iconic representations and the *sameness* of the users of diagrams regarding prior knowledge.

© Springer Nature Switzerland AG 2021
A. Basu et al. (Eds.): Diagrams 2021, LNAI 12909, pp. 315–330, 2021.
https://doi.org/10.1007/978-3-030-86062-2_32

2   Assumptions in the Study of Learning with Diagrams

Within an objectivist perspective, researchers consider diagrams to be external representations which allow learners to interact with some content domain in the absence of the objects and phenomena of interest [20, 32, 36]. So called "representational pictures" [7] are deployed to learn about physical objects, e.g. bicycle pumps, gears, and toilet flushes, as well as natural phenomena, e.g. the movement of tectonic plates, volcano outbursts, and thunder storms. Furthermore, formal diagrams, such as concept maps, visual programming languages, and inquiry diagrams, allow learning in advanced conceptual domains. Research within a cognitively oriented perspective rests on Palmer's definition of representation as "something that stands for something else" [14]. Moreover, following Peirce, Schnotz [30] distinguished between descriptions (symbols, texts, propositions) which represent by virtue of some arbitrary rule, habit, or convention and depictions (icons, pictures, images, diagrams) which represent by virtue of a resemblance relation between the represented and the representing world. This distinction between symbolic and iconic representations is pervasive throughout the literature albeit with different terms: auditory verbal and visual spatial, propositional and imagistic for internal mental representations, text and pictures [25], linguistic and graphical [34], sentential and diagrammatic [24], descriptive and depictive [31], textual and pictorial for external representations. Thus, within cognitive perspectives, the design of instructional material should capitalize on both channels: the literal arbitrary meaning of words (force, lever, acid, cell, cylinder, square, etc.) and the similarity of pictures and diagrams to real world objects and phenomena. Although this conceptual framework is straightforward and parsimonious and produced many design principles for instructional materials, it builds on two assumptions that may be subject to debate.

2.1   The Directness of Visual Representations

Iconicity is the ground for qualifying external visual representations as more natural, direct and intuitive than verbal representations. In contrast with text, which may be ambiguous regarding spatial information (e.g. "the fork is next to the plate"), pictures more effectively show the phenomena at hand. Thus, relevant information can simply be read off a diagram. However, the sheer existence of ambiguous pictures shows that, in principle, visual representations can have more than one interpretation. For example, knowing both animals, I can see Jastrow's drawing alternatively as a duck and as a rabbit. Indeed, we should not overlook that prior knowledge is a prerequisite: knowing only one animal would make the drawing quite ineffective for learning about the other. As argued by Wittgenstein, multiple ways of seeing flip back and forth in the mind but a drawing cannot be seen in multiple ways at the same point in time [39].

According to Giardino and Greenberg [14], pictures and diagrams are varieties of iconicity: whereas pictures include a perspective, diagrams do not. These authors also show that pictures and diagrams do not differ in the syntactical elements (the marks on paper or on screen), but need to be distinguished by the semantics (the content). As an illustration, it is impossible to attribute a language to each of the marks in Fig. 1 in isolation, i.e. without referring to a context or a content. The marks may have a natural cause, such as a fossil or a form in the sand, or a human cause, such as a drawing, a character, a figure, or a symbol. They may be iconic, i.e. a ribbon, a fish, a ground, a closed eye, a ring, or symbolic, i.e. the characters a, o, k in Latin or Oe with diaeresis in Cyrillic script. They might have denotative (a ribbon, a heart, a fish) or connotative meaning (love, aids awareness, military decoration, Jesus, Christianity, unity, rewind). A rectangle may represent a frame, a brick, or a cylindrical chamber in a picture, but also an action, a concept, an argument, a substance in a diagram [37]. The marks can be geometrical forms or symbols in STEM domains (proportionality, angular acceleration, change). The point is that categorization as pictorial or textual, iconic or symbolic, perspectival or diagrammatic, denotative or connotative, *follows* a particular interpretation in terms of a content (see also Rastier [29]).

Fig. 1. Visual marks may be naturally occurring or artificially created, symbolic or iconic, perspectival or diagrammatic, and follow a literal, figurative or metalinguistic signification mode.

Diagrammatic representations differ in the degree to which they comply to formal rules and the degree to which these are known in a given society [38]. For example, freehand drawings and diagrams most often do not comply with formal rules of composition. They are ruled by polysemy: a given graphical element represents something as a function of the configuration of inscriptions in which it is inserted [4]. Freehand drawings may have a local character, invented on the fly, relevant during some activity, a game, or in a vocational context [38]. Furthermore, many instructional graphics, such as structural diagrams of the human circulatory system, are of this kind.

Fig. 2. Two alternative diagrams of a process model. The right-hand diagram shows action verbs in rectangles, which suggests compliance with flowchart conventions.

In contrast with freehand drawings, graphical representations, such as line graphs, bar and pie charts do comply with formal rules of composition. These are widely used in today's society and their conventions are taught in school. All kinds of graphical models disseminated in Western society (boxes-and-arrows, flowcharts, concept maps, electrical circuit diagrams) are ruled by monosemy: a graphical element represents something irrespective of the configuration in which it is inserted [4]. A legend suffices to explicit the rules and conventions that might be unknown in a specific community [34]. In fact, most conventions are firmly established in today's society and graphical representations are robust so that transgressions of the rules do not result in a breakdown of the communication. However, the apparent directness of graphical representations conceals the interrelatedness of content and graphical convention: prior knowledge of the content and graphical language is required to assert compliance to a particular convention. In the examples in Fig. 2, knowledge of process models allows the inference that the left-hand diagram complies with rules for boxes-and-arrows models and the right-hand diagram with rules for flowcharts.

Finally, experts in highly specialized scientific domains use very elaborate visual representations tailored to their needs, developed and sanctioned by scientific communities themselves [8]. Indeed, diagrams, as the tools of the trade, are powerful reasoning tools in logic, in mathematics, and in science. Of course, experts have the knowledge in their domain of scientific endeavor as well as the skills in producing and using adequate domain-specific visual languages and representational conventions. Since these are not known by the public, learning in highly specialized domains involves acquiring the content as well as the form of representational discourse [2].

2.2 The Sameness of the Producers and Users of Instructional Materials

The aim of research within a cognitive perspective is to produce scientific knowledge of learning with diagrams, texts, and pictures, taking into account the properties of the human cognitive architecture. Scientific knowledge ideally is independent of a particular culture, language, context or content domain (except in science education and didactics). Likewise, when diagrams are the object of study in their own right, the context of producing or using diagrams and the specificity of a content domain play a minor role. Both endeavors aim at generalizable knowledge of learning and problem solving with diagrams. Thus, experimental research in these fields requires participants to be relatively homogeneous regarding culture, majority language, and prior knowledge and most results are obtained with Western, Educated, Industrialized, Rich and Democratic participants (WEIRD, [16]). The question arises as to whether results hold universally for any human being in any culture, content, context or language setting. In today's society, teaching and learning may take place at a distance, across contexts, cultures, and languages and with learners varying in prior knowledge and dominant language.

Another reason for questioning the sameness of producers and users of instructional materials in educational settings is the fundamental knowledge gap between teachers and learners, lecturers and students. In educational settings, producers and users of diagrams in principle differ in prior knowledge of both the content domain and the relevant visual languages. Moreover, an obstacle in learning research is that once expert in a domain, it is virtually impossible to "see" diagrams as a novice, i.e. as an individual lacking prior knowledge. In other words, it is difficult to consider text, pictures and diagrams from the learner's standpoint[1].

In distinguishing a representing, which may be internal mental or external, and a represented world, cognitive perspectives on representation are predominantly dyadic [35, 38]. For example, learners may construct a mental model of a bicycle pump through the study of a diagram. Both the mental model (internal representation) and the diagram (external representation) represent a bicycle pump as an object in the material world. Researchers then evaluate the amount of learning through the similarity between the two, discarding any alternative interpretation of the diagram, consistent with prior knowledge

---

[1] To experience learning with a diagram out of context, without knowing either the content or the language, one might try to learn how a *kaasschaaf* works from some text and pictures. http://www.degebruiksaanwijzingen.nl/Html/gebruiksaaanwijzingen/gebruiksaanwijzingen27.htm.

of the learner, as a failure to learn. Nevertheless, Palmer [14] claimed that a representation exclusively contains information for which there is an operation extracting it. According to Palmer, representation presupposes a third entity, human or artificial, exploiting the correspondence relations between a representing and a represented world. Recognizing three entities, rather than just two, corresponds to semiotic perspectives going back to Peircian definitions of the sign[2]. Within a triadic approach, representation involves a viewpoint or interpreting entity with prior knowledge.

In conclusion, critical inspection of the two assumptions reveals that learning with diagrams, as studied within an objectivist perspective, relies on shared cultural and linguistic background, strong contextualization, as well as on prior knowledge of both content and graphical conventions. We claim that the lack of such common ground would put the learner in the position of a cryptanalyst or an archeologist. A cryptanalyst comes into possession of a message with no prior knowledge of the underlying code and must break the code through dexterous manipulations of the message [19]. Consequently, we set out to study learning with diagrams in the hypothetical situation in which a learner has knowledge of neither the content nor the type of diagram.

3 The Learner as an Archeologist

We use the term *inscription* to refer to the way archaeologists, as third-party observers, characterize patterns in a physical medium apart from any reference to how they might be used, understood, or perceived, and, apart from any structure they might embody [22]. In effect, archeologists do not know *in advance* whether a pattern is the result of some natural phenomenon or whether it has a human cause. Moreover, artificial patterns (created by a human) do not necessarily serve communicational purposes; patterns may be sheer embellishments, i.e. seductive details or ornaments, or still interface controls. We already showed that inscriptions cannot be categorized as depictive (iconic) or descriptive (symbolic) *without* knowledge of the context or the intended meaning. Finally, in the presence of more than one potentially relevant code, convention, or language, archeologists proceed by postulating multiple language-content combinations. We propose, as a method of study, to act *as if* there is no common ground amongst the producers and users of diagrams. Thus, we adopt an archeologist's approach to the multiple different inscriptions that a learner may come across in instructional settings.

Archeologists rely on abductive reasoning, the third mode of reasoning or "inference to the best explanation" [9–11, 27, 33]. Abductive reasoning goes as follows: encountering some fact (q) triggers an existing rule (if p, then q) and leads to *conjecturing*, not asserting, the antecedent (p). In fact, affirming p as the consequent would correspond to a logical fallacy. Abduction is ubiquitous for generating hypotheses, in science and technology, in everyday life, and in activities that involve finding an explanation for some symptom or some phenomenon, such as in medical diagnostics and mechanical trouble shooting. Text comprehension also proceeds by abduction since the attribution of

---

[2] The semiotic triangle involves three entities: the signifier or sign-vehicle (mark, sound, etc.), the signified or *interpretant* (idea created in the mind), and the referent or object in the world. In our notation, slashes indicate the signifier, double quotes indicate the signified, and no typographic marks indicate the referent, such that /dog/ evokes "dog" in the absence of a dog.

meaning requires hypothesizing a language as a frame of reference [11]. Monolinguals are unlikely to be aware of abduction, or interpretative moment, when comprehending text in their language because the intended meaning leaps off the page or the screen. However, multilinguals may consciously experience abduction when they encounter so-called *non-cognate inter-lingual homographs* [13]. For example, the inscription /sale/ denotes selling at a reduced price in English and dirty in French. Upon perceiving /sale/ on a store at a certain time of year, an English-French bilingual understands "selling at a reduced price", which in turn leads to the abduction, or best explanation, that the sign is probably in English. However, the perception of /sale/ on a laundry basket evokes two possible denotations in the mind of a bilingual ("reduced price" and "dirty") and thus two competing explanations (English and French). Another real world example is the inscription /pain/ on an aisle sign in a Canadian supermarket. The aisle sign may indicate the presence of either pain relief pills (in English) or bread (in French). Upon understanding the sign as one or the other, an English-French bilingual infers a language as the best explanation. A glance at the actual products under the sign then either confirms or refutes the hypothesis. In the multilingual mind, multiple meanings of the same inscription compete for selection and non-cognate inter-lingual homographs are seen as verbal equivalents of ambiguous pictures [5, 23].

Several abductive steps (see Table 1) take place when learning from a textbook or computer screen full of inscriptions. In the following, we will illustrate these with the help of the patterns in Fig. 3.

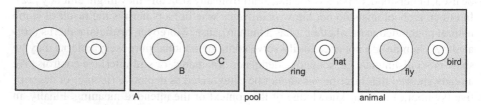

Fig. 3. Some inscriptions, patterns, pictures, and/or diagrams

First, as Hofstadter [17] noted, one has to be aware of the presence of a content before paying attention to an inscription amongst the multitude of patterns and marks available. Encountering the first pattern in Fig. 3 triggers an inference to the best explanation, i.e. that someone tries to signify something (first abduction in Table 1). Of course, this can safely be assumed in an instructional document or on a computer screen, but alternative explanations cannot be excluded formally. In fact, any pattern may also result from some naturally occurring phenomenon or from some technical problem in printing or displaying. Moreover, rectangles and circles may embellish rather than carry meaning. According to Peirce, in reference to Descartes, abduction allows inferring, from our understanding of an inscription, that we know the code or the language used (second abduction in Table 1). In effect, we seem to understand something in all four patterns in Fig. 3, so the best explanation is that these inscriptions are probably meant to convey meaning and that we know their code or language.

As mentioned above, identical marks, characters, and geometrical figures are used in different contexts, in drawings, pictures and diagrams, and in different visual languages. The marks in Fig. 3 may be geometrical figures, characters, shapes of portrayed objects, Euler circles, Venn or Peirce's diagrams. The sheer recognition of a mark is not formal proof of a specific graphical code or language. Thus, in principle, we may wrongly conjecture a language for a pattern (third abduction in Table 1). An archeologist in fact considers several plausible meaning-language combinations for a particular mark. For example, for a circle, we consider a ring in a top view, a figure in a geometry lesson, and a set in a Euler diagram. For /pool/, we consider a swimming pool in English and a pole in Dutch. For /hat/, we consider a hat in English and the third person form of the verb to have in German (/has/ in English). For /animal/, we hesitate between English and French.

Table 1. Three examples of abductive reasoning in understanding inscriptions

Fact (q)	Rule ($p \rightarrow q$)	Hypothesis ($\Diamond p$)	Other explanations
These are marks	If someone wants to signify, then marks arise	Someone probably wants to signify	Natural phenomena, unintentional behavior
I understand these marks	If I know a language, then I understand marks	I probably know this language	Squiggles, interface controls, decorations
This circle is a set to me	If a pattern is a Euler diagram, then circles are sets	This is probably a Euler diagram	Other graphical codes or visual languages

The third abduction (Table 1) shows how we infer the language from the signification attributed to an inscription. Let us take a closer look at the third pattern of Fig. 3. If I see a swim ring and a sun hat floating around, I infer that it is probably a picture of the top view of a pool scene. Alternatively, if I see sets of entities, I conjecture that it is a Euler diagram. I can even flip both interpretations back and forth in my mind. As Wittgenstein told us, the mind can only focus on one perception at a time. So-called visual puzzles, such as those depicting large hats (sombreros, cowboy hats), exploit this phenomenon. The puzzle solver conjectures possible interpretations of a drawing until he or she is told the intended one (cf. a Mexican/cowboy frying an egg).

In the fourth pattern, I see sets of entities, so again, I conjecture that it is a Euler diagram. In fact, English as the likely language gives two competing interpretations for the inscription /fly/. It could be either a verb (to fly) or a noun (a fly). Since I already know that birds fly, I hypothesize that /fly/ probably has to be read as a noun. The example shows how prior knowledge of birds and flies guides disambiguation. Finally, although denotation (literal meaning) is the primary signification mode in instructional material, an archeologist cannot exclude connotation (figurative meaning) a priori. For example, to some, swing rings and sun hats evoke "holiday", "free time", or "leisure", to others, flies may suggest "nuisance" or "death", birds may suggest "freedom" or still "peace", etc. Potential significations are infinite, solely limited by context, cultural linguistic background and prior knowledge. Upon encountering an inscription, either only one interpretation of a given inscription is available in the mind or the content itself allows deciding amongst different interpretations in alternative languages.

4   A Formal Framework with Recursive Rules

In acknowledging different categories of observers, such as teachers and learners or experts and novices, with different cultural and/or linguistic backgrounds, the above showed that a particular configuration of inscriptions might represent or evoke a number of different things by virtue of an individual's prior knowledge of the domain and of disciplinary representational systems. The resulting situation deviates from the straight-forward denotative semiotics described in the introduction. In this section, we develop a formal system for exploring such a situation. A formal system consists of axioms and rules for producing new propositions from existing ones. It guarantees the validity of a reasoning chain independently of the semantic content of the propositions.

4.1   Representation as a Three-Place Predicate

We adopt a triadic semiotic perspective in which a representation requires an *interpretant* in Peirce's terminology: an entity to which something represents something else. Peirce (1935–1985) gives a number of examples: a weathercock represents the direction of the wind to the conception of a person understanding it, a barrister represents a client to a judge and a jury. Such a form of *personal representation* requires three places for the representing x, the represented y, and the interpretant a, a three-place predicate noted $P(a, x, y)$. In words: x stands for y to a (or for a). It precisely allows for varying viewpoints due to differences in culture, language, and prior knowledge.

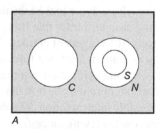

Fig. 4. In the set of all entities A, C is the set of mental entities; N is the set of publicly available physical objects and phenomena; S (within N) is the set of entities intentionally carry meaning

Our universe of discourse is the set of entities A (Fig. 4). We define C as the set of entities that are purely mental. Such entities in the mind can be concrete, such as "chair" and "pump", or abstract, such as "love" and "freedom". We conceive of them as necessarily individual, in someone's mind, but we assume that a group of individuals may share similar mental entities across languages. For example, the idea of "freedom" maybe similar within a group even if evoked by /freedom/ in English, /vrijheid/ in Dutch, and /liberté/ in French. The complement of C in A is N, that is the set of physical objects and phenomena that are publicly available, perceivable by the senses (see also [14]). We do not distinguish between natural entities, such as sand, rocks, water, plants, lightning, etc., and human-made ones, such as cars, factories, books, computers, and marks. Within N, we define S as the set of entities which are intentionally produced to carry meaning, such as language, notational systems, graphics, tables, images, schemas, diagrams, paintings, and statues. We also define M as the set of entities that are natural or manufactured and publicly available, but that did *not* come into existence for intentionally carrying meaning (in N and not in S).

Whereas emblematic instances of representation involve an element of S (the word /tree/) that represents an element of M (a tree) for an element of C (the idea "tree"), infinite semiosis formally permits elements of the three sets M (material), S (intentionally signifying) and C (mental) to take up the positions of x, y or a. Hence, 27 different triads (3*3*3) can be distinguished. For example, in the design of consumer goods, a material object such as a Ferrari ($\in M$) may represent something such as "wealth" ($\in C$) in some society ($\in C$) regardless of the intention of the designer of the car. Barthes [3] called them function-signs to explicitly designate their utilitarian, functional *raison d'être*. Furthermore, elements of M might also occupy the place of the interpreting entity, such as virtual agents (computer programs) interpreting external symbolic representations. Table 2 shows examples of the nine types relevant in this contribution: those with purely mental entities as the interpreting entity ($a \in C$).

Table 2. Nine types of representation $P(a, x, y)$ with $a \in C$ and $x, y \in C, S$, or M

	Types	Examples
1	$x \in C, y \in C$	A theory in mathematics, psychology, or sociology
2	$x \in C, y \in S$	A theory in linguistics, or semiotics
3	$x \in C, y \in M$	A theory in physics, chemistry, or biology
4	$x \in S, y \in C$	A diagram of an argument
5	$x \in S, y \in S$	A graphical representation of a data set
6	$x \in S, y \in M$	A drawing of a tree, a flowchart of a process in a factory
7	$x \in M, y \in C$	A dove representing peace, marbles representing a number
8	$x \in M, y \in S$	A toy-train with letter-wagons, a car representing its CAD drawing
9	$x \in M, y \in M$	A ball-and-rod model of a molecular structure, a sand castle

For the types including the mental domain, the direction of representation is most often from the mental to the material or symbolic, such as conceptions of the world of concepts, signs and things (types 1, 2, 3). Furthermore, types 4 through 6 correspond to the traditional view of representation as a three-place predicate, with a viewpoint, involving the intentionally carrying meaning and representing something abstract or concrete. Types that involve the mental domain for neither x nor y (types 5, 6, 8, 9) are most frequent in instructional material. Indeed, to Jorna and van Heusden [21], ignoring the mental domain entails a positivist view: a unique interpretation for each representation by a shared decoding mechanism. Finally, a mental entity in the place of the represented world is found in semantic networks (type 4) or marbles (type 7).

For the examples in Table 2, interpretation by a human is a necessary and sufficient condition for representation. At the same time, this might seem excessive since only one interpreting entity (to the conception of which something is a representation of something else) is sufficient for affirming the existence of a particular representation. As an illustration, let us read the examples in Table 2 by inversing the roles of x and y. Although ideosyncratic, such as for "peace" to represent a dove, or a process in a factory to represent a flowchart, the formal framework should not formally exclude the existence of such a representation in the mind of at least one individual. For instance, to an experienced Computer Assisted Design (CAD) operator, a consumer object, such as a plastic bottle, represents the CAD model from which it was manufactured (see [21] for this counterintuitive example). Whereas Bunge's [6] system excludes interchanging the roles (representation is a non-symmetrical relation), our relativist account of representation accepts the reversibility of the roles of the represented and the representing.

4.2 A Set of Test Patterns

For illustration purposes, we generated a collection of content-free patterns by combining four graphical and three textual elements in two spatial layouts (Fig. 5). Textual elements suggest different things depending on the attribution of a language. For example, the inscription /pain/ evokes "bread" in the mind of a French and "pain" in the mind of an English monolingual. The inscription /bread/ evokes "bread" in English, but is a non-word in many other languages. The inscription /pooq/ is a non-word in English and French and presumably does not evoke anything. The graphical elements were very basic forms featuring in freehand pictorial drawings, as well as in formal diagrams such as flowcharts and electrical circuit diagrams. Thus, multiple personal representations can be

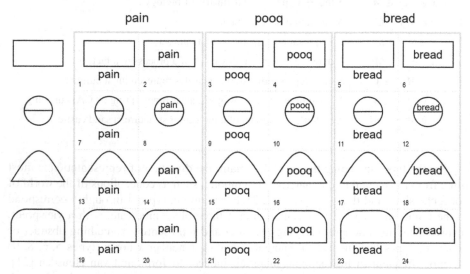

Fig. 5. All 24 combinations of two spatial layouts of four graphical and three textual inscriptions

generated depending on context, prior knowledge, and cultural or linguistic background (the reader is invited to look for individual subjective interpretations).

4.3 Denotation, the Regular Case of Instructional Material

Denotative semiotics [3], or literal meaning, relates to the representational or referential function of language [18]. Indeed, in the regular case, pictures or words such as /force/, /lever/, and /acid/ in instructional material stand for their respective objects or concepts in physics, technology, and chemistry. Likewise, in mathematics, a drawing of a rectangle represents nothing else than a quadrilateral polygon with right angles. Thus, we first define the base case in rule (1).

$$\forall x \forall y \in A \ \exists a \in C \ P(a, x, y) \Rightarrow R(x, y) \tag{1}$$

In words, for any x and y, if there is at least one mental entity a, to the conception P of which x stands for y, then x is a representation R of y. Rule (1) states that one personal representation $P(a, x, y)$ suffices to affirm the existence of a representation. The base case satisfies the obligation to stop a recursive function and constitutes the most frequent situation in which learners are not supposed to go beyond the immediate dictionary definition or resemblance in a predefined cultural context.

Let us look for denotation in Fig. 5. From a Western English point of view, examples of denotation include a picture of a pain relief pill (pattern 8), a loaf of bread (patterns 23 and 24), and maybe of a bun (pattern 11). However, in the absence of the inscription /douleur/, none of the patterns in Fig. 5 evokes pain in a French mind, but denotation includes a picture of a bun (pattern 7) and of a loaf of bread (patterns 19 and 20). For a French-English bilingual, /pooq/ does not evoke anything, but we cannot exclude that there is a language in which /pooq/ has a denotation. From an archeologist's standpoint, /pooq/ could be some meaningless squiggle, a word in an unknown language or still the to-be-learned topic precisely because we do not understand it yet. Upon encountering pattern 15 or 16 in instructional material, the denotative mode could make us infer that a pooq is a kind of haystack shaped object. Hence, paradoxically, prior knowledge of pooq is required to learn about it from a visual representation.

4.4 Connotation or Figurative Meaning

Connotation or figurative meaning produces alternative interpretations of inscriptions depending on context and cultural background. For example, one of the marks in Fig. 1 represents a fish (denotation), but to some, this representation in turn evokes Jesus, which in turn evokes Christianity. In our relativistic frame, if the fish shape represents Christianity to at least one entity, then we decide that indeed that fish shape is a representation of Christianity. Recursive rules allow modelling the fact that there may be representations of representations. It corresponds to Peirce's infinite semiosis: "The idea of representation involves infinity, since a representation is not really such unless it be interpreted in another representation" [27]. There are several ways of defining recursive rules. With Barthes [3], we argue that the simple scheme of transitivity as in rule (2) does *not* fit the domain of representation.

$$\forall x \forall y \exists z \in A \ R(x, z) \wedge R(z, y) \Rightarrow R(x, y) \tag{2}$$

Such a transitive rule holds in the domain of ancestorship. In words, for any x and y, if there is a z, so that x is an ancestor of z and z is an ancestor of y, then x is also an ancestor of y. In contrast to ancestorship, which involves separate entities, a representation as a *reference* of an x to a y taken as a whole plays a role in a higher-order representation. Hence, we must consider two forms of complex recursive rules: a given representation can replace either the representing x or the represented y in another representation [3]. In other words, while we acknowledge a certain degree of isomorphism between the two domains of representation and ancestorship, we do not formulate new propositions on the basis of the similarity (a case of active interpretation [17]). The semantic domain of representation, *not* the domain of ancestorship, should furnish an interpretation relation that makes the formal system true, i.e. it is a model in model-theoretic terms [34]. For intelligibility, we do not insert the base case into the recursive rules (necessary in an implementation), because according to rule (1), any $P(a, x, y)$ gives rise to a $R(x, y)$.

The first complex scheme of recursion is *connotation* foreseen by Barthes [3] in which a representation replaces the representing entity rule (3).

$$\forall x \forall y \exists z \in A \; R(R(x, z), y) \Rightarrow R(x, y) \tag{3}$$

In words, for any x and y, if there exists a z, so that x is a representation of z, and this representation of itself is a representation of y, then x is also a representation of y. The connotative recursive role indefinitely expands the number of things that can be seen in something depending on prior knowledge. For example, to an observer from Western culture, a circle (Fig. 3) through connotation represents a ring, unity, and marriage by virtue of prior knowledge of weddings. As an illustration, rectangles such as in Fig. 5 provide multiple connotations in particular for patterns 3 and 4 since the inscription /pooq/ does not provide a clue in English, French, or Dutch.

$$R(R(R(R(\Box, \text{rectangle}), \text{cylinder}), \text{chamber}), \text{pump}) \Rightarrow R(\Box, \text{pump}) \tag{4}$$

$$R(R(R(R(\Box, \text{rectangle}), \text{cuboid}), \text{case}), \text{shoebox}) \Rightarrow R(\Box, \text{shoebox}) \tag{5}$$

The connotative recursive examples in (4) and (5) show how a pooq might be seen as a pump or a shoebox, but multiple other possibilities are available for a rectangle, such as a photo frame, a brick, a dough roller, or still a ramp. Abductive reasoning goes backwards from the consequent to the antecedent: if you see a rectangle as a pump, then you infer that the rectangle probably has to be interpreted as a hollow cylinder. The important point is that prior knowledge is required to disambiguate between competing interpretations of a rectangle in a drawing.

4.5 Metalanguage

The second form of recursion follows Barthes' [3] scheme of *metalanguage*, which occurs when a representation of itself is the represented entity of another representation. Thus, whereas in connotation, a representation replaces x, here it replaces y rule (6).

$$\forall x \forall y \exists z \in A \; R(x, R(z, y)) \Rightarrow R(x, y) \tag{6}$$

In words, for any x and y, if there exists a z, so that x is a representation of another representation involving a z standing for y, then x is also a representation of y. As an example, we examine Euler circles as a metalanguage for reasoning about sets of entities. A logician reasoning in set theory has an internal representation involving a mental entity m_l that stands for a set. Since we cannot look inside a mind, we use the notation m_l (instead of "set") to designate a mental entity in a logician's mind. The circle in a Euler diagram is a second-order representation of that representation (7). In other words, circles in Euler diagrams are a metalinguistic shorthand for sets within the community of logicians. The same circle stands for other representations in other metalanguages (a connector in flowcharts or a voltmeter in electrical circuit diagrams). Experts with shared knowledge of a metalanguage have similar mental entities, e.g. m_e representing a meter in the minds of electricians (8).

$$R\big(\circ, R(\text{m\_l}, \text{set})\big) \Rightarrow R(\circ, \text{set}) \tag{7}$$

$$R\big(\circ, R(\text{m\_e}, \text{meter})\big) \Rightarrow R(\circ, \text{meter}) \tag{8}$$

These instances exemplify that the choice of a particular geometrical form (a circle, a rectangle, a square) is an arbitrary convention within the community of experts in a field independent of the context in which it is inserted: a case of monosemy [4].

4.6 Mutual Recursion

Finally, we distinguish between internal representations (in the head, $x \in C$) and external representations (outside the head, $x \in N$). In mutual recursion, two rules are defined in terms of one another (indices are introduced to label representations).

$$\forall x \in C \; \forall y \in A \; \exists z \notin C \; R_i(x, R_e(z, y)) \Rightarrow R_i(x, y) \tag{9}$$

$$\forall x \notin C \; \forall y \in A \; \exists z \in C \; R_e(x, R_i(z, y)) \Rightarrow R_e(x, y) \tag{10}$$

Rule (9) put in words: if x is an internal representation of an external representation z of y, then x is also an internal representation of y. The rules mutually calling internal and external representations necessarily follow the scheme of metalanguage, not connotation. These rules allow understanding propositional networks as researchers' external representations of a cognitive psychological model of participants' internal representations (11).

$$R_e\left(\boxed{\text{pooq}}, R_i\left(\text{m\_r}, R_e(/\text{pooq}/, R_i(\text{m\_p}, \text{pooq}))\right)\right) \Rightarrow R_e(\boxed{\text{pooq}}, \text{pooq}) \tag{11}$$

A participant has a first-order internal representation that involves the participant's mental entity (m_p) standing for a pooq. The mark /pooq/ written by this participant is a second-order external representation of the reference of the mental entity to a pooq. The researcher in turn has third-order internal representation in which the researcher's mental entity (m_r) stands for the nested representation. Finally, the box $\boxed{\text{pooq}}$ in a concept map is a researcher's fourth-order external representation of a pooq. Thus, such a node in a propositional network stands for a profound stack of mutually calling internal and external representations. Note that the fourth-order representation evicts both the researcher's and the participant's mental entities. The rules of the framework illustrate how propositional networks as scientific models shortcut the representing mental entities of both the researcher and the participant and exemplify the symbol grounding problem [15]. Fodor [12] referred to this phenomenon: since we do not have a language to form genuine structural descriptions of internal representations, we take natural language structural descriptions to be internal representations.

5 Conclusion

Piaget [28] asserted the impossibility to construct knowledge by copying from the world, since knowledge of the objects and of the phenomena is necessary in order to decide what aspects to copy in the first place. Our investigation shows the impossibility to learn from visual representations, since prior knowledge of the represented objects and phenomena is necessary to recognize them in a picture, drawing or diagram in the first place. The framework gives a very pessimistic outlook on teaching and learning with diagrams because of the knowledge disparity between teachers and learners, professors and students, instructors and trainees, experts and novices.

Three directions for evaluating the framework can be sketched. First, the framework aids in establishing the boundary conditions of mainstream research in the field. In particular, the framework should be gauged on its merits in providing alternative explanations for failures to show effects of certain kinds of instructional materials. For example, the lack of previous exposure to a graphical code or language constitutes an obstacle in learning with diagrams. Contrary to natural language, one cannot reflect on a graphical language in that graphical language, i.e. flowcharts cannot be used to explain the use of flowcharts. A graphical code requires textual explanations because graphical languages are non-reflexive [40]. Second, although we explored an idiosyncratic highly unlikely imaginary situation, we think the developed framework contributes to our understanding of multicultural and/or multilingual cognition. Corroborating evidence from research on cultural differences could be sought to strengthen and enrich the current frame. A third direction for evaluation is to empirically test the existence of different semiotics (denotation, connotation, metalanguage) in learning with diagrams.

Fortunately, most educational contexts can rely on common ground in the form of shared cultural background, dominant language, and general prior knowledge. In any case, whether advocate or skeptic about multiple external representations, in the event of learners successfully interpreting pictures and diagrams, the representational equivalent of the frame problem still stands today: How do learners deal with the multiplicity of codes and languages that potentially apply to a given inscription?

References

1. Ainsworth, S.: DeFT: a conceptual framework for considering learning with multiple representations. Learn. Instr. **16**(3), 183–198 (2006). https://doi.org/10.1016/j.learninstruc.2006.03.001

2. Airey, J., Linder, C.: A disciplinary discourse perspective on university science learning: achieving fluency in a critical constellation of modes. J. Res. Sci. Teach. **46**(1), 27–49 (2009). https://doi.org/10.1002/tea.20265
3. Barthes, R.: Elements of Semiology. Hill and Wang, New York (1968)
4. Bertin, J.: Semiology of Graphics: Diagrams, Networks, Maps. The University of Wisconsin Press Ltd, Madison (1983)
5. Bialystok, E., Shapero, D.: Ambiguous benefits: the effect of bilingualism on reversing ambiguous figures. Dev. Sci. **8**(6), 595–604 (2005). https://doi.org/10.1111/j.1467-7687.2005.00451.x
6. Bunge, M.: Analogy, simulation, representation. Rev. Int. Philos. **87**, 16–33 (1969)
7. Carney, R.N., Levin, J.R.: Pictorial illustrations still improve students' learning from text. Educ. Psychol. Rev. **14**(1), 5–26 (2002)
8. diSessa, A.A.: Metarepresentation: native competence and targets for instruction. Cogn. Instr. **22**(3), 293–331 (2004). https://doi.org/10.1207/s1532690xci2203_2
9. Douven, I.: Abduction. In: Zalta, E.N. (ed.) The Stanford Encyclopedia of Philosophy. Stanford University (2017)
10. Eco, U.: A Theory of Semiotics. Indiana University Press, Bloomington (1976)
11. Eco, U.: Semiotics and the Philosophy of Language. Indiana University Press, Bloomington (1986)
12. Fodor, J.A.: Computation and reduction. In: Fodor, J. (ed.) RePresentations, Essays on the foundations of cognitive science, pp. 146–174. MIT Press, Cambridge (1983)
13. French, R.M., Ohnesorge, C.: Using non-cognate interlexical homographs to study bilingual memory organization. In: Moore, J.D., Lehman, J.F. (eds.) Proceedings of the Seventeenth Annual Conference of the Cognitive Science Society, pp. 31–36 Erlbaum Associates, Mahway, Lawrence (1995)
14. Giardino, V., Greenberg, G.: Introduction: varieties of iconicity. Rev. Philos. Psychol. **6**(1), 1–25 (2014). https://doi.org/10.1007/s13164-014-0210-7
15. Harnad, S.: The symbol grounding problem. Physica D **42**(1), 335–346 (1990)
16. Henrich, J., et al.: Most people are not WEIRD. Nature **466**(7302), 29 (2010)
17. Hofstadter, D.R.: Gödel, Escher, Bach: An Eternal Golden Braid. Basic Books, New York (1979)
18. Jakobson, R.: Closing statement: linguistics and poetics. Style Lang. **350**(377), 570–579 (1960)
19. Jakobson, R., Halle, M.: Fundamentals of Language. Walter de Gruyter (2010)
20. Jonassen, D.H.: Objectivism versus constructivism: do we need a new philosophical paradigm? Educ. Tech. Res. Dev. **39**(3), 5–14 (1991)
21. Jorna, R., van Heusden, B.: Why representation (s) will not go away: crisis of concept or crisis of theory? Semiotica **143**(1/4), 113–134 (2003)
22. Kaput, J.J.: Representations, inscriptions, descriptions and learning: a kaleidoscope of windows. J. Math. Behav. **17**(2), 265–281 (1998)
23. Kroll, J.F.: Juggling two languages in one mind. Psychological Science Agenda, American Psychological Association, vol. 22, p. 1 (2008)
24. Larkin, J.H., Simon, H.A.: Why a diagram is (Sometimes) worth ten thousand words. Cogn. Sci. **11**(1), 65–100 (1987). https://doi.org/10.1111/j.1551-6708.1987.tb00863.x
25. Mayer, R.E.: Multimedia Learning. Cambridge University Press, New York (2001)
26. Pande, P.: Learning and expertise with scientific external representations: an embodied and extended cognition model. Phenomenol. Cogn. Sci. **20**(3), 463–482 (2020). https://doi.org/10.1007/s11097-020-09686-y
27. Peirce, C.S.: Collected Papers, vols. 1–8. Harvard University Press, Cambridge (1931)
28. Piaget, J.: Genetic Epistemology. Columbia University Press, New York (1970)

29. Rastier, F.: On signs and texts : cognitive science and interpretation. In: Perron, P. et al. (eds.) Semiotics as a Bridge between the Humanities and the Sciences, pp. 409–450. Legas Press, New-York/Toronto (2000)

30. Schnotz, W.: Sign systems, technologies and the acquisition of knowledge. In: Rouet, J.-F., et al. (eds.) Multimedia Learning: Cognitive and Instructional Issues, pp. 9–30. Elsevier, Oxford (2001)

31. Schnotz, W., Bannert, M.: Construction and interference in learning from multiple representation. Learn. Instr. **13**(2), 141–156 (2003). https://doi.org/10.1016/S0959-4752(02)000 17-8

32. Schuh, K.L., Barab, S.A.: Philosophical perspectives. In: Spector, J.M. et al. (eds.) Handbook of Research on Educational Communications and Technology, pp. 67–82 (2007)

33. Shelley, C.: Visual abductive reasoning in archaeology. Philos. Sci. **63**(2), 278–301 (1996)

34. Stenning, K., Oberlander, J.: A cognitive theory of graphical and linguistic reasoning: logic and implementation. Cogn. Sci. **19**(1), 97–140 (1995)

35. de Vries, E., et al.: External representations for learning. In: Balacheff, N. et al. (eds.) Technology-Enhanced Learning, pp. 137–153. Springer, Heidelberg (2009). https://doi.org/10.1007/978-1-4020-9827-7_9

36. de Vries, E.: Learning with external representations. In: Seel, N.M. (ed.) Encyclopedia of the Sciences of Learning, pp. 2016–2019. Springer, New York (2011)

37. de Vries, E.: What's in a rectangle? An issue for AIED in the design of semiotic learning tools. In: Looi, I.C.-K. et al. (eds.) Artificial Intelligence in Education. Supporting Learning Through Intelligent and Socially Informed Technology, pp. 938–940. IOS Press, Amsterdam (2005)

38. de Vries, E., Masclet, C.: A framework for the study of external representations in collaborative design settings. Int. J. Hum Comput Stud. **71**(1), 46–58 (2013). https://doi.org/10.1016/j.ijhcs.2012.07.005

39. Wittgenstein, L.: Philosophical Investigations. Basil Blackwell, Oxford (1968)

40. Wittgenstein, L.: Tractatus Logico-Philosophicus. New York (1922)

41. Zhang, J.: The nature of external representations in problem solving. Cogn. Sci. **21**(2), 179–217 (1997)

The Fall and Rise of Resemblance Diagrams

Mikkel Willum Johansen(✉) and Josefine Lomholt Pallavicini

Department of Science Education, University of Copenhagen,
Universitetsparken 5, 2100 Copenhagen, Denmark
{mwj,jlp}@ind.ku.dk
https://www.ind.ku.dk

Abstract. A recent investigation of the changes in the use of diagrams in published mathematics papers shows that diagrams were frequently used at the end of the 19th century and the beginning of the 20th. They then largely disappeared in the period 1910–1950, whereafter they reappear [1]. Although this story is unsurprising considering the dominance of formalist ideology in the first half of the 20th century, the detailed investigation of the development points out several interesting open questions. Especially, we do not know if the diagrams that disappeared with the advent of formalism are the same as those that are used today.

In this paper, we will focus on so-called "resemblance" diagrams, which are one of three general categories of diagrams covered in the investigation in [1]. We will analyze and compare resemblance diagrams used in the late 19th century with those used in the early 20th century to determine if there have been substantial changes. The comparison shows that even though the diagrams can be said to belong to the same general category and share certain general features, the resemblance diagrams used today are very different from those used before the advent of formalism. The criticism raised by the formalist movement of the diagrams used in the late 19th century can be seen as a possible explanation of this change.

Keywords: Mathematical diagrams · Publication practice · Corpus study

1 Entering the Valley of Formalism

Looking at research papers published in mathematics journals during the last century, one gets the impression that there are substantial differences in the frequency and types of diagrams published in different time periods. In [1], the overall trends and changes in the use of diagrams in the period 1885–2015 are investigated by coding all papers published in the *Bulletin of the AMS*, *Acta Mathematica*, and *Annals of Mathematics* in years separated by five-year intervals beginning with 1885. The investigation revealed that diagrams were relatively frequently used until 1910. Then they disappeared for several decades before they reappeared during the 1950s and 1960s. The disappearance can, in

© Springer Nature Switzerland AG 2021
A. Basu et al. (Eds.): Diagrams 2021, LNAI 12909, pp. 331–338, 2021.
https://doi.org/10.1007/978-3-030-86062-2_33

part, be ascribed to the influence of what we will loosely call "formalist ideology", e.g., the nexus of ideas expressed by David Hilbert [3], Mouriz Pasch [4], Bertrand Russell [5], and others, that mathematics should be formalizable and that diagrammatic reasoning should be confined to the heuristics of mathematics. In other words, we see a half-a-century-wide "valley of formalism," where diagrams almost disappear from mathematics publications.

The investigation reported in [1], however, also shows that the diagrams that reappeared in the 1950s were not the same as those that disappeared half a century earlier. To track the overall changes in the types of published diagrams, the investigation operates with a rough distinction between three general categories of diagrams: resemblance, algebraized, and abstract diagrams. Here "resemblance diagrams" are diagrams with a direct resemblance to the objects being represented, "algebraized diagrams" represent objects in an algebraized domain, and "abstract diagrams" essentially depend on a conceptual map (see [2] for details). The resurge of diagrams in the 1950s and 1960s was mainly due to the advent of abstract diagrams, especially commutative diagrams, and closely related diagram types. As diagrams of this type were rarely used at the beginning of the 20th century, it appears that the formalist ban on diagrams was not broken by the reappearance of the "old" diagrams, but rather by the introduction of a completely new kind of diagram that conforms better to the specific formalistic demands for rigor.

The data in [1], however, also show that resemblance diagrams and algebraized diagrams eventually reappeared as well, although slightly later. As the criticism of the use of diagrams in mathematics made by the formalist movement was mainly aimed at resemblance diagrams, especially diagrams involving geometric intuition, this is particularly interesting. Does the reappearance of resemblance diagrams indicate that diagrams attacked by formalism are once again used in mathematical publication practice?

Unfortunately, the three categories used to classify diagrams in [1] are very broad, and each category includes several sub-categories of diagrams. Thus, although resemblance diagrams reappeared in the late 20th century, we do not know if the resemblance diagrams that are used today are of the exact same type as those used before the advent of formalism, or if there has been an internal development within the category. In this paper, we will investigate this question by analyzing and comparing the resemblance diagrams that were published at the end of the 19th century with those published at the beginning of the 21st century. The aims of this investigation are 1) to give a more detailed picture of the influence formalist ideology has had on the use of diagrams and 2) to strengthen our understanding of the role certain types of diagrams play in modern mathematical practice.

2   Methods

The sampling and coding strategy for the full investigation is described in detail in [1,2]. In the full corpus, 1,143 diagrams were coded as resemblance diagrams.

For this paper, all diagrams coded as resemblance diagrams in the years 1885–1895 and 2005–2015 were revisited with a qualitative and grounded approach aimed at identifying relevant sub-classes of diagrams within the general category. Each subclass was given a short description, and a prototypical example of the class was picked out and will be presented below.

The number of resemblance diagrams and the distribution over the six years in question can be seen in Table 1. As the number of diagrams is relatively small, we did not attempt a structured quantitative analysis (e.g., counting the exact number of diagrams in each subclass) but restricted ourselves to a purely qualitative analysis describing the types of diagrams present in the two periods supplemented with rough estimates describing the relative frequency of the most common types.

Table 1. Resemblance diagrams in the years 1885–1895 and 2005–2015.

Year	1885	1890	1895	2005	2010	2015
Diagrams	1	21	17	114	25	439

3 Results

3.1 Resemblance Diagrams 1885–1895

In the analysis of the resemblance diagrams published in the early period, that is, 1885, 1890, and 1895, three major sub-categories (what we will call "types") emerged. We will call the first of these *Euclidian construction diagrams*. Diagrams of this type are used to anchor a sequence of geometric constructions involving basic geometric objects, such as circles, triangles, and straight lines. A prototypical example is shown in Fig. 1.

The second subcategory is *object illustrating diagrams*. This type of diagram is closely related to Euclidian construction diagrams, but instead of anchoring a geometric construction, these diagrams simply represent a constellation of idealized geometric shapes, often modelling real life objects. A prototypical example, where different ways of stretching a string between two objects is modeled, is shown in Fig. 2.

The final subcategory identified in the early period is *diagrams illustrating an operation*. Here, geometric shapes are used to illustrate the effects of an operation. Thus, in a sense, such diagrams do not illustrate the geometric shapes per se but rather something more abstract, such as an operation. A prototypical example is shown in Fig. 3.

In the period in question, Euclidian construction diagrams are the most common of the three types, closely followed by diagrams illustrating objects. Only one paper contains diagrams illustrating operations. From this analysis, it thus follows that in our corpus, the resemblance diagrams used in the period 1885–1895 had a close connection to specific geometric objects either by anchoring compass and straightedge constructions or by modeling (metric properties of) constellations of geometric shapes.

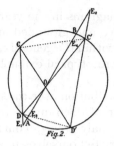

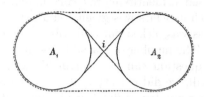

Fig. 1. Euclidian diagram. Reproduced with kind permission from [6, p. 176].

Fig. 2. Diagram illustrating an object (a string stretched between the two figures A_1 and A_2). Reproduced with kind permission from [7, p.189].

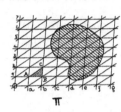

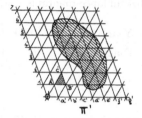

Fig. 3. Diagram illustrating the effect of an operation on a geometric shape. Reproduced with kind permission from [8, p.4].

3.2   Resemblance Diagrams 2005–2015

Diagrams illustrating objects and those illustrating operations are also present in the late period (viz. 2005, 2010 and 2015). Yet, most of the diagrams illustrating objects are of a slightly different nature than the similar diagrams in the early period. In the early period, the resemblance between the object and the diagram illustrating the object is established through direct metric likeness, whereas in the late period, the resemblance is most often established through topological likeness. We consider this to be a new type of diagram that we will call a *topological illustration diagram*. The objects illustrated in this kind of diagram can be certain kinds of knots, graphs, and abstract objects, such as manifolds. As an example, the diagram represented in Fig. 4 is used to illustrate a proof of why the neck of a dumbbell pinches before the bells become extinct. Notice, however, that although the objects are generalized manifolds, the diagram draws heavily on intuitions based on sensory-motor experience.

Among the 578 resemblance diagrams in the late part of the corpus, we did not encounter any Euclidian construction diagrams. However, another kind of construction diagrams are very common. These diagrams can be called *topological construction diagrams*, and they can be characterized as diagrams representing constructions with or on objects of topological illustration diagrams. As a prototypical example, Fig. 5 illustrates how two abstract surfaces can be glued together. In this period, there is a thin line between representing objects and constructions, as the representation diagrams often suggest movement or manipulation of the objects represented, although this is not explicitly stated, as the implied pinching of the surfaces in Fig. 4 illustrates.

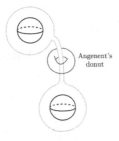

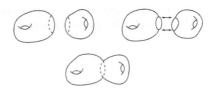

Fig. 4. Diagram illustrating an object through topological likeness. In this case the diagram illustrates a vital step in a proof by Angenent. Reproduced with permission from [9, p. 303].

Fig. 5. A topological construction diagram). Reproduced with permission from [10, p. 62].

Thus, in short, in both periods, we see resemblance diagrams illustrating objects and diagrams representing constructions. However, there is a crucial difference between the two periods in the sense that in the early period the resemblance is established through metric likeness, whereas in the late period it is (mostly) established through topological likeness. We will explain why this distinction is crucial in the discussion below.

Apart from the move from metric to topological likeness, we also saw some qualitatively new types of diagrams in the late period. Especially, a diagram type we will call *syntactic manipulation diagrams* is widely used in the late period. In these diagrams, the objects represented are also transformed, but not as freely as in the topological construction diagrams, but rather in accordance with strict syntactic rules. This in effect turns this type of diagram into a hybrid between diagrams and symbols (as pointed out in [11]). A prototypical example is Reidemeister moves representing manipulations of idealized knots (Fig. 6). This general type of diagram, however, is widespread in the late part of the corpus, and the specific diagrams belonging to this type are often innovative and appear to play a central role in the arguments of the papers in which they appear. As an example, in [13], a particular type of "puzzle piece" is introduced along with rules of manipulation, and diagrams constructed with these puzzle pieces are then used to perform complex calculations within the particular area of mathematics at hand (see Fig. 7 and Fig. 8). We have coded all representations involving these puzzle pieces in [13] to be diagrammatic, although there may be little difference between the arrangement seen in Fig. 8 and an arrangement of algebraic symbols in the sense that calculations can (seemingly) be performed as purely syntactic manipulations with both representational types. The hybrid nature of syntactic manipulation diagrams thus makes it difficult to maintain a clear distinction between algebraic and diagrammatic representations.

Fig. 6. Diagram illustrating a particular Reidemeister move on an idealized knot. Reproduced with kind permission from [12, p. 500].

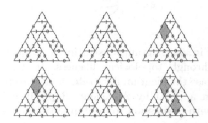

Fig. 7. Partial explanation of the syntax of the "puzzle piece" diagrams. Reproduced with kind permission from [13, p. 179].

Fig. 8. Puzzle pieces needed for a particular computation. Reproduced with kind permission from [13, p. 180].

There is extensive variety in the types of diagrams used in the late part of the corpus—both in general and within the category of resemblance diagrams, which is our focus point in this paper. The sub-categories we have presented here do not cover all the resemblance diagrams in the late part of the corpus, only what we believe to be the most common types of diagrams. This explosion in the variety of diagram types and designs is by itself an interesting result worth noticing; the fact that contemporary mathematicians are willing to spend the effort needed to design and typeset new—often complex—diagram types suggests that diagrams play a relatively central role in their practice (as also pointed out in [1]).

4   Discussion

In summary, we see considerable differences between the resemblance diagrams present in the two parts of the corpus under investigation here. The type of diagrams most commonly used in the years 1885–1895, Euclidian construction diagrams, was not at all present in the late period, and the two other diagram types used in the early period are rarely used in the same form in the years 2005–2015. As a clear trend, the use of metric likeness in the early period is generally (although not completely) replaced with topological likeness in the late period. We thus see diagrams illustrating objects and anchoring constructions in both parts of the corpus, but in contrast to the early period, these diagrams most often take departure in topological likeness in the late period rather than metric likeness. Furthermore, syntactic manipulation diagrams are extensively used in the late period, whereas they are not present in the early period (in the part of our corpus under investigation here).

Following this analysis, it is especially worth noting that the resemblance diagrams used in the late period are better suited to answer at least part of the criticism raised against the use of diagrams in mathematical reasoning. A typical point of criticism centers on the idea that diagrams are over-specific and thus do not allow general conclusions (although this can be contested, see e.g. [14]). Another typical criticism points out that reasoning with the exact features of a diagram may lead to false conclusions (e.g., in the famous example of the "proof" that all triangles are isosceles used by Hilbert [3, p.541]). This kind of criticism is much easier to raise against diagrams that depend on metric likeness, such as Euclidian construction diagrams, than against those depending on topological likeness, as the latter type is more abstract than the former (since metric features are abstracted away in topological diagrams). The move from metric to topological likeness can thus, in part, be seen as a response to the criticism raised by the formalist movement. Similarly, syntactic manipulation diagrams are well aligned with formalist ideology's understanding of rigor, as the diagrams can (in part) be operated syntactically without the use of geometric intuition in a manner similar to algebraic symbols.

Thus, the narrative that diagrams disappeared from published mathematical material at the beginning of the 20th century and reappeared half a century later is much too simple. The development in the attitude toward and use of diagrams in the 20th century not only led to the introduction of a qualitatively new category—abstract diagrams—but, as the in-depth analysis of resemblance diagrams above has shown, it also led to substantial revisions of other categories, leading to a diagrammatic practice that is better aligned with formalist ideology. The return of diagrams thus does not indicate that formalism has been abandoned. Rather, some of its basic ideas seem to have been embedded in current diagrammatic practice.

Finally, it should be noted that the development in diagrammatic practice cannot be understood solely as the product of ideological development. Other factors such as technological development and changes in research interests, should also be taken into consideration. These are not disconnected; it would be difficult to pursue a research interest in an area of mathematics that depends heavily on diagrams if you do not have the technical means to produce them and if current ideology forbids their publication. We have chosen to focus on ideology as an explanatory factor in this short paper, but the story could clearly be nuanced taking other factors into consideration as well.

References

1. Johansen, M.W., Pallavicini, J.L.: Entering the valley of formalism. Trends and changes in mathematicians' publication practice 1885 to 2015 [In review]

2. Johansen, M.W., Misfeldt, M., Pallavicini, J.L.: A typology of mathematical diagrams. In: Chapman, P., Stapleton, G., Moktefi, A., Perez-Kriz, S., Bellucci, F. (eds.) Diagrams 2018. LNCS (LNAI), vol. 10871, pp. 105–119. Springer, Cham (2018). https://doi.org/10.1007/978-3-319-91376-6_13

3. Michael, H., Ulrich, M. (eds.) David Hilbert's Lectures on the Foundations of Geometry 1891–1902. Springer, Berlin (2004)

4. Pasch, M., Dehn, M.: Vorlesungen über neuere Geometrie. Die Grundlehren der mathematischen Wissenschaften, vol. 23. Springer, Berlin (1882/1926)

5. Russell, B.: Mathematics and the metaphysicians. bertrand russel: mysticism and logic and other essays, London: George Allen and Unwin 1917, pp. 74–96. (First published as "Recent Work on the Principles of Mathematics"). In: International Monthly, vol. 4, pp. 83–101 (1917/[1901])

6. Candy, A.L.: A general theorem relating to transversals, and its consequences. Ann. Math. **11**(1/6), 175–190 (1895)

7. Sylvester, J.: On a funicular solution of Buffon's "problem of the needle" in its most general form. Acta Mathematica **14**, 185–205 (1890)

8. Emch, A.: On the fundamental property of the linear group of transformation in the plane. Ann. Math. **10**(1/6), 3–4 (1895)

9. Colding, T.H., Minicozzi II, W.P., Pedersen, E.K.: Mean curvature flow. Bull. (new Series) Am. Math. Soc. **52**(2), 297–333 (2015)

10. Morgan, J.W.: Recent progress on the Poincaré conjecture and the classification of 3-manifolds. Bull. (new Series) Am. Math. Soc. **42**(1), 57–78 (2005)

11. De Toffoli, S.: Chasing the diagram–the use of visualizations in algebraic reasoning. Rev. Symb. Logic **10**(1), 158–186 (2017)

12. Lackenby, M.: A polynomial upper bound on Reidemeister moves. Ann. Math. **182**(2), 491–564 (2015)

13. Buch, A.S.: Mutations of puzzles and equivariant cohomology of two-step flag varieties. Ann. Math. **182**(1), 173–220 (2015)

14. Giaquinto, M.: The epistemology of visual thinking in mathematics. In: Edward, N.Z. (ed.) The Stanford Encyclopedia of Philosophy (Spring 2020 Edition) (2020). https://plato.stanford.edu/archives/spr2020/entries/epistemology-visual-thinking/

The Science of Seeing Science: Examining the Visuality Hypothesis

Lisa Best[1] (ID) and Claire Goggin[2](✉) (ID)

[1] University of New Brunswick, 100 Tucker Park Road, Saint John, NB E2L 4L5, Canada
lbest@unb.ca
[2] St. Thomas University, 55 Dineen Drive, Fredericton, NB E3B 5G3, Canada
cgoggin@stu.ca

Abstract. Fundamental disciplinary differences may be traceable to the use of visual representations, with researchers in the physical and life sciences relying more heavily on visuality. Our goal was to examine how inscriptions are used by scientists in different disciplines. We analyzed 2,467 articles from journals in biology, criminology and criminal justice, gerontology, library and information science, medicine, psychology, and sociology. Proportion of page space dedicated to graphs, tables, and non-graph illustrations was calculated. A Visuality Index was defined as the proportion of page space dedicated to visual depictions of data and non-data information. An ANOVA indicated a statistically significant difference between disciplines, interaction between inscription type and discipline, with articles published in biology journals dedicating more page space to graphs. The significant overlap in inscription use and visuality indicates imperfect disciplinary demarcation, suggesting similar methodological and data analytic practices within a discipline and between subdisciplines.

Keywords: Disciplinary differences · Graphs and tables · Data analysis · Non-inferential analyses

1 Graph Use in Science

Whatever relates to extent and quantity may be represented by geometrical figures. Statistical projections which speak to the senses without fatiguing the mind, possess the advantage of fixing the attention on a great number of important facts.

Alexander D Humboldt, 1811 [1].

Graphical representations allow researchers to summarise data using numerical information in a form that is easily understood by researchers across disciplines [2–6]. The centrality of graphs to scientific inquiry implies that they are at the heart of scientific communication and the construction of scientific facts [7] and have a larger impact on the public than do written descriptions and mathematical proofs [8]. Visual images play an integral role in the comprehension of scientific theories and their constituent data.

© The Author(s) 2021
A. Basu et al. (Eds.): Diagrams 2021, LNAI 12909, pp. 339–347, 2021.
https://doi.org/10.1007/978-3-030-86062-2_34

Kevles [9] asserted that images allow scientifically dense results to be rendered accessible to non-scientists because, "the viewer transforms the static image into an active intellectual experience" [10, p. 9]. Illustrations afford description, classification, order, analysis, and comprehension [10] and modern scientific depictions often mirror early visualization strategies [9].

Latour's [2] classic essay about the importance and usefulness of graphs laid out several fundamental graphical attributes, noting that they enable the transformation of transitory data into a stable, enduring depiction, and facilitate the detection of relationships between variables. Practical benefits of graphs include their easy transportability and reproducible nature, and their ability to be scaled depending upon the magnitude of the data that they represent. Furthermore, they greatly enhance scientific discussion and production of knowledge. Latour claimed that graphs were often central to a lucid and coherent argument, as, without them, scientists "stuttered, hesitated and talked nonsense and displayed every kind of political or cultural bias" (p. 22).

1.1 Disciplinary Differences in Graph and Table Use

Cleveland [4] analyzed graph use in 57 natural, mathematical, and social science journals that represented the hard-soft science continuum. Cleveland's underlying hypothesis was that researchers include visual inscriptions only if they are considered central to the message conveyed in the paper. He calculated the proportion of total page area in an article dedicated to graphical displays (FGA) and found that natural scientists included more graphs in their publications than mathematical and social scientists. Although Cleveland's results support the contention that "harder" sciences make use of more graphs than "softer" ones, he noted that differences in graph use could not be attributed to amount of data but to differences in data representation.

Smith and his colleagues [11] asked psychology faculty and graduate students to rate the scientific hardness of each of Cleveland's [4] disciplines using a 10-point Likert scale. This subjective measure of hardness strongly correlated ($r = 0.97$) with Cleveland's FGA ratings, with physics rated as the hardest and sociology rated as the softest. Thus, harder, more codified disciplines (i.e., natural sciences) used more graphs than softer, less codified disciplines (i.e., social sciences). Arsenault et al. [3] analyzed articles sampled from the same journals as Cleveland and argued that the relationship between FGA and disciplinary hardness could be generalized to other forms of visual displays, supporting the 'visuality hypothesis'. Taken together, Smith and his colleagues [3; 6; 11; 12] have shown that graph use differs according to perceived scientific hardness both between and within a discipline; that is, researchers in harder areas use more graphs than those in softer areas. From a Latourian [2] point of view, these findings illustrate that graph use is proportional to the codification of disciplines, supporting the idea that graphs are a powerful communication device but that they are used differentially by researchers in different disciplines.

1.2 Purpose of the Current Study

This project is the result of 20 years of research, which began with the work of Larry Smith and Alan Stubbs examining inscription use in psychology (PSYC). The research has expanded to include journals from biology (BIO), criminology and criminal justice (CCJ), gerontology (GERO), library and information sciences (LIS), medicine (MED), and sociology (SOC).[1] Our purpose was to examine the use of inscriptions in high impact natural and social science journals. Following Arsenault et al. [3], we were interested in examining disciplinary differences in the use of visual inscriptions (i.e., graphs + non-graph illustrations) and data presentation (i.e., graphs + tables). Further, we examined whether articles published in high impact journals include a wider variety of scientific inscriptions.

2 Method

2.1 Selecting and Coding the Sample

Journal titles per discipline were selected for the current study based on impact (i.e., h-index)[2] and/or journal prestige rankings and were sampled at 5 year intervals from 1980 to 2015, inclusive. Using an on-line random number generator, four issues per year and four articles per issue per journal were identified and selected for inclusion. We analyzed a total of 2,467 articles published in BIO ($k = 11, n = 324$)[3] [13]; CCJ ($k = 16, n = 397$) [5]; GERO ($k = 25, n = 360$) [14]; LIS ($k = 11, n = 524$) [15]; MED ($k = 10, n = 300$); PSYC ($k = 12, n = 339$) [6; 12]; and SOC ($k = 7, n = 223$) journals. For each article, bibliographic factors were recorded and the proportion of page spaced dedicated to graphs, tables, and non-graph illustrations was calculated.

A graph was defined as a figure with a scale that displayed quantitative information [4]. Specific graph types (i.e., bar, scatter, line, 3D) were coded. Numbers and types of graphs, plus total graph area, and fractional graph area (FGA) [see 4] were recorded for each article. A table was defined as information presented in a series of rows and columns distinct from the main body of text [see 6] and were classified as a data or non-data table. Non-data tables typically presented qualitative information (i.e., lists, models). Numbers and types of tables, plus total table area and fractional table area (FTA) were recorded for each article [see 6]. A non-graph illustration was defined as any visual inscription (i.e., photograph, schematic, methodological illustration) that did not meet the criteria of a graph or table [5, 6]. Numbers and types of illustrations as well as total non-graph illustration area and fractional non-graph illustration area (FIA) were recorded for each article [see 6].

3 Results

3.1 Inscription Use Per Discipline

Following Cleveland [4] and Smith et al. [6], differences per inscription type per discipline were examined by comparing mean FGA, FTA, and FIA values. Overall, more

---

[1] Data from BIO, CCJ, GERO, LIS, and PSYC were previously published or presented.

[2] Hirsch's [16] h-index is a measures of scientific impact, based on the number of times journal articles are cited. Data were downloaded in April 2021 and are available from L. Best.

[3] k = number of journals per discipline; n = number of articles per journal.

342   L. Best and C. Goggin

page space was devoted to tables (8.99%) than to graphs (5.20%) or non-graph illustrations (2.13%). Collectively, 14.23% of page space was dedicated to the presentation of data (i.e., graphs + tables). Mean FGA ranged from 1.64% for CCJ journal articles to 10.36% for PSYC journal articles. With respect to table use, mean FTA ranged from 6.08% for BIO journal articles to 12.69% for LIS journal articles. Mean FIA ranged from 0.21% for articles in CCJ journals to 6.52% for those in BIO.

To test specific differences in mean FGA, FTA, and FIA across disciplines, a mixed model ANOVA was conducted. The interaction between inscription type and discipline was statistically significant, $F(12, 4894) = 13.02$, $p < .001$, and post hoc analyses indicated that articles published in BIO and PSYC dedicated significantly more page space to graphical displays than did those in CRIM, GERO, and SOC ($ps < .001$); articles in PSYC had higher graph use than those in LIS and MED ($ps < .001$). Articles published in LIS journals used more tables than those in journals from all other disciplines ($ps < .001$). GERO, MED, and SOC articles consistently included more tables than articles published in BIO and PSYC. Articles in BIO journals dedicated more page space to non-graph illustrations (M = 6.52% vs. < 3% for each of the other disciplines; $p < .001$).

3.2 Data Presentation and Visuality Indices

To examine differences between data presentation techniques and overall use of visual inscriptions, a Visuality Index (VI = FGA + FIA) and a Total Data Presentation Index (DPI = FGA + FTA) were calculated [see 3].

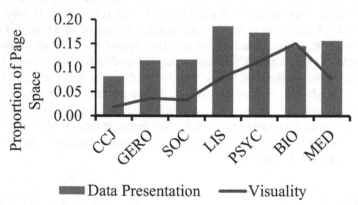

Fig. 1. Disciplinary differences in Data Presentation (FGA + FTA) and Visuality (FGA + FIA).

As indicated in Fig. 1, there was more variability in VI than in DPI. The mean score (SD) on the DPI was 14.23% (0.22) with scores ranging from 8.18% for CCJ articles to 17.35% for LIS articles. A one-way ANOVA indicated disciplinary differences in DPI, $F(6, 2452) = 11.78$, $p < .001$, with CCJ articles having the lowest DPI. Articles published in BIO, LIS, and PSYC had consistently higher DPI than those published in CCJ, GERO, and SOC. The mean (SD) score on VI was 7.34% (0.21) with scores ranging from 1.86% for CCJ articles to 15.06% for articles in BIO journals, with statistically significant disciplinary differences, $F(6, 2457) = 17.38$, $p < .001$. Articles published in BIO had a higher VI than other disciplines ($ps < .001$), with LIS, MED, and PSYC having a greater VI than CCJ, GERO, and SOC ($ps < .001$). To examine differences in VI and DPI for each discipline, a series of paired samples t-tests were conducted. With the exception of BIO, there were statistically significant differences between VI and DPI, indicating that researchers dedicate more space to data presentation than to visualization.

A Heterogeneity Index (HI), defined as the number of inscription types used in different articles, was calculated [see 3]. The HI ranged from 0 to 3, with HI $= 0$ indicating that an article included no graphs, tables, or illustrations and HI $= 3$ indicating the inclusion of at least one of each type of inscription. Overall, mean HI ranged from 0.92 (SD $= 0.70$) in CCJ articles to 2.14 (SD $= 0.72$) in BIO articles (see Fig. 2). As expected, there were statistically significant disciplinary differences, $F(6, 2452) = 91.17, p < .001$. Overall, BIO had the highest HI and CCJ had the lowest ($ps < .001$), with MED, PSYC, and SOC using, on average, approximately 1.6 different types of inscriptions per article. Finally, an examination of the relationship between average HI and h-index revealed a statistically significant correlation, $r(90) = 0.34$, $p < .001$. Interestingly, HI was more strongly correlated with h-index than mean DPI, $r(90) = 0.27, p = .032$, or VI, $r(90) = 0.26, p = .04$, suggesting that journal impact factor is associated with using a variety of scientific inscriptions.

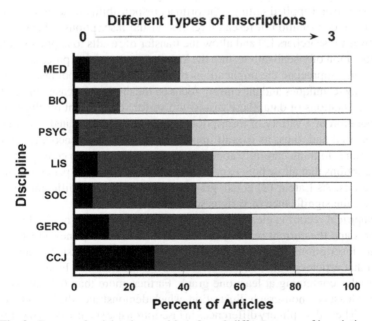

Fig. 2. Percent of articles that used 0, 1, 2, or 3 different types of inscriptions.

4 Discussion

The current results summarise how scientists in different disciplines use scientific inscriptions and are generally consistent with previous research. This sample of articles included a variety of different types of inscriptions and, across all disciplines, almost 15% of article page space was dedicated to data presentation. As expected, there were statistically significant differences in graph use, with articles published in BIO, MED, and PSYC dedicating proportionally more page space to graphical displays [4; 6; 17]. It is important to note that, although, on average, articles published in the harder areas of science (BIO, MED) had higher FGAs than those published in the softer areas (CCJ, SOC), there was considerable disciplinary overlap.

Overall, tables were the most common scientific inscription, with approximately 14% of page space dedicated to tabular data presentations. With the exception of BIO, most researchers in most disciplines included fewer non-graph illustrations in their articles. Although there were statistically significant differences, the page space dedicated to data presentation ranged from approximately 8% in CCJ journals to 19% in LIS journals. This variability is likely attributable to how a graphical representation was defined (i.e., a display that contains quantitative information on a scale). For example, many articles in LIS included architectural plans, which served to increase the FGA in this discipline's journals. In light of disciplinary differences, we suggest that our data supports the assertion that all scientific disciplines have a large amount of data but that researchers convey their results using different types of inscriptions [4].

In both basic and applied settings, the primary responsibility of scientists is to ensure the meaningful contribution of research results. Visual inscriptions enhance communication between researchers [2] and allow the transfer of results to applied settings [18]. The incorporation of visualizations into the knowledge generation process allows stakeholders to fully appreciate the implications of specific research findings. Given the rise in data display techniques and the current focus on the creation of easily understandable visual representations of data, these results are useful. The current results suggest that the incorporation of a variety of inscriptions (see Fig. 2) is associated with the h-Index [16] of a journal and, thus, its overall impact factor. We would encourage researchers in all disciplines to think carefully about the message they want to convey and to create visual inscriptions that allow both experienced researchers and laypersons to appreciate research results. As Latour [2] noted, graphs allow people from different backgrounds to appreciate the significance of various scientific discoveries.

The empirical record is clear that graphical displays enhance our ability to discern data patterns and the current results illustrate that, across disciplines, most researchers incorporate inscriptions. They also support the link between visuality and scientific impact, in that articles published in BIO and PSYC journals had higher VI, with over 70% of articles containing at least one graph. Further, more than 60% of BIO articles contained at least one non-graph illustration, which demonstrates the importance placed on visuality. These disciplinary differences in inscription use speak to specific differences in theory development and codification [11]. Appropriately constructed illustrations augment the persuasiveness of research findings and, over the long term, help increase a discipline's codification. Through such efforts, the goals of knowledge cumulation and transfer – translating empirical results into practical applications – will be met and yield benefits for both applied and research contexts.

4.1 Implications of Current Study

The current study adds to the body of literature concerning good data representation, analysis, and communication. To highlight the importance of graphs as a powerful supplement to inferential statistics, Wilkinson and the Task Force on Statistical Inference [19] advocated graphing one's data prior to statistical analyses. As these authors claimed, "graphics broadcast; statistics narrowcast" and plotting one's data prior to analysis may also help uncover coding errors, or threats to the integrity of the data [20, 21]. In science, measurement is central to discovery, and integrating graphical analyses and displays allows researchers to better understand the phenomena that they are studying.

4.2 Conclusions

The current results highlight several disciplinary differences in inscription use, with articles published in BIO and PSYC dedicating more page space to graphical presentation of data. Although some researchers in the social sciences include a variety of visual inscriptions in their publications, many researchers focus almost solely on tabular data presentation. Given the explosion of social media and news sites, we suggest that researchers should focus on overall communication strategies that extend beyond publication in academic journals and could include infographics designed for social media sites and well-designed videos that clearly communicate important information. The spread of misinformation is best countered by the spread of accurate information. Visual displays of data and, in fact, scientific methodologies, can slow the spread of false information. As Helen Purchase [22] noted, there are many different types of diagrams and, when designed properly, all serve to aid in the communication of science.

References

1. Funkhouser, H.G.: Historical development of the graphical representation of statistical data. Osiris 3, 269–404 (1937)
2. Latour, B.: Drawing things together. In: Lynch, M., Woolgar, S. (eds.) Representation in Scientific Practice, pp. 19–68. MIT Press, Cambridge (1990)
3. Arsenault, D.J., Smith, L.D., Beauchamp, E.A.: Visual inscriptions in the scientific hierarchy: mapping the treasures of science. Sci Commun. **27**, 76–428 (2006). https://doi.org/10.1177/1075547005285030
4. Cleveland, W.S.: Graphs in scientific publications. Am Stat. **38**, 261–269 (1984). https://doi.org/10.1080/00031305.1984.10483223
5. Goggin, C., Best, L.A.: The use of scientific inscriptions in criminology and criminal justice journals: An analysis of publication trends between 1985 and 2009. J. Crim. Just. **24**, 517–535 (2013). https://doi.org/10.1080/10511253.2013.841971
6. Smith, L.D., Best, L.A., Stubbs, D.A., Archibald, A.B., Roberson-Nay, R.: Constructing knowledge: the role of graphs and tables in hard and soft psychology. Am. Psychol. **57**, 749–761 (2002). https://doi.org/10.1037/0003-066X.57.10.749

 7. Zuckerman, H.A., Merton, R.K. Institutionalization patterns of evaluation in science. In: Storer, N. (ed.) The sociology of Science, pp. 460–496. University of Chicago Press, Chicago (1973)
 8. Myers, G.: Every picture tells a story: illustrations in E O. Wilson's sociobiology. Hum. Stud. **11**, 235–269 (1988). https://doi.org/10.1007/BF00177305
 9. Kevles, D.J.: Historical foreword. In: Robin, H. (eds.) The Scientific Image: From the Cave to the Computer. WH Freeman & Co, New York, NY (1992)
10. Robin, H.: The Scientific Image: From Cave to Computer. WH Freeman & Co, New York, NY (1993)
11. Smith, L.D., Best, L.A., Stubbs, D.A., Johnston, J., Archibald, A.B.: Scientific graphs and the hierarchy of the sciences: a Latourian survey of inscription practices. Soc. Stud. Sci. **30**, 73–94 (2000). https://doi.org/10.1177/030631200030001003
12. Best, L.A., Smith, L.D., Stubbs, D.A.: Graph use in psychology and other sciences. Behav. Process. **54**, 155–165 (2001). https://doi.org/10.1016/S0376-6357(01)00156-5
13. Best, L.A., Goggin, C., Buhay, D.N., Caissie, L.T., Boone, M., Gaudet, D.: Describing data using pictures: the use of visual inscriptions in science. In: Pracana, C., Wang, M. (eds.) International Psychological Applications Conference and Trends: Proceedings, Lisbon, Portugal, pp. 296–298. World Institute for Advanced Research and Science (2016)
14. Caissie, L.T., Goggin, C., Best, L.A.: Graphs, Tables, and Scientific Illustrations: visualisation as the Science of Seeing Gerontology. Can. J. Aging **36**, 536–548 (2017). https://doi.org/10.1017/S0714980817000447
15. Buhay, D., Best, L.A., Goggin, C., McPhee, R.: Use of Scientific Inscriptions in Library Sciences: An Empirical Study of Publication Practices. Poster presented at the 83rd IFLA World Library and Information Congress, Wroclaw, Poland (2017)
16. Hirsch, J.E.: An index to quantify an individual's scientific research output. Proc. Nat. Sci. **102**(46), 16569–16572 (2005)
17. Kubina, R.M., Kostewicz, D.E., Datchuk, S.M.: An initial survey of fractional graph and table area in behavioral journals. Behav. Anal. **31**, 61–66 (2008). https://doi.org/10.1007/BF03392161
18. Ahmed, M., Boisvert, C.M.: Enhancing communication through visual aids in clinical practice. Am. Psychol. **58**, 816–819 (2003). https://doi.org/10.1037/0003-066X.58.10.816
19. Wilkinson, L.: Task Force on Statistical Inference, American Psychological Association, Science Directorate. Statistical methods in psychology journals: Guidelines and explanations. Am. Psychol. **54**, 594–604 (1999). https://doi.org/10.1037/0003-066X.54.8.594
20. Brand, A., Bradley, M.T., Best, L.A., Stoica, G.: Accuracy of effect size estimates from published psychological research. Percept. Mot. Ski. **106**, 645–649 (2008). https://doi.org/10.2466/pms.106.2.645-649
21. Brand, A., Bradley, M.T.: The precision of effect size estimation from published psychological research: surveying confidence intervals. Psychol. Rep. **118**, 154–170 (2016). https://doi.org/10.1177/0033294115625265
22. Purchase, H.C.: Twelve years of diagrams research. Vis. Lang. Comput. **25**, 57–75 (2014). https://doi.org/10.1016/j.jvlc.2013.11.004

Open Access This chapter is licensed under the terms of the Creative Commons Attribution 4.0 International License (http://creativecommons.org/licenses/by/4.0/), which permits use, sharing, adaptation, distribution and reproduction in any medium or format, as long as you give appropriate credit to the original author(s) and the source, provide a link to the Creative Commons license and indicate if changes were made.

The images or other third party material in this chapter are included in the chapter's Creative Commons license, unless indicated otherwise in a credit line to the material. If material is not included in the chapter's Creative Commons license and your intended use is not permitted by statutory regulation or exceeds the permitted use, you will need to obtain permission directly from the copyright holder.

Can Humans and Machines Classify Photographs as Depicting Negation?

Yuri Sato[1]([✉])[iD] and Koji Mineshima[2]

[1] The University of Tokyo, Tokyo, Japan
satoyuri0@g.ecc.u-tokyo.ac.jp
[2] Keio University, Tokyo, Japan
minesima@abelard.flet.keio.ac.jp

Abstract. How logical concepts such as negation can be visually represented is of central importance in the study of diagrammatic reasoning. To explore various ways in which negation can be visually represented, this study focuses on photographs as instances of purely visual representations. We use real-world photographic image data and study how well humans can classify those images as depicting negation. We also compare the human performance with a state-of-the-art machine (deep) learning model on this classification task. The present paper gives some preliminary results on our data-driven analyses.

Keywords: Negation · Photograph · Machine learning · Cognitive science

1 Introduction

How logical concepts can be visually expressed has been of central importance in the study of logical and cognitive study of visual representations. Barwise and Etchemendy [1] famously claimed that pictures are not suitable for depicting indeterminate information such as negation and disjunction. This view can be seen as one of the motivations behind the study of diagrammatic logic, whose aim is to build heterogeneous systems combining visual representations with conventional symbolic devices (e.g., [4]).

However, it remains unclear whether visual representations in the real world could express negation without any aid of symbolic devices (see [3] for some discussion). In previous studies, we investigated this issue by focusing on comic illustration and the role of iconic conventional devices to express negation such as "effect lines" that trace disappearing figures [6,7]. As a sequel to this study, the present paper focuses on photographs as instances of pure visual representations and explores whether and how they can visually represent negation. Photographs have been widely studied in recent AI studies [2], yet their cognitive functions to depict negated information are understudied. The present study aims to fill this gap by taking a data-driven approach: we use real-world photographic image data and study how well humans can classify those images as depicting negation.

© Springer Nature Switzerland AG 2021
A. Basu et al. (Eds.): Diagrams 2021, LNAI 12909, pp. 348–352, 2021.
https://doi.org/10.1007/978-3-030-86062-2_35

We also compare human performance with a state-of-the-art machine (deep) learning model on this classification task. We give some preliminary results on our data-driven analyses.

2 Experiments

2.1 Negation and Negation-Free Photograph Data Collection

We use photographic image data from MS-COCO [5] and their Japanese captions from STAIR Captions [10]. We randomly selected a set of 345 images (called "Image Pool 1") whose captions contain a negation word "*nai (not)*". Each image in MS-COCO is annotated with a set of object categories (e.g., *person, umbrella*, etc.). To extract negation-free images, we selected images that do not contain the negation word and share at least one category with those in Image Pool 1. We excluded those images that differ in more than two categories. We collected the same number of negation-free images (called "Image Pool 2").

For the total 690 images, three people manually annotated the gold standard label, negation or negation-free. They were also asked to write down the reason why they think it is natural or appropriate to use (or not to use) negation to describe the image (e.g., *because there is no ...*). For the negation images in Image Pool 1, there were 65 images to which two or more of the three annotators gave the "negation" label and their reasons mention the same object. We chose the same number of the corresponding negation-free images from Image Pool 2 that were judged as "negation-free" by two or more of the three annotators. We used these 65 negation images and 65 negation-free images in our analysis.

2.2 Machine Learning on the Negation Image Classification Task

We ran experiments on the negation image classification task using a machine (deep) learning model. We divided the 65 negation images into 35 training images, 10 validation images, and 20 test images. We used a convolutional neural network (CNN) model with a fine-tuning technique (VGG16 [8]). Following the previous studies suffering from the small sample learning dilemma in fields that deal with hard-to-find data [9], we used techniques of data-argumentation. More specifically, we used techniques in OpenCV (https://opencv.org/) such as reverse turning and contrast adjustments to augment the training and validation images, which generated 630 and 180 images, respectively. An equal number of negation-free images were also provided in the same way. For implementation, we use a standard library (Keras) in Python. Key parameters were the following: sequential model; activation function for intermediate layer = relu; dropout rate = 0.5; activation function for output layer = softmax; VGG16 model weights for up to 14 layers; loss function = crossentropy; batch size = 18; epochs = 3.

The results showed that the CNN model correctly classified 11 out of 20 negation images (55%) as negation. For the negation-free images, we obtained the same accuracy (55%). The performance is almost identical to the chance level of 50%, indicating that the model is generally not able to classify negation and negation-free images.

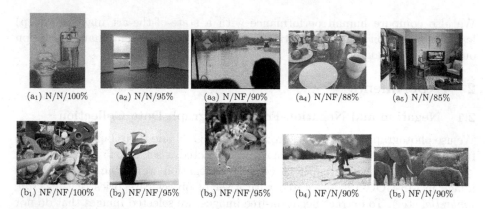

(a₁) N/N/100%   (a₂) N/N/95%   (a₃) N/NF/90%   (a₄) N/NF/88%   (a₅) N/N/85%

(b₁) NF/NF/100%   (b₂) NF/NF/95%   (b₃) NF/NF/95%   (b₄) NF/N/90%   (b₅) NF/N/90%

Fig. 1. Examples of image classification results. "N/NF/90%" means that the gold label is "negation (N)", the label predicted by CNN is "negation-free (NF)", and the average human accuracy is 90%. For each type, the top five examples with the highest accuracy were selected.

2.3   Human Performance on the Negation Image Classification Task

Two hundred and three people over the age of 20 participated in the experiment. The mean age of participants was 39.8 (SD = 9.39). The experiment was conducted in Japanese, online, and participants were given informed consent and paid for their participation with the approval of the local ethics committee. Participants were given photographic images and were asked to classify whether or not they involve negation (i.e., it is natural or appropriate to use negation to describe the given image). The instruction specifies that negation is typically expressed by such phrases as *"there is no __"*, *"__ does not exist"*, *"__ is not __"*, *"__ disappeared"*, *"__ is empty"*, *"__ cannot do __"*, *"__ does not move"*.

In the first "training" phase, the participants were given 18 classification task items (half for negation, the other half for negation-free) and were shown the correct/incorrect answers for their classification choices. This training was conducted twice. In the third phase (final test phase), the participants were given four new classification task items (half for negation, the other half for negation-free). 18 images for training were randomly chosen from 45 images in the training and validation set for the deep learning experiment, while 4 images for the test were randomly chosen from the 20 images used for the test set.

The average accuracy rate of humans was 67.6% for negation images and 73.6% for negation-free images, showing that the human performance outperformed the performance of the CNN model in classifying negation images.

3   Discussion

Figure 1 shows some examples of the image classification results. Type (a) shows images whose gold label is negation. The reasons provided by the human annotators were "There is no furniture in the room" for (a₂) and "the road was flooded with water, making it impossible to move." for (a₃). Note here that the target of each description is not directly depicted in the image (*furniture* and *movement of a car*). To recover what is negated in the content of an image, one needs to use background knowledge, such as "Usually, there is furniture in a room" and "A car is usually stuck in the water." The same is not true in the case of the negation-free images in (b). In (b₁), for example, the annotated description is "there are a ukulele and food". No special external knowledge is required to extract this information from the image.

In the case of machine learning, judging from the overall average accuracy of 55%, it would be fair to say there is room for improvement in classifying negation automatically, though it is premature to draw any conclusive claim from this experiment given the size of the data. Given that a classification criterion that is not related to negation works here, it might be reasonable to conjecture that the difficulties may come from the failure to infer background knowledge from image features only [2].

Although we used the classification task for negation images, it is interesting to consider a generation task where a human or a machine learning model is asked to answer a question like "What is missing (or does not happen) in the image?". This could be the next step towards explaining human cognitive capacity to understand negated information from photographs, as well as to improving the ability of current machine learning models. In [6], we compared three forms of visual representations: photographs, comics, and videos. How such representations as videos that involves sequential pictures, changes, and communicative intention can depict negated information is also left for future study.

Acknowledgments. This work was supported by JSPS KAKENHI Grant Number JP20K12782 to the first author.

References

1. Barwise, J., Etchemendy, J.: Hyperproof: logical reasoning with diagrams. In: Reasoning with Diagrammatic Representations, pp. 77–81. AAAI Press (1992)
2. Bernardi, R., et al.: Automatic description generation from images: a survey of models, datasets, and evaluation measures. J. Artif. Intell. Res. **55**, 409–442 (2016). https://doi.org/10.1613/jair.4900
3. Grzankowski, A.: Pictures have propositional content. Rev. Phil. Psych. **6**, 151–163 (2015). https://doi.org/10.1007/s13164-014-0217-0
4. Howse, J., Stapleton, G., Taylor, J.: Spider diagrams. LMS J. Comput. Math. **8**, 145–194 (2005). https://doi.org/10.1112/S1461157000000942
5. Lin, T.Y., Maire, M., Belongie, S., Hays, J., Perona, P.: Microsoft COCO: common objects in context. In: Fleet, D., Pajdla, T., Schiele, B., Tuytelaars, T. (eds.) ECCV 2014. LNCS, vol. 8693, pp. 740–755. Springer, Cham (2014). https://doi.org/10.1007/978-3-319-10602-1_48
6. Sato, Y., Mineshima, K.: Depicting negative information in photographs, videos, and comics: a preliminary analysis. In: Pietarinen, A.-V., Chapman, P., Bosveld-de Smet, L., Giardino, V., Corter, J., Linker, S. (eds.) Diagrams 2020. LNCS (LNAI), vol. 12169, pp. 485–489. Springer, Cham (2020). https://doi.org/10.1007/978-3-030-54249-8_40

7. Sato, Y., Mineshima, K., Ueda, K.: Visual representation of negation: real world data analysis on comic image design. In: CogSci 2021, pp. 1166–1172 (2021)
8. Simoyan, K., Zisserman, A.: Very deep convolutional networks for large-scale image recognition. In: ICLR 2015 (2015)
9. Vabalas, A., Gowen, E., Poliakoff, E., Casson, A.J.: Machine learning algorithm validation with a limited sample size. PLoS ONE, **14**(11), e0224365 (2019). https://doi.org/10.1371/journal.pone.0224365
10. Yoshikawa, Y., Shigeto, Y., Takeuchi, A.: Stair captions: constructing a large-scale Japanese image caption dataset. In: ACL 2017, pp. 417–421 (2017). https://doi.org/10.18653/v1/P17-2066

The Presence of Diagrams and Problems Requiring Diagram Construction: Comparing Mathematical Word Problems in Japanese and Canadian Textbooks

Mari Fukuda[1]([✉]), Emmanuel Manalo[2], and Hiroaki Ayabe[3]

[1] Graduate School of Education, The University of Tokyo, Tokyo, Japan
mari_fukuda@p.u-tokyo.ac.jp
[2] Graduate School of Education, Kyoto University, Kyoto, Japan
manalo.emmanuel.3z@kyoto-u.ac.jp
[3] National Institute for Physiological Sciences, Okazaki, Aichi, Japan
ayabe@nips.ac.jp

Abstract. It is generally considered beneficial for learners to construct and use appropriate diagrams when solving mathematical word problems. However, previous research has indicated that learners tend not to use diagrams spontaneously. In the present study, we analyzed textbooks in Japan and Canada, focusing on the possibility that such inadequacy in diagram use may be affected by the presence (or absence) of diagrams in textbooks, the kinds of diagrams that are included, and whether problems requiring the construction of diagrams are provided in those textbooks. One set each of Japanese and Canadian elementary school textbooks were analyzed, focusing on the chapters dealing with division. Results revealed that the Japanese textbooks contain worked examples and exercise problems accompanied by diagrams more than the Canadian textbooks. Furthermore, the Japanese textbooks often use line diagrams and tables that abstractly represent quantitative relationships and they include more problems that require students to use diagrams. However, to encourage students to use diagrams spontaneously, it may be necessary to include problems that scaffold the use of diagrams in a step-by-step manner in both the Canadian and the Japanese textbooks.

Keywords: Mathematical word problem · Textbooks analysis · Utilizing diagrams

1 Introduction

Diagrams are one of the effective tools students can use in solving mathematical word problems [1]. For example, in the problem-solving process by Mayer [2], pictures representing the situation or scene of the problem may be effective in understanding the context of the problem, and diagrams such as arrays, blocks, and line diagrams representing quantities and their relationships may be effective in forming a representation of the whole problem and in facilitating the planning stage of problem-solving.

© The Author(s) 2021
A. Basu et al. (Eds.): Diagrams 2021, LNAI 12909, pp. 353–357, 2021.
https://doi.org/10.1007/978-3-030-86062-2_36

However, previous studies have revealed that students do not spontaneously use diagrams [3]. Since the use of learning strategies, such as utilizing diagrams in mathematics word problem solving, requires knowledge of strategies, it is necessary to provide scaffolding for developing the ability to use diagrams in the math classroom. One of the critical factors that influence teachers' teaching and students' learning are textbooks. Although textbooks and their systems vary from country to country, they are considered to be the physical tools most closely linked to teaching and learning [4], that provide opportunities for teacher professional development, and that are essential for both teachers and students to use [5]. If diagrams were not presented with word problems in textbooks, instruction that promotes the use of diagrams could be inhibited.

The present study comprised an initial exploratory analysis of textbook sets from two countries to determine the extent to which diagrams accompany mathematics word problems in those textbooks. The following questions were addressed:

RQ1. How many problems are accompanied by diagrams in the textbooks?
RQ2. What kinds of diagrams are presented with the problems?
RQ3. How many problems explicitly require the use of diagrams (e.g., construction or fill-in-the-blank) in the textbooks?

2 Method

We analyzed textbooks for elementary school students from one Japanese textbook company and one Canadian textbook company. Japanese textbooks are certified by the government and are required to be used in the classroom. The textbooks of Dainippon Tosho Publishing [6] were selected out of the six available companies. In Canada, the education system differs between provinces, and the contents of textbooks differ accordingly. Also, there is no requirement to use textbooks, and their selection and use depend on schools and teachers. In this study, "Math Makes Sense" by Pearson [7] was selected and analyzed. The units analyzed were about "Division," but the learning contents of the two countries' grade levels do not correspond perfectly. Based on these differences, fractional division was excluded from the analysis (due to mismatches), and only the units on whole numbers and decimal division were analyzed. Textbooks for grades 3–5 in Japan and grades 3–6 in Canada were used for the analysis.

The word problems were coded according to the problem type (i.e., worked example, exercise, review), the presence of diagrams, the number and kinds of diagrams, and the presence of problems requiring construction or filling in of parts of diagrams. Worked examples contain a problem sentence, formula, and answer, and exercises and reviews are problems for practicing and mastering skills. Exercises and reviews differ in that the exercises are included in each chapter for contents to be learned, and the reviews are aimed at applying the contents that have been learned (including multiple chapters).

Diagrams were categorized into one of five kinds: (1) Pictures (images representing/relating to the problem situation), (2) Concrete diagrams (illustrations/pictures, like counters and blocks, depicting quantitative relationships), (3) Schematic diagrams (arrows, lines, figures showing procedures and quantitative/functional relationships), (4) Line diagrams (line or tape diagrams, segments of which indicate quantities or show relationships between quantities), (5) Tables (arrays of numbers and words/letters).

3 Results and Discussion

3.1 Math Word Problems Accompanied by Diagrams in Textbooks

Table 1 shows the number of mathematical word problems accompanied by diagrams in Canadian and Japanese textbooks. Chi-square tests revealed that Japanese textbooks contained worked examples and exercises accompanied by diagrams more than Canadian textbooks (worked examples: $\chi^2(1) = 6.41$, $p = .011$, *Cramer's V* $= 0.30$; exercises: $\chi^2(1) = 4.46$, $p = .035$, *Cramer's V* $= 0.14$).

Next, we examined the number of diagrams included per question to determine whether the books differed. Table 2 shows the number of diagrams in word problems accompanied by diagrams. We found that the number of diagrams contained in each problem in the Japanese textbooks was higher, with an average of just over two diagrams per problem ($t(175) = 3.20$, $p = .002$, $d = 0.50$). Considering that Japanese textbooks are required to be used in classroom teaching, it is possible that diagrams are utilized more frequently in Japanese classes when teaching word problems.

Table 1. The number of mathematical word problems with/without diagrams

		With diagrams	Without diagrams	Total
Worked example	Canada	20 (87%)	3 (13%)	23
	Japan	47 (100%)	0 (0%)	47
Exercise	Canada	37 (41%)	54 (59%)	91
	Japan	69 (55%)	56 (45%)	125
Review	Canada	5 (28%)	13 (72%)	18
	Japan	8 (15%)	45 (85%)	53
Total	Canada	62 (47%)	70 (53%)	132
	Japan	124 (55%)	101 (45%)	225

Table 2. The number of diagrams in word problems containing diagrams

	Problems with diagrams	Total diagrams	Per problem	
			Mean	SD
Canada	62	88	1.42	1.28
Japan	124	273	2.20	2.07

3.2 Types of Diagrams Presented with Math Word Problems

Table 3 shows the number of each diagram type included in the Canadian and Japanese textbooks. The first and second coders independently coded 20% of the total diagrams.

Interrater reliability was satisfactory ($\kappa = .88$), discrepancies were resolved after discussion, and all remaining coding was done by the first coder. Chi-square test showed that the Canadian textbooks contain more picture and concrete diagrams, while the Japanese textbooks contain more line diagrams and tables ($\chi^2(4) = 39.81, p < .001$, *Cramer's V* $= 0.33$). The results suggest that the Canadian textbooks use more concrete diagrams to clarify the problem context, while the Japanese textbooks use more abstract diagrams to clarify the quantity relationships.

Table 3. The number of each diagram type included in word problems

	Picture	Concrete	Schematic	Line	Table	Total
Canada	51 (58%)	34 (39%)	0 (0%)	3 (3%)	0 (0%)	88
Japan	106 (39%)	62 (23%)	17 (6%)	80 (29%)	9 (3%)	274
Total	157 (43%)	96 (27%)	17 (5%)	83 (23%)	9 (2%)	362

3.3 Problems Requiring the Use of Diagrams

The number of problems that require students to construct a diagram or fill in parts of provided diagrams were counted. Using the same procedure as before, two coders coded 20% of the problems independently, and high interrater reliability was confirmed ($\kappa = .94$); discrepancies were discussed and resolved. The Japanese textbook contained 19 of such problems (8.4% of total problems), while the Canadian textbook contained five of such problems in total (3.8% of total problems). The questions in the Canadian textbook required students to construct a picture from scratch (e.g., "Draw a picture. Use grid paper"), whereas many of the problems in the Japanese textbook required students to write only numbers that can be read from the problem text in the blanks of a given figure (e.g., "Let's complete the number line").

Such scaffolding of diagram use is possibly effective considering the common difficulties students manifest in constructing diagrams, but the construction of diagrams from scratch may be difficult for many students. Therefore, it may be necessary to divide the construction of diagrams into steps, starting with filling in the blanks and gradually fading the scaffolding by having students construct more of the diagrams by themselves. Further investigation is needed to see if such scaffolding and fading support of diagram construction is effective. As this study is preliminary and focused only on division, it is necessary to expand to other topics and problem types to examine whether there are differences there, and to investigate the relationship between teachers' use of textbook diagrams in class and the students' use of diagrams and textbooks at home.

Acknowledgment. This research was supported by a grant-in-aid (20K20516) received from the Japan Society for the Promotion of Science.

References

1. Hembree, R.: Experiments and relational studies in problem-solving: a meta-analysis. J. Res. Math. Educ. **23**, 242–273 (1992)
2. Mayer, R.E.: Thinking, Problem Solving, Cognition, 2nd edn. W H Freeman, New York (1992)
3. Uesaka, Y., Manalo, E., Ichikawa, S.: What kinds of perceptions and daily learning behaviors promote students' use of diagrams in mathematics problem solving? Learn. Instruct. **17**, 322–335 (2007)
4. Valverde, G.A., Bianchi, L.J., Wolfe, R.G., Schmidt, W.H., Houang, R.T.: According to the book: Using TIMSS to investigate the translation of policy into practice through the world of textbooks. Springer (2002)
5. Ball, D.L., Cohen, D.K.: Reform by the book: What is – or might be – the role of curriculum materials in teacher learning and instructional reform? Educ. Res. **25**, 6–14 (1996)
6. Souma, K., et al.: Tanoshii Sansuu Grades 3-5, 1st edn. Dainippon Tosho Publishing (2020)
7. Appel, R., et al.: Math Makes Sense, Grades 3-6. Pearson, London (2009)

Open Access This chapter is licensed under the terms of the Creative Commons Attribution 4.0 International License (http://creativecommons.org/licenses/by/4.0/), which permits use, sharing, adaptation, distribution and reproduction in any medium or format, as long as you give appropriate credit to the original author(s) and the source, provide a link to the Creative Commons license and indicate if changes were made.

The images or other third party material in this chapter are included in the chapter's Creative Commons license, unless indicated otherwise in a credit line to the material. If material is not included in the chapter's Creative Commons license and your intended use is not permitted by statutory regulation or exceeds the permitted use, you will need to obtain permission directly from the copyright holder.

Diagrams and Computation

Extracting Interactive Actor-Based Dataflow Models from Legacy C Code

Niklas Rentz[1]([✉])[iD], Steven Smyth[1][iD], Lewe Andersen[2],
and Reinhard von Hanxleden[1][iD]

[1] Department of Computer Science, Kiel University, Kiel, Germany
{nre,ssm,rvh}@uni-kiel.de
[2] Scheidt & Bachmann System Technik GmbH, Melsdorf, Germany
Andersen.Lewe@scheidt-bachmann-st.de

Abstract. Graphical actor-based models provide an abstract overview of the flow of data in a system. They are well-established for the model-driven engineering (MDE) of complex software systems and are supported by numerous commercial and academic tools, such as Simulink, LabVIEW or Ptolemy. In MDE, engineers concentrate on constructing and simulating such models, before application code (or at least a large fraction thereof) is synthesized automatically. However, a significant fraction of today's legacy system has been coded directly, often using the C language. High-level models that give a quick, accurate overview of how components interact are often out of date or do not exist. This makes it challenging to maintain or extend legacy software, in particular for new team members.

To address this problem, we here propose to reverse the classic synthesis path of MDE and to synthesize actor-based dataflow models automatically from source code. Here functions in the code get synthesized into nodes that represent actors manipulating data. Second, we propose to harness the *modeling-pragmatic* approach, which considers visual models not as static artefacts, but allows interactive, flexible views that also link back to textual descriptions. Thus we propose to synthesize actor models that can vary in level of detail and that allow navigation in the source code. To validate and evaluate our proposals, we implemented these concepts for C analysis in the open source, Eclipse-based KIELER project and conducted a small survey.

Keywords: Actor-based dataflow · Program comprehension · Interactive Documentation

1 Introduction

Precise and up to date documentation is a key aspect of quality maintenance of software systems [4,19]. Good documentation does not only help the developer

This work has been supported by the project Visible Code, a cooperation between Kiel University and Scheidt & Bachmann System Technik GmbH.

© The Author(s) 2021
A. Basu et al. (Eds.): Diagrams 2021, LNAI 12909, pp. 361–377, 2021.
https://doi.org/10.1007/978-3-030-86062-2_37

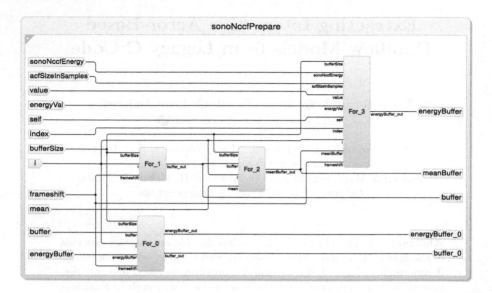

Fig. 1. An example actor model, illustrating the data flow in a signal processing component named sonoNccfPrepare.

to keep a better overview about the project, but also enables users to gain a perception about the usage, functionality, and connections inside the project.

Regardless of the advantages of a well-documented project, the documentation for many projects is outdated, as stated by different surveys, e.g., by Lethbridge et al. [14] and Singer [24]. Singer states that even if most developers appreciate good documentation, the time needed for its creation or maintenance leads to inconsistency, which further leads to a lack of trust from the developers.

We here propose to enhance the documentation of existing codebases by the usage of diagrammatic *actor-based dataflow models*, or actor models for short. Actor models are already commonly used in model-driven engineering and are supported by numerous commercial and academic tools, such as Simulink, LabVIEW or Ptolemy. Lee et al. [13] describe *actors* as components that can execute and communicate with other actors in a model. Their *ports* represent an interface for communication with other actors and their environment. For example, Fig. 1 shows an actor named sonoNccfPrepare, with input ports sonoNccfEnergy, acfSizeInSamples, etc., and output ports energyBuffer, meanBuffer, etc. That actor includes other actors; For_1, for example, receives the inputs bufferSize, i, and frameshift from the environment, and provides output buffer to actor For_2.

Unlike the MDE setting, where developers create visual (actor) models and textual code is synthesized from the models, we here propose to reverse the synthesis path and to automatically create models from existing code. We thus propose to extract documentation directly and automatically from the source, which has been shown to be helpful and wanted by developers [6]. Rugaber [21] states that program comprehension is the biggest bottleneck in development time-wise and is mainly done manually, so automation and simplifications in the comprehension can cut down on that bottleneck. A key benefit of visual code comprehension tools is that users are not burdened anymore with the manual creation and maintenance of such visualizations, and that such visualizations are more up to date. This is not meant to completely replace manual documentation, as documentation extracted from code can only be a description and not define a specification or the thoughts that went into design decisions, as discussed by Parnas [19]. The generated visualizations proposed here are meant to complement and structure other documentation.

Contributions and Outline

- We propose an approach to automatically generate actor models from C programs, where the actors match the program structure and their interconnections reflect the data flow (Sect. 2).
- We have prototyped and integrated this model extraction and visualization in an open source, Eclipse-based modeling environment, which allows flexible diagram views that link back to the source code (Sect. 3).
- We have conducted a small user experiment that, for a given set of tasks, compares the effectiveness of source code analysis vs. the inspection of visual models (Sect. 4).

Section 5 discusses related work, we conclude in Sect. 6.

2   Actor-Based Dataflow Visualization

Most developers define the behavior of a program with imperative programming languages [15]. In programs split up into many different functions it may be unclear where data in functions come from and where they are used. Most commonly used IDEs provide some support for tracing the data, through highlighting or function usage trees, but in general they do not give an overview of the intraprocedural dataflow. We propose an actor-based dataflow view, akin to Ptolemy [5] and SCCharts dataflow [29]. This section describes how to visualize such a dataflow model for imperative programming languages. We illustrate this with the example of C code.

2.1   Actor-Based Dataflow

The example actor model shown in Fig. 2b, which represents the code shown in Fig. 2a, shows a possible mapping from elements of an imperative language to dataflow elements. We show this for the C language, the principles, however, are applicable to other higher-level imperative and functional languages as well. As they might use more complex constructs such as classes, generics, etc., these points have to be addressed in future work. The actor surrounding the dataflow model may have some declarations to define variables used as in- or outputs for the main *dataflow region*. These declarations show the interface of the actor

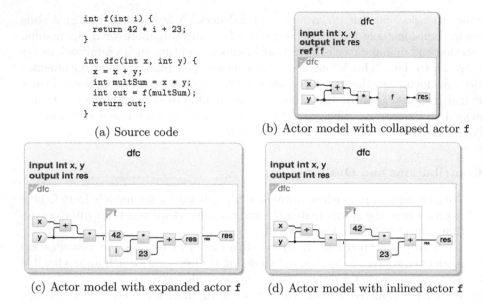

```
int f(int i) {
    return 42 * i + 23;
}

int dfc(int x, int y) {
    x = x + y;
    int multSum = x * y;
    int out = f(multSum);
    return out;
}
```

(a) Source code

(b) Actor model with collapsed actor f

(c) Actor model with expanded actor f

(d) Actor model with inlined actor f

Fig. 2. Automatically generated dataflow views.

regarding the data it reads from and gives back to the environment. This is also visible in the dataflow region, as the in- and outputs of the actor are visualized as named flags on the left or the right of the dataflow, respectively.

The dataflow region itself visualizes a collection of *assignments*. All assignments connect their inputs to their outputs through *simple actors*, which are the basic operations available for numbers, such as +, −, *, etc.., or *complex actors* with inner behavior, such as the actor f. We also make use of configurable *views*, to show the dataflow only at top-level, more detailed by expanding any complex actor, or completely inlined with connected interface, see Figs. 2b, 2c and 2d.

2.2   Constructing Actor Models

For the translation of the source code into an actor model, the Abstract Syntax Tree (AST) is analyzed and a view is presented based on its constructs. The following explains how we propose to visualize the different language constructs.

Assignment Statements. To start with the basics, each assignment of an expression is represented by the connection of its simple or complex actors via an edge or *wire* in the view. These wires can connect to further assignments or other uses in complex actors.

Compound Statements. The translation of a compound statement is the core of the dataflow extraction. It represents the body of a function, but it is also

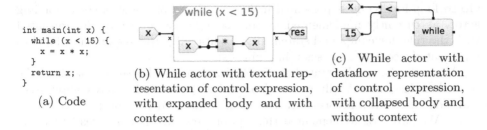

```
int main(int x) {
    while (x < 15) {
        x = x * x;
    }
    return x;
}
```

(a) Code

(b) While actor with textual representation of control expression, with expanded body and with context

(c) While actor with dataflow representation of control expression, with collapsed body and without context

Fig. 3. Alternative actor views of a `while` statement.

used to represent the body of control statements such as the `if` statement. To translate this into dataflow, all statements within are distinguished as detailed in this section and visually added to a dataflow region.

Function Definitions. Figure 2a shows the definitions of the functions `f` and `dfc`. The representation of the function `dfc` in actor-based dataflow is shown in Fig. 2b. It shows the function as an actor which can be referenced by other actors to represent function calls. The parameters are represented by variable declarations. Furthermore, the actor declares an output variable matching the return type of the function, shown with the name `res`. The compound statement of the function is then represented by the dataflow region of the actor.

Function Calls. In Fig. 2a, `out = f(multSum)` is an assignment statement that makes a function call. A complex actor is created that references the actor for the called function. The parameters of the function and its return value are linked to the corresponding variables in the dataflow region. This results in the diagrams shown in Figs. 2b to 2d. If the called function is also defined in the given source code—and is not for example a library function—the referenced actor is expandable to show the behavior of the called function. This way, the resulting diagram can be navigated without showing every detail from the beginning, so that the users can choose by themselves which details they want to see.

While and Do-While Statements. Control statements, such as loops, have no direct representation in classical dataflow views. However, they are an important part of the program structure that we want to preserve in our visualization. We therefore propose to represent each control flow statement as a complex actor referenced in the dataflow region, as it was shown for function calls. That complex actor shows control flow in an abstract way or in a state machine fashion to keep the main focus on the dataflow while giving the control flow a natural counterpart. We explain the loop representations in detail for the `while` statement only, as `do-while` and `for` statements are similar.

A `while` loop contains a conditional expression that controls how often the loop is repeated, and the loop body that represents its behavior. An example `while` loop in C code is presented in Fig. 3a. Figure 3b shows a corresponding actor, as currently implemented in our tooling. The loop body is translated with the rules for compound statements. Additionally, this region has a label to represent the control expression in plain text.

Alternatively, as illustrated in Fig. 3c, one might choose to not show the control expression textually but to also synthesize it including any computation and side effects into actors, with the result connected to a port of the `while` actor. We have not implemented this option yet since it might lead to many additional elements in the resulting graphic. However, for future extensions it might be worth considering to offer this alternative visualization as an option.

The definition of the in- and outputs of the `while` actor is done by searching for any variable defined outside of the `while` loop that is read or modified inside the loop. In the resulting dataflow region, each read variable is connected with the inputs, and each written variable is connected with the outputs of the actor, as in Fig. 3b.

For Statements. The translation of the `for` statement is very similar to the other loop statements. The difference is that it does not only contain one expression for the condition. The `for` statement contains two more expressions for an initialization and an update of the loop, also shown as text.

If and Switch Statements. These control flow statements are again visualized as complex actors with the analysis of read and written variables. The control flow can be modeled and displayed as simple state machines. This is different to other approaches, as discussed in Sect. 5, since we use state chart visuals to represent the branching control flow that is not directly translatable into traditional dataflow views. Keeping this as a dataflow-only view with combination actors similar to multiplexers would be another viable approach allowing for non-trivial control expressions that we have not implemented yet.

The `if` statement shown in Fig. 4a can be translated as shown in Fig. 4b, with a branching initial state that hands the control either to the `then` or the `else` branch, depending on whether the condition expression results as `true` or `false`. The compound statements of both branches are then put into the dataflow regions of their respective states as described above. The numbers next to the transitions to the branches indicate their priority, where the lower number stands for a higher execution priority. So the `then` transition with the higher priority 1 will only be taken if the condition is true, otherwise the always-true transition with the lower priority 2 is taken, matching the semantics of the `if` statement. The connection of this actor to the outside is, as for all complex actors, via the ports for all in- and outputs as shown before and omitted here.

The `switch` statement follows a very similar strategy, as it corresponds to multiple chained up `if` statements. The code and resulting visualization are shown in Fig. 5. Each `case` statement is equivalent to an equality check of the variable in the `switch` with the value in the `case` statement. The translation to the state chart visualization therefore is as for the `if` translation, where now

```
int main(int x) {
  int a;
  if (x > 42) {
    a = 1337 * x;
  } else {
    a = 420 / x;
  }
  return a;
}
```

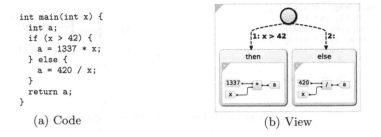

    (a) Code   (b) View

Fig. 4. The actor representation of an `if` statement.

```
int main(int x) {
  int a = 0;
  switch (x) {
    case 1:
      a = 42;
      break;
    case 2:
      a = 14;
    case 3:
      a = a + x;
    case 4:
      a = a * x;
      break;
    case 5:
      a = x - a;
    default:
      a = x;
  }
  return a;
}
```

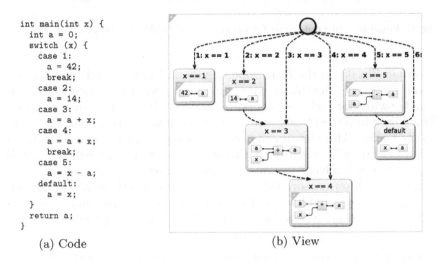

    (a) Code   (b) View

Fig. 5. The actor representation of a `switch` statement.

instead of a single **then** branch a conditional transition to a **case** state is added for each **case** statement with decreasing priority. The **else** branch is equal to the **default** case, as it will be the transition with the lowest priority and no condition attached to it. Additionally, every **case** gets an outgoing transition into the **case** state with the next lower priority if it is missing the final **break** statement, as the control flow will continue into the next **case** block in that case.

2.3 Data Types

Many use cases are already covered when using primitive data types such as `int`, `float`, and so on. These represent single values and are shown to be held by wires in the view. Other data types, however, are also widely used in C and other imperative languages. Their representations are presented here.

Arrays. Arrays are used in most languages, and we propose two visual counterparts. A code example swapping two array indices with their possible views

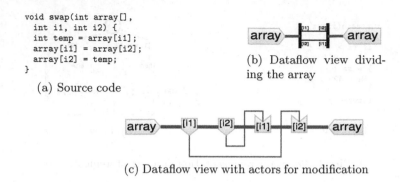

```
void swap(int array[],
    int i1, int i2) {
    int temp = array[i1];
    array[i1] = array[i2];
    array[i2] = temp;
}
```

(a) Source code

(b) Dataflow view dividing the array

(c) Dataflow view with actors for modification

Fig. 6. Possible representations for arrays, which could also be used for structs.

are shown in Fig. 6a. In the first proposed view in Fig. 6b, which we have opted for in our prototype, the thicker wire representing the array is divided and split into further wires for each read operation, labeled with the read index, and then combined back into the thicker array wire for each write operation, labeled with the written index. This view is compact and shows all array accesses in a fashion where only the data is relevant. Another possible view, not implemented yet but illustrated in Fig. 6c, has a continuous thicker array wire and visualizes every array access as an own actor. This representation may be less tidy, due to additional edge crossings, and has a strict dependency on the order of operations on the array, even if they may be interchangeable in the dataflow sense.

Structured Data. Structured data such as in structs in C or classes in other languages can be visualized similar to arrays, where the thick wire represents said structured data and its field accesses map to the index accesses in Fig. 6.

Pointers. Pointers to structured data and pointers in general can be visualized as wires as described above, where the wires carry the data that is pointed to. However, this only applies to pointers which are not modified. Pointer arithmetic can be used in C, though that disconnects the pointer variable from the data it points to and cannot be visualized easily with the concepts shown here anymore, as it is may depend on actual runtime data and cannot be analyzed statically as described.

Structs. A common practice to pass data around among functions is to collect these as structured data such as a struct in C and to pass a pointer to that data to the functions. We propose an alternative way to show each field of a central struct like individual variables together with the usual variables to flow through the algorithm. An example of a dataflow graphic using this *struct flow abstraction*, based on signal processing code provided by an industrial partner, can be seen in Fig. 9.

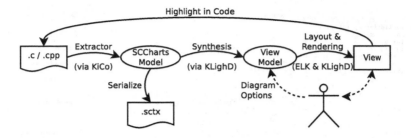

Fig. 7. The workflow for visualizations in KIELER.

3 Implementation and Validation

For the extraction and visualization of dataflow from source code we make use of pre-existing technologies for model-based design from the Kiel Integrated Environment for Layout Eclipse Rich Client (KIELER) project. The typical workflow and user interaction in the Eclipse-based tool explained in this section is presented in Fig. 7 and described below. A key aspect is the separation of *model* and customizable *views*, which serve as abstract documentation and navigation aid, as advocated in *modeling pragmatics* [7].

KIELER uses the model-based framework KIELER Compiler (KiCo) [26] to create compiler chains to and from modeling languages, executable code, visual models, or intermediate models in multiple configurable steps, so-called *processors*. We use this compiler framework to build a new compiler chain compiling from source code, in this case C code, to the visual modeling language SCCharts. Our synthesis chain can also parse C++ code, and can, e.g., synthesize state machines from a state pattern based on C++ templates, as discussed further in Andersen's thesis [2]. We use the C/C++ Development Tooling (CDT) for parsing the source code and a novel extraction to generate SCCharts models from the parsed code.[1]

The Sequentially Constructive StateCharts (SCCharts) language presented by von Hanxleden et al. [9] provides determinate concurrency using a graphical statechart notation and also supports the use of dataflow. We use it as a modeling language for its support of modeling dataflow combined with its possibility to create configurable and interactive visuals using KLighD.

The KIELER Lightweight Diagrams (KLighD) framework, as presented by Schneider et al. [22], generates interactive views from arbitrary models using an abstract *view model* to describe node-link diagrams. As SCCharts was designed as a visual language using KLighD to automatically visualize modeled instances, the use of SCCharts as our basis creates a shortcut to generating views. This provides filtering and configurability of the view out of the box that are useful for our use case.

---

[1] The code for the extraction is available at https://git.rtsys.informatik.uni-kiel.
de/projects/KIELER/repos/semantics/browse/plugins/de.cau.cs.kieler.c.sccharts?
at=refs%2Ftags%2Fdiagrams21.

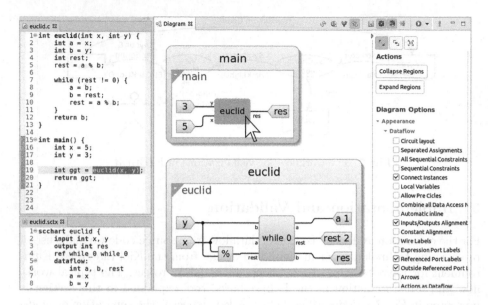

Fig. 8. Automatic highlighting from a dataflow view to the source code and interaction in the tool.

All figures shown of dataflow use the KIELER infrastructure with SCCharts. Here we use the visual aspect of the language, ignoring the semantics of the language itself. To provide more than static diagrams, this infrastructure allows the user to browse the models freely and interactively. The user can expand and collapse regions and actors, such as shown in Figs. 2b and 2c. One may also use *diagram options* to show or hide labels for the inputs and outputs of referenced complex actors, as also shown in Figs. 2b and 2c, and other visual options. Fuhrmann and von Hanxleden [7] presented how view management is conceptualized and implemented in KIELER and can be used for SCCharts.

Furthermore, when having the extracted diagram and the source code of an algorithm next to each other, we allow the user to interactively trace the origin of diagram elements. For example, clicking on a complex function call actor leads the user directly to the function call expression in the source code for investigating the diagram and the source code itself in parallel. An impression on how this looks in the KIELER environment is shown in Fig. 8.

Figure 7 visualizes how to combine these modes of interaction with the tool to get from the source code to a visualization and other artifacts. KIELER employs the Eclipse Layout Kernel (ELK) and KLighD to compute concrete views from the view models. One aspect not to be underestimated when using visual models is how the proper usage of *secondary notation* affects readability [20]. To that end, we employ the *layer-based layout* computed by ELK [23], which facilitates a natural left-to-right reading direction; for dataflow diagrams, this means that information flows from inputs to outputs.

Figure 8 also shows the user interface in KIELER. There are two open editors, one with the base C source code and one with the extracted and serialized SCChart model, together with the view of that SCChart model with tools for interactivity, such as a side bar for the diagram options, and the possibility to highlight and navigate to elements from the view back to the source code.

For practical validation, we implemented a prototype following the concepts as described in the previous sections. The dataflow view does not fully support the pointers as discussed in Sect. 2.3 yet, the basic functionality for passing pointers and writing to them in function calls, however, is already implemented. The struct flow abstraction is used as a proof of concept of the view for **structs**. In it, we require the algorithm to be separated into different functions that all may manipulate one central struct. If they manipulate the struct, it is expected that each of these functions take it as the first parameter. Finally, as the struct flow abstraction does not support arrays as the default dataflow view does, those elements are not wired up in Fig. 9 used for the survey, leaving the complex actor named **For_8** unconnected in the view. Having full support for structs as described above in our main dataflow view and adding that to the visual model will be part of future work. This will also allow us to present the tool with larger programs than the examples used here and in part improve the scalability and understandability. For example, Fig. 9 relies heavily on the use of structs and would get clearer.

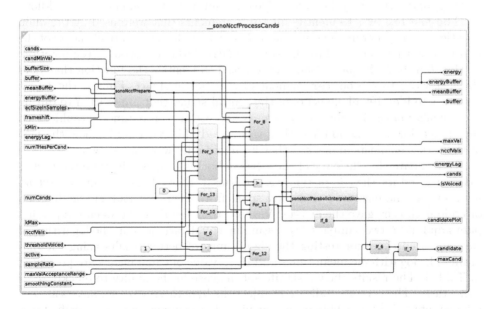

Fig. 9. Top level struct flow view filtered to display the main struct. Hierarchies are all collapsed to show a first overview.

Table 1. Survey results for the tasks.

Task	Code	Diagram	p-value
Used for statements	13.6 s	3 s	0.020
Read struct elements	23.7 s	5.6 s	0.039
Last write	22.4 s	7.6 s	0.054
Written elements	125.9 s	5.1 s	<0.001

4   Survey

As a first assessment of the benefit of the resulting dataflow diagrams, we conducted a preliminary survey with six PhD students from the working group. The participant group was split in half, with each group being asked to complete the same tasks regarding the dataflow example in Fig. 9, utilizing only the core concepts. Half of the participants got to view the source code in an Eclipse IDE with the installed CDT for the tasks for the dataflow example. The other group had the same tasks, using the extracted diagram in the interactive view in KIELER. All participants had previous knowledge of the C programming language, the usage of the Eclipse IDE, and the diagram view with SCCharts, thus mitigating any skill-based bias in the results. They were allowed to use all features of the IDE and the interactive diagram freely to simulate real workflow.

The *first task* was to identify all for statements that were used by the main function of the program. The *second task* asked for the number of elements of the main struct that were used by the sonoNccfParabolicInterpolation function. In the *third task* the participants were asked to identify the function that does the last write to the buffer element of the central struct. Finally, the *fourth task* was to identify all elements of the central struct that are written to during the calculation of the main function. We note that the dataflow graphic used the struct flow abstraction and thus was already filtered to only show the flow of the struct elements alone, so other local variables were not shown.

To compare the efficiency of the tasks in both scenarios, the time the participants needed for the completion was measured. Everything was already set up for them, so the measured time only spans from their first look at the diagram or code until they gave the answer. We then took the times between the participants and did a two-sample *t-test* assuming equal variances on the logarithm of the measured times for testing the significance of better results using the diagram. The logarithmic mean times and the p-value of equal means are presented in Table 1. The results show a significant increase in the ability to quickly solve these tasks. The p-values indicate to reject the hypothesis of equal means at confidence intervals of over 90% each, meaning that completing these specific tasks is most likely faster. In this sample all of these tasks were solvable on average more than three times as fast when using the extracted diagram than without, while almost all tasks were solved correctly. Only in the fourth task none of the participants was able to give the correct answer given the code alone and one participant even gave up and stated that there was no way to solve the task in the two minutes that were suggested as a soft time limit. They said that they needed at least a piece of paper to take notes and that the IDE's functionality to highlight the name of the central struct was essential to make it even possible.

This indicates that the generated diagrams will make these tasks quicker and easier to solve than with the code and usual IDE tooling alone. Surely, these tasks are constructed in such a way that they should be solvable faster with the diagram, for example showing what was asked for in the questions on a single screen in the first task compared to needing to scroll through the code, but that confirms the assumption that a diagram like this can present information in a more accessible way than typical IDE features alone, and that can be harnessed by using the extraction and interactivity presented in this paper. Furthermore, the small sample size, the current state of the tool, and the tailored questions only allow for preliminary study results that cannot be generalized and need to be validated with a more complete tool. But it is not the focus to replace the code, but rather to provide a tool to help the developer comprehend the code alongside it, so these results already show potential to improve specific tasks.

5   Related Work

Analyzing dataflow is common practice, often used for low level compiler optimizations [10,17]. However, these are typically low-level analyses, and not meant to help program comprehension. To quote Ishio et al. [11], *while developers have to investigate dataflow paths, existing source code viewers focus on method calls as a main relationship*. In this paper, however, we focus on a higher level analysis for the comprehension of developers. Ishio et al. [11] have investigated interprocedural dataflow for Java programs. They propose Variable Data-Flow Graphs (VDFGs) that represent interprocedural dataflow, but abstract from intraprocedural control flow. Unlike our work, their analysis ignores sequential control flow, thus in program fragments like x = y; y = z; it (falsely) assumes that x depends on z. For method calls, they assume that these produce data only via their return value. They provide an interactive viewer, but their interaction consists of selecting specific nodes (e.g., a variable) in the VDFGs, for which then the neighboring nodes are indicated. Our work instead aims for providing high-level overviews of whole program (regions) that can be explored interactively.

Namballa et al. [18] use VHDL and create a control and dataflow graph (CDFG) for the program, which is an integral part during the synthesis from the behavioral specification of a program into an electronic circuit using logic gates. This CDFG described by Amellal and Kaminska [1] focuses on a control flow graph with hints for the variables used in the dataflow. These graphs are not extracted to help programmers comprehend the code but are specific to the hardware synthesis. Furthermore, the graphs focus on showing the concrete control flow, while we focus on displaying the concrete dataflow, while hinting at the control flow for non-linear flow of data.

Beck et al. [3] and Gèvay et al. [8] present methods to execute imperative control flow on dataflow machines and graphical representations for those translations. The use of their visualizations as a program comprehension tool is restricted as they mainly focus on the execution on parallel dataflow machines. Moreover, they do not describe any interactive features for the diagrams.

There are further tools and frameworks to reverse engineer diagrams from C code. CPP2XMI is such a framework as used by Korshunova et al. [12] for extracting class, sequence, and activity diagrams, or MemBrain for analyzing method bodies as presented by Mihancea [16]. UML class models are extracted from C++ by Sutton and Maletic [27], and another framework for the analysis of object oriented code is presented by Tonella and Potrich [28]. All these tools and frameworks show the importance of reverse engineering and presenting views to programmers, whereas these do not cover dataflow like in our approach.

Commercial tools such as the McCabe tool suite or SCITools' Understand also focus on visualizing code metrics for an easier comprehension, but they do not include such specific means of visualizing intraprocedural dataflow. The DMS Software Reengineering Toolkit is another tool supporting some dataflow analysis framework, that, however, also concentrates on the control flow and the statements themselves, with data dependencies between each statement added to the view, similar to the graphs shown in Namballa et al. [18].

Smyth et al. [25] implemented a generic C code miner for SCCharts. The focus was to create semantically valid models from legacy C code, which can then be compiled to modern code for various platforms. The paper only considers a small subset of C, which is translated into control flow constructs. The work also tried to find appropriate means for visualizing common C patterns in control flow, which still is a difficult question for larger models.

6 Conclusions and Outlook

As argued by others before, visual models may be a valuable documentation of textual programs. Diagrams leverage the human perception capabilities in ways that program text typically does not use, and visual models typically entail some level of abstraction. Thus visual models are—at least here—not meant to replace code, but rather to augment them. In terms of visual documentation of source code, a main novelty presented here is the synthesis of actor-based dataflow diagrams. Not all C language features are fully supported yet our implementation, but in our experience, the dataflow and struct flow visualizations appear to be valuable for the user and to provide an easy way to analyze the flow of data within functions. A first user experiment indicated the advantages of the visual models over the original program text for answering certain questions.

The synthesis of diagrams for code documentation also shows the broad applicability of visual models such as SCCharts. They can be used for programming in the SCCharts language directly, but also for the automatic generation of informative graphics. A compiler framework such as KiCo in the IDE makes the mapping to a visual model easy to complete.

As future work SCCharts and the dataflow extraction could be extended with support for the remaining constructs of the C language to enlarge the set of programs that can be shown in dataflow in their entirety. With that, evaluating the views and their scalability on production codebases is a next step in verifying this approach. Also, further filtering could be applied to the elements shown in the visualization, adding to the existing filtering possibilities in SCCharts. Finally, formalizing the notation and a more representative survey would be worthwhile future work.

References

1. Amellal, S., Kaminska, B.: Scheduling of a control and data flow graph. In: 1993 IEEE International Symposium on Circuits and Systems, vol. 3, pp. 1666–1669. IEEE (1993)
2. Andersen, L.: Dataflow and Statemachine Extraction from C/C++ Code. Master thesis, Kiel University, Department of Computer Science (December 2019). https://rtsys.informatik.uni-kiel.de/~biblio/downloads/theses/lan-mt.pdf
3. Beck, M., Johnson, R., Pingali, K.: From control flow to dataflow. J. Parallel Distrib. Comput. **12**(2), 118–129 (1991)
4. Cook, C., Visconti, M.: Documentation is important. CrossTalk **7**(11), 26–30 (1994)
5. Eker, J., et al.: Taming heterogeneity-the Ptolemy approach. Proc. IEEE **91**(1), 127–144 (2003)
6. Forward, A., Lethbridge, T.C.: The relevance of software documentation, tools and technologies: a survey. In: Proceedings of the 2002 ACM Symposium on Document Engineering, DocEng 2002, pp. 26–33. Association for Computing Machinery, New York (2002)
7. Fuhrmann, H., von Hanxleden, R.: On the pragmatics of model-based design. In: Choppy, C., Sokolsky, O. (eds.) Monterey Workshop 2008. LNCS, vol. 6028, pp. 116–140. Springer, Heidelberg (2010). https://doi.org/10.1007/978-3-642-12566-9_7
8. Gévay, G.E., Rabl, T., Breß, S., Madai-Tahy, L., Markl, V.: Labyrinth: compiling imperative control flow to parallel dataflows. CoRR (2018). http://arxiv.org/abs/1809.06845
9. von Hanxleden, R., et al.: SCCharts: Sequentially Constructive Statecharts for safety-critical applications. In: Proceedings of ACM SIGPLAN Conference on Programming Language Design and Implementation (PLDI 2014), pp. 372–383. ACM, Edinburgh (June 2014)
10. Hecht, M.S.: Flow Analysis of Computer Programs. Elsevier Science Inc., New York (1977)
11. Ishio, T., Etsuda, S., Inoue, K.: A lightweight visualization of interprocedural dataflow paths for source code reading. In: Beyer, D., van Deursen, A., Godfrey, M.W. (eds.) IEEE 20th International Conference on Program Comprehension (ICPC), pp. 37–46. IEEE, Passau (June 2012)
12. Korshunova, E., Petković, M., van den Brand, M.G.J., Mousavi, M.R.: CPP2XMI: reverse engineering of UML class, sequence, and activity diagrams from C++ source code. In: 13th Working Conference on Reverse Engineering (WCRE 2006), pp. 297–298. IEEE Computer Society, Benevento (October 2006)
13. Lee, E.A., Neuendorffer, S., Wirthlin, M.J.: Actor-oriented design of embedded hardware and software systems. J. Circuits Syst. Comput. (JCSC) **12**(3), 231–260 (2003)
14. Lethbridge, T.C., Singer, J., Forward, A.: How software engineers use documentation: the state of the practice. IEEE Softw. **20**(6), 35–39 (2003)

15. Meyerovich, L.A., Rabkin, A.S.: Empirical analysis of programming language adoption. In: Proceedings of the 2013 ACM SIGPLAN International Conference on Object Oriented Programming Systems Languages & Applications (OOPSLA 2013), pp. 1–18. Association for Computing Machinery, New York (2013)

16. Mihancea, P.F.: Towards a reverse engineering dataflow analysis framework for Java and C++. In: Negru, V., Jebelean, T., Petcu, D., Zaharie, D. (eds.) 2008 10th International Symposium on Symbolic and Numeric Algorithms for Scientific Computing (SYNASC), pp. 285–288. IEEE Computer Society, Timisoara (September 2008)

17. Muchnick, S.S., Jones, N.D.: Program Flow Analysis: Theory and Applications. Prentice-Hall Inc., Englewood Cliffs (1981)

18. Namballa, R., Ranganathan, N., Ejnioui, A.: Control and data flow graph extraction for high-level synthesis. In: IEEE Computer Society Annual Symposium on VLSI, pp. 187–192. IEEE (2004)

19. Parnas, D.L.: Precise documentation: the key to better software. In: Nanz, S. (ed.) The Future of Software Engineering. Springer, Heidelberg (2011). https://doi.org/10.1007/978-3-642-15187-3_8

20. Petre, M.: Why looking isn't always seeing: readership skills and graphical programming. Commun. ACM **38**(6), 33–44 (1995)

21. Rugaber, S.: Program comprehension. Encycl. Comput. Sci. Technol. **35**(20), 341–368 (1995)

22. Schneider, C., Spönemann, M., von Hanxleden, R.: Just model! - Putting automatic synthesis of node-link-diagrams into practice. In: Proceedings of the IEEE Symposium on Visual Languages and Human-Centric Computing (VL/HCC 2013), pp. 75–82. IEEE, San Jose (September 2013)

23. Schulze, C.D., Spönemann, M., von Hanxleden, R.: Drawing layered graphs with port constraints. J. Vis. Lang. Comput. **25**(2), 89–106 (2014). Special Issue on Diagram Aesthetics and Layout

24. Singer, J.: Practices of software maintenance. In: Proceedings of the International Conference on Software Maintenance (Cat. No. 98CB36272), pp. 139–145. IEEE (1998)

25. Smyth, S., Lenga, S., von Hanxleden, R.: Model extraction for legacy C programs with SCCharts. In: Proceedings of the 7th International Symposium on Leveraging Applications of Formal Methods, Verification and Validation (ISoLA 2016), Doctoral Symposium. Electronic Communications of the EASST, vol. 74. Corfu, Greece (October 2016). poster

26. Smyth, S., Schulz-Rosengarten, A., von Hanxleden, R.: Towards interactive compilation models. In: Margaria, T., Steffen, B. (eds.) ISoLA 2018, Part I. LNCS, vol. 11244, pp. 246–260. Springer, Cham (2018). https://doi.org/10.1007/978-3-030-03418-4_15

27. Sutton, A., Maletic, J.I.: Mappings for accurately reverse engineering UML class models from C++. In: 12th Working Conference on Reverse Engineering (WCRE 2005), pp. 175–184. IEEE Computer Society, Pittsburgh (2005)

28. Tonella, P., Potrich, A.: Reverse Engineering of Object Oriented Code. Springer Science+Business Media Inc., New York (2005). https://doi.org/10.1007/b102522

29. Wechselberg, N., Schulz-Rosengarten, A., Smyth, S., von Hanxleden, R.: Augmenting state models with data flow. In: Lohstroh, M., Derler, P., Sirjani, M. (eds.) Principles of Modeling. LNCS, vol. 10760, pp. 504–523. Springer, Cham (2018). https://doi.org/10.1007/978-3-319-95246-8_28

Open Access This chapter is licensed under the terms of the Creative Commons Attribution 4.0 International License (http://creativecommons.org/licenses/by/4.0/), which permits use, sharing, adaptation, distribution and reproduction in any medium or format, as long as you give appropriate credit to the original author(s) and the source, provide a link to the Creative Commons license and indicate if changes were made.

The images or other third party material in this chapter are included in the chapter's Creative Commons license, unless indicated otherwise in a credit line to the material. If material is not included in the chapter's Creative Commons license and your intended use is not permitted by statutory regulation or exceeds the permitted use, you will need to obtain permission directly from the copyright holder.

Visualising Lattices with Tabular Diagrams

Uta Priss$^{(\boxtimes)}$

Ostfalia University, Wolfenbüttel, Germany
http://www.upriss.org.uk

Abstract. Euler and Hasse diagrams are well-known visualisations of sets. This paper introduces a novel type of visualisation, Tabular diagrams, which is essentially a type of Euler diagram where lines have been omitted or a 2-dimensional Linear diagram. Tabular diagrams are utilised to visualise lattices in comparison to Euler and Hasse diagrams. For that purpose, lattice terminology is applied to all three types of diagrams.

1   Introduction

Formal Concept Analysis (FCA) is a mathematical method for knowledge representation with many applications that uses lattice theory [5]. A challenge for FCA is that users need to be trained in order to be able to read Hasse diagrams of lattices. Euler diagrams tend to be perceived as more "intuitive" to read than Hasse diagrams but also have certain disadvantages compared to Hasse diagrams [8]. An experiment of Chapman et al. [3] demonstrates that Linear diagrams are more effective for certain retrieval tasks than Euler diagrams. But Linear diagrams often require repetition of attributes in order to represent the data of an application. Tabular diagrams as suggested in this paper are essentially 2-dimensional Linear diagrams which can represent more attributes without repetitions than Linear diagrams. The usability of Tabular diagrams should be similar to Linear diagrams because reading tables is a skill that most users are accomplished in. But the focus of this paper is on structural aspects, not on usability which will be left for future research.

Euler diagrams are a form of graphical representation of set theory that is similar to Venn diagrams but leaves off any regions that are known to be empty. Figure 1 shows three Euler diagrams with corresponding Hasse diagrams (which are explained in the next Section). Euler diagrams consist of closed curves with labels representing sets. The smallest undivided areas in an Euler diagram are called *minimal regions*. Regions are defined as sets (or unions) of minimal regions. Zones are maximal regions that are within a set of curves and outwith the remaining curves. Thus for a set of sets, $A := \{a_1, ..., a_n\}$, zones can be described as $(a_1 \cap ... \cap a_k) \backslash (a_{k+1} \cap ... \cap a_n)$. The reason for distinguishing between minimal regions and zones is that zones are the smallest mathematical meaningful areas of an Euler diagram whereas minimal regions are the smallest visibly undivided areas. In Euler diagrams that are not *well-formed*, it is possible that zones and minimal

© Springer Nature Switzerland AG 2021

A. Basu et al. (Eds.): Diagrams 2021, LNAI 12909, pp. 378–386, 2021.
https://doi.org/10.1007/978-3-030-86062-2_38

regions do not coincide and some zones consist of several minimal regions. A number of other criteria for being well-formed can be specified (cf. [4]), for example, allowing at most two curves to meet in any point and disallowing curve edges to meet in more than one adjacent point. In this paper we are only discussing Euler diagrams that fulfil a "zones=minimal regions" condition. Shading of a zone is sometimes used in order to indicate that a zone must be empty. In some cases, it is not possible to generate a well-formed Euler diagram without using shading.

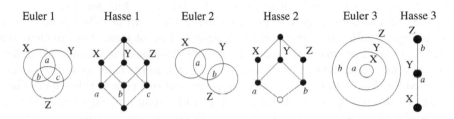

Euler 1   Hasse 1   Euler 2   Hasse 2   Euler 3   Hasse 3

Fig. 1. Euler and Hasse diagrams

2   Euler Diagrams and FCA

Because the focus of this paper is on FCA applications, the normal Euler diagram terminology is amended in this paper using notions from FCA. In this paper, the labels of curves are called *attributes*. Elements of sets denoted by attributes are called *objects*. These two notions allow a distinction between the sets and elements of an application, consisting of objects and attributes, from other sets and elements encountered in the discussion about Euler diagrams. The notions have purely a structural meaning. Thus objects and attributes can correspond to any kind of data in an application. Regions that can be described as intersections of attributes without using union or set difference as operations are called *i-regions*. The following lemma summarises well-known structural properties of zones and i-regions:

Lemma 1. *There is a 1-to-1 correspondence between zones and i-regions in an Euler diagram: for each zone $(a_1 \cap ... \cap a_k) \setminus (a_{k+1} \cap ... \cap a_n)$ its corresponding i-region is $a_1 \cap ... \cap a_k$ or denoted as an element of a powerset: $\{a_1, ..., a_k\}$. A set of i-regions together with $\subseteq$ forms a partially ordered set. This ordering can be isomorphically transferred to an ordering amongst zones if a zone a is called below another zone b (written as $a \leq b$) if $a_1 \subseteq b_1$ holds for the i-regions a_1 and b_1 corresponding to a and b.*

It is know from lattice theory that a subset of a powerset forms a lattice if it is closed with respect to intersections. Therefore a set of i-regions forms a lattice if it is closed with respect to intersections of sets of attributes. Thus in Fig. 1 Euler diagram 2, $\{\{\}, \{X\}, \{Y\}, \{Z\}, \{X, Y\}, \{Y, Z\}\}$, the intersection of the sets $\{X, Y\}$ and $\{Y, Z\}$ is required ($\{Y\}$) but not intersections of attributes such as $X \cap Z$. The i-regions in Euler diagrams 1 and 3 in Fig. 1 form lattices, the i-regions in Euler diagram 2 do not because the intersection over the empty set, i.e. $\{X, Y, Z\}$, is missing. If a set of i-regions does not form a lattice, it can be embedded into a lattice by adding the missing intersections.

The remainder of this section translates Euler diagram terminology into lattice terminology as provided by FCA. It is assumed in this paper that all sets of i-regions are embedded into lattices. An i-region $\{a_1, ..., a_k\}$ then determines a *concept* as a pair of sets of objects and attributes. The set $\{a_1, ..., a_k\}$ of attributes of a concept is called the *intension* of the concept. The set of objects of a concept is called an *extension* and consists of all objects that are elements of $a_1 \cap ... \cap a_k$. The ordering amongst i-regions discussed in Lemma 1 is called *conceptual ordering*, the lattice a *concept lattice*. Concepts that are added during an embedding into a lattice are called *supplemental concepts* in this paper. They correspond to missing or shaded zones of an Euler diagram. Supplemental concepts and thus concepts corresponding to missing or shaded zones of an Euler diagram do not have any *immediate objects* in their extension that is objects that belong to them but not to any lower concepts.

Hasse diagrams (as in Fig. 1) are a well-known diagrammatic representation of partially ordered sets and thus of lattices. The ordering is visualised as a transitive reduction because all edges that are implied by the transitivity of the ordering are omitted. For concept lattices, the ordering represents the conceptual ordering. Hasse diagrams are directed graphs where all edges are read in the direction from the visually lower to the higher end. Nodes in a Hasse diagram correspond to concepts. In order to read the extension of a concept in a Hasse diagram, all objects at the concept or at any concepts below it (according to the ordering) need to be collected. In order to read the intension of a concept, all attributes at a concept or at any concepts above it need to be collected. The nodes of supplemental concepts are drawn as unfilled circles in the Hasse diagrams (as in Hasse diagram 2 in Fig. 1). One advantage of using FCA is that it provides a variety of existing software[1] for generating lattices and their Hasse diagrams from data.

For the purposes of this paper, it should not be necessary to explain more details about FCA. Priss [8] provides a slightly more detailed introduction to FCA and its relationship to Venn, Euler and Hasse diagrams. Priss [8] concludes that lattice-theoretical properties can provide some further clues about when Euler diagrams are well-formed and discusses some advantages and disadvantages of Hasse diagrams compared to Euler diagrams.

3   Tabular Diagrams

Apart from Euler and Hasse diagrams, a further visualisation of concept lattices called *Tabular diagrams* is introduced in this paper. Tabular diagrams are essentially a 2-dimensional version of the "Linear diagrams" invented by Leibniz

---

[1] cf. https://upriss.github.io/fca/fcasoftware.html.

and discussed by Chapman et al. [3]. Tabular diagrams are best characterised as matrix-based diagrams (according to [2]) and appear under many different notions (mosaic plots/displays, contingency tables, Karnaugh maps) often with additional purposes, for example displaying frequencies within each zone. So far we have not been able to find a more general notion for or discussion of Tabular diagrams in the literature.

Figure 2 shows an example of a lattice where the attributes are the numbers 2, 3, 5, and 7 and the objects are numbers that are products of prime numbers. Objects have an attribute if they are divisible by that number. For example, 30 has the attributes 2, 3 and 5, but not 7. Such lattices as on the left side of Fig. 2. are well-known as examples of Boolean algebras. For the Tabular diagram in the middle, the set of attributes is partitioned into two sets. Intersections amongst attributes that belong to the same partition are indicated by overlapping brackets and a separate column. Intersections amongst attributes that belong to different partitions correspond to regions of the table. Objects are written into the zones. The Tabular diagram in Fig. 2 contains as many zones as there are concepts, but it is possible to have fewer zones than concepts if there are supplemental concepts. It is not possible to add another attribute to the Tabular diagram in Fig. 2 which intersects with all previous attributes. The strategy for Linear diagrams (as in Fig. 2, on the right) is to repeat attributes if a diagram is impossible to construct otherwise. Attributes can also be repeated for Tabular diagrams. But not all Tabular diagrams with more than four attributes require repetitions. Shading can be used for zones that do not belong to concepts at all if the Tabular diagram cannot be constructed otherwise. Supplemental concepts correspond to empty or missing zones but not to shaded zones. The bottom concept can be omitted if it has an empty extension.

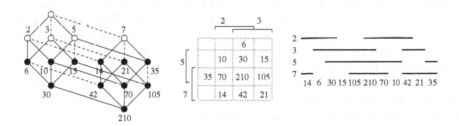

Fig. 2. Hasse, Tabular and Linear diagrams of a Boolean lattice with 4 attributes

All types of diagrams (Hasse, Euler, Linear and Tabular) are difficult to visually parse for large sets of data. Therefore, the fact that Tabular diagrams have some limitations for larger data sets does not necessarily disadvantage them compared to the other types. The dashed lines in the Hasse diagram in Fig. 2 indicate pairs of zones which are not neighbours in the Tabular diagram

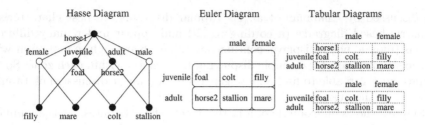

Fig. 3. Hasse, Euler and 2 Tabular diagrams of a linguistic example

even though they could be neighbours if the rows and columns were permuted differently. In the Tabular diagrams in this paper, the top is usually the left upper corner and the bottom is close to the centre or omitted. Thus some relationships are more easily visible in a Hasse diagram than in a Tabular diagram. Tracing lines can become difficult in a large Hasse diagram. Determining which attributes belong to an object seems to be easier in a Tabular diagram because it only involves reading the row and column headings.

Figure 3 contains another example modelled as isomorphic Hasse, Euler and Tabular diagrams. Brackets are not needed in Tabular diagrams for attributes that only span one row or column. The bottom element is omitted in all four diagrams. Figure 3 demonstrates that Tabular diagrams correspond to Euler diagrams where curves are rectangular and arranged in rows and columns. Euler diagrams with parallel curves are usually considered not well-formed. We would argue that in the case of Tabular diagrams the parallel curves are not a limitation because rows and columns of a table are easy to read. The Euler diagram in Fig. 3 cannot be drawn in a well-formed manner. The word "horse" has two meanings in this example: "horse1" refers to the species, "horse2" to the adult animal. The outer zone may be less obvious to see in the Euler diagram. The object "horse1" which would have to be placed into the outer zone has therefore been omitted in the Euler diagram. If the object "horse1" is deleted from the example, then the Tabular diagram can be reduced to the version shown in the lower corner. The lower Tabular diagram still contains 9 concepts: 4 labelled rows and columns, 4 intersections of rows and columns and the table as a whole, but three supplemental concepts are omitted.

From a structural viewpoint the question arises as to which sets of data can be represented by Tabular diagrams without shading or repetition. With respect to repetitions Petersen's [7] analysis provides some clues. Translated into the terminology of this paper, Petersen provides a characterisation with FCA of when a Linear diagram can be represented without repetitions. Her characterisation essentially checks whether a planar lattice exists for the data in which a line can be drawn from each object to the bottom concept without crossing the edges of the Hasse diagram. If a Hasse diagram is tree-like after omitting its bottom node, then it fulfils Petersen's condition. Furthermore if a Hasse diagram contains a cycle, the cycle must contain the bottom node or be at the side of the diagram, but not in the middle in order to fulfil the condition.

For Tabular diagrams the question is then whether the set of attributes can be partitioned into two sets which are representable as Linear diagrams without repetitions. In that case, it is known from FCA that the resulting lattice can be embedded into the direct product of the two lattices corresponding to the Linear diagrams. Most likely providing a characteristic or algorithm for producing Tabular diagrams in this manner is non-trivial. Because there are $2^n/2$ possible ways to split a set of n attributes into two partitions, calculating all possibilities is not feasible for large sets. But as mentioned before, diagrams are mainly of interest for fairly small sets of data and some heuristics can be applied to reduce the number of possibilities, such as:

- check whether omitting the bottom concept generates a more tree-like structure,
- look for partitions that split the set of attributes approximately in half,
- check whether the attribute set can be simplified (for example if an attribute and its negation exist in the data, only one of them may be required),
- determine if certain lattice properties exist in the data using FCA software and use them to partition the data (for example, the attributes of a chain or antichain should be kept in one partition).
- According to FCA: if a lattice is a direct product (possibly minus the bottom concept), then the number of rows and columns of the Tabular diagram should be a divisor of the number of concepts (possibly minus 1). Figure 2 and Fig. 3 show examples for this case (with 4×4 and 3×3 concepts).

4 More Examples of Tabular Diagrams

This section discusses some slightly more complex examples that pertain to data from applications. The Hasse diagram in Fig. 4 is a well-known FCA example of a lexical field of bodies of water based on linguistic "componential analysis" which determines semantic components of words [6]. Each semantic component relates to a positive and a negative attribute depending on whether it exists or not in the word. This results in each object in the lattice (with the exception of "channel") having either the positive or the negative counterpart of each of the four attributes (such as "inland" or "maritime"). For "channel" it is not specified within the provided data whether it is natural or artificial. The supplemental concepts correspond to more general concepts (such as "stagnant natural body of water") which are not lexicalised in the data.

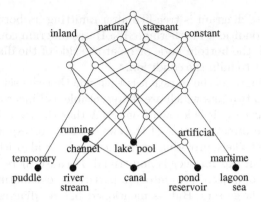

Fig. 4. Hasse diagram of a lexical field

The left side of Fig. 5 shows a Tabular diagram for the lattice in Fig. 4. It contains 23 zones, corresponding to 23 concepts except the bottom concept. The 16 empty zones represent supplemental concepts. The shaded zones belong to combinations which do not exist in the data at all, for example, there is no concept for maritime and artificial because no object has both those attributes. If all the non-shaded zones of one attribute are adjacent, it is possible to represent the non-shaded zones by drawing a curve around them as shown in the right side of Fig. 5. In a sense this corresponds to adding a third dimension to the Tabular diagram. The diagram on the right side of Fig. 5 also contains 23 zones, but the zone "constant, artificial, inland" might be overlooked because it is empty, quite small and not rectangular in shape. The top row and left column can be omitted if it is not desired to show all supplemental concepts. One might be tempted to simply add the negative attributes as row and column headings in the right Tabular diagram in Fig. 5. But that would change the data. For example, the information that temporary bodies of water are always inland and stagnant would be lost.

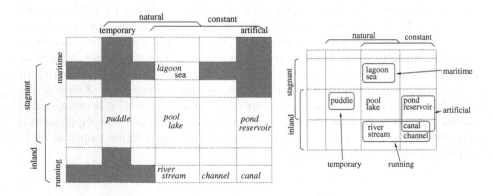

Fig. 5. Tabular diagrams for the lexical field in Fig. 4

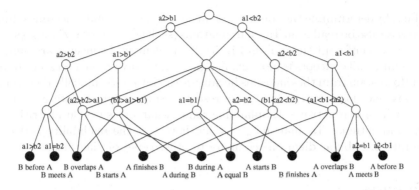

Fig. 6. Hasse diagram of Allen's [1] temporal relations

Figure 6 and Fig. 7 display a further example containing Allen's 13 temporal relations [1]. These relations are used in formal ontologies in order to express all possibilities of how two temporal intervals (A and B) can be related to each other, such as one occurring before the other one or both overlapping based on the relationships of their start points (a1 and b1) and end points (a2 and b2). The concept lattice in Fig. 6 contains 29 concepts (without the bottom concept). Allen himself uses a table with 144 fields to display the transitivity relationships amongst the 13 temporal relations. A lattice representation summarises the relationships because each field in Allen's table is the intension of a concept. The lattice contains all possible logical combinations. In this example, all concepts other than the neighbours of the bottom concept are supplemental yielding a Tabular diagram in Fig. 7 without empty zones. Because of the missing supplemental concepts it is more difficult in the Tabular diagram to count the total number of concepts in the lattice because all possible intersections of attributes need to be considered.

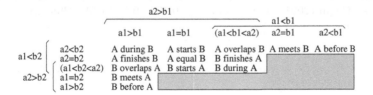

Fig. 7. Tabular diagram of Allen's [1] temporal relations

5   Conclusion

Tabular diagrams provide a concise representation of the information contained in a concept lattice. It is straightforward to determine the attributes for each object by retrieving the row and column headings and to determine the objects for each attribute by determining the zones belonging to the row or column headings. Implications can also be read from Tabular diagrams (such as "temporary" implies "natural", "stagnant" and "inland" in Fig. 5). If supplemental concepts are omitted, then counting all concepts is more challenging but it may not be necessary to read such information from the diagrams because it can be algorithmically determined.

Many questions are left for future research, such as which data can be represented as non-repetitive Tabular diagrams, what are suitable algorithms to create and optimise Tabular diagrams and how does the readability and visual scalability of Tabular diagrams compare to other types of diagrams.

References

1. Allen, J.: Maintaining knowledge about temporal intervals. Commun. ACM **26**(11), 832–843 (1983)
2. Alsallakh, B., Micallef, L., Aigner, W., Hauser, H., Miksch, S., Rodgers, P.: The state-of-the-art of set visualization. Comput. Graphics Forum **35**(1), 234–260 (2016)
3. Chapman, P., Stapleton, G., Rodgers, P., Micallef, L., Blake, A.: Visualizing sets: an empirical comparison of diagram types. In: Dwyer, T., Purchase, H., Delaney, A. (eds.) Diagrams 2014. LNCS (LNAI), vol. 8578, pp. 146–160. Springer, Heidelberg (2014). https://doi.org/10.1007/978-3-662-44043-8_18
4. Flower, J., Fish, A., Howse, J.: Euler diagram generation. J. Vis. Lang. Comput. **19**(6), 675–694 (2008)
5. Ganter, B., Wille, R.: Formal Concept Analysis. Mathematical Foundations. Springer, Heidelberg (1999). https://doi.org/10.1007/978-3-642-59830-2
6. Kipke, U., Wille, R.: Formale Begriffsanalyse erläutert an einem Wortfeld. LDV-Forum **5**, 31–36 (1987)
7. Petersen, W.: Linear coding of non-linear hierarchies: revitalization of an ancient classification method. In: Fink, A., et al. (eds.) Advances in Data Analysis, Data Handling and Business Intelligence, pp. 307–316. Springer (2010). https://doi.org/10.1007/978-3-642-01044-6_28
8. Priss, U.: Set visualisations with euler and hasse diagrams. In: Cochez, M., Croitoru, M., Marquis, P., Rudolph, S. (eds.) GKR 2020. LNCS (LNAI), vol. 12640, pp. 72–83. Springer, Cham (2021). https://doi.org/10.1007/978-3-030-72308-8_5

Understanding Scholarly Neural Network System Diagrams Through Application of VisDNA

Guy Clarke Marshall[1]([✉]) [iD], Caroline Jay[1] [iD], and André Freitas[1,2] [iD]

[1] Department of Computer Science, University of Manchester, Manchester, UK
guy.marshall@postgrad.manchester.ac.uk,
{caroline.jay,andre.freitas}@manchester.ac.uk
[2] Idiap Research Institute, Martigny, Switzerland

Abstract. We utilise VisDNA as a tool for understanding neural network system architecture diagrams. Through examples from scholarly proceedings, we find that the application of the framework to this ecological and complex domain is effective for reflecting on these diagrams. We argue for additional vocabulary to describe semiotic variability and internal inconsistency or misuse of visual encoding principles in diagrams. Further, for application to system diagrams, we propose the addition of "Grouping by Object" as a new visual encoding principle, and "Emphasising" as a new visual encoding type.

Keywords: Neural networks · Visual encoding · Graphic language · System diagrams

1 Introduction

The VisDNA framework proposed by Engelhardt and Richards [3, 9] is designed to be applicable to "all types of visualizations". VisDNA provides a method and vocabulary for analysing diagrams and diagram components, which has been tested on a large variety of visualization types. We aim to advance the understanding of the heterogeneous diagrammatic representations of neural network (NN) systems found in scholarly Natural Language Processing (NLP) conference proceedings by applying this framework. Scholarly NN systems diagrams have been chosen due to their heterogeneity, importance and prevalence in communicating about NN systems research. Whilst VisDNA has been tested on a wide variety of information visualisation resources, it has not yet been applied to (a) system diagrams, or (b) "ecological" diagram examples not conforming to an established standard or grammar. Our method is to manually select a variety of NN system diagrams from recent NLP conference proceedings, to apply VisDNA in order to analyse these examples, and to discuss and make suggestions to improve the applicability of VisDNA. In the course of this analysis, we necessarily comment on existing diagrams. Our goal is not to criticise the authors of these diagrams, nor to provide suggestions for effective diagrams, but to explore the application of VisDNA. Our contribution is to:

© Springer Nature Switzerland AG 2021
A. Basu et al. (Eds.): Diagrams 2021, LNAI 12909, pp. 387–394, 2021.
https://doi.org/10.1007/978-3-030-86062-2_39

- Analyse scholarly NN system diagrams
- Qualitatively evaluate VisDNA as a tool for reflecting on scholarly neural network systems diagrams
- Suggest further requirements for VisDNA, including emphasis, graphical object schematicity, and semiotic considerations

2   Background

2.1   Neural Network Systems

Neural networks are software which utilise a series of mathematical operations, conceptually arranged in layers, in order to process an input space to an output space. The NN learns functions to transform different input by converting it to numeric data and optimising the transformation function. It is utilised to recognise patterns, usually by classifying, ranking, or predicting the next item in a sequence. A definition of NN systems is provided by Goodfellow et al. [4].

The representation of NN systems in conference proceedings is done through natural language, in pseudocode, and in diagrams which often appear to describe the system beyond the NN itself, including inputs, outputs, and the relationship between components. Code may be shared as a supporting artifact external to the formal proceedings. System outputs are often shared as tables, charts, and metrics. In order to consider diagrams separately from the text, we focus on *self-contained diagrams* which were deemed to meaningfully exist with assumed knowledge but without reference to other resources.

2.2   Frameworks for Analysing Diagrammatic Representations

We are not aware of research applying existing diagram analysis frameworks to scholarly communication. "Physics of Notations" [8] is a highly cited framework for designing visual notations, and could potentially be re-purposed as an analytic lens to describe diagrammatic phenomena. However, Physics of Notations is fairly abstract, with categories such as "include explicit mechanisms for dealing with complexity". Similarly, Cognitive Dimensions [1] is designed at a high level, and as a consequence many NN diagrams would be indistinguishable.

There are also frameworks which analyse diagrams from a cognitive perspective. Hegarty [5] lists cognitive advantages, including external storage and organisation of information, and suggests practices for effective graphics grounded in empirical research. Cheng [2] provides a taxonomy for analysing representations. "A1.1 One token for each type", for example, is violated within many diagrams and certainly within the corpus, where vectors are represented in a plethora of visual encodings. It would be possible to analyse these diagrams with Cheng's framework, however in our domain much groundwork is missing. For example, to discuss suitability of diagram components against reading and inference operations such as "A2.2. Prefer Low Cost Operators" appears to require context-specific evidence.

In this paper we utilise the VisDNA framework due to its concreteness, its task-independence and its potential for conducting systematic analysis.

3   Application of VisDNA "Mode of Visual Encoding" to Scholarly Neural Network Systems Diagrams

3.1   Methodology

VisDNA consists of three modes, five types, and 15 principles of visual encoding, and provides a systematic method for analysing diagrams and visualisations [3]. The framework focuses on the principles of visual encoding, which we also focus on in our analysis. Discussion, with additional examples, can be found in the extended pre-print of this paper [7]. Within the visualisation design space, system diagrams could broadly be considered flow diagrams, though they do not conform to a consistent or standard form. Examining self-contained diagrams found at ACL 2019, a top NLP conference, we found a visually and semantically heterogeneous set of diagrams across the proceedings.

3.2   Applying VisDNA

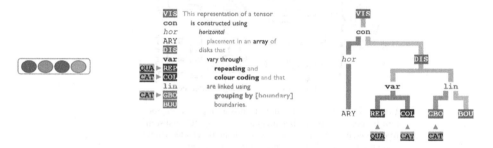

Fig. 1. Left: Tensor (multi-dimensional data used in a NN system) visually encoded as disks. Centre: VisDNA natural language description. Right: VisDNA parse tree

Figure 1 shows the VisDNA tree of a tensor, indicating it may be possible to apply VisDNA to sub-figures of NN diagrams. However, even this simple component requires a larger number of VisDNA blocks than a complete Scatter Plot [9], making it impractical to capture VisDNA for entire NN diagrams to a high specificity. We proceed to describe visual encoding types for NN diagrams:

Scaling. This is variable between and within diagrams. Within a diagram, size can be used to represent dimensional differences, often indicating a binary "bigger or smaller than," rather than precise scaling.

Ordering. Often these diagrams are (broadly) read linearly left-right, in chronological order of data processing at training-time. The diagrams are not usually of the system at run-time. Some information important for the creation and operation of the system, including chronological information such as "parameter training process," intervals, epochs, and updating of parameters are often omitted.

Grouping. Varied, often multiple different encodings are used within each diagram. Often, multiple visual encodings are applied to perform the same grouping (such as colour, proximity, alignment, and boundary, plus a label and a caption).

Linking. This is perhaps the most important part of the NN system diagrams, since along with components themselves, the relationship between components determines the architecture. Arrows are often used.

3.3   Example Diagram

Figure 2 can be thoroughly described using VisDNA, which we will briefly do here, avoiding minutiae sub-figure discussion.

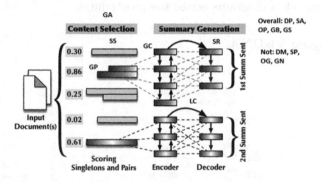

Figure 2: System architecture. In this example, a sentence pair is chosen (red) and then merged to generate the first summary sentence. Next, a sentence singleton is selected (blue) and compressed for the second summary sentence.

Fig. 2. An example diagram [6], with VisDNA labels overlaid

There is grouping by alignment, by colour and by position. It uses two different visual components for linking (dotted lines and arrows). The content is mostly schematic, with omission of details not important to the core contribution, such as layer labels and operations, which are commonly seen in other diagrams. Moody et al. [8] would describe this figure as demonstrating good "graphic economy," whilst in Gestalt theory this would be "Prägnanz" [10], language that VisDNA appears to lack. The framework does not facilitate discussion of the schematic and relatively minimalist content style, as the mode of depiction has a physical definition. Currently, the framework has the mode of depicting being schematic or realistic/precise, but this does not support the examination of individual visual components which is necessary for comparing between diagrams.

There are some important features that the framework does not currently include, such as the bracket object being used for grouping, which might be termed *Grouping by Object*. Unusual styling is present, in the graded colouring within the rectangles, and it is not clear if this serves an encoding purpose. The rotated text also is visually notable, and may require head-tilting which physically changes the readers interaction with the diagram. The use of an icon to accompany "Input Document(s)" is a dual-coding unexpected given the schematic nature of the rest of the diagram. Note that the system inputs are data rather than physical pieces of paper, so the mode of correspondence remains non-literal. Moody [8] might describe this dual-coding as "semiotically unclear," as there is not a 1:1 mapping between semantic constructs and visual components. The precision of the diagram is also arbitrary; the numbers are indicative rather than meaningful, as are the number of rectangles. Emphasis by Colour is used as well as Grouping by Colour.

Semiotically, the object being signified is variable within this diagram, with sentence-specific numeric information alongside the system-level "document(s)." Indeed more broadly this diagram appears to be representing the "contribution" rather than the "system." Where the authors reference the diagram in the text, it is described as an "illustration" [6], perhaps reflecting that the mode of depiction is schematic rather than precise.

4   Results and Discussion

4.1   Strengths of Applying the Framework

We found it extremely useful to have the vocabulary and breadth of considerations provided by the framework, particularly around the visual components. It has successfully and usefully described how different principles are used to group and link, within diagrams in this domain. In general, we found the visual encoding principles unambiguous and straightforward to identify.

Applying this framework has also allowed us to discover internal inconsistency in some of the diagrams, and a number of errors in the creation of the diagrams appear to have been found. Perhaps authors and reviewers would benefit from examining diagrams with a critical eye, using Visual Encoding Principles.

4.2   Refinements and Extensions to Improve Utility of VisDNA for (Scholarly NN) System Diagrams

There are requirements for VisDNA to have additional utility in our domain. It would be helpful to have standard terminology for some additional attributes of diagrams.

Author Assumed Knowledge and Conventions. Use of context-specific visual encoding conventions is not captured by VisDNA.

Internal Consistency. For complex diagrams, "consistency" may be important. This does not only apply to how visual encoding principles are used, but also in natural language, consistency of capitalisation or use of symbols. Regardless of whether internal inconsistency is cognitively problematic, it is necessary to capture this in order to describe the diagram, and would provide an entry point to discussing differences in representational choice within the diagram.

Semiotics of Examples vs Systems. As part of the future work, we suggest including semiotic or representational considerations (keeping within the framework's scope of "meaning represented in a diagram"). For example, if a systems diagram includes an example, this shifts the diagrammatic representamen to being a specific instantiation of the system.

Common Domain-Specific Symbols. Within either "mode of depicting" or "visual encoding: depicting," it would be useful to have visual components categorised to describe symbol choices. For visual encoding principles, quantitative, ordinal and nominal attributes are mentioned, and "Depicting: picturing" could be extended to include other properties of the object that are in a general sense iconic.

Sub-figures. In this domain, sub-figures (such as the layers) sometimes contain different representational choices to the "macroscopic" diagram. VisDNA terms these "nested visual structures," which we abbreviate as sub-figures. This indicates the utility of this framework at different levels of granularity.

Schematics. The framework does not capture the schematic (or otherwise) nature of the diagram, making comparison or discussion on "How much content is included or omitted" difficult.

Physical Emphasis. For non-physical systems, dimensions of discussion are reduced to being "non-literal" correspondence with "realistic/precise" depiction. It may be useful to formalise additional vocabulary to describe the use of non-physical metaphor and models. This could be part of the content work.

Quantitative Analysis. Allowing for quantitative discussion of diagrams at scale is time consuming, particularly for complex diagrams. This capability would be useful in order to do justice to the heterogeneity of visual encoding techniques employed, and allow for quantitative comparison (say, between different domains or media).

Grouping by Object. In encoding by visual appearance, we found visual components typified by "{," and natural language labels, being used to group other visual components. The right brackets of Fig. 2 is one such example. In this example, it is to group together individual rows by "Input Document(s)," "1st Summ Sent" or "2nd Summ Sent," where spatial grouping is used to link the coloured rectangular objects with the descriptions. A bracket should not be considered a "linking" symbol as it applies a common property to a set of other objects and is therefore about "category," which is defined to be the goal of grouping [3, p. 206].

Emphasis Principle. Additionally we propose "Emphasis" as a worthwhile extension to the visual encoding principles. It is most similar to the principle "Ordering," but fulfils a different function:

- **Emphasising** is used to make visually salient particular aspects. *Emphasising* answers questions such as *What is most important?* and *What should the reader look at first?*

- **Emphasis by colour** is a visual encoding principle often utilising bright or high contrast colours.
- **Emphasis by position** is done by placing visual primacy to certain elements, for example with a prominent position (extreme left, right or central), or providing lower spatial clustering with respect to other visual components or sub-figures.
- **Emphasis by uniqueness** is done by using a unique visual encoding principle for an element (or sub-figure). This could be a different boundary, colour, or visual component.

Ambiguities and Other Observations

- It is unclear whether a caption should be considered as "part of" the diagram.
- VisDNA lends itself to diagrams that are consistently assembled. However, it lacks the vocabulary to describe errors or misleading visual encoding, which would be useful as part of validating this framework and providing feedback to diagram authors, a potential use case of this framework.
- The vocabulary enables standard terminology to describe visual features of diagrams, and has been useful in describing figures in our domain. However, it does not yet provide guidance for semantic-content-focused vocabulary, and borderline topics such as graphic economy, and semiotics. The definition of "mode of depiction" also appears to need clarification in order to disambiguate precise/realistic and describe schematics and conventions.
- VisDNA has facilitated description and discussion of complex systems diagrams, including sub-figures. However, the framework is not yet optimised for sub-figures, or for describing the layout of the diagram. Sometimes we have to infer the intention or meaning of the author, and clarification is only possible outside of the diagram (e.g., in text or speech).
- For "positioning along an axis," we have assumed this axis does not need to be explicitly drawn, though for the infographics domain perhaps this distinction is useful. More generally, we felt that more precise definitions would aid meaningful and unambiguous application.
- In our domain, authors sometimes seem to use visual encoding mechanisms in unconventional ways (e.g. colour gradient). The framework would benefit from vocabulary to describe this.

4.3   Limitations

The manual selection of four diagrams (in the extended paper) is not statistically representative. However, we did not feel this was necessary in the first instance, as these examples faciliate discussion of the application of the framework. We have only used one coder, though the three authors of this paper were in agreement about the assessments. With the ambiguity in applying the framework as an analysis tool, this is a risk not only for this paper but also for the VisDNA framework itself. Further application in other domains would be an aid to making refinements to VisDNA and increasing practical clarity.

5   Conclusion

We have applied VisDNA to describe and discuss scholarly neural network systems diagrams, identifying unconventional visual encoding choices and internal inconsistencies in the diagrams. We have suggested additions to VisDNA, including the Emphasis Principle, contextual factors, and assumed knowledge. We also support creation of a method for encoding sub-figures. These modifications would make the application of this framework more useful in our domain, and perhaps for system diagrams more generally.

Acknowledgement. The authors would like to thank Clive Richards and Yuri Engelhardt for useful discussions about VisDNA, and anonymous reviewers for their feedback on an earlier version of this paper.

References

1. Blackwell, A., Green, T.: Notational systems-the cognitive dimensions of notations framework. In: HCI Models, Theories, and Frameworks: Toward an Interdisciplinary Science. Morgan Kaufmann, San Francisco (2003)
2. Cheng, P.C.-H.: What constitutes an effective representation? In: Jamnik, M., Uesaka, Y., Elzer Schwartz, S. (eds.) Diagrams 2016. LNCS (LNAI), vol. 9781, pp. 17–31. Springer, Cham (2016). https://doi.org/10.1007/978-3-319-42333-3_2
3. Engelhardt, Y., Richards, C.: A framework for analyzing and designing diagrams and graphics. In: Chapman, P., Stapleton, G., Moktefi, A., Perez-Kriz, S., Bellucci, F. (eds.) Diagrams 2018. LNCS (LNAI), vol. 10871, pp. 201–209. Springer, Cham (2018). https://doi.org/10.1007/978-3-319-91376-6_20
4. Goodfellow, I., Bengio, Y., Courville, A., Bengio, Y.: Deep Learning. MIT press, Cambridge (2016)
5. Hegarty, M.: The cognitive science of visual-spatial displays: implications for design. Topics Cogn. Sci. 3(3), 446–474 (2011)
6. Lebanoff, L., et al.: Scoring sentence singletons and pairs for abstractive summarization. arXiv preprint arXiv:1906.00077 (2019)
7. Marshall, G.C., Jay, C., Freitas, A.: Understanding scholarly natural language processing system diagrams through application of the Richards-Engelhardt framework. arXiv preprint arXiv:2008.11785 (2020)
8. Moody, D.: The "physics" of notations: toward a scientific basis for constructing visual notations in software engineering. IEEE Trans. Softw. Eng. 35(6), 756–779 (2009)
9. Richards, C., Engelhardt, Y.: The DNA of information design for charts and diagrams. Inf. Des. J. 25(3), 277–292 (2019)
10. Wertheimer, M.: Untersuchungen zur Lehre von der Gestalt. II. Psychologische Forschung 4(1), 301–350 (1923). https://doi.org/10.1007/BF00410640

A Universal Grammar for Specifying Visualization Types

Yuri Engelhardt[1](✉) 📷 and Clive Richards[2] 📷

[1] University of Twente, Enschede, The Netherlands
yuri.engelhardt@utwente.nl
[2] Birmingham City University, Birmingham, UK
clive.j.richards@me.com

Abstract. A 'universal grammar' for the full spectrum of visualization types is discussed. The grammar enables the analysis of any type of visualization regarding its syntactic constituents, such as the types of visual encodings and visual components that are used. Such an analysis of a type of visualization, describing its compositional syntax, can be represented as a specification tree. Colour coded tree branches between constituent types enforce the combination rules visually. We discuss how these specification trees differ from linguistic parse trees, and how visual statements differ from verbal statements. The grammar offers a basis for generating visualization options, and the potential for formalization and for machine-readable specifications. This may serve as a basis for a system providing computer-generated visualization advice.

Keywords: Visualization types · Visual encodings · Syntax of diagrams · Specification trees of visualization types · Visually enforced syntax rules · Comparing visualization types · Comparing verbal and visual statements

1 Purpose of This Work

The grammar presented here is the most recent addition to the 'DNA of visualization', a framework that may help designers to generate visualization options. Descriptions of this framework can be found in Engelhardt and Richards 2018, Richards and Engelhardt 2020, and at VisDNA.com. The framework also offers a tool for research, a basis for formalization, and the potential for computer-based visualization advice. Put another way, by defining the fundamental building blocks of visualization, their interrelationships and the grammar for their combination, as discussed in this paper, the framework provides a method for deconstructing visualizations – which in turn provides a toolkit for exploring design choices. The system may even support combinations of visual encoding possibilities that result in entirely novel visualization types. The 'DNA of visualization' can thus be thought of as a *compositional* taxonomy of visualization, which goes beyond what other taxonomies offer. This work can also be framed as a *pattern language* for

Y. Engelhardt and C. Richards—Both authors contributed equally to the work.

© The Author(s) 2021
A. Basu et al. (Eds.): Diagrams 2021, LNAI 12909, pp. 395–403, 2021.
https://doi.org/10.1007/978-3-030-86062-2_40

visualization, or as a system to describe 'the structure of the information visualization design space' – in the sense of the paper with that title by Card and Mackinlay (1997, IEEE VIS 'Test of Time Award' in 2017).

We use both a linguistic analogy of 'parts of speech' and 'grammar' (see Sects. 2 and 5), and the biological metaphor of 'DNA' and 'species' (see Sect. 3) when discussing types of visualizations, and commonalities and differences between them.

Like academic work in linguistics, the work presented here is primarily not prescriptive but descriptive, in the sense that it facilitates the understanding, modelling and creation of (visual) language.

2　A Grammar for 'Parts of Graphical Speech'

We share with Fred Lakin the view that, "When a person employs text and graphic objects in communication, those objects have meaning under a system of interpretation, or 'visual language'." (Lakin 1987, p. 683). The constituents of visualizations – which include *visual components* as well as *visual encodings* – can be conceived of as "parts of graphical speech", in the sense that Graham Wills is suggesting: "a visualization can be defined by a collection of 'parts of graphical speech', so a well-formed visualization will have a structure, but within that structure you are free to substitute a variety of different items for each part of speech" (Wills 2012, p. 22).

We offer a 'universal grammar' that describes how the constituents of a visualization – Graham Wills' 'parts of graphical speech' – can be combined into a specification tree[1] describing that type of visualization. This grammar is expressed visually, through colour-coded couplings between constituents, enforcing the combination rules in a visual way (see Sect. 6 and Fig. 1). This facilitates the systematic and detailed analysis of commonalities and differences between one type of visualization and another, as well as the exploration of visualization options.

This system can be applied to the full diverse spectrum of different types of visualization. While Wilkinson's 'grammar of graphics' (2005) or the *Vega-Lite* visualization grammar (Satyanarayan et al. 2016) can be used to describe many statistical visualizations, these frameworks are unable to deal with, for example, most non-statistical visualizations. The framework presented here covers statistical visualizations as well as non-statistical visualizations, such as family trees, Venn diagrams, flow charts, texts using indenting, technical drawings and scientific illustrations.

3　'DNA' and 'Species' – A Metaphor for Visualizations

In their 'Tour through the Visualization Zoo', Jeffrey Heer et al. (2010) say that "all visualizations share a common 'DNA' – a set of mappings between data properties and visual attributes such as position, size, shape, and color – and that customized species of visualization might always be constructed by varying these encodings." (ibid. p. 60). We use this metaphorical idea of the 'DNA of visualization' in a similar vein, taking

---

[1] Engelhardt (2006) proposed a grammar-driven analysis of graphics with tree diagrams. The paper included an earlier version of the specification tree shown here in Fig. 1.

it to the extent of identifying a comprehensive set of individual DNA building blocks of visualization, and the rules for combining them. This allows for the construction of a broad range of different types of visualizations – Heer's "customized species of visualization". We will refer to the DNA of visualization as 'VisDNA'. VisDNA building blocks are shown as colour coded three-letter abbreviations, see Fig. 1.

We offer a grammar for combining VisDNA building blocks. The grammar rules are presented in Sect. 5 and at VisDNA.com. We refer to a 'well-formed' combination of VisDNA building blocks, i.e. one that follows the rules, as a **visualization species**. Many common *visualization species* have been given a name (e.g. 'pie chart') and are generally referred to as 'chart types', while novel or rare visualization species often do not have a name (yet). As Heer et al. (2010, p. 67) write, "many more species of visualization exist in the wild, and others await discovery." We have analyzed a large number of *visualization species* using our system, including most of the corpus at datavizproject.com plus many other examples. Examples are shown in Fig. 1 and Fig. 2. Many more examples can be found at VisDNA.com.

4 Visual Encodings: Arranging, Varying, Linking

Visual encodings are at the centre of the VisDNA system (other building blocks include *types of information*, *visual components*, *layout principles* and *directions* – see Richards and Engelhardt 2020, and our accompanying website VisDNA.com). *Visual encodings* can be divided into three subgroups. By applying *visual encodings*, visual components can be spatially *arranged* in order to con·struct visualizations, *var·ied* regarding their visual properties, or *lin·ked* by adding *configurator components*.

- **Arranging**: All the *types of information* (grey DNA) in the VisDNA system can be represented by how visual components are spatially *arranged* into a meaningful configuration. Examples of *arranging* are positional encodings such as *grouping by position*, *positioning on an axis*, *nesting*, or *coupling by adjacency* (red DNA).
- **Varying**: Quantity, order and category membership can be represented by how visual components are visually *var·ied*. Examples of *var·ying* are visual encodings such as *colour coding* or *sizing* (blue DNA).
- **Linking**: Relationships between entities, and in some cases category membership, can be represented by *lin·king* visual components using *configurator components* (such as connector lines or boundaries). These visual encodings are *connecting* and *grouping by boundary* (pink DNA).

A visual component can be involved in *several different* visual encodings, simultaneously representing different types of information.

5 A Visual Grammar for Combining VisDNA Building Blocks

We are proposing a visual grammar for combining VisDNA building blocks, the key rules of which are given in Fig. 1. This grammar includes colour-coded representations of VisDNA building blocks and their couplings. The colour codings enforce the key

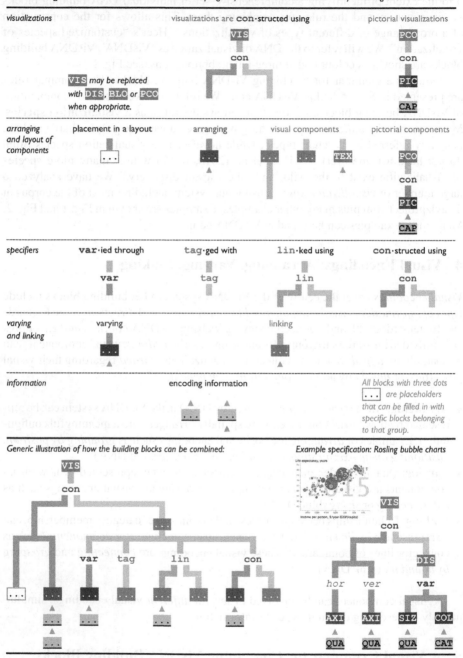

Fig. 1. The visually enforced key grammar rules for combining VisDNA building blocks, and an example visualization with the VisDNA specification tree that defines this visualization species. (Color figure online)

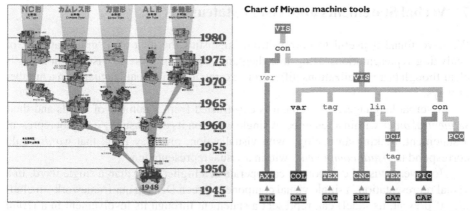

Fig. 2. Example visualization (courtesy of Citizen Machinery Miyano Co. Ltd.) with the VisDNA specification tree that defines this visualization species. (Color figure online)

combination rules in a *visual* way – only couplings of matching colours can connect, forming the branches of a VisDNA specification tree.

Various further rules for combining VisDNA, for example regarding the constraints for connecting visual encodings to types of information, to directions, and to visual components, are given at VisDNA.com.

A VisDNA specification tree specifies a *visualization species*. All individual specimens belonging to the same visualization species can thus be described by the same VisDNA specification tree. A visualization species can be transformed into another visualization species by adding, replacing or removing one or more VisDNA building blocks.

6   VisDNA Specification Trees

VisDNA specification trees are drawn so that they have a layer at the bottom showing the *types of information* (grey DNA) that are represented. The layer directly above that includes the *visual encodings* (red/blue/pink DNA) that are used to represent those types of information, plus any *layout principles* (black-on-white DNA) that may be involved. The remaining layers above the visual encodings show the **specifiers** '*var·ied* through', *tag·*ged[2] with, '*lin·ked* using' and/or '*con·structed* using' (in that left-to-right order). These *specifiers* characterize *visual components* (green DNA) by connecting them to the *visual encodings* in which they are involved, and when needed, to other visual components (either subcomponents, configurator components or tags), integrating any *directions* that may apply. The branches come together in a single node at the top.

---

[2] *Tagging* refers to the identification or annotation of visual components with either text, symbols or embedded visualizations. Tagging is a feature of most visual representations of information.

7   Verbal Statements and Visual Statements

We have found it useful to borrow from linguistics the idea of a 'grammar' and of analyzing representations using tree diagrams, applying these to visualizations. It is clear though that visualizations differ from expressions in verbal languages in a number of ways.

In verbal languages, a statement is constructed from a number of words and their *sequential order* within a sentence. A single word, on its own, usually does not represent a statement. Making an analogy with visualization, one may think that *words* could correspond to *visual components* within a visual representation.

However, a visual component can operate at a higher level than a single word. In a visual representation, a single visual component (green DNA in our framework – including, if present, its label) can represent a statement through its involvement in a *visual encoding*. This may include its position in a meaningful *spatial arrangement* (usually non-sequential – red DNA), its *visual properties* (blue DNA), and any relationship to a *configurator component* (e.g. to a *boundary* or to a *connector line* – pink DNA). For example, a symbol together with its colour gradient may represent the statement 'this measuring station records a *medium* level of pollution'.

A single visual component can even be involved in *several* visual encodings simultaneously (regarding the component's position, its visual properties, and its relationships to configurator components), thus representing *several* statements through a *single* visual component.

Let us consider, for example, a dot in a scatter plot. Such a single dot may represent that 'the UK has an average life expectancy of 81 years and a GDP per capita of 40,000 dollars'. This single dot thus makes one or more statements. To take another example, a single symbol within the intersecting circles of a Venn diagram may represent that 'dolphins are mammals, and not fish'. Every single symbol added within the circles of a Venn diagram thus makes additional separate statements. As a final example, a bar in a bar chart may represent that '2 cm of rain fell on April 1st'. Every single bar added to the bar chart likewise makes an additional separate statement.

In summary, in verbal languages, in order to represent a statement, one can combine words into a sentence. In visualizations, in order to represent one or more statements, one can apply one or more *visual encodings* to a *single* visual component. A visualization usually contains a number of visual components that are following the same visual encoding rules, and thus visualizations represent *sets* of statements – statements that are characterized by the same syntactic structure.

8   VisDNA Specification Trees and Linguistic Parse Trees

For spoken and written languages, linguistic frameworks have been developed that enable the parsing of expressions. Parsing a linguistic expression is based on a grammar of the concerned language, and involves dividing an expression into its (syntactic) constituents and the identification of the (syntactic) relationships between these constituents. The result of parsing an expression can be represented in a parse tree. The development of grammars and parsing are regarded as key accomplishments of the field of linguistics.

Syntactic constituents of a sentence are also referred to as 'parts of speech'. At the beginning of this paper we quoted Graham Wills: "a visualization can be defined by a collection of 'parts of graphical speech', so a well-formed visualization will have a structure, but within that structure you are free to substitute a variety of different items for each part of speech" (Wills 2012, p. 22).

Linguistic parse trees and VisDNA specification trees both identify syntactic constituents of a representation – 'parts of speech' or 'parts of graphical speech' – and the syntactic relationships between those constituents, thus describing the representation's compositional syntax. This is what linguistic parse trees and VisDNA specification trees have in common – in other regards they are quite different.

In linguistics, a parse tree describes the syntactic categories of words and their combination into a *sequential* order in a specific sentence.

In visualization, due to its very nature, syntactic constituents include not only different types of *visual components,* but also *visual encodings* – different types of spatial *arranging,* visual *varying,* and *linking* with configurator components. A VisDNA specification tree describes the syntactic categories of such constituents of a visualization species and their *simultaneous* combination. For example, a symbol's position in a visualization, its size and its colour, all exist simultanously, rather than in a sequential order. If applicable, a VisDNA specication tree also indicates the nesting of smaller visual components within larger ones – which is also non-sequential.

Another difference is that, in linguistic analysis, usually every individual word is featured in a parse tree, while in VisDNA specification trees only sets of constituents are specified. The number of instances in each set (e.g. the number of *visual components,* or the number of colours used in a *colour coding)* is not part of the specification of a *visualization species.*

9   Future Work

This framework is an evolving programme. Because of the flexible structure of the framework, further types of VisDNA building blocks may be added, in order to accommodate any additional visualization species that one may want to describe and that cannot be fully specified using the current scheme. Examples may be the addition of VisDNA building blocks for animation or interactivity in visualizations.

VisDNA specification trees offer a potential research tool for exploring various kinds of commonalities, family resemblances and differences between visualization species within collections of a wide range of visual representations. For example, an application of the VisDNA framework to neural network system diagrams has been described in Marshall et al. (2021, in this volume).

Future work in using this framework to compare visualization species may lead to a better understanding of the structure of the visualization design space. The VisDNA building blocks and the grammar proposed here, also offer the potential for formalization and for machine-readable specifications. This may serve as a basis for a system providing computer-generated visualization advice, which could be linked to an application, such as, for example, a future version of the grammar-based tool *Vega-Lite,* in order to produce actual visualizations and variants of them.

Acknowledgements. The authors would like to thank Guy Clarke Marshall for a number of valuable discussions about the VisDNA framework, and three anonymous reviewers for their thorough and useful comments.

References

Card, S.K., Mackinlay, J.: The structure of the information visualization design space. In: Proceedings of the 1997 IEEE Symposium on Information Visualization (InfoVis 1997), pp. 92–99 (1997). https://doi.org/10.1109/INFVIS.1997.636792

Engelhardt, Y., Richards, C.: A framework for analyzing and designing diagrams and graphics. In: Chapman, P., Stapleton, G., Moktefi, A., Perez-Kriz, S., Bellucci, F. (eds.) Diagrams 2018. LNCS (LNAI), vol. 10871, pp. 201–209. Springer, Cham (2018). https://doi.org/10.1007/978-3-319-91376-6_20

Engelhardt, Y.: Objects and spaces: the visual language of graphics. In: Barker-Plummer, D., Cox, R., Swoboda, N. (eds.) Diagrams 2006. LNCS (LNAI), vol. 4045, pp. 104–108. Springer, Heidelberg (2006). https://doi.org/10.1007/11783183_13

Heer, J., Bostock, M., Ogievetsky, V.: A tour through the visualization zoo. Commun. ACM **53**(6), 59–67 (2010). https://doi.org/10.1145/1743546.1743567

Lakin, F.: Visual grammars for visual languages. In: Proceedings of the sixth National Conference on Artificial Intelligence (AAAI 1987), vol. 2, pp. 683–688 (1987). https://dl.acm.org/doi/10.5555/1856740.1856793

Marshall, G.C., Jay, C., Freitas, A.: Understanding scholarly neural network system diagrams through application of VisDNA. In: Basu, A., et al. (Eds.): Diagrams 2021, LNAI, vol. 12909, pp. 1–17 (2021)

Richards, C., Engelhardt, Y.: The DNA of information design for charts and diagrams. Inf. Des. J. **25**(3), 277–292 (2020). https://doi.org/10.1075/idj.25.3.05ric

Satyanarayan, A., Moritz, D., Wongsuphasawat, K., Heer, J. Vega-Lite: A grammar of interactive graphics. IEEE Trans. Vis. Comput. Graph. **23**(1), 341–350 (2016). https://doi.org/10.1109/TVCG.2016.2599030, https://vega.github.io

Wilkinson, L.: The Grammar of Graphics, 2nd edn. Springer, New York (2005). https://doi.org/10.1007/0-387-28695-0

Wills, G.: Visualizing Time. Springer, New York (2012). https://doi.org/10.1007/978-0-387-779 07-2

Open Access This chapter is licensed under the terms of the Creative Commons Attribution 4.0 International License (http://creativecommons.org/licenses/by/4.0/), which permits use, sharing, adaptation, distribution and reproduction in any medium or format, as long as you give appropriate credit to the original author(s) and the source, provide a link to the Creative Commons license and indicate if changes were made.

The images or other third party material in this chapter are included in the chapter's Creative Commons license, unless indicated otherwise in a credit line to the material. If material is not included in the chapter's Creative Commons license and your intended use is not permitted by statutory regulation or exceeds the permitted use, you will need to obtain permission directly from the copyright holder.

Visualizing Program State as a Clustered Graph for Learning Programming

Oleg Sychev(✉) (ID)

Volgograd State Technical University, Volgograd, Russia
o_sychev@vstu.ru

Abstract. The poster presents a method of visualizing the program state as a diagram of data elements that can both include nested elements and reference other elements. It differs from the known methods because it allows nested objects and arrays; the method also acknowledges high-level data structures like containers and iterators. Layout methods for graphs of this complexity are discussed, and examples of the resulting graphs are provided. The approach was used in computer science courses; the survey showed that it was popular among middle- and low-performing students while high-performing students preferred a professional tool, giving more compact images. Educational uses of the generated diagrams are discussed.

Keywords: Mapping · Visualization · Program state · Education

1 Introduction

Novice programmers often have difficulties with understanding program dynamics as an implementation of fundamental programming concepts. Working with program code, they may have little knowledge of program state during its execution and how their code affects it [5]. Modern programming paradigms like object-oriented programming lead to programs with complex states that are hard to grasp: objects can reference each other; they may contain values, other objects, and arrays. Keeping track of program state during debugging is challenging to novices [4], while regular debuggers are mostly aimed at experienced programmers. This led to the widespread development of program visualization tools [6], representing the program state as a diagram of objects.

Most of these tools have significant limitations. Among widespread object-oriented programming languages, C++ has the most complex program state model because it supports object aggregation by value – i.e. objects can contain other objects and arrays. Most of the other languages allow aggregating only scalar values, while arrays and objects can only be referenced. C++ programmers need to choose between storing objects by value and by reference, which requires an understanding of the data structures it entails. Sorva et al. [6] put references and

The reported study was funded by RFBR, project number 20-07-00764.

© Springer Nature Switzerland AG 2021
A. Basu et al. (Eds.): Diagrams 2021, LNAI 12909, pp. 404–407, 2021.
https://doi.org/10.1007/978-3-030-86062-2_41

pointers at the top of the list of topics about which students develop misconceptions. While some program visualizers like jGRASP [1] and HDPV [7] claim C++ support, they mostly concentrate on visualizing references, ignoring object aggregation. The site of the widely-used visualization tool Jeliot says that supporting C++ remains a project to complete despite attempts by Kirby et al. [3] to create a variation of Jeliot for visualizing C programs. Visualizing aggregated objects is not easy because it makes diagram vertices non-atomic: they can contain other vertices, and edges can connect nested vertices across their parents' boundaries. Data Display Debugger (DDD) [8], originally developed for C, allows node nesting, but it does not show the edges from and to nested vertices correctly, significantly decreasing the value of the resulting diagrams.

An important problem of program state visualization is choosing the layout algorithm. For example, the force-based algorithm implemented in HDPV often causes diagram edges to cross vertices and other edges, making the resulting diagram unreadable. Another problem is parallel edge sections that are often displayed by visualizers like Jeliot. It is too easy to confuse edges in such groups.

2   Method

Given that the resulting diagram should support nested vertices, it can be represented as a clustered graph [2]. The layered graph drawing strategy was chosen because it is well-suited to show common data structures like linked lists and trees. There are two ways to generalize layered graph drawing to clustered graphs. (1) Place all the vertices (including nested vertices) at once, grouping

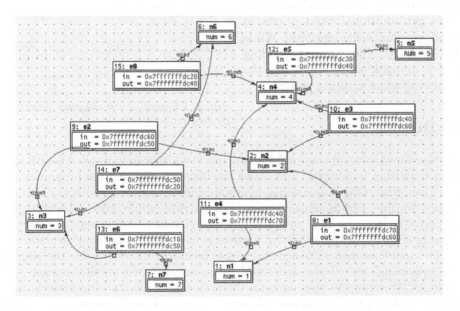

Fig. 1. An example of program state diagram generated by DDD.

them by clusters. This allows creating more compact diagrams but does not allow changing the layout for vertices of different clusters, which is bad for showing complex data like nested arrays and objects. (2) Consider clusters as vertices and apply the algorithm recursively to place nested vertices. This method creates bigger graphs but allows adjusting layouts depending on the kind of cluster.

The recursive approach was chosen because it is better suited to showing data structures (e.g. clusters for one-dimensional arrays can arrange their vertices horizontally, while clusters for objects - vertically). For making the graph acyclic, which is an NP-complete problem, the following heuristic is used: the graph vertices are numbered using depth-first search and the edges that lead from vertices with bigger numbers to vertices with smaller numbers are reversed. This heuristic was chosen after experimenting with different methods because it improves the layout of canonical data structures: doubly linked lists and trees. Vertices within each layer are permuted using the barycentric approach. After determining the coordinates of each vertex and edge, the edges get additional bend points to avoid crossing other vertices. While this increases the sum of bending angles, it significantly decreases the number of edges crossing vertices.

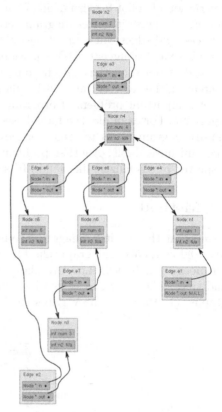

Fig. 2. Program state diagram generated by SeePP.

3   Results

The developed method was implemented in a visualization tool called SeePP that takes data from a debugger and draws the diagram, showing the program state at a given time. One-dimensional arrays are shown as clusters with horizontally aligned vertices, two-dimensional arrays as matrices, while objects use vertically aligned vertices. The tool supports multi-level nesting of objects and arrays, drawing edges from and to nested vertices across cluster borders if necessary.

The tool was used in Volgograd State Technical University for the courses "Programming" and "Data structures" for undergraduate computer science (CS) students. According to the teachers, the majority of the students actively used SeePP to debug their programs. A survey was conducted among 43 volunteer undergraduate CS students, asking them to choose the best and the worst visual representation of program data among diagrams for 6 data structures (tree, single-linked list with an iterator, double-linked list, tree containing lists, hash table, and graphs with edge objects pointing to nodes) produced by DDD (see Fig. 1), SeePP (e.g. Fig. 2) and the prototype of SeePP. The majority of the students chose the diagrams generated by SeePP as the best (mean 28.17, st. dev. 5.91) and DDD as the worst (mean 25.5, st. dev. 4.46). More students answered in favor of SeePP for complex diagrams (trees, hash-table, and graph) than for lists. A small group of 8 the best-performing students consistently chose DDD as the best (except hash table), commenting that compact diagrams are better for them.

4   Conclusion

The developed tool SeePP is capable of generating diagrams of program states including pointers and references, arrays, and nested objects. Edges can link vertices belonging to different clusters; the algorithm avoids drawing close parallel edges that are hard to trace. Most of the students preferred SeePP as a learning tool over the visual debugger DDD. The developed method can also be used as a part of an intelligent tutor for constructing access expressions both to visualize the problem and display feedback that is considered for further work.

References

1. Cross II, J.H., Hendrix, T.D., Barowski, L.A.: Combining dynamic program viewing and testing in early computing courses. In: 2011 IEEE 35th Annual Computer Software and Applications Conference, pp. 184–192. IEEE (2011). https://doi.org/10.1109/COMPSAC.2011.31
2. Eades, P., Feng, Q.W.: Multilevel visualization of clustered graphs. In: North, S. (ed.) Graph Drawing, pp. 101–112. Springer, Heidelberg (1997). https://doi.org/10.1007/3-540-62495-3_41
3. Kirby, M.S., Toland, M.B., Deegan, D.C.: Visualisation tool for teaching programming in c. In: Proceedings of the International Conference on Education, Training and Informatics (ICETI'10) (2009). http://www.iiis.org/CDs2010/CD2010IMC/ICETI_2010/PapersPdf/EB134TP.pdf
4. Shinners-Kennedy, D.: The everydayness of threshold concepts: state as an example from computer science, pp. 119–128. Brill — Sense, Leiden, The Netherlands Jan 01 2008). https://doi.org/10.1163/9789460911477_010
5. Sorva, J.: Notional machines and introductory programming education. ACM Trans. Comput. Educ. 13(2) (2013). https://doi.org/10.1145/2483710.2483713
6. Sorva, J., Karavirta, V., Malmi, L.: A review of generic program visualization systems for introductory programming education. ACM Trans. Comput. Educ. **13**, 1–64 (2013). https://doi.org/10.1145/2490822
7. Sundararaman, J., Back, G.: HDPV: interactive, faithful, in-vivo runtime state visualization for C/C++ and Java. In: Proceedings of the 4th ACM Symposium on Software Visualization, SoftVis 2008, pp. 47–56. Association for Computing Machinery, New York (2008). https://doi.org/10.1145/1409720.1409729
8. Van Hoey, J.: Data display debugger, pp. 51–55. Apress, Berkeley, CA (2019). https://doi.org/10.1007/978-1-4842-5076-1_6

Dynamic Flowcharts for Enhancing Learners' Understanding of the Control Flow During Programming Learning

Mikhail Denisov⊙, Anton Anikin$^{(\boxtimes)}$⊙, and Oleg Sychev⊙

Volgograd State Technical University, Lenin Ave, 28, Volgograd 400005, Russia
anton.anikin@vstu.ru

Abstract. In introductory programming learning, flowcharts are often used to help students comprehend patterns of behavior of control flow statements like alternatives, loops, and switches. The use of flowcharts in assignments along with lectures can increase their learning effect. In some areas, the potential of flowcharts remains largely underestimated. Flowcharts are rare in intelligent learning systems for learning programming, although class, object, and sequence diagrams are quite popular. This poster describes the implementation of a non-editable flowchart widget in our intelligent programming tutor on control flow structures (hereinafter referred to as the exerciser). In normal use, the student chooses the next correct action in the program text (algorithm), i.e., makes one transition through the flowchart. Our exerciser allows making a wrong choice and gives a text hint explaining the reason for incorrectness. The wrong or inexistent transition is also displayed on the flowchart to help localize the error.

Keywords: Flowchart · Introductory programming learning · Intelligent tutoring systems

1 Introduction and Related Work

According to the review of the tools for introductory programming courses [4], typical visualizations include a) UML class diagrams (statically), object (dynamic), and sequence diagrams; b) pointers and references linking instances (dynamic); c) step-by-step expression evaluation; d) call stack and more.

Flowchart is closely related to students' understanding of what a program does and thus is useful in introductory programming courses. For example, the supporting tool called FRIMAN [8] allows a student to create functions or methods as abstract flowcharts instead of writing code in a concrete programming language. Learner-oriented programming environment jGRASP [6] provides the "control structure diagram" feature that shows flowchart-like pictograms in the code; the authors discuss that such markup enhances readability and protects

The reported study was funded by RFBR, project number 20-07-00764.

© Springer Nature Switzerland AG 2021
A. Basu et al. (Eds.): Diagrams 2021, LNAI 12909, pp. 408–411, 2021.
https://doi.org/10.1007/978-3-030-86062-2_42

against the misreading of nested constructs in indent-free languages (as opposed to, e.g., Python). Jsvee (JavaScript Visual Execution Environment) library with Kelmu toolkit [3,9] lets instructors visualize code execution (in particular, using execution stack visualization) and create corresponding program animations for online courses with additional annotations like textual explanations and arrows on top of the animation. These tools are language-independent; they were used in introductory programming courses. The research shows that learners found the animations useful, and annotated animations change the student behavior. Flowchart-based Intelligent Tutoring System (FITS) [5] uses Bayesian networks and automatic text-to-flowchart conversion to support novice programmers in flowchart development and improving their problem-solving skills. It generates appropriate reading sequences and suggests learning goals; the students who use FITS are assisted in navigating the online learning materials. So, it uses flowcharts for learning process management and not for program execution modeling. The tools like code2flow [1], Flowgorithm [2] allow to create of flowcharts based on code (or specific pseudo-code) or generate code based on user-created diagrams for programming learning.

To the best of our knowledge, no work uses flowcharts to illustrate program execution.

2 Method

Visual patterns for typical control structures are simple and straightforward; they use a common notation (see, for example, [7]). To preserve the benefits of recognizable canonical visual patterns of algorithm structure, we used a template-based layout with templates fixing the relative positions of the nested elements of every control-flow statement. Our diagram does not show the names of conditions and simple actions; only compound statements (i.e. loops and alternatives) are named. This increases readability by avoiding overloading the diagram with small graphic elements and encourages the student to look for the necessary details in the code and the trace, which helps to understand the relationship between the diagram and the code and reinforces the mental images of control structures.

3 Results

When using our "How It Works: Algorithms" exerciser[1] intended for improving an understanding of program execution, the student is offered a small teacher-defined algorithm, composed of loops, alternatives, and sequences [10]. The student's goal is to build the execution trace by activating the algorithm elements in the correct order. While this is trivial for sequential sections, transitions related to control conditions (in loops and alternatives) often cause difficulties for novice programmers. An incorrect choice produces a message with a text explanation

---

[1] https://howitworks.app/en/algorithms.

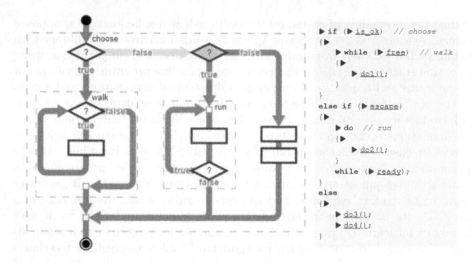

Fig. 1. Complex example of control-flow structures (if-elseif-else, while, do-while): flowchart and algorithm in C language with clickable actions (on the right).

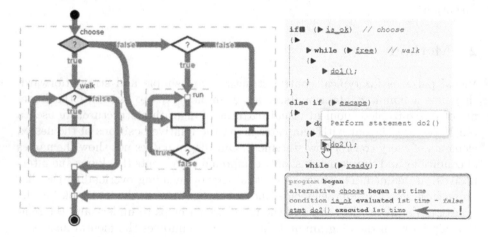

Fig. 2. Example of an error: flowchart shows the incorrect transition (on the left) made by clicking on 'do2()' action in the algorithm (on the right) right after the condition 'is_ok' was evaluated (see the trace under the code).

of the error that helps the student to realize the error cause and fix their mental models. The values of the control conditions at each step are set by the teacher; the student learn them only after choosing to evaluate the condition.

We have extended this exerciser with a dynamic flowchart designed to show all the steps that the student makes. This helps navigating the code and complements error messages. The correct choice highlights the current action and the transition in green (Fig. 1). In case of an error, the red arrow connects the previous action to the selected one (Fig. 2) showing what was done and helps to see how serious the error is (e.g. whether there is a valid path between the ends of the red arrow) and what the correct answer in this position can be. The preloaded example is available at https://howitworks.app/diagram-demo.

4 Conclusion

In this poster, we described our implementation of a dynamically updated flow-chart widget in an intelligent programming exerciser "How It Works: Algorithms" to enhance the understanding of control structures in introductory programming courses, in addition to context-sensitive explanatory messages about incorrect steps. Unlike text messages, the diagram provides a more intuitive way of error explanation. Designing the graphics, we focused on the textbooks-like appearance of the canonical structures like sequences, loops, and alternatives. Further work implies adding support for more types of loops, a specific appearance for the "if-else" case of selection statement, adding animations, and possibly the user customization of the look and feel.

References

1. code2flow - interactive code to flowchart converter. http://www.code2flow.com/
2. Flowgorithm - flowchart programming language. http://www.flowgorithm.org/
3. Learning + technology research group (2021). https://github.com/Aalto-LeTech
4. Earwood, B., Yang, J., Lee, Y.: Impact of static and dynamic visualization in improving object-oriented programming concepts. In: 2016 IEEE Frontiers in Education Conference (FIE), vol. 2016, pp. 1–5 (2016). https://doi.org/10.1109/FIE.2016.7757639
5. Hooshyar, D., Ahmad, R., Yousefi, M., Yusop, F., Horng, S.J.: A flowchart-based intelligent tutoring system for improving problem-solving skills of novice programmers. J. Comput. Assist. Learn. **31**(4), 345–361 (2015). https://doi.org/10.1111/jcal.12099
6. Miller, A., Reges, S., Obourn, A.: jGRASP: a simple, visual, intuitive programming environment for CS1 and CS2. ACM Inroads **8**(4), 53–58 (2017). https://doi.org/10.1145/3148562
7. Prasad, B.: On mapping tasks during product development. Concurrent Eng. Res. Appl. **24**(2), 105–112 (2016). https://doi.org/10.1177/1063293X15625098
8. Sedlacek, P., Kvet, M., Václavková, M.: Development of FRIMAN - supporting tool for object oriented programming teaching. Open Comput. Sci. **11**(1), 90–98 (2021). https://doi.org/10.1515/comp-2020-0117
9. Sirkiä, T.: Jsvee & kelmu: creating and tailoring program animations for computing education. J. Softw. Evol. Process **30**(2), e1924 (2017). https://doi.org/10.1002/smr.1924
10. Sychev, O., Denisov, M., Anikin, A.: Verifying algorithm traces and fault reason determining using ontology reasoning. In: ISWC 2020 Posters, Demos, and Industry Tracks, vol. 2721, pp. 49–53. CEUR-WS (2020). http://ceur-ws.org/Vol-2721/paper495.pdf

Cognitive Analysis

Cognitive Properties of Representations:
A Framework

Peter C.-H. Cheng[1](✉) [iD], Grecia Garcia Garcia[1] [iD], Daniel Raggi[2] [iD],
Aaron Stockdill[2] [iD], and Mateja Jamnik[2] [iD]

[1] University of Sussex, Brighton, UK
{p.c.h.cheng,g.garcia-garcia}@sussex.ac.uk
[2] University of Cambridge, Cambridge, UK
{daniel.raggi,aaron.stockdill,mateja.jamnik}@cl.cam.ac.uk

Abstract. We present a framework for assessing the relative cognitive cost of different representational systems for problem solving. The framework consists of 13 *cognitive properties*. These properties are mapped according to two dimensions: (1) the time scale of the cognitive process, and (2) the granularity of the representational system. The work includes analyses of those processes that are relevant to the internal mental world, and those that are relevant to the external physical display too. The motivation for the construction of this framework is to support the engineering of an automated system that (a) selects representations, (b) that are suited for individual users, (c) and works on specific classes of problems. We present a prototype implementation of such an automated representation selection system, along with an evaluation.

Keywords: Representational systems · Cognitive cost · External and internal representation

1 Introduction

The motivation for (yet) another analysis of the nature of representations stems from our project that is building an automated approach to the *selection* of appropriate representations for solving problems. The motivation and goals of the project are described more fully in [9]. Representation selection must take into account: (a) the type of problem, (b) the specific representational system in which the problem may live, and (c) the users' abilities and familiarity across various representational systems. Expert teachers are able to pick alternative representations to suit each individual students' ability for specific classes of problems; thus, our aim is to design and build a system that can make similar selections. So, what kind of aspects do we need to take into account when building an automated system? In our project, we identified *formal properties* and *cognitive properties* of representational systems. Further, we are developing methods to combine those properties with information about individual users in order to suggest candidate representations for them as well as rank them according to their efficacy for each individual. In this paper, we focus on the cognitive properties. (We describe formal properties in

© Springer Nature Switzerland AG 2021
A. Basu et al. (Eds.): Diagrams 2021, LNAI 12909, pp. 415–430, 2021.
https://doi.org/10.1007/978-3-030-86062-2_43

detail in [18, 19, 23]). Fundamental to our approach is the assessment of the relative *cognitive cost* of alternative representations. Therefore, this requires us to state what the cognitive properties are and to formulate cost measures for them, which will be used in the calculations of an overall cost of a representational system.

Many empirical studies have been conducted on the relative benefit of selected representations for specific tasks, such as [5, 11, 27]. However, it is unclear how these findings can be applied to the assessment of the cognitive costs of representations in general: they only address particular isolated factors. In contrast, we aim to address the following research questions:

1. What are cognitive properties and where do they come from?
2. How should cognitive properties' relative importance be assessed in the context of their multitude and diversity?
3. How can cost measures of the properties be combined to give the relative order of the effectiveness of representations?

Our aim in this paper is to provide the foundational framework from which to address these questions. To be clear, we are not pursuing a general psychological theory of representational systems, but aim to engineer a system to reason about representations; in other words, we want to explore how to give computers the ability to select effective representations for humans. Give the scope of this goal, it is not possible to cover all relevant areas of the literature within this paper, so we have necessarily been selective.

The framework is presented in the next section. This is followed by the presentation of three sample solutions to one problem in three different representational systems. The five sections that then follow describe classes of cognitive properties identified by the framework. We then present an example on how the framework has been used in a prototype of an automated system for representation selection. The final discussion section reflects on the scope and limitations of our framework.

2 Analysis Framework

We use these abbreviations: R – representation; RS – representational system[1]; ER – external representation[2]; IR – internal (mental) representation; CP – cognitive property.

A *cognitive property* is a feature of a representational system that influences how information is processed, and is thus likely to affect the cognitive cost of using the representation (e.g., the number of symbols in a R can affect its cognitive cost).

By *cognitive cost*, we mean the cognitive load that a user experiences using a representational system. This might be measured empirically in terms of: the time taken to complete a problem; the number of operations or procedures used; a rating of the moment to moment subjective effort that the user perceived; the amount of unproductive

---

[1] Following [18], a *representational system* is an abstract entity from which many distinct individual *representations* may be created.

[2] Following [27], ERs are information and objects that exist in the external environment and can be perceived; while IRs are knowledge and structures in memory (p. 180).

effort due to errors or the pursuit of unproductive solution paths. At the level of cognitive processes, some of the factors that are known to underpin cognitive cost include (e.g., [4, 17, 20]): instantaneous working memory load; less accessible information; operators that take more effort to select or to apply; reduced ability to anticipate the consequence of applying operators; the possession of poor problem-solving heuristics; the lack of externalised memory or free-ride inferences [20]. In contrast to Sweller's [24] notation of cognitive load, our approach broadens the idea to wider temporal and granularity scales, rather than whole instructional tasks, but narrows the focus specifically on representational systems, rather than instructional interventions in general.

Table 1. Cognitive properties framework.

		Notation granularity		
		Symbol	Expression	Representational system
Cognitive type or level	0. General			Sub-RS-variety
	1. Registration	Registration-process Number-of-symbols/expressions Variety-of-symbol/expressions		–
	2. Semantic encoding	Concept-mapping ER-semantic-process IR-semantic-process		–
	3. Inference	Quantity-scale	Expression-complexity Inference-type	–
	4. Problem solution	–	–	Solution-depth Solution-branching-factor Solution-technique

A framework for cognitive properties has stringent requirements. First, it should systematically identify cognitive properties without neglecting important high impact properties. Second, the CPs should directly relate to established cognitive phenomena and accepted theoretical cognitive constructs associated with representational systems (e.g., [13]). Third, it should identify unique CPs that overlap minimally in scope.

So, to define the framework, three distinct primary cognitive dimensions have been adopted, guided by insights from [1, 16, 21]. The space is represented in Table 1. The dimensions are:

(1) The *granularity of components* of the ER: column headings in Table 1.
(2) The *type and temporal level* of cognitive processing: row headings in Table 1.
(3) Whether the component or the process is primarily *associated with the ER or IR*: see the names of some CPs in the cells of Table 1.

The framework embodies the idea that, as CPs are manifestations of interactions between cognitive and representational systems, both are conceptualised as nearly-decomposable hierarchical systems [21] that function over large ranges of spatial and temporal scales [1, 16] and are distributed between the IR and ER.

Granularity of Components. This is a dimension ranging across the size of cognitive objects that encode meaning. The *Symbol*[3] level is for elementary, non-decomposable carriers of concepts. Expressions are assemblies of elementary symbols, which occur at different hierarchical levels. The *Representational System* level concerns the complete notational system that is used in a particular case (the representation) for problem solving, which may include distinct sub-representational systems (*sub-RSs*).

Type and Temporal Level of Cognitive Processing. This dimension has two parts. The first part is composed of four temporal levels at which cognitive processes operate, ranging from 100 ms to tens of minutes (e.g., from the time to retrieve a fact from memory, to the time to develop a problem solution). These levels are: (1) *registration*, (2) *semantic encoding*, (3) *inference*, and (4) *problem solution.* Registration refers to the process of acknowledging the existence and location of objects. The encoding level considers the cost of associating symbols with concepts. The inference level considers the cost of the arguments and difficulty of inferences. The problem solution level captures the complexity of the problem state and goal structure. Relatively strong interactions occur between processes at a particular time scale, and relatively weak interactions between different time scales [1, 16]. So, for the sake of analysis, cognitive processes at scales, differing by an order of magnitude, may be treated as nearly independent. Nevertheless, short processes impact long processes cumulatively.

The second part of this dimension is a further level zero, *general*, in Table 1, which accommodates a CP that is not covered by the four temporal levels, but it is a feature that affects how information is processed too.

Association with the ER or IR. This third dimension is recognised because the nature of some processes that serve the same cognitive function may actually differ substantially between IR and ER, and so, they need to be explained in terms of different CPs.

The CP framework builds upon the taxonomy of characteristics of effective RSs compiled by [4], but diverges from that work by providing an underpinning cognitively motivated theoretical justification for the framework's structure. CPs are included in the framework on the basis that a theoretical argument can be made that the CP impacts the cost of using a representation. Inclusion makes no claim that a simple measure of cognitive cost or practical means to compute the cost is necessarily available; this issue is discussed below. As will be noted, some of our proposals need additional empirical support. Before considering the CPs named in Table 1, we present the solutions to a problem in alternative representations to provide running examples.

3 Sample Representations and Problem

We selected probability problems as one target domain for our project because they are knowledge-rich and can be solved using a large variety of alternative representations.

---

[3] Across disciplines, different terminology is used for *symbols* and *expressions*. From a computational perspective, [18] refers to *primitives* instead of *symbols*, and *composites* instead of *expressions*. These differences partially rise from different perspectives on what is understood by a basic/elementary unit, whether it is considered decomposable or not. As this paper focuses on cognitive aspects of RSs, we have adopted cognitive-oriented terminology.

Probability tests are a good exemplar: they are an important class of problems that have wide application in many disciplines, but are known to be challenging for problem solvers and learners. Consider this medical problem:

1% of the population has a disease D. There is a test, T, such that: (i) if you have the disease, the chance that T comes out positive is 98%; (ii) if you don't have the disease, the chance that T comes out positive is 3%. Suppose Alex takes the test and it comes out positive. What's the probability that Alex has the disease?

Figure 1 shows ideal solutions in two conventional representations: algebraic *Bayesian representation* and *contingency table*; a third representation uses *Probability Space (PS) diagrams* [3]. PS diagrams exemplify how our framework can be applied to novel representations for which analysts do not have established intuitions. Also, PS diagrams provide an interesting test case as they integrate information about sets and probability relations using a coherent diagrammatic scheme that has been shown to substantially enhance problem solving and learning with little instruction [3]. The green text in Fig. 1 shows values given in the problem statement shown above.

The problem is a fairly canonical test situation, but has a complication. The test is not an independent trial, but depends on whether the disease is actually present or not. Thus, the five-line Bayesian solution (Fig. 1a) employs steps that are beyond school level probability: (1) Bayes' theorem; (2) law of total probability applied to the denominator; (3) De Finetti's axiom of conditional probability. Clearly, this solution requires a high degree of mathematical sophistication.

The contingency table solution (Fig. 1b) assumes that the user knows the arithmetic rules governing continency tables; the formulas in smaller letters at the bottom right of the cells. The solution progresses by successively entering given values of the problem statement into the cells, taking into account the arithmetic constrains. It is completed by selecting the values from the cells that correspond to the target condition probability and calculating the answer, as captured by the line below the table. Since the user must be proficient at using contingency tables, they should be able to handle the impact of lack of independence of the test and to complete only germane cells.

Students, who do not have mathematical instruction beyond 16 years of age, can solve the medical problem by drawing a diagram like Fig. 1c, after just two hours of instruction on PS diagrams [3]. A typical solution using PS diagrams might proceed by sketching the sub-diagram for a binary outcome trial first: this is the horizontal line *D* in the diagram, which consists of the slightly misaligned 'no' and 'yes' subsegments. Then, two more sub-diagrams are drawn within line *T* (below line *D*); each one covers the two test outcomes of each state of *D*. For example, the left sub-diagram of *T* (consisting of two slightly misaligned segments on the left) covers the test outcome when the person '*does not have the disease*' (since it is under the 'no' sub-segment of *D*) and it shows a sub-segment for when '*the chance that T comes out positive is 3%*,' which is labelled with the '+' sign and the '0.03' value; thus, the sub-segment labelled with the sign '−' and value '0.97' represents the chance that T comes out negative. With kinder numbers, the diagram could be drawn to scale, nevertheless, the (green) numbers record the information given in the problem statement. Knowing the probability of each space (or sub-space), proceed to review the full diagram vertically. As required, we focus on the positive (+) outcomes of the test (the two middle segments labelled '+' within T), which gives us a conditional sub-space that is represented with the horizontal line '*Ans.*'. Using one of the basic rules of PS diagrams, we can calculate the probability of the outcomes in that sub-space, by multiplying the values of the no_D and yes_T outcomes ($0.99 * 0.03 = 0.0297$), and the values of yes_D and yes_T outcomes ($0.01 * 0.98 = 0.0098$). Now, the probability of "*Alex has the disease*" is given by the portion of the conditional space that is yes_D within line '*Ans.*' (thicker sub-segment) which by an approximate mental calculation, is about a quarter.

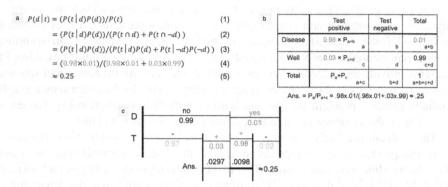

Fig. 1. (a) Bayesian representation, (b) contingency table, and (c) probability space diagram solutions to the medical problem. (Color figure online)

The comparison of these examples will informally support the claims below about cognitive cost of different CPs.

4   General Cognitive Property – Sub-RS-Variety

Much of the literature on representational systems has typically focused on RSs with a single format and made comparisons between such unitary RSs. However, all but the simplest RSs are heterogeneous mixtures comprised of *sub-RSs*. Thus, the **sub-RS-variety** is a CP, because sub-RSs are systems which must work in a coordinated fashion. This entails matching information between the sub-RSs or translating information from the format of one into another. Impacts of multiple sub-RSs include, for instance: increased frequency of attention switches between sub-RSs, with all of the attendant delays in reactivating propositions associated with each sub-RSs; greater number of inference rules to handle; more opportunity for potential errors. Thus, high heterogeneity of sub-RSs incurs a heavy cognitive cost [25].

Obviously, an RS is heterogeneous when it is composed of sub-RSs that would be independently considered as RSs in their own right. For example, in Fig. 1a, the Bayesian notation operates on the quantities of probability, $P(\ldots)$, separately from the set theory notation embedded within the parentheses. More formally, sub-RSs may be distinguished in four related ways. (1) A part of the RS is governed by an exclusive set of syntactic rules, likely applied to distinct operator symbols (i.e., in [12]'s terms, it possesses a different *format* compared to the rest of the RS). (2) A part of the RS encodes a distinct set of domain concepts, so it may be a separate sub-RS: in the contingency table representation, rows and columns encode relations among sets, whereas the cell entries are formulas involving magnitudes of probabilities. (3) An RS has an indexing system that serves to coordinate between sub-RSs, but that does not directly encode domain concepts: for example, the cell labels and subscripts within the contingency table. (4) A part of the RS is a sub-RS and is spatially remote from the RS: for instance, in Fig. 1b the equation below the contingency table.

Numeration systems are in themselves RSs [28], so any RS that includes numbers has at least two sub-RSs. This is the case in our three representations in Fig. 1. However, numbers may be set aside in the count of sub-RSs because every one of our representations uses them in a similar fashion. So, the *differential* cost of their presence across the three representations will be small compared to other CPs.

The Bayesian and contingency table representations are likely to have a similar cognitive cost in terms of the number of sub-RS-variety CP. In contrast, the PS diagram does not meet many of the criteria for the existence of other sub-RSs; in fact, it may be a special case of a representation without instances of other sub-RSs, and thus, its cognitive cost is predicted to be less than the cost of the other two representations.

5 Registration Cognitive Properties

Registration is the first of the four main temporal levels of cognitive process in the framework. An RS has a vast number of possible features that might serve as symbols because any part of a feature of a graphical element could be selected arbitrarily, such as the 'l' or the '–' in a '+', or even their point of intersection. Registration process establishes what particular objects, features, or groups of objects are taken to be a potential symbol (or expression), by acknowledging their existence and noting their location in the representation.

Registration occurs when we seek a symbol in the ER to match a concept (in the IR). Alternatively, we may examine an ER to find symbols in at least two ways. (1) We may use our knowledge about the RS. For example, the answer to a problem, in a problem solution, is likely to be found at the bottom of the solution – as in Fig. 1a. (2) If we are not familiar with an RS, then those features that vary with the RS are potential symbols or expressions, but constant features are not. For instance, the size of the font in the Bayesian example in Fig. 1 is fixed, so it is not meaningful, but it would be if the formulas included subscripts (as in Fig. 1b).

The **registration-process** CP concerns the various types of cognitive processes that are used to register symbols or expressions. The purpose of this CPs is to specify the relative cost arising from those processes. The processes, in order of increasing cost, are: (a) *iconic*, (b) *emergent*, (c) *spatial-index*, (d) *notational-index*, and (e) *search*. (a) The iconic registration process rapidly focuses attention upon 1 object or 1 group that is highly recognisable to the user due to its familiarity. For example, following instruction, students familiar with PS diagrams will perceive the main space (*D* and *T* lines) in Fig. 1c as a single object; or the symbol '≈' in Fig. 1a can be rapidly recognised given its location and shape. (b) Emergent registration processes occur when a group of symbols are arranged so that they form a perceptual Gestalt (e.g., continuity, closure). For example, the numbers in parentheses in Fig. 1a, which are not part of the solution, but can be used to refer to the different algebraic statements. (c) Spatially-indexed registration processes exploit the spatial organisation in the RS, as described by [12]. (d) Notational-index registration processes exploit some alphanumeric system to organise or index objects, such as the reference letters in the contingency table of Fig. 1b. (e) Lastly, the registration process may default to mere search, perhaps using heuristics or just exhaustively, when the other processes are unavailable (e.g., find 'tl¬d' in Fig. 1a). Although we consider our proposed order for these processes to be sensible, further empirical evidence is needed to confirm this order.

The other pair of CPs at the registration level address (a) the **number-of-symbols** or **expressions** and (b) the **variety-of-symbols** or **expressions**. An elementary *symbol* is a non-decomposable carrier (representation) of a concept. For example, in our three sample representations, symbols include: variables and mathematical operators, table cells, and labelled line segments, respectively. The notion of *symbols* also encompasses graphical properties of ER tokens that in themselves may encode particular concepts; for example, the thickness of a line segment in the PS diagram denoting the solution. *Expressions* are assemblies of elementary symbols, which occur at different hierarchical levels; such as algebra formulas or their parts, rows and columns of the contingency table, or the horizontal lines for a particular trial in the PS diagram. In some circumstances we may treat expressions as single objects; e.g., dividing throughout by one side of an equation to obtain a form equal to unity. So just as the number of symbols will impact the cost of using a representation, so will the number of expressions.

It is unlikely that the cognitive cost of the number-of-symbols CP will be a simple linear function of the number-of-symbols, because of the propensity of the mind to chunk information [14]. The same is likely to be true for number-of-expressions, as chunking is a hierarchical process [21]. In the Bayesian representation, the number of symbols including 'P(…)' is 14. However, the cognitive cost is more likely to be a count of the variety-of-symbols/expressions, as chunking does not operate directly on categories. For the contingency table representation, the varieties (types) include the table cells, variable names, and numbers.

6   Semantic Encoding Cognitive Properties

This set of CPs considers the cost of associating symbols and expressions with concepts, that is, the establishment of meaning (not just mere existence and location as in the registration level). Two aspects are considered. One addresses the relation between concepts and things encoding them in a representation, and the other concerns the cognitive processes.

The first CP of the first aspect is **concept-mapping**, which applies both to symbols and expressions. This CP draws upon the literature on the nature of possible matches between symbols (tokens) and expressions in the ER and concepts in the IR [7, 15]. There are five ways in which matches may occur, which are described next in likely order of cognitive cost. As our focus is cognitive, we propose a slightly different ranking to [15]. (1) *Isomorphic*: Matching occurs when each concept precisely matches one symbol; this entails the lowest cognitive cost. (2) *Symbol-excess*: It occurs when some symbols do not represent any domain concept, they only add noise to the representation. Normally, when a user is familiar with the representation, such noise (junk) symbols can be ignored without undue effort. (3) *Symbol-redundancy*: It occurs when one concept maps to many symbols. For example, as in the Bayesian representation in Fig. 1a, the symbol '*d*' appears several times. In terms of cost, some effort is required to handle this, but since we are naturally able to deal with duplicated symbols and synonyms, the cost may not be too high. (4) *Symbol-deficit*: The cost increases in this case because there is no symbol for a concept, so the benefits of externalising memory are not available. Thus, effort must be expended to place a mental pointer to where its symbol would have appeared in the ER. (5) *Symbol-overload*: This is the worst kind of match. It occurs when multiple concepts map to one symbol. This has the grave potential of propagating error due to confusion. To avoid such errors, laborious inferences exploiting contextual information must be executed to mitigate such ambiguities. The contingency table and the PS diagram are largely isomorphic, in part because the numerical contents of cells of the *Test negative* column have been omitted from the table and the negative test values have been greyed out in the PS diagram, specifically to reduce symbol-excess for the medical problem. Finally, regarding the proposed order for these processes, we are currently working on supporting these claims with empirical evidence.

The next pair of CPs deal with cognitive processing costs. The **ER-semantic-process**, which applies both to symbols and expressions, refers to five cognitively different types of processes that associate symbols or expressions in the ER to concepts in the IR; these are listed here in our proposed rank order of cost. (1) The easiest, *known-association* encoding, depends on the familiarity of the user with the RS (e.g., people are typically familiar with numbers, such as the numbers in Fig. 1). (2) *Visual-properties* can be used to represent quantities. This generally has a low cognitive cost, but there are variations among properties that may increase the cost, such as position, length or angle for instance [5]. (3) The *linear-order* in one spatial dimension can readily encode information. For example, temporal sequencing of events D and T in the PS diagram, or placing the result of a computation to the right side (instead of the left) of an equal sign in a Bayesian solution (Fig. 1a). (4) Encoding the meaning of a symbol due to its *spatial-arrangement* in 2D is more challenging and uses devices such as: coordinate systems or arrays (e.g., the contingency table), hierarchical assemblies (e.g., the PS diagram), or networks (e.g., trees or lattices). (5) The costliest encoding is for *arbitrary* unstructured list of collections.

IR-semantic-process is the other in the pair of CPs and applies to symbols and expressions. We identify five processes within this CP, which are presented in our proposed rank order of cost (c.f., [13]). (1) The lowest are known *cases*, or prototypes, such as our understanding of the general format of a contingency table. (2) More complex and costly are *schemas*, whose slots and fillers require more processing (e.g., PS diagrams are diagrammatic configuration schemes [10]). (3) IRs based on *rules* are next, which are more costly because they have fewer constraints, so effort must be expended just to identify categories and track concepts. (4) *Mental-imagery* is more costly still, because the imagery system's limited functionality and resolution will tend to demand multiple iterations of procedures [6]. (5) *Propositional-networks*, such as analogies, are the costliest because they are largely built on simple associations, which place little constraint on valid inferences. The form of a given RS may suggest what IR a user will likely adopt (e.g., for Fig. 1a: rules; for Fig. 1b: schema; for Fig. 1c: diagrammatic schema). So, the ordering provided by these processes provides means to estimate the relative cost of the CP.

Note that the order of our proposed processes for ER- and IR-semantic-process CPs, although sensible, is something that needs to be demonstrated empirically too.

7   Inference Cognitive Properties

This penultimate group of CPs concerns costs at the level of making inferences. One of the properties in this group is **quantity-scale**, which concerns the type of quantity or measurement scale that dominates an RS, specifically, *nominal, ordinal, interval* or *ratio* [22]. Zhang [26] considered the role of quantity scales in the design of representational systems, and the scale hierarchy is well documented [29]. Here, we claim, further, that as the more sophisticated scales have more information content, they will impose greater cognitive cost. However, it is unlikely that RSs will differ in their use of quantity scales, because this is substantially determined by the content of the problem. For example, all three of our examples in Fig. 1 involve quantities related to nominal (manipulation of sets) and ratio (manipulation of probability quantities) scales. Rather, this CP is included because users' degree of experience in reasoning with more sophisticated scales is likely to have cost implications. For this CP, we are currently conducting empirical studies about the relative costs of the scales.

The next CP in the inference group is **expression-complexity**. Obviously, the longer an expression, the more components it possesses or the more tortuous it is, the greater the costs of using it to generate new information. For instance, it is easier to understand how each part of a PS diagram constrains the size of other parts than it is to work out how the magnitudes of variables vary in relation to each other in the Bayesian representation. Expression-complexity may be decomposed into particular factors such as the depth of relations and the arity of relations. The former is the number of levels of nesting of relations. The latter is the number of arguments that relations take. The more arguments, the more information must be handled, so the greater the cost [8]. For instance, the calculation of the final answer in the Bayesian and the contingency table solutions take six numbers, whereas only two are used in the PS diagram solution.

Not all inferences have the same difficulty, so the **inference-type** CP considers various types, for which we propose this rank ordering cost: (1) *symbol-selection* (e.g., lookup a table cell entry); (2) *assign/substitute* a symbol or concept (let the top-left sub-segment line in the PS diagram in Fig. 1c stand for no_*D*); (3) *compare/match* symbols or concepts; (4) *select-expression*; (5) *substitute-expression*; (6) *calculate*; and (7) *transform-expression*, which re-arranges the structure, resulting in a new relation (e.g., writing a new line in Fig. 1a; drawing a new sub-space in Fig. 1c). The Bayesian representation in Fig. 1a is dominated by the costliest of the 7 inference-types (e.g., transform-expression), but not so for Fig. 1b and 1c. Again, some empirical evidence will be needed to support our proposed order of processes for this CP.

8 Problem Solution Cognitive Properties

To capture the impact at the overall level of problem solutions, three CPs are proposed [17]. The first two are **solution-depth** and **solution-branching-factor**, which consider the overall topology of the hierarchical problem state space that users of a representation generate when solving problems. Solution-depth is the number of steps on the most direct path between the initial state and solution. The solutions to the medical problem in Fig. 1 are ideal solutions, with no back-tracking nor branching, so the number of operations that generate the solutions is also the solution depth. The solution-branching-factor addresses the likely width of the problem space experienced by a problem solver. For example, the branching factor from step 1 to 2 in Fig. 1a is higher than in Fig. 1c: a problem solver using a Bayesian representation may need to consider several theorems to move from step 1 to 2; while a problem solver using the PS diagram just needs to draw the different events for each of those steps. A problem state space given by an RS offers the problem solver alternative paths to follow and it will increase costs in at least two ways. First, it is the simple challenge of choosing which path to follow; and second, many alternative paths may lead to impasses rather than solutions. Clearly, the heuristics possessed by a problem solver will influence the solution-depth and the solution-branching-factor.

The **solution-technique** CP considers problem solution approaches that depend on the nature of the problem, which are distinct from general heuristics, and focuses on the nature of the procedures that are used for solutions. Two problem solutions might have the same breadth and depth but may vary in the variety of operators that are used to generate expressions. For example, a solution in a PS diagram typically involves iterative applications of finding a subspace in the diagram and drawing further subdivisions of them, whereas algebraic solutions invoke a larger range of operations that vary with the changing structure of the expressions [3]. As teachers of programming know, iterative processes are typically easier to grasp and to implement than recursive processes. Hierarchical processes also tend to be more complex than iterative processes, because they require nested sub-procedures and the management of sub-goals.

9 Example of Application

One can envisage many uses for the CP framework [9]. It may serve as a checklist of factors that instructors might consider when they develop a curriculum in order to determine the order in which to introduce different representations. More ambitiously, we are using the framework to develop an AI engine that will automatically select representations that are suited to particular problems and users with different levels of familiarity of a target pool of representations. This section of the paper summarises the role of the CPs framework in the development of our first prototype of a representation selection system called rep2rep [18] as a concrete illustration of the framework's utility. In [18], the main focus is on the formal properties and the application of our framework, whereas the underpinning cognitive rationale is the main contribution of this paper.

The general challenge is to develop computational mechanisms that formalise the CPs described by the framework in such a way that their associated cognitive costs can be accurately calculated – to enable the selection of effective RSs for problem solving.

In order to meet this challenge, we need to cover two levels of abstraction. At the lower level we have questions such as 'how do we count the number of symbols in a representation?' and 'what is the expected cost of reading any of the symbols in Fig. 1a?' – which requires a prediction of how the physical components would be chunked into discrete symbols and how much time and effort it would take. And at the higher level, we have questions such as 'how does the number of symbols affect the cost?'. For our computational formalisation, we *assume-as-given* some answers to the lower-level type of questions. We only address computationally the higher level. To be clear, this does not mean that we *have* concrete answers to the lower level-type of questions. It only means that to turn our implementation into a full computational formalisation of the framework we need to *plug in* mechanisms that yield the lower-level values.

Computationally, we encode representations abstractly as collections of *primitive terms*, *patterns*, *laws*, and *tactics*. We call these the *formal components* of a representation. Terms (or symbols) are assigned *types*, and patterns capture the idea that higher-granularity items (*composite terms*) in a representation are formed from lower-granularity items, all the way down to the primitive terms. Specifically, a pattern describes the structure of composite terms (of a certain type) which are made up from more basic terms of certain types. This abstraction – of patterns as the glue of composite terms – can capture the complexity of various grammars: from natural language, to formal mathematics, to graph-theoretic or geometric diagrams [18, 19]. Analogous to the way in which patterns describe the structure of composite terms from more basic terms, tactics encode the structure of inferences from more basic knowledge, all the way down to laws[4]. Moreover, the links between different representations (e.g., how the same problem is encoded in multiple RSs) is captured by the concept of *correspondence*. Lastly, the user's general *expertise* is captured simply as a value between 0 (novice) and 1 (expert).

Given the abstraction of representations into their formal components, the question now is how the CP framework is applied. For the work in [18], we formalised a version of each of: sub-RS variety, registration (of primitives and composite terms), concept-mapping, quantity-scale, expression-complexity, inference-type, solution-depth, and solution-branching-factor[5]. As stated above, the formalisation of these properties relies on some low-level assumed-as-givens. These take either of the following forms:

1. Given a problem-solution representation, its abstraction into formal components is assumed. This means, for instance, that the question of which terms are considered primitive (in practice, a question of chunking) must be given. Furthermore, a value of *importance* is assigned to each component, encoding its relevance with respect to the solution (e.g., a component that plays no role in the solution is considered unimportant and given a value of 0).

2. The assignment of cognitive attributes to components is assumed. This means, for instance, that whether a tactic is assigned the attribute of being a substitution or a calculation (see Sect. 7), must be given. Furthermore, the parameter values for basic

---

[4] In formal, sentential mathematics these would be called axioms, but we do not want to give the impression that either (i) our system only applies to axiomatic systems or that (ii) laws have to be as low level as axioms typically are.

[5] Other CPs, e.g., IR & ER-semantic-process and solution-technique, are yet to be implemented.

costs, associated with these attributes, are assumed. This means, for instance, that the cost of a single inference which is a calculation is assumed to be twice as costly as that of a simple substitution. Lacking specific and accurate empirical data, ratios such as this one were chosen arbitrarily with the simple constraint that they must preserve the rank order specified by the framework.

Given these low-level assumed values, we assign a cognitive cost for each CP using a variety of methods. For example, registration and inference-type costs are similarly computed as a sum of the basic parameter values for the given components modulated by importance and expertise (expertise is assumed to reduce the impact of noisy components, as these can be ignored). Expression-complexity and solution-branching-factor, on the other hand, are computed from the branchiness and nestiness of patterns and tactics, respectively, with a similar effect from expertise. Quantity-scales is computed via the correspondences of components to arithmetic operators, and concept-mapping is computed via the type of relation given by the correspondence map to a fixed representation. Sub-RS-variety is simply computed from the number of *modes* (a given) which are intended to capture individual formats used in the representation.

Once the cognitive cost associated to each CP is computed, they are combined in a weighted sum, with CPs in higher cognitive level and higher notation granularity being assigned greater weights. Moreover, expertise is assumed to have a stronger impact on the cost of CPs of higher notation granularity components.

Our prototype engine for representation selection can also be used to produce an informational suitability score, which estimates the likelihood that a given RS can be used to represent and solve a problem. An interesting question for future research is how the informational and cognitive computations can be used synergistically. It is clear that it depends on the application in which our framework is employed. Precise formulae for informational suitability and cognitive costs, and details of their implementation can be found in [18].

9.1 Evaluation

In [18], we presented an evaluation of the effectiveness of the implementation, which is summarised here. Since there are no other systems to compare against, the evaluation was done by comparing computed measures of informational suitability[6] (IS) and cognitive cost against data obtained from surveying expert analysts. That is, was our system producing similar rankings as expert humans? The evaluation focused on the domain of probability and the medical problem presented in Sect. 3, albeit using different values. The RS used were Natural Language (NL), Bayes, Areas, and Contingency Table. The computation of IS was done as stated at the start of this section, and the cognitive cost function was computed considering 3 user profiles, which were set through the general expertise function described above.

Eleven analysts with strong mathematical background completed an online questionnaire, which contained 2 tasks. In Task 1, participants were first shown the description

---

[6] Information suitability measures how well a representation encodes the informational content of a problem and is computed using the formal properties of representations.

of the medical problem. Then they were asked to give feedback on how informationally sufficient (descriptions of) RSs were using a 7-point Likert scale. In Task 2, participants were asked to rank the same RS descriptions, but for novice, expert and average users.

The mean Likert score given to different RSs in Task 1 was used to derive IS ranking, and the mean of the rank scores across different RS was used to derive the ranking of different RSs for different user profiles. In terms of IS, the rank order produced from the rep2rep system and the analysts was similar for the most and least IS RSs (Bayes and NL, respectively), but different on the Areas and Contingency Table RS. Although the correlation was not significant, it was considered that the overall ranking produced by the system was sensible. In terms of cognitive costs, the rankings given by the analysts and the rep2rep system for the expert and average profiles showed high and statistically significant correlations at $p < .05$ ($r = 0.9$), but not for the novice profile. A possible explanation of this result is that users' familiarity with the RS is not yet modelled in the system. Details can be found in [18].

Overall, the results are promising in terms of the AI system being able to recommend effective representations – although more empirical work still needs to be done.

10   Discussion

To identify cognitive properties that contribute to the cognitive cost of an RS, we formulated the analysis framework, as summarised in Table 1. We proposed 13 diverse CPs. Some relate cost to counts of instances found, some require the calculation of an average to represent some commonly occurring factor, and others propose ranking of processes as guides to relative cost. Although 13 CPs are postulated, we make no claim that they are exhaustive, and note that some are applicable at multiple levels of granularity of RSs. A key feature and potential benefit of the framework is its differentiation of CPs within a two-dimensional space of cognitive level and notation granularity. Given a particular problem-solving process, one can use the dimension to locate its position within the space and, hence, the CPs that are likely to be important factors that impact the cost of the process in different representations. Nonetheless, CPs are not perfectly orthogonal. For example, the number-of-symbols will likely increase with the number of sub-RSs. However, the distinction between these RSs is important, not just because they span very different ranges in the framework, but because we can imagine a situation where one RS A is comprised of two sub-RSs, and a second RS B without sub-RSs has an equal number-of-symbols. In that case, the RS A will have a higher cost because of the challenges related to multiple sub-RSs.

Whilst more extensive justification and rigorous definition could be made about the values of CPs and the rank order of the costs of particular CPs, we consider that the given notions and orders are reasonable.

Note that the three example representations in Fig. 1 encode equivalent sets of concepts. If this were not the case, then fair comparisons could not be made [2]. However, the framework does permit comparisons where the ERs of two RSs are not equivalent, as long as any difference is remedied in the IR content of the RS in deficit.

As part of our ongoing work with the framework, we are investigating how to combine the CPs into a single cost measure for whole RSs. Three critical issues will need to be addressed.

1. How can the disparate measures, with their different scales, be normalised so that they can be reasonably combined?
2. What weighting should be given to those normalised CPs, as they naturally have different levels of impact?
3. How should the weights of each CP be moderated given differences in individual's expertise with alternative RSs?

Our first prototype representation selection engine rep2rep, described in Sect. 9, provides one tentative solution to the first two issues, at least for selected CPs. More broadly and fortunately, the framework supports our analyses of the questions, because it acknowledges the range of granularity scales applicable in the use of RSs. For instance, we have some basis to examine trade-offs between changes to CPs at the lower levels (registration, semantic encoding), which have small impacts on numerous symbols and expressions, versus changes to CPs at higher levels (inference and solution), which impact just a few large-scale procedures.

Acknowledgements. We thank Gem Stapleton, from Cambridge University, for her comments and suggestions for this paper. This work was supported by the EPSRC grants EP/R030650/1, EP/T019603/1, EP/R030642/1, and EP/T019034/1.

References

1. Anderson, J.R.: Spanning seven orders of magnitude: a challenge for cognitive modeling. Cogn. Sci. **26**, 85–112 (2002)
2. Cheng, P.C.-H.: Electrifying diagrams for learning: principles for effective representational systems. Cogn. Sci. **26**(6), 685–736 (2002)
3. Cheng, P.C.-H.: Probably good diagrams for learning: representational epistemic re-codification of probability theory. Top. Cogn. Sci. **3**(3), 475–498 (2011)
4. Cheng, P.-H.: What constitutes an effective representation? In: Jamnik, M., Uesaka, Y., Elzer Schwartz, S. (eds.) Diagrams 2016. LNCS (LNAI), vol. 9781, pp. 17–31. Springer, Cham (2016). https://doi.org/10.1007/978-3-319-42333-3_2
5. Cleveland, W.S., McGill, R.: Graphical perception and graphical methods for analysing scientific data. Science **229**, 828–833 (1985)
6. Finke, R.A.: Principles of Mental Imagery. The MIT Press, Cambridge (1989)
7. Gurr, C.A.: On the isomorphism, or lack of it, of representations. In: Marriott, K., Meyer, B. (eds.) Visual Language Theory, pp. 293–306. Springer, New York (1998). https://doi.org/10.1007/978-1-4612-1676-6_10
8. Halford, G.S., Baker, R., McCredden, J.E., Bain, J.D.: How many variables can humans process? Psychol. Sci. **16**, 70–76 (2005)
9. Jamnik, M., Cheng, P.C.-H.: Endowing machines with the expert human ability to select representations: why and how. In: Muggleton, S., Chater, N. (eds.) Human Like Machine Intelligence. Chapter 18. Oxford University Press, Oxford (2021). (in press)
10. Koedinger, K.R., Anderson, J.R.: Abstract planning and perceptual chunks: elements of expertise in geometry. Cogn. Sci. **14**, 511–550 (1990)
11. Kotovsky, K., Hayes, J.R., Simon, H.A.: Why are some problems hard? Cogn. Psychol. **17**, 248–294 (1985)

12. Larkin, J.H., Simon, H.A.: Why a diagram is (sometimes) worth ten thousand words. Cogn. Sci. **11**, 65–99 (1987)
13. Markman, A.B.: Knowledge Representation. Lawrence Erlbaum, Mahwah (1999)
14. Miller, G.A.: The magical number seven plus or minus two: some limits on our capacity for information processing. Psychol. Rev. **63**, 81–97 (1956)
15. Moody, D.L.: The "physics" of notations: toward a scientific basis for constructing visual notations in software engineering. IEEE Trans. Softw. Eng. **35**(6), 756–779 (2009)
16. Newell, A.: Unified Theories of Cognition. Harvard University Press, Cambridge (1990)
17. Newell, A., Simon, H.A.: Human Problem Solving. Prentice-Hall, NJ (1972)
18. Raggi, D., Stapleton, G., Stockdill, A., Jamnik, M., Garcia Garcia, G., Cheng, P.C.-H.: How to (re)represent it?. In: 2020 IEEE 32nd International Conference on Tools with Artificial Intelligence (ICTAI), pp. 1224–1232. IEEE, Baltimore (2020)
19. Raggi, D., Stockdill, A., Jamnik, M., Garcia Garcia, G., Sutherland, H.E.A., Cheng, P.-H.: Dissecting representations. In: Pietarinen, A.-V., Chapman, P., Bosveld-de Smet, L., Giardino, V., Corter, J., Linker, S. (eds.) Diagrams 2020. LNCS (LNAI), vol. 12169, pp. 144–152. Springer, Cham (2020). https://doi.org/10.1007/978-3-030-54249-8_11
20. Shimojima, A.: Semantic Properties of Diagrams and their Cognitive Potentials. CSLI Press, Stanford (2015)
21. Simon, H.A.: Sciences of the Artificial, 2nd edn. MIT Press, Cambridge (1981)
22. Stevens, S.S.: On the theory of scales of measurement. Science **103**(2684), 677–680 (1946)
23. Sockdill, A., et al.: Correspondence-based analogies for choosing problem representations. In: 2020 IEEE Symposium on Visual Languages and Human-Centric Computing (VL/HCC), pp. 1–5. IEEE, Dunedin (2020)
24. Sweller, J.: Cognitive load during problem solving: effects on learning. Cogn. Sci. **12**(2), 257–285 (1988)
25. van Someren, M.W., Reimann, P., Boshuizen, H.P.A., de Jong, T.: Learning with multiple representations. Advances in Learning and Instruction Series. ERIC (1998)
26. Zhang, J.: A representational analysis of relational information displays. Int. J. Hum. Comput. Stud. **45**, 59–74 (1996)
27. Zhang, J.: The nature of external representations in problem solving. Cogn. Sci. **21**(2), 179–217 (1997)
28. Zhang, J., Norman, D.A.: A cognitive taxonomy of numeration systems. In: Proc. of the 15th Annual Conference of the Cognitive Science Society, pp. 1098–1103. Lawrence Erlbaum, Hillsdale (1993)
29. Zhang, J., Norman, D.A.: A representational analysis of numeration systems. Cognition **57**(3), 271–295 (1995)

Intentional Diagram Design: Using Gestalt Perceptual Grouping in Cladograms to Tackle Misconceptions

Jingyi Liu$^{(\boxtimes)}$ and Laura R. Novick

Vanderbilt University, Nashville, TN 37203, USA
{j.liu, laura.novick}@vanderbilt.edu

Abstract. People have many incorrect beliefs about evolutionary relationships among living things, in part due to the prominence people place on observable similarities as an indicator of such. Consider two examples: (a) a highly detrimental misconception surrounding the current pandemic—that COVID-19 is like the common flu—is based on the similarity of the visible symptoms; (b) people think mushrooms are more closely related to plants than to animals because mushrooms and plants both grow in the ground. Our research asked whether it is possible to combat misconceptions using compelling visual representations. Previous research found that the Gestalt principles of perceptual grouping affect reasoning with evolutionary trees. We explored the potential of designing such trees as a "myth buster" tool to target biological misconceptions. More specifically, we tested the hypothesis that students would judge misconception-based inferences to be weaker when the perceptual grouping of the branches looked less consistent, as opposed to more consistent, with the misconception. The results showed that our manipulation of perceptual grouping affected students' propensity to make inferences consistent with their misconceptions, in the direction predicted, for six separate misconceptions.

Keywords: Evolutionary trees · Diagram design · Gestalt grouping · Inference

1 Introduction

During the early stage of the COVID-19 pandemic, there was a popular but highly detrimental misconception that COVID-19 is just like the flu. People infer that items of the same category tend to share characteristics [1]. The misconception that COVID-19 and the common flu are related misled some people to believe that no safety precautions were needed. However, a phylogenetic tree of viruses shows that coronaviruses are only distantly related to flu viruses [2]. The tendency to treat COVID-19 like the flu is a good example of how misconceptions can come from observable similarities that are not grounded in underlying structural commonalities. Could COVID-related misconceptions be prevented or mitigated with a diagram that powerfully contradicts the misconceptions?

© Springer Nature Switzerland AG 2021
A. Basu et al. (Eds.): Diagrams 2021, LNAI 12909, pp. 431–438, 2021.
https://doi.org/10.1007/978-3-030-86062-2_44

Graphic design plays a critical role in science education and science communication [3, 4]. Many design principles are inspired by principles of cognitive psychology, specifically Gestalt perception [5, 6]. In the present study, we explored whether a diagram designed with the principles of Gestalt perceptual psychology in mind has the potential to perceptually promote the diagram-based inferences intended by the designer and attenuate viewers' misconceptions. The following sections will introduce cladograms, a type of phylogenetic tree used in the current study, and cognitive principles involved in reasoning with cladograms. We targeted common misconceptions college students have about living things, and we tested the effect of intentional design of the structures of diagrams on disabusing students of their evolutionary misconceptions.

1.1 Understanding Evolutionary Trees

Cladograms (see Fig. 1) are hypotheses about nested groups of taxa that share characters due to descent from a most recent common ancestor (MRCA). A *taxon* is a category of like organisms (e.g., manatees, mammals, vertebrates). A group of taxa consisting of a most recent common ancestor and that ancestor's descendants is called a *clade*. The MRCA is the first taxon in evolutionary history to have a particular character that will be shared by its descendants. In the cladograms shown in Fig. 1a and 1b, porpoises and whales form a clade. Bison, porpoises, and whales also form a clade. Thus, the clades are nested. Whales are more closely related to bison than to manatees because whales share a more recent common ancestor with bison than with manatees. The "X" in Fig. 1 represents the MRCA of all the taxa in the cladogram.

Research shows that students find interpreting cladograms to be difficult [7–9]. They often misconstrue physical distance between the perceptual grouping of taxon branches as indicators of evolutionary relatedness [10]. However, the only relevant indicator is relative recency of common ancestry.

1.2 Conception and Perception

Comprehension is the result of the interaction between top-down (i.e., knowledge driven) processes and bottom-up (i.e., data-driven) processes [11, 12]. Diagram comprehension similarly involves interaction between the top-down prior knowledge (conception) one brings to the interpretation and the bottom-up structure (perception) one sees in the diagram.

Conception Affects Perception. Students often have incorrect beliefs about the evolutionary relationships among living things because they tend to make inductive inferences based on folk-biological categorizations (e.g., "fish," "trees," "mammals that live in the sea"). Folk-biological categorizations are derived from observable morphological and behavioral similarities such as similar appearances and habitats [1]. In Fig. 1b, manatee, despite being a sea animal, is more closely related to elephant than to whale.

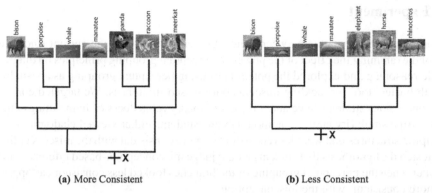

(a) More Consistent   (b) Less Consistent

Fig. 1. The targeted misconception is that porpoises and whales are more closely related to mana-tees than to bison because the former taxa all live in the ocean. (a) A cladogram that is more consistent with the misconception (MC). (b) A cladogram that is less consistent with the misconception (LC). The taxon photos were printed in color in the experiment.

Previous research found that students tend to rely on inaccurate prior knowledge (i.e., their misconceptions) when reasoning with cladograms that include taxa about which they have evolutionary misconceptions [13]. In that study, students saw clado-grams containing either familiar taxa or unfamiliar taxa on top of the same branching structures. The familiar taxa were living things about which students were likely to have evolutionary misconceptions, while the unfamiliar taxa were living things about which students had insufficient prior knowledge to harbor misconceptions (e.g., spiders, types of strep bacteria). The results showed that students were more successful at tree thinking when the branches were labeled with unfamiliar taxa.

Perception Affects Conception.   A bottom-up cognitive process that heavily influences diagrammatic reasoning is Gestalt grouping. The Gestalt principles of grouping govern how viewers of complex visual scenes perceive patterns, form perceptual organizations, and segment the scenes into meaningful units [14, 15]. Visual elements that are grouped together not only form a visual unit, but, more importantly, they are the units of cognition. The principle of element connectedness is an important grouping principle for inferring evolutionary relatedness among taxa in cladograms. Element connectedness describes the tendency for two or more distinct elements to be grouped into a single unit because they are connected [16, 17]. Some may question whether element connectedness is suit-able for cladograms because all elements (taxa) are connected. Cladograms scientifically visualize the theory of evolution, and all species are related if one goes far enough back in evolutionary time. However, the more proximally connected taxa that form a clade (e.g., *porpoise* and *whale*) also form a stronger perceptual group compared to distantly con-nected taxa (e.g., *whale* and *manatee*), as shown in Fig. 1. Previous research found that students' perceptual interpretations of groupings in cladograms agreed with the Gestalt principles, particularly the principle of element connectedness. Students were also more likely to infer that two taxa used the same enzyme to help regulate cell functions if they were in the same perceptual group than if they were in different perceptual groups [18].

2   Experiment

Incorporating the dynamic between conception and perception, the current study went beyond examining the effect of the powerful perceptual grouping principles on diagrammatic reasoning and explored the potential of using perceptual grouping as a visualized "myth buster" tool for tackling misconception-based inferences. We targeted common misconceptions students have about the evolutionary relatedness of living things due to their visual similarity and/or common habitat. Students either viewed cladograms with grouping structures that were less consistent or more consistent with the misconceptions. We tested the hypothesis that students would judge misconception-based inferences to be weaker when the perceptual grouping of the branches looked less consistent, as opposed to more consistent, with the misconception.

2.1   Method

Subjects. Vanderbilt University undergraduates (N = 76; 58 females, 16 males, 2 preferred not to respond) participated in partial fulfillment of requirements for their introductory psychology class. Students also provided their race/ethnicity in a free response format. Their responses were coded into four categories: White (51.3%), Asian (22.4%), Hispanic or Latinx origin (14.5%), Black or African American (5.2%), and mixed ethnicity (11.8%).

Design. The primary independent variable was the consistency of the cladogram structure with a misconception about the evolutionary relationships among a subset of three taxa included in the cladogram. There were six pairs of experimental cladograms that involved misconceptions. Participants saw one cladogram from each pair, with three cladograms inducing perceptual groupings that were more consistent with the misconceptions and three cladograms inducing perceptual groupings that were less consistent with the misconceptions. Thus, consistency was manipulated within-subjects when considered across misconceptions but between-subjects for each individual misconception. Participants provided ratings on a scale from 1 to 4 for how likely two taxa, often misconceived to be closely related, share a character, which evaluated their propensity to make inferences based on their misconceptions.

Materials. There were 18 cladograms in total. In addition to six pairs of experimental cladograms that involved misconceptions, there was a practice problem and five filler problems.

Manipulation Checks. Manipulation checks using bare tree structures without labeled branches confirmed that the pair of trees used for each misconception differed with respect to whether the perceived grouping structure was more versus less consistent with the misconception. In each diagram, the branches that would be assigned misconception taxa were labeled with the letters *J, K,* and *L*. From left to right, half of the diagrams had *JKL,* and the other half had *LJK.* For example, the bare structures of both manatee cladograms shown in Fig. 1 had *J, K,* and *L* corresponding to *porpoise, whale,* and *manatee.* Participants rated the extent to which it looks like branch *L* is in the same group as branches *J* and *K* on a scale of 1 to 5. For all six pairs of misconceptions structures, we found that participants gave higher ratings for the designated *more consistent* structures than the *less consistent* structures ($M_{MC} = 2.89$ vs. $M_{LC} = 1.77$).

Misconceptions. The misconceptions were primarily based on students' tendency to believe that similar habitat is a good indicator of evolutionary relatedness [19]. Categorization based on folk-biology [20, 21] often contradicts scientific taxonomy. In Fig. 1, the misconception is that porpoises and whales are more closely related to manatees because they all live in the ocean, but in fact porpoises and whales are more closely related to bison. The cladogram in Fig. 1b presents the target misconception taxa (*porpoise, whale, manatee*) within a structure that looks like porpoise and whale are in a different perceptual group than manatee, which makes the cladogram structure less consistent with the misconception. In contrast, the cladogram in Fig. 1a presents those three taxa within a structure that looks more like porpoise and whale are in the same perceptual group as manatee, which is more consistent with the misconception. Contrary to the misconception that *porpoise, whale,* and *manatee* form a valid biological group, the correct biological group in Fig. 1 is: *(bison + (porpoise + whale)*.

The other misconceptions were: (a) mushroom: mushrooms are more closely related to plants than to animals because they grow in the ground; (b) fish: salmon and bass are more closely related to sharks and stingrays than to moose because those taxa all live in the ocean and are part of the folk-biological category of *fish*; (c) sand dollar: sand dollars are more closely related to clams and scallops than to birds and mammals because they all live in water and (seemingly) have hard shells; (d) tree: palm trees are more closely related to maple trees and oak trees than to corn and bamboo because the former taxa all belong to the folk-biological category of *trees*; (e) crustacean: shrimp are more closely related to starfish and sea urchins than to insects because the former taxa all live in water and (seemingly) have hard shells.

Inference Questions. For each cladogram, participants provided a rating for how likely two taxa, often misconceived to be closely related, share a character [8]. For example, the inference question for the manatee problem was: "Manatees have a circumferential placenta. How likely is it that whales also have a circumferential placenta?" Participants gave a rating from 1 to 4, with 1 being very unlikely and 4 being very likely. The characters used for the inference questions were scientifically accurate but unfamiliar to ensure that participants made inferences rather than retrieved answers from memory.

Procedure. Participants were randomly assigned to one of two sets of cladograms and one of two stimulus orders. Each set contained the more consistent structure for three experimental cladograms, the less consistent structure for the other three experimental cladograms, and the practice and filler problems. The two orders were the reverse of each other. Preliminary analyses indicated there were no order effects. Participants completed the experiment through an online program using REDCap [22]. After providing informed consent and instructions, participants were asked to use the information presented in the diagram to answer an inference question to the best of their ability. Participants worked at their own pace to answer the questions. Finally, participants answered demographic and biology background questions.

2.2 Results

The inference question assessed students' propensity to make a character inference from one taxon to another taxon that is incorrectly believed to be closely related. We hypothesized that students would provide lower ratings for the cladograms whose structures were less consistent with the misconception than for the cladograms whose structures were more consistent with the misconception.

We conducted a 2 (set) X 2 (consistency of structure) repeated-measures ANOVA on students' mean ratings. As predicted, students gave significantly lower ratings for cladograms that were less consistent with their misconceptions than for cladograms that were more consistent with their misconceptions: $F(1, 74) = 84.05, p < .001, \eta_p^2 = 0.53$ ($M_{MC} = 2.23$ vs. $M_{LC} = 1.51$). In other words, students were significantly less likely to believe that the misconceptions taxa shared a character if the perceptual grouping of the cladogram branches was less consistent with the misconception. There was no main effect of set, $F(1, 74) = 0.05, p > .80, \eta_p^2 = 0.00$, and there was no interaction between set and consistency of grouping structure, $F(1, 74) = 3.23, p > .07, \eta_p^2 = 0.04$.

We then conducted a univariate ANOVA for each pair of experimental cladograms using consistency of grouping structure with the targeted misconception as the between-subjects factor. We found a significant difference in the mean ratings, in the predicted direction, for all six cladogram pairs: (a) manatee: $F(1, 74) = 20.23, p < .001, \eta_p^2 = 0.22$ ($M_{MC} = 2.31$ vs. $M_{LC} = 1.51$); (b) mushroom: $F(1, 74) = 17.51, p < .001, \eta_p^2 = 0.19$ ($M_{MC} = 2.28$ vs. $M_{LC} = 1.57$); (c) fish: $F(1, 74) = 23.93, p < .001, \eta_p^2 = 0.24$ ($M_{MC} = 2.11$ vs. $M_{LC} = 1.28$); (d) sand dollar: $F(1, 74) = 14.72, p < .001, \eta_p^2 = 0.17$ ($M_{MC} = 2.41$ vs. $M_{LC} = 1.60$); (e) tree: $F(1, 74) = 6.75, p < .05, \eta_p^2 = 0.08$ ($M_{MC} = 2.04$ vs. $M_{LC} = 1.59$); (f) crustacean: $F(1, 74) = 23.10, p < .001, \eta_p^2 = 0.24$ ($M_{MC} = 2.34$ vs. $M_{LC} = 1.45$). A previous study that used the mushroom, manatee, and fish cladogram pairs found the same pattern of results.

3 Discussion

Students struggle with tree thinking. They may inappropriately rely on their prior knowledge when interpreting cladograms, which is problematic if the cladograms include taxa about which students have misconceptions [13]. Students are also influenced by the apparent perceptual groups in cladograms, a factor that is irrelevant: When taxa are in the same perceptual group, they are interpreted as being more closely related than when they are in different perceptual groups [18].

Previous research has also provided insights on how well-designed curricula can improve tree thinking [9, 23]. The present research adopted a different strategy. The involuntary and powerful perceptual grouping of cladogram branches could be misleading if the grouping structure unintentionally appears to support viewers' misconceptions. Accordingly, we used Gestalt perceptual grouping to design cladogram structures that challenge students' misconceptions about evolutionary relationships among taxa. Across six pairs of cladograms, students' ratings that two "misconception" taxa share a character were significantly lower when those taxa were in different perceptual groups than when the taxa were in the same perceptual group. Thus, perceptual grouping might help disabuse students of their misconceptions when making biological inferences and encourage reasoning based on the evidence presented in the diagrams.

The results from the current study have the potential to inform diagram design decisions in educational settings. A diagram is (sometimes) worth ten thousand words because a diagram is more efficient in illustrating complex relationships [24]. However, the bottom-up segregation of diagrammatic information into separate units can interact with top-down prior knowledge to affect students' inferences. When creating educational cladograms, designers should avoid creating potentially misleading cladograms by (a) anticipating students' biological misconceptions and (b) making sure the grouping structure of cladogram branches aids rather than hinders appropriate interpretation of evolutionary relationships. Future research can explore how these strategies could be paired with tree-thinking curricula to have a stronger impact on student learning.

Diagram designers should also explore the potential of creating goal-directed diagrams. By intentionally implementing perceptual grouping in cladograms, designers can create cladograms that not only accurately depict evolutionary relatedness among the taxa included in the tree but also graphically highlight the relationships among a few taxa for communicative purpose, such as tackling a misconception. Future research can explore the general applicability of intentional design, particularly using Gestalt perceptual grouping, in other types of diagrams to promote more impactful dissemination of scientific information.

References

1. Lopez, A., Atran, S., Coley, J.D., Medin, D.L., Smith, E.E.: The tree of life: universal and cultural features of folkbiological taxonomies and inductions. Cogn. Psychol. **32**(3), 251–295 (1997)
2. Nasir, A., Caetano-Anollés, G.: A phylogenomic data-driven exploration of viral origins and evolution. Sci. Adv. **1**(8), e1500527 (2015)
3. Cook, M.P.: Visual representations in science education: the influence of prior knowledge and cognitive load theory on instructional design principles. Sci. Educ. **90**(6), 1073–1091 (2006)
4. Murchie, K.J., Diomede, D.: Fundamentals of graphic design—essential tools for effective visual science communication. FACETS **5**(1), 409–422 (2020)
5. Chang, D., Dooley, L., Tuovinen, J.E.: Gestalt theory in visual screen design — a new look at an old subject. In: Selected Papers from the 7th World Conference on Computers in Education 2001. Australian Computer Society, vol. 8, pp. 5–12. Australian Topics, Melbourne (2002)
6. Moore, P., Fitz, C.: Gestalt theory and instructional design. J. Tech. Writ. Commun. **23**(2), 137–157 (1993)
7. Meir, E., Perry, J., Herron, J.C., Kingsolver, J.: College students' misconceptions about evolutionary trees. Am. Biol. Teach. **69**(7) 71–76 (2007)
8. Novick, L.R., Catley, K.M., Funk, D.J.: Inference is bliss: using evolutionary relationship to guide categorical inferences. Cogn. Sci. **35**(4), 712–743 (2011)
9. Novick, L.R., Catley, K.M.: Reasoning about evolution's grand patterns: college students' understanding of the tree of life. Am. Educ. Res. J. **50**(1), 138–177 (2013)
10. Novick, L.R., Catley, K.M.: Fostering 21st-century evolutionary reasoning: teaching tree thinking to introductory biology students. CBE Life Sci. Educ. **15**(4), ar66 (2016)
11. Neisser, U.: Cognition and Reality: Principles and Implications of Cognitive Psychology. W H Freeman/Times Books, Henry Holt & Co., New York (1976)

12. Norman, D.A., Rumelhart, D.E.: The LNR approach to human information processing. Cognition **10**(1–3), 235–240 (1981)
13. Novick, L.R., Catley, K.M.: When relationships depicted diagrammatically conflict with prior knowledge: an investigation of students' interpretations of evolutionary trees. Sci. Educ. **98**(2), 269–304 (2014)
14. Koffka, K.: Principles of Gestalt Psychology. Lund Humphries, London (1935)
15. Wagemans, J., et al.: A century of gestalt psychology in visual perception: I. Perceptual grouping and figure–ground organization. Psychol. Bull. **138**(6), 1172 (2012)
16. Palmer, S.E., Beck, D.M.: The repetition discrimination task: an objective method for studying perceptual grouping. Percept. Psychophys. **69**(1), 68–78 (2007)
17. Palmer, S., Rock, I.: Rethinking perceptual organization: the role of uniform connectedness. Psychon. Bull. Rev. **1**(1), 29–55 (1994)
18. Novick, L.R., Fuselier, L.C.: Perception and conception in understanding evolutionary trees. Cognition **192**, 104001 (2019)
19. Morabito, N.P., Catley, K.M., Novick, L.R.: Reasoning about evolutionary history: post-secondary students' knowledge of most recent common ancestry and homoplasy. J. Biol. Educ. **44**(4), 166–174 (2010)
20. Osherson, D.N., Smith, E.E., Wilkie, O., Lopez, A., Shafir, E.: Category-based induction. Psychol. Rev. **97**(2), 185 (1990)
21. Goldberg, R.F., Thompson-Schill, S.L.: Developmental "roots" in mature biological knowledge. Psychol. Sci. **20**(4), 480–487 (2009)
22. Harris, P.A., Taylor, R., Thielke, R., Payne, J., Gonzalez, N., Conde, J.G.: Research electronic data capture (REDCap)—a metadata-driven methodology and workflow process for providing translational research informatics support. J. Biomed. Inform. **42**(2), 377–381 (2009)
23. Novick, L.R., Catley, K.M.: Understanding phylogenies in biology: the influence of a gestalt perceptual principle. J. Exp. Psychol. Appl. **13**, 197–223 (2007)
24. Larkin, J.H., Simon, H.A.: Why a diagram is (sometimes) worth ten thousand words. Cogn. Sci. **11**(1), 65–100 (1987)

Cognitive Style's Effects on User Task Performance in Network Visualisations

Nikita Dev Lomov[1] (ID), Weidong Huang[2] (ID), Jing Luo[3](✉),
and Quang Vinh Nguyen[1] (ID)

[1] Western Sydney University, Parramatta, NSW, Australia
[2] University of Technology Sydney, Ultimo, NSW, Australia
[3] Chongqing Jiaotong University, Chongqing, China
luojing0923@sina.com

Abstract. With the increasing importance of visualisation in being able to understand large sets of data, there is a growing body of research on how individual differences can influence a user performance in tasks using network visualisations. Individual differences in how users interact with and respond to visualisations presents an opportunity to inform how we construct visualisations. In this study, we chose to explore the effect of cognitive style on users' performance in network visualisations. Three psychological constructs were used to account for individual differences: the Verbal-Imagery Cognitive Style, Rational-Experiential Inventory and Wholist-Analytic Cognitive Style. Using a sample of university students, we measured participants accuracy, effort, time, and efficiency to complete three separate tasks on network visualisations. Overall, the results of the study show evidence that cognitive styles account for some individual differences in user's visualisation performance

Keywords: Cognitive style · Network visualisation · Evaluation · Individual difference

1 Introduction

Through the increase in the importance of visualisation to communicate information, there is a growing interest in how individual differences in people affect their perception of visualisations [5]. Conventional means of measuring visualisation performance such as time and accuracy only capture a part of individual difference, therefore using a psychometric measure allows us to see if there are any additional factors influencing visualisation perception [6, 11].

1.1 Current Study

As there are not many studies in the field visualisation that have utilized cognitive styles as a measure of individual differences we seek to explore if there is any influence cognitive styles can have on comprehension, accuracy, speed, or efficiency across the three

© Springer Nature Switzerland AG 2021
A. Basu et al. (Eds.): Diagrams 2021, LNAI 12909, pp. 439–442, 2021.
https://doi.org/10.1007/978-3-030-86062-2_45

visualisation tasks [5]. Based on the three cognitive styles and prior research, there are three hypotheses to explore. First, in the Rational Experiential Inventory (REI), participants who score higher in experiential cognitive style will correlate to lower accuracy in the visualisation task [5, 11, 13]. Second, in the Verbal Imagery Cognitive Style (VICS), the visualizer cognitive style will correspond to a better performance in the visualisation tasks [4, 5, 10]. Thirdly, in the Cognitive Styles Assessment-Wholistic Analytic (CSA-WA), participants who score higher in the analytic style will also correlate to higher accuracy and performance in visualisation tasks [7, 14].

2   Method

2.1   Participants

Participants (n = 28) were recruited from information technology university students with ages ranging from 22 to 31 (M = 27.29, SD = 2.14), out of the 28 participants only one was female.

2.2   Cognitive Style Measures

Csa-Wa. The CSA-WA is 80 items that measure or survey both analytic and wholistic cognitive styles, recording the time and accuracy of the participants' responses [8].

VICS Test. The VICS contains 232 survey items with half being verbal and the other being imagery stimuli [8].

REI. The REI-40 is comprised of a 40 item survey with 20 questions weighted towards experiential cognitive style and the other towards rational cognitive style [1].

2.3   Visualisation Efficiency

Combining the response time (RT), response accuracy (RA) and the mental effort (ME) in Formula (1) we can obtain a measure of visualisation efficiency. As suggested by Huang, Eades [2], the measure of visualisation efficiency was used as a measure of cognitive load.

$$E = \frac{zRA - zME - zRT}{\sqrt{3}} \tag{1}$$

2.4   Procedure and Visualisation Tasks

Participants were asked to give their consent, then three cognitive style tests were administered, and the three visualisation tasks were given after a brief tutorial example. The stimuli were 120 node-link diagrams that were drawn with a commonly used force-directed algorithm from 120 randomly generated networks [3, 14]. Network visualisations, when drawn to minimize crossings and possess a low to moderate density, can more effectively measure the user's performance visualisation [2]. The three tasks chosen were:

1. Shortest Path: Involved the participants finding the shortest path between the two highlighted nodes in the figure.
2. Common Neighbour: Tasked participants to find how many common neighbours the highlighted nodes have.
3. Degree: Asked the participants to compute the largest degree of the three highlighted nodes

3   Results

Table 1. Pearson correlation

Measure	Wholistic analyst style	Rational cognitive style	Experiential cognitive style	Wholistic cognitive style	Imagery cognitive style
Degree task efficiency	−.483*	−.304	−.492**	.294	.396*
Degree task effort	.238	.312	.557**	−.318	−.230
Degree task time	−.146	−.209	−.307	.316	.268
Degree task accuracy	−.472*	−.289	−.461*	.284	.392*
Common neighbour effort	.284	.482**	.525**	−.378*	−.245
Common neighbour time	−.123	.155	−.163	.456*	.246

$^*p < .05. ^{**}p < .01$
Tasks or measures with non-significant correlations were removed from the table.

4   Discussion and Conclusion

The findings of the study show that there is some correlation between the participants' visualisation task performance and their cognitive style, with evidence for the first two hypotheses. The three cognitive styles tested were compared as combined measure such as rational-intuitive, or individually as either rational or intuitive. The results also show that the individual cognitive styles showed a greater correlation to performance than the combined measures (See Table 1). Cognitive styles also demonstrated greater relationship to time, accuracy, and effort measures with only on Degree Task Efficiency providing any significant correlations to the cognitive styles. Furthermore, cognitive styles such as wholist or experiential which operate in opposition to visualisation performance showed a greater negative correlation in time and effort spent on some tasks. The cognitive styles such as imagery which operate in the same domain as visualisation showed a greater accuracy in certain tasks.

The results of the study are encouraging as they show that cognitive styles play a role in understanding how individual differences in users can impact how they interact with a visualisation. There are several implications in computing, if the sample of users is known to prefer using visual cognitive style, it allows us to design visualisation or learning material that will provide greater engagement and show a greater degree of understanding. Although the current study is exploratory, the results may provide an opportunity for a study with a larger sample which would allow more comprehensive statistical analysis.

References

1. Epstein, S., et al.: Individual differences in intuitive–experiential and analytical–rational thinking styles. J. Pers. Soc. Psychol. **71**(2), 390 (1996)
2. Huang, W., Eades, P., Hong, S.H.: Beyond time and error: a cognitive approach to the evaluation of graph drawings. In: Proceedings of the 2008 Workshop on Beyond Time and Errors: Novel Evaluation Methods for Information Visualization (2008)
3. Huang, W., et al.: Effects of curves on graph perception. In: 2016 IEEE Pacific Visualization Symposium (PacificVis). IEEE (2016)
4. Koć-Januchta, M., et al.: Visualizers versus verbalizers: effects of cognitive style on learning with texts and pictures–an eye-tracking study. Comput. Hum. Behav. **68**, 170–179 (2017)
5. Liu, Z., Crouser, R.J., Ottley, A.: Survey on individual differences in visualization. arXiv preprint arXiv:2002.07950 (2020)
6. Ottley, A., et al.: Manipulating and controlling for personality effects on visualization tasks. Inf. Vis. **14**(3), 223–233 (2015)
7. Parkinson, A., Redmond, J.A.: Do cognitive styles affect learning performance in different computer media? In: Proceedings of the 7th Annual Conference on Innovation and Technology in Computer Science Education (2002)
8. Peterson, E.R., Deary, I.J., Austin, E.J.: A new measure of verbal–imagery cognitive style: VICS. Pers. Individ. Differ. **38**(6), 1269–1281 (2005)
9. Price, L.: Individual differences in learning: cognitive control, cognitive style, and learning style. Educ. Psychol. **24**(5), 681–698 (2004)
10. Reyna, C., Ortiz, M.V.: Psychometric study of the rational experiential inventory among undergraduate Argentinean students. Revista de Psicología **34**(2), 337–355 (2016)
11. Sheidin, J., et al.: The effect of user characteristics in time series visualizations. In: Proceedings of the 25th International Conference on Intelligent User Interfaces (2020)
12. Witteman, C., et al.: Assessing rational and intuitive thinking styles. Eur. J. Psychol. Assess. **25**(1), 39–47 (2009)
13. Xu, K., et al.: A user study on curved edges in graph visualization. IEEE Trans. Visual Comput. Graph. **18**(12), 2449–2456 (2012)
14. Yuan, X., Liu, J.: Relationship between cognitive styles and users' task performance in two information systems. Proc. Am. Soc. Inf. Sci. Technol. **50**(1), 1–10 (2013)

Concentrating Competency Profile Data into Cognitive Map of Knowledge Diagnosis

Viktor Uglev[1]([⊠])[iD] and Oleg Sychev[2][iD]

[1] Siberian Federal University, Zheleznogorsk, Russia
vauglev@sfu-kras.ru
[2] Volgograd State Technical University, Volgograd, Russia

Abstract. The paper describes the process of aggregating primary learning data in the form of learning digital footprints for managing educational process using cognitive visualization techniques. The over-arching process of transition from the learning data to competency diagrams and concentrate them into Cognitive Maps of Cnowledge Diagnosis that. The competency profile, represented as a radar-chart diagram, is compressed into a cognitive map, allowing interpreting this information during decision making by faculty and administrative staff. This allows generalizing results for groups of students. The results of using competency profiles and respective cognitive maps to analyze the results of summative assessments, final, and cross-curricular exams are provided as an illustration of the proposed approach.

Keywords: Radar-chart diagrams · Cognitive visualization · Learning digital footprint · Competency profile · Cognitive map of knowledge diagnosis · Information concentration

1 Introduction

Cognitive visualization is used intensively by scientists of various fields to speed up data interpretation [4], including education [1,8] that is the focus of this paper. The more complex the required analysis is, the more concentrated the information should be. Diagrams are an important tool of generalization and visualization [5] that must be used in conjunction with more complex analysis methods [6]. Data, represented in modern learning management systems (LMS) [7] as learning digital footprint are generalized from raw data to a complex object [9]. The results of methods that analyze only student's knowledge like [3] are not enough even for tactical pedagogical decisions because they also require taking into account other factors like competencies development levels, semantic links between learning units, learning trajectory, trends in indicators, etc. To help teachers and tutors solve these complex problems, the authors propose using cognitive visualization methods: visualize student's competency profile as a radar chart diagram, and overlay the information about selected competencies on Cognitive Map of Knowledge Diagnosis (CMKD) for managing the learning process.

© Springer Nature Switzerland AG 2021
A. Basu et al. (Eds.): Diagrams 2021, LNAI 12909, pp. 443–446, 2021.
https://doi.org/10.1007/978-3-030-86062-2_46

2   Method

The method of estimating competency development level (CDL) described in [10] allows building competency profiles for particular courses, shown in Fig. 1 (a) and (b). Each axis represents an estimate of competency development, ranging from −1 (high certainty that the competency is not developed) to 1 (high certainty that the competency is developed), with 0 representing full uncertainty. These levels are calculated based on certainty factor Buchanan and Shortliffe [2] from the student's answers and other elements of digital learning footprint. Generalizing information from several courses, it is possible to create a cross-curricular competency profile, shown in Fig. 1 (c). But these estimates do not provide enough information to manage the learning process.

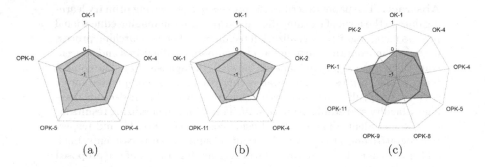

(a)   (b)   (c)

Fig. 1. Competency profiles of student: (a) for course "Simulation modeling", (b) for course "Artificial Intelligence", (c) cross-curricular profile. The poles are competencies. Red lines show zero certainty factor (i.e. nothing is known about this competency development); blue polygons show actual development. (Color figure online)

To concentrate learning data further and show different factors on one diagram, it is proposed to use the method of Cognitive Maps of Knowledge Diagnosis, described in [9]. CMKD is a map that represents the information summarised by LMS to simplify the expert analysis of learning digital footprint, evaluate learning situation, and determine an efficient reaction to learner's actions. The interpretation of the learning situation is based on the course's cognitive map, individualized by the data from the learning footprint, e.g. student's performance, goals, competency development, etc.

Visual representation of CMKD for the given course includes several units di, ordered into the regular sequence of their learning. The semantic links between the components are shown inside the circle. The teacher chooses a competency from the competency profile and can see it overlaid on the map (see Fig. 2). The color of the unit represents the competency development according to the results of summative assessments for this unit: red means the competency is not developed, green – the competency is developed, white shows uncertainty about competency development, while gray shows that the unit does not affect developing the competency in question.

To interpret a CMKD and making an informed decision, a teacher or tutor need to analyze the following factors:

- concentration and sequence of semantic links for the problematic units;
- the trend of competency development, shown by the font of "d" letter in the unit (bold shows that the competency development rises; italics – the competency development lowers; regular font – no significant changes);
- the units' importance for learning goals and student's goals, represented by the unit's figure, changing from circle (low importance) to square (high importance).

The complex analysis of the learning situation, using a set of different maps, allows the teacher (or tutor) to make better-informed decisions.

3   Results

Figure 2 shows an example of CMKD, visualizing the data about competency OPK-4 "Ability to generalize and structure information and make conclusions" from the competency profile shown in Fig. 1 (a). The map is visualized for the course "Simulation modeling" (Siberian Federal University, master's degree program "Computer Science and Engineering"). It shows the units requiring attention – d8, d9, d12, d17, and d18 – in three different sections of the course. Taking into account the semantic links between units, the best way to further develop this competency is working on unit d10, then on units d6 and d8. I.e.

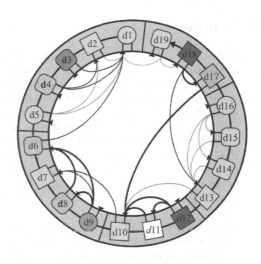

Fig. 2. CMKD of e-learning course "Simulation modeling", showing competency development for OPK-4

unit d10, according to its importance to the student (square form), and the number and boldness of outgoing semantic links is the key unit for topic 3, and the student still didn't master it (light-green color). Direct work on the poorly-mastered units d11 and d12 is not recommended because they both heavily depend on unit d10. For the complex analysis of learning situations (e.g. assessing the development of all the required competencies) the teacher can use several maps showing different aspects, creating an atlas of cognitive maps.

Our proposed method of raw data concentration and visualization was used in Siberian Federal University for teaching 10 courses in computer science and mathematics to graduate students (more than 400 exam blanks total). Six teachers used CMKDs during one semester to give individualized advice to their students about their competency development, registering the time spent on making the decision and finding arguments to support it. One course was taught traditionally, with teachers providing advice after reading LMS reports. In 89% of cases, the teachers using CMKDs made the decision faster (up to 4 times) than when working with raw data from LMS. The teacher's survey showed that in 94% of cases the teachers became surer in their decisions when using CMKD.

4   Conclusion

Building radar-chart diagrams representing Competency Development Levels, while analyzing learning digital footprint allows concentrating multidimensional data on the course and cross-curricular levels. As the experience of Siberian Federal University shows, in conjunction with methods of building Cognitive Maps of Knowledge Diagnosis, these techniques of visualization increase the effectiveness of learning analytics in the modern educational process.

References

1. Aldowah, H., Al-Samarraie, H., Fauzy, W.: Educational data mining and learning analytics for 21st century higher education: a review and synthesis. Telematics Inform., 13–49 (2019). https://doi.org/10.1016/j.tele.2019.01.007
2. Buchanan, B., Shortliffe, E.: Rule-based Expert System: The MYCIN Experiments of the Stanford Heuristic Programming Project. Addison-Wesley, New York (1984)
3. Erdemir, A., Atar, H.Y.: Simultaneous estimation of overall score and subscores using MIRT, HO-IRT and bi-factor model on TIMSS data. J. Measur. Eval. Educ. Psychol. **11**, 61–75 (2020). https://doi.org/10.21031/epod.645478
4. Han, J., Kamber, M., Pei, J.: Data mining concepts and techniques third edition. Morgan Kaufmann Ser. Data Manag. Syst. **5**(4), 83–124 (2011)
5. Moktefi, A.: Diagrams as scientific instruments, pp. 81–89. Peter Lang, Frankfurt (2017)
6. Moreno-Marcos, P.M., Martínez de la Torre, D., González Castro, G., Muñoz-Merino, P.J., Delgado Kloos, C.: Should we consider efficiency and constancy for adaptation in intelligent tutoring systems? In: Kumar, V., Troussas, C. (eds.) ITS 2020. LNCS, vol. 12149, pp. 237–247. Springer, Cham (2020). https://doi.org/10.1007/978-3-030-49663-0_28
7. Nicolay, R., Malotky, N., Martens, A.: Centralizing the teaching process in intelligent tutoring system architectures. In: 19th International Conference on Advanced Learning Technologies, ICALT (2017)
8. Niemiec, J.: Visualizing curricula. In: Pietarinen, A.-V., Chapman, P., Bosveld-de Smet, L., Giardino, V., Corter, J., Linker, S. (eds.) Diagrams 2020. LNCS (LNAI), vol. 12169, pp. 544–547. Springer, Cham (2020). https://doi.org/10.1007/978-3-030-54249-8_53
9. Viktor, U., Kirill, Z., Ruslan, B.: Cognitive maps of knowledge diagnosis as an element of a digital educational footprint and a copyright object. In: Silhavy, R., Silhavy, P., Prokopova, Z. (eds.) CoMeSySo 2020. AISC, vol. 1295, pp. 349–357. Springer, Cham (2020). https://doi.org/10.1007/978-3-030-63319-6_31
10. Uglev, V.A., Ustinov, V.A.: The new competencies development level expertise method within intelligent automated educational systems. In: Bajo Perez, J., et al. (eds.) Trends in Practical Applications of Heterogeneous Multi-Agent Systems. The PAAMS Collection. AISC, vol. 293, pp. 157–164. Springer, Cham (2014). https://doi.org/10.1007/978-3-319-07476-4_19

Diagrams as Structural Tools

Interactivity in Linear Diagrams

Peter Chapman[(✉)]

Edinburgh Napier University, Edinburgh, UK
p.chapman@napier.ac.uk

Abstract. Linear diagrams have been shown to be an effective method for representing set-based data. Moreover, design principles have been empirically developed that, when followed, improve the efficacy of linear diagrams. These principles are task-independent. However, linear diagrams may be produced to aid with a variety of tasks, for which different representations may be more effective. In this paper, we introduce simple interactivity into linear diagrams. Namely, we gave users: the ability to move sets; the ability to move overlaps (set-intersections); and the ability to focus the diagram on a particular group of sets. Whether these interactions improved cognition was investigated via two empirical studies. In the first, we observed that interactivity improved participants' accuracy, confidence and speed. In the second, we observed that these improvements were based on the diagrams participants produced in the first study, rather than being an artefact of interactivity itself. We conclude that adding simple interactivity is useful in the case of linear diagrams.

Keywords: Linear diagrams · Interaction · Empirical evaluation · Set-based representation

1 Introduction

Linear diagrams can be effective at representing set-based data [3]. In contrast to region-based diagrams such as Euler and Venn diagrams, they are also easy to draw. We interpret a linear diagram in the following way: where two horizontal lines exist in the same vertical space, the intersection between the represented sets is non-empty. For example, were a linear diagram to contain some vertical space where lines for set A and B were present, but C was absent, then the diagram would represent that the intersection $A \cap B \cap \bar{C}$ is non-empty.

In essence, a linear diagram can be seen as a matrix, where each matrix entry is either an empty space, or contains a line segment. The columns of the matrix are the *overlaps* of the diagram, and the rows of the matrix are the *sets* of the diagram. With this conceptual model in mind, matrix operations can be recontextualised as linear diagram manipulations, and easily implemented as interactive elements. Making use of these interactive elements could cause the linear diagram to violate drawing guidelines in [15]. Specifically, the number of line segments present in the diagram could increase. The guidelines were

© Springer Nature Switzerland AG 2021
A. Basu et al. (Eds.): Diagrams 2021, LNAI 12909, pp. 449–465, 2021.
https://doi.org/10.1007/978-3-030-86062-2_47

developed for static diagrams, and so have to apply when the task for which the diagram is being used is not known. By contrast, through interaction the user can manipulate the diagram to produce a layout that is helpful for the specific task they have in mind. In this paper, we evaluate whether or not interaction can aid users, when compared to using static diagrams which follow best practice.

In Sect. 2, we investigate interaction, and explain the specific interactive elements added to linear diagrams in Sect. 2.1. The first user-study is explained in Sect. 3, including discussion of the diagrams generated through the interaction process (Sect. 3.5). A question arising from the results of study 1 is investigated in study 2, in Sect. 4. Finally, we conclude and outline further research areas in Sect. 5. The study materials, including diagrams, datasets, and statistical analyses can be found at https://doi.org/10.17869/enu.2021.2748492.

2   Interaction Design

Despite being of central importance to the usability of information visualisations, interaction has until recently received relatively little attention in the literature [7]. Where new visualisations are introduced, the interactive aspects are treated as of secondary importance, if at all [18]. Interaction is being increasingly studied, however, in the literature [13]. Interactions themselves need not be complex, as Dix and Ellis state: "virtually any static representation can become more powerful by the addition of simple interactive elements" [6]. We seek to test this (theoretical) assertion in an empirical manner in this paper.

The literature on interaction in the InfoViz field which does exist is concerned with classifying broad types of interaction users might want. These types are categorised differently by different authors. For example, Shneiderman [16] provides seven tasks that users may wish to undertake: *overview, zoom, filter, details-on-demand, relate, history* and *extract*. Dix and Ellis [6], meanwhile, suggest that the necessary interactions are: *highlighting and focus; accessing extra information; overview and context; same representation, changing parameters; same data, changing representation;* and *linking representations*.

The approach of [7] is different. Rather than focussing on what tasks interaction should support, Elmqvist *et al.* provide a set of eight interaction guidelines for fluidity. When followed, the authors contend that they will produce "effective information visualizations that support fluid interaction." In the following section, we explain the interaction elements added to linear diagrams, and relate them to the theoretical guidelines from the literature.

2.1   Interactive Elements in Linear Diagrams

We introduce four interactive controls to a linear diagram, extended from [4]. Two concern the horizontal ordering of the overlaps, and two concern the vertical ordering of the sets. All alter the diagrams in a sound manner: the underlying information represented is not changed through these interactions [5].

Swap Two Adjacent Overlaps. An example of a button which controls this interaction can be seen in the circle labelled 1 in Fig. 1.

Force Order. This element is the most complex interaction. For this paper, we use one that can prioritise up to 2 sets. The 2-set prioritisation works as follows. The user supplies up to two sets, for example A and B. The overlaps are then split into four mutually exclusive groups, namely those containing: A but not B; both A and B; B but not A; and finally neither A nor B. If only one set is selected, then the overlaps are split into two groups: those containing the set selected, and those not containing the set selected. The diagram is then drawn by separately ordering the four groups of overlaps using the drawing algorithm of [15] and then concatenating them together. In this way we can guarantee that the lines A and B (in the two-set case) will be drawn using single line segments. How this interactive element appears is seen at 2 in Fig. 1. An example of the outcome can be seen in Fig. 2, where the order was forced on the circled sets.

Swap Two Adjacent Sets. As an example, if the button in the circle labelled 3 in Fig. 1 was clicked, then the sets Economics and Food would be switched. The sets would keep their original colours: i.e. Economics would still be purple after the switch, whilst Food would still be brown. By maintaining the original colours, a user's mental map can be preserved [12].

Move Set to Top. This interaction is encoded as repeated application of the "swap two adjacent sets" functionality. The user does not see the intermediate steps, however: the new set simply appears at the top. The remainder of the sets keep their colours and relative order. As an example, if the button in the circle labelled 4 in Fig. 1 was clicked, then the set Movie would move to the top, and all other sets would move down by one.

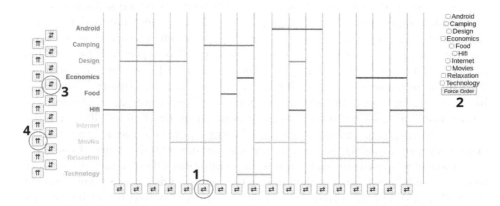

Fig. 1. Interactive elements for linear diagrams: 1. swap overlaps; 2. force order; 3. swap sets; 4. move set to top

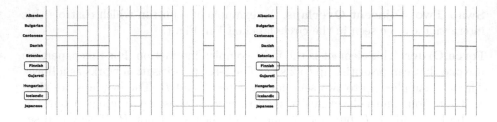

Fig. 2. Original (left) vs. forced (right): forcing on Finnish and Icelandic.

Discussion. The interactions outlined above change the representation of the data, but not the data itself. From the framework of Shneiderman [16], the tasks that will be possible with these interactions are *relate* (view relationships among items) and *details-on-demand* (select an item or group and get details when needed). To some extent, the interactive elements also support *zoom* (zoom in on items of interest) and *filter* (filter out uninteresting items); however the altered diagrams do not remove the uninteresting information (or zoom onto the interesting information) but can rather partition the diagram into interesting and uninteresting regions.

From the framework of Dix and Ellis [6], meanwhile, the kinds of interaction supported are *highlighting and focus* along with *same representation, changing parameters*. In the former, by bringing the overlaps/sets of interest into close proximity to each other, we exploit that "interaction is good, but eye movements [...] are faster still" [6]. For the latter, forcing the drawing order of the diagram can possibly allow more effective representation of parts of the dataset.

Of the eight design guidelines outlined by Elmqvist [7], some were not attempted. As this work is the first step towards an empirically validated tool, it was determined that the use of animated transitions between diagrams was not of high importance. Thus design guidelines 1 (use smooth animated transitions between states) and 5 (reward interaction, through "the use of animations, sounds and pretty graphics") were not followed. (This aspect will be revisited in Sect. 5.) However, the interactions endeavoured to follow design guidelines 2 (provide immediate visual feedback on interaction); 3 (minimize indirection in the interface) and 4 (integrate user interface components in the visual representation) by having the buttons for manipulating overlaps be directly adjacent to the diagram, and not separated by any border etc.; and 7 (reinforce a clear conceptual model) by maintaining the first-presented colours of each set, regardless of where the set ends up after manipulation. Of secondary importance were the guidelines 6 (ensure that interaction never 'ends') and 8 (avoid explicit mode changes). For the former, there is always the ability to continue interacting with the diagram, whereas for the latter there was only one representation type present, and so changing mode (i.e. representation type) was impossible.

Our interactive elements are relatively narrow in their scope, but the representation itself is also limited to displaying set-based information. As such, the interactive elements should help the user complete most of the tasks associated with set-based data (i.e. determining the intersections, unions, disjointness, containment etc. between various groups of sets). The interactions follow suggested guidelines, with the biggest omission that of animated transitions. In the next section, we determine whether the tools which were designed to help users complete a range of tasks are actually helpful with one specific task.

3   First Study: Interactive vs. Static Diagrams

The research question we will be answering is:

> [**RQ1**]: Does interaction aid users in performing set-based tasks with linear diagrams?

Owing to space limitations, we restrict the number of independent variables considered. We always use diagrams that contain 10 sets, and either 19 or 20 overlaps. We also only focus on one particular task, with two variants: identifying intersections between a set and a given combination of two other sets. The variants come from considering two different given combinations of sets. These choices necessarily limit the direct applicability of the results to other tasks, and other size diagrams. However, we can still address our research question.

Examples of the two task variants are to identify the sets where "**some** of the set is also in common with **either** set A **or** set B", and to identify the sets where "**some** of the set is also in common with **both** set A **and** set B". In other words, in the first we are asking participants to identify those sets which intersect with $A \cup B$, and in the second we are asking participants to identify those sets that intersect with $A \cap B$. Given that the use of multiple line segments to represent a set is permissible in linear diagrams, both $A \cup B$ and $A \cap B$ may be relatively concentrated in one small area of the diagram, or spread widely across it. In order to control for this variance, we investigate the concept of *question shape*.

Question Shape. Participants will be required to interrogate a diagram to determine whether or not a given combination of sets has elements in common with remaining sets. In order to do this participants *must* find the given combination of sets described in the task. Once this region of the diagram is discovered, the participant should focus their attention on this region, and expand their focus if needed. By controlling the size of this task region (hereafter referred to as *question shape* or just *shape*), we can determine whether various measures of diagram performance are affected by interactivity and shape.

Shape is determined in two dimensions. Within the context of linear diagrams, a tall shape would represent a situation where the question sets are separated by some vertical distance. By contrast, a short shape would represent a situation where the question sets are relatively close vertically. A wide shape would represent a situation where the question overlaps span a large horizontal distance. Finally, a narrow shape would represent a situation where the question overlaps are relatively close horizontally. Taken together, these two dimensions give rise to four shapes: a short-wide shape (hereafter represented in the paper as ▭); a short-narrow shape (□); a tall-narrow shape (▯); and a tall-wide shape (▢). The four question shapes can be seen highlighted in Fig. 3.

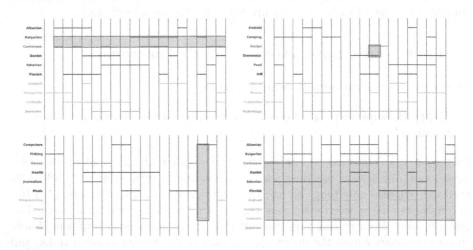

Fig. 3. Question shapes: Top-left the question refers to Bulgarian ∪ Cantonese; top-right Design ∩ Economics; bottom-left Fishing ∩ Travel; bottom-right Cantonese ∩ Icelandic.

We characterise a narrow shape as one which spans less than half of the horizontal space of the diagram, and a wide shape as one which spans greater than half of the horizontal space of the diagram. Similarly, a short shape is one which spans less than half of the vertical space of the diagram, and a tall shape is one which spans greater than half of the vertical space of the diagram. Three tasks were presented to participants for each shape, yielding 12 tasks in total.

3.1   Study Design and Materials

Our first study has two independent variables: **group** (interactive group vs. static group), and **shape**. We use a between-groups design, in that participants either saw only diagrams they could interact with (interactive group), or only static diagrams (static group). Participants are presented with 12 tasks, with 3 tasks for each shape. An example of how a task was presented is given in Fig. 4, which is task 5 from the study. This figure shows a union variant of a task, with wording "Tick the checkboxes where **some** of the people are also interested in **either** Food **or** Internet". The wording for the task for the other variant would be (for example) "Tick the checkboxes where **some** of the people are also interested in **both** Android **and** Design." Two contexts were used: the interests of a group of people (as seen in Fig. 4), and the languages spoken by a group of people (as seen in the top-left panel of Fig. 3). Because all diagrams in the study have 10 sets, and all tasks follow the pattern seen in Fig. 4 with two given sets, there are always 8 checkboxes available for selection. The response variables collected for both groups were the number of check-boxes correctly filled in (giving a score from 0 to 8); the self-reported confidence level of the participant (from 0 to 5); and the time taken to answer the question (in milliseconds). In addition, the diagram created by the interactive group was also collected.

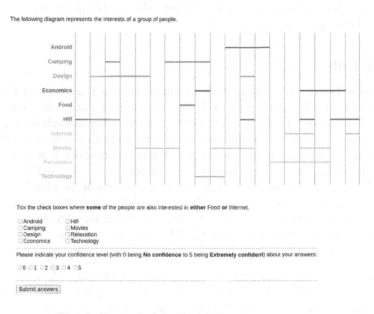

Fig. 4. How a task in the static group appears.

3.2 Hypotheses

The interactive elements will allow participants to redraw a diagram so that the area they must focus their attention on is (a) contiguous, in that the question sets are drawn as single lines, and (b) in a known location on the diagram (at the left-hand side). These two facts should allow participants to find the intersecting sets more easily, and to be more certain that they have found all intersecting sets. However, interacting with the diagram takes time: users must perform two sub-tasks. They must first re-arrange the diagram, and then interrogate the new diagram. We thus have our first hypothesis (in three parts):

[H1.i/ii/iii] the use of interaction leads participants to answer *significantly more accurately/confidently/slower* than when using static diagrams.

When considering shape, the more space a user needs to interrogate, the more potential there is for mistakes. Further, it could take longer. There is a possible interaction between shape and interaction: as noted above the use of interaction can reduce the search space for the user. However, we could still reasonably expect larger question shapes to be more challenging to users. We thus have our second hypothesis (in three parts):

[**H2.i/ii/iii**] diagrams with a small question shape permit users to answer tasks *significantly more accurately/confidently/quickly* than with diagrams with a large question shape.

3.3   Methods

Data Collection. Participants were recruited from a university in the southeast of Scotland. Owing to the pandemic, face-to-face data collection was not possible. Participants were invited to download the study materials (when packaged as a zip), and open the first page (`start.html`) in a browser. The study would then run locally on their machine. A random number was generated on the start page: this was used to assign participants to a group, and to generate a unique identifier. Only the participant knows both parts of the identifier-identity pair. The responses were automatically collected in a text file, which was produced at the end of the study. Participants were then asked to submit the text file to an `ftp` server. Once collected, the files were deleted from the server. This study received ethical approval from Edinburgh Napier University.

Training. There were two phases to the study: a training phase and a main phase. In the training phase, participants were first given a description of a linear diagram itself, and how to interpret them. For the interactive group, they were further given a description of what each interactive control did. Note that they were not given any information of *when* to use a control, but rather only *how* to use a control. In other words, the interactive controls were not linked to the specifics of completing a task.

Participants were then presented with three training questions to familiarise them with the task. In the first and second, the correct answers were pre-selected, and explained. In the third, the correct answers were not pre-selected. Participants were then taken to a holding page, which contained information about how many questions were remaining, and given the opportunity for a rest.

Main Phase. Participants were presented the 12 tasks in a random order. Participants were free to select no sets as their answer. However, participants were not able to submit their answer without selecting a confidence-level. Between consecutive tasks, users were taken to a holding page where they could have a rest. The number of questions remaining was presented on this page. The holding page also fulfilled a technical role: various timers and variables were reset on the holding page. After all tasks were completed, participants were asked to indicate their age (in bands) and whether or not they had any colour-blindness. Responses of "prefer not to say" were available for both questions.

Statistical Analysis. Multiple responses were collected from each participant. The dataset therefore exhibits clustering, meaning that the assumption of standard approaches (ANOVA, χ^2-test for goodness of fit, etc.) of independence of observations was violated, possibly leading to overstated statistical significance and underestimated standard errors [2]. We thus used the approach of generalised estimating equations [8], used in conjunction with generalised linear models (GLM). The statistical software R was used for the analysis, with the package geepack [9]. This approach still yields p-values, and so is appropriate for hypothesis testing. The models fitted are explained in each section. Initially, a model with both main terms (here shape and group) and interaction terms was attempted, and simplified if no interaction effect was present.

3.4 Interaction vs. Static Results

A pilot study was conducted, with 3 participants in each group (interactive, and static). No issues were identified during the pilot, and no changes to the study design were made, and so these 6 participants were included in the main results. A total of 53 participants were recruited, of which 29 were randomly assigned to the static group, and 25 to the interactive group. As the participants were students on Computing-related courses, the sample skewed male and young.

It was observed that, of the 25 participants in the interactive group, not all of them used the interactive tools available to them. 17 of the 25 made changes to the diagrams presented (the details of these changes will be discussed in Sect. 3.5). Of note is that participants either altered all 12 diagrams they saw, or altered none of them. From an analysis point of view, a decision thus needed to be made. Either the groups remained an independent variable in the model (i.e. the distinction between participants is whether they had the *potential* to interact with the diagrams, regardless of whether they *did*), giving a group split of 25 to 29, a new independent variable is introduced to encode whether a participant interacted with the diagrams or not, giving a group split of 17 to 37; or the non-interacting participants in the interactive group were discarded, giving a group split of 17 to 29. Because we are interested in whether or not the interactions participants make are useful, rather than whether the interface encouraged interaction, we have split the data according to the middle choice. Thus, a new independent variable (changed vs. unchanged) was introduced.[1]

Accuracy. Both groups had high accuracy rates. Each question contributed a score of 0 to 8: the average question score in the changed group was 7.83, and in the unchanged group was 7.61 (giving accuracy rates of 98% and 95% respectively). Whilst high, these rates are not unexpected. High rates were found in [3,15]. For each shape, the following accuracy rates were observed: for the shape ⌐⌐, the rates were 94.8% changed vs. 95.7% unchanged; for ◻, 99.5% changed vs.

---

[1] The results remain true whichever choice was made. The other models can be seen in https://doi.org/10.17869/enu.2021.2748492. The p-values differ, but the models do not give remove (or introduce) significance at 95%.

96.3% unchanged; for ⌶, 99.5% changed vs. 94.7% unchanged; and for ☐, 97.9% changed vs. 93.6% unchanged.

A GLM with an ordinal response variable was initially anticipated. However, owing to the high accuracy rates some categories did not have enough responses for robust analysis [10].[2] Thus, responses were combined in the following way: question responses were recoded as either the participant was completely correct (i.e. a score of 8 on a question), or made at least one mistake (i.e. a score of 7 or lower). In this way, the model became a GLM with a binomial response. It is these results which we report. The model gives an interaction effect between the group and shapes ⌶ ($p = 0.0195$) and ☐ ($p = 0.0153$). We can thus conclude that for shapes ⌶ and ☐, *participants who changed the diagrams performed significantly better than those who did not*. For the other two shapes ▫ and ⊏⊐, there was no difference between the two groups.

Confidence. Both groups had high confidence rates. The changed group reported an average confidence level of 4.7 (out of 5), and the unchanged group reported an average confidence level of 4.4. By shape, we observed the following confidence levels. For ⊏⊐, 4.7 for changed vs. 4.2 for unchanged; for ▫, 4.9 for changed vs. 4.5 for unchanged; for ⌶, 4.8 for changed vs. 4.5 for unchanged; and for ☐, 4.6 for changed vs. 4.2 for unchanged.

As with the accuracy results, there were not enough responses in some categories to perform robust ordinal regression, and so a GLM with a binomial response was fitted. The model thus compares those who had full confidence (i.e. level 5) with those who had confidence levels of 4 or lower.

There was no interaction effect between group and shape, and so a more simple model was fitted with only main terms for group and shape. The changed group were found to be *more confident* in their responses than the unchanged group ($p = 0.035$). Further, for the narrow shapes ⌶ ($p = 0.026$) and ▫ ($p < 0.001$), participants were more confident with their responses than with the wide shapes ☐ and ⊏⊐. No differences were found between ⌶ and ▫, or between ☐ and ⊏⊐.

Time. The average time taken for the changed group to answer a question was 45.0 s, compared with 63.3 s for the unchanged group. Note that this time *includes* any time re-arranging the diagram for the changed group. When comparing by shape, we have: for ⊏⊐, 52.0 s for changed vs. 80.6 s for unchanged; for ▫, 39.3 s for changed vs. 47.6 s for unchanged; for ⌶, 33.4 s for changed vs. 47.2 s for unchanged; and for ☐, 55.3 s for changed vs. 77.8 s for unchanged.

A GLM with a normal response was fitted to the data. There was found to be no interaction effect between shape and group, and so a simpler model with just main terms for group and shape was fitted. It was found that the changed group were *significantly faster* ($p = 0.0073$) than the unchanged group, and that shapes ▫ ($p < 0.0001$) and ⌶ ($p < 0.0001$) were *significantly faster* than the

---

[2] These models were developed, and broadly show the same results as the binomial model. However, the more conservative model is reported as it is more robust. The ordinal logistic model can be found with the rest of the study materials at https://doi.org/10.17869/enu.2021.2748492.

shapes ⊏⊐ and ☐. There were no differences within each pair of shapes: neither the narrow shapes (◻ and ◫) nor the wide shapes (⊏⊐ and ☐).

Discussion. We can answer RQ1: interaction can help users in performing set-based tasks with linear diagrams. We have seen that hypotheses **H1.i** and **H1.ii** can be supported by the data. However, **H1.iii** can be rejected. The evidence is more mixed for **H2**. The tall shapes produced worse accuracy results (evidence for **H2.i**), whereas the narrow shapes gave higher confidence **H2.ii** and quicker responses **H2.iii**. Overall, larger shapes were more problematic, but the dimension of size increase was important.

The high levels of accuracy amongst participants, as mentioned, was not unexpected. We also gained further (anecdotal) evidence that participants struggle with broken lines, as the wide shape questions produced lower confidence. This phenomenon has been observed before: in [15], which recommended minimising the number of line segments in a diagram; in [1], more line segments equated to a higher perceived clutter; and in [17] participants reported that "broken lines were problematic". Of interest here, however, is that whilst participants were less confident with finding information across a wide area of the diagram, there was no significant lowering of accuracy in these cases. Indeed, what was found to increase the accuracy was to reduce the *vertical* size of the question, not the horizontal. In some cases, it would not be possible to change the horizontal shape (from wide to narrow), but it was always possible to change the vertical shape (from tall to short). We will return to this theme in Sect. 3.5.

We see that confidence levels for those who made changes are higher even when there is no corresponding improvement in accuracy (i.e. for shapes ⊏⊐ and ◻). This finding could be seen as an example of the illusion of control [11], where increased confidence in performance is not related to an improved performance when participants have perceived control. In general, however, it is re-assuring that the overall high levels of accuracy translate into high levels of confidence.

The time findings were surprising. Participants were *faster* when they had to perform two tasks rather than one: first alter the diagram, and then interrogate the new diagram to produce the required answer. Of course, participants who changed the diagram then knew where the information they were looking for was to be found. Thus, there would be little need to search the diagram, or to spend time checking other regions of the diagram that were known to be outside the area of interest.[3] However, that no searching outside a given region would be necessary is known only to the user who altered the diagram. It is an open question as to whether the benefits of the layout are available to users who did not generate the layout. We attempt to answer this question in the second study, in Sect. 4.

---

[3] This was alluded to as the zoom-like or filter-like behaviour of [16] in the Discussion of Sect. 2.1.

3.5   User-Generated Diagrams

There were 17 participants who changed diagrams, as detailed in Sect. 3.4. In this section, we examine those altered diagrams. Owing to space restrictions, we focus on the diagrams created for shapes ⨅ and ⬜; these are the shapes where participants saw an accuracy improvement. However, the themes hold for the other shapes, too.

Two main patterns for redrawing were observed. The first can be seen in the left panel of Fig. 5, which was a diagram created by a participant to help complete task 4 (originally shown in the lower-right panel of Fig. 3): *tick the check boxes where* **some** *of the people also speak* **either** *Cantonese* **or** *Icelandic*. The participant has used the "force order" functionality, applied to the lines Cantonese and Icelandic, but has left those two lines in their original vertical location. As such, they have transformed what was a ⬜ diagram (the lower-right figure of Fig. 3) into a ⨅ diagram (Fig. 5, left).

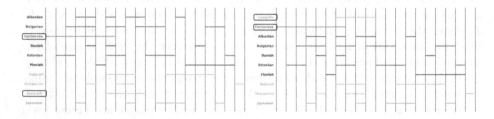

Fig. 5. User-generated diagrams: forced order only (left), forced order followed by vertical re-ordering (right). The sets of interest are circled.

The second main pattern observed can be seen in the right panel of Fig. 5. Again, this diagram was collected from a participant's interaction for the same task. The participant has used the "force order" functionality, but has then subsequently moved the two question lines to the top of the diagram. As such, they have transformed what was a ⬜ diagram into a ▢ diagram.

For the six tasks concerning ⨅ and ⬜ diagrams, the frequency of each pattern is given in Table 1. The category "Other" includes: instances where force-order was applied to only one set; where force-order was applied to both and some vertical re-arrangement made the question sets be adjacent but not at the top; and one instance some individual overlaps had been moved. Within the "force order, move to top" category, there were two subcategories. The first is seen in the right panel of Fig. 5: the line which starts left-most is *below* the other. The second is not shown for space reasons, but transposes the top two sets. Participants produced the former over the latter at a ratio of roughly 3:1.

Discussion. We can infer from the frequencies given in Table 1 exactly which functionality participants used. The force order functionality was used in all but one of the 102 diagrams created. The ability to move a set to the top of the

Table 1. Frequency of user-generated diagrams

Task	Force order	Force order, move to top	Other
3	3	13	1
4	4	13	0
7	3	13	1
8	5	12	0
11	4	13	0
12	3	12	2

diagram was used in 75% of instances. However, the ability to move a single overlap left and right, and the ability to move an individual set up and down one step at a time, were rarely, if ever used. It would be beneficial to remove that functionality: the buttons are creating visual clutter on the screen, but not performing any useful function for this task.

4   Second Study: User-Generated vs. Original Diagrams

The results of Sect. 3.5 lead us to pose the following research question:

[**RQ2**]: are the improvements in the interactive group owing to the user-generated layout, or the process of interaction itself?

In other words, if person A created a diagram through interaction, would person B gain the performance benefits if shown that diagram over the original? We could address this question partially through the analysis of Sect. 3.4. If the process of interaction was driving the accuracy improvements, then we would have expected to see uniform improvements for all shapes. This uniform improvement did not occur, although there was a confidence improvement.

We can also address the question through a secondary study, however. By presenting each participant with both the original and the most common user-generated diagram (i.e. Fig. 5, right), we can observe whether accuracy, confidence and time follow the same changes as for those who created the diagrams. We only present participants with the diagrams that were found to improve accuracy in Sect. 3.4, namely ⊔ and ☐ shaped diagrams, in order to keep the anticipated time per participant to complete the set of tasks low enough so as not to cause fatigue [14]. Further, we changed the context of the altered diagrams so that participants would not know they were seeing each task twice: i.e. if the original diagram was about languages, then the user-generated diagram was altered to be about interests. The alphabetical order was retained, however.

We thus conducted a second study with a *within* factor (the **type** of the diagram, with two levels user-generated and original). In addition, we had the **shape** factor, with two levels ⊔ and ☐ within the model. As in Sect. 3.3, there was also an interaction term in the model type × shape. Where there was found to be no interaction effect, the model was simplified to only include the main terms. The number of correct responses, the confidence level, and the time taken to answer a question was recorded for each participant.

4.1 Hypotheses

The user-generated diagrams contain more line segments than the originals (on average 11.7% more line segments). However, the region of interest for the task is contiguous. A further confounding issue is that the order of the sets is no longer alphabetical: it may be more difficult to find the particular question sets than for the original diagrams. We still anticipate that the contiguous task region will be the most important factor, leading to our hypotheses **H3** (shown later).

With regards to shape, we are only looking at those shapes which caused more problems from the first study. However, we still have two different sizes in the horizontal direction: narrow and wide shapes. Given the results of the first study, we can hypothesise that a similar effect will be present, leading to hypotheses **H4**.

> **[H3.i/ii/iii]** user-generated diagrams permit participants to answer *more accurately/confidently/quickly* than the original diagrams.
> **[H4.i/ii/iii]** narrow shapes permit participants to answer *more accurately/confidently/quickly* than the wide shapes.

4.2 Results

A pilot study was conducted, with 3 participants. No issues were identified during the pilot, and no changes to the protocols were made, and so the pilot participants' data were included in the main dataset. A further 17 participants were recruited. As with the first study, the participants were students enrolled on Computing-related courses, and so skewed male and young.

Accuracy. Participants again showed high accuracy rates. Overall, the user-generated diagrams returned an accuracy rate of 92.4%, whereas the original diagrams had an accuracy rate of 89.3%. When considered by shape: for ⬭, the accuracy was 97.3% for user-generated vs. 94.0% for the original diagrams. For ⬜, the rates were 87.5% for user-generated vs. 84.6% for the original diagrams.

As with study 1, the high accuracy rates in some categories meant that responses were recoded as either completely correct, or at least one mistake. The results reported are for a GLM with a binomial response variable. There was no interaction effect between shape and type ($p = 0.778$) and so a simpler model with no interaction terms was fitted. There was a main effect for user-generated vs. original diagrams ($p = 0.0033$), but there was no effect owing to shape ($p = 0.20$). We can thus conclude that *the user-generated diagrams performed significantly better than the original diagrams*.

Confidence. Participants had high confidence scores. The average confidence was 4.5 for the user-generated diagrams, versus 4.2 for the original diagrams. By shape, the confidence for ⬚ was 4.8 for user-generated vs. 4.4 for original. For ⬜, the scores were 4.1 for user-generated vs. 4.0 for the original diagrams.

As with study 1, it was necessary to recode confidence as either full confidence (level 5), or a confidence level of 4 or lower. A GLM with a binomial response was fitted to these data. An interaction effect between type (user-generated vs. original) and shape was present ($p = 0.0041$), and so each type was analysed separately. For the user-generated diagrams, there was significantly lower confidence with shape ⬜ than with shape ⬚ ($p = 0.0003$), whilst for the original diagrams there was no difference in confidence levels between the shapes ($p = 0.18$). Overall, there was a significant decrease in confidence between the user-generated and the original diagrams ($p = 0.0004$). Again, we can conclude that *the user-generated diagrams gave participants higher levels of confidence than the original diagrams.*

Time. One participant took roughly 100 times longer on one task than the next highest time. That response was then removed as an obvious outlier. The summary statistics and analysis are described for the reduced dataset. Overall, the average time taken to complete a task was 50.6 s for the user-generated versus 78.8 s for the original diagrams. By shape, we have for ⬚: 33.6 s for the user-generated vs. 60.4 s for the original diagrams. For c, we have 67.4 s for the user-generated vs. 97.2 s for the original diagrams.

A GLM with a normal response variable was fitted to the data. There was no interaction between type and shape ($p = 0.80$), and so a simpler model with no interaction terms was fitted. Participants answered significantly faster ($p < 0.0001$) when answering tasks with user-generated diagrams than with the original diagrams. In addition, ⬜ shaped-questions took significantly longer to answer than ⬚ shaped-questions ($p = 0.0085$). We can again conclude that *the user-generated diagrams outperformed the original diagrams.*

Discussion. Participants were more accurate, faster, and more confident, when using user-generated diagrams than the original diagrams, giving supporting evidence for hypothesis **H3**. We can then answer research question **RQ2**: the user-generated layouts give performance enhancements to those who did not create them. Of course, the only diagrams examined are those which produced an improvement, namely ⬚ and ⬜. As with the first study, the hypotheses regarding shape are more subtle. For accuracy, we can reject **H4.i**; for confidence, the interaction effect means that we only have partial evidence for **H4.ii**; and for time we have supporting evidence for **H4.iii**.

5 Conclusions and Further Work

We have implemented simple interactive controls for linear diagrams, and shown that they are useful for one type of set-based task. Further, we have identified that the question shape affects user accuracy, confidence and speed. However, the type of shape change is important: confidence varies with horizontal changes, whereas accuracy and speed vary with vertical. In either instance, interactions which allow the shape of the question to be altered are beneficial.

There is much scope for further work. On an immediate level, the interactive elements themselves could be incorporate smooth transition effects, allowing more guidelines from [7] to be implemented. In more general terms of interaction and linear diagrams, more task types could be investigated. For example, we only looked at intersection. Whether interactivity also helps with containment or disjointness would be an obvious extension. Similarly, moving to a different representation could yield different results. The force order algorithm works by letting the user, rather than a heuristic, specify the drawing order. Where set representations allow a set-by-set drawing algorithm, order forcing through interaction can be implemented.

Region-based representations can draw more sets before known problems occur: whilst linear diagrams can only draw Venn-2 using a single line for each set, it is possible to draw Venn-3 using circles, and Venn-5 using ellipses, without using duplicate curves. A force order algorithm applied to these representations can guarantee reasonable representations for a higher number of sets, and interaction could therefore be highly useful to them.

References

1. Alqadah, M., Stapleton, G., Howse, J., Chapman, P.: The perception of clutter in linear diagrams. In: Jamnik, M., Uesaka, Y., Elzer Schwartz, S. (eds.) Diagrams 2016. LNCS (LNAI), vol. 9781, pp. 250–257. Springer, Cham (2016). https://doi.org/10.1007/978-3-319-42333-3_20
2. Cameron, A.C., Miller, D.L.: Robust inference with clustered data. Technical Report, Working paper 10–7, University of California, Davis (2010)
3. Chapman, P., Stapleton, G., Rodgers, P., Micallef, L., Blake, A.: Visualizing sets: an empirical comparison of diagram types. In: Dwyer, T., Purchase, H., Delaney, A. (eds.) Diagrams 2014. LNCS (LNAI), vol. 8578, pp. 146–160. Springer, Heidelberg (2014). https://doi.org/10.1007/978-3-662-44043-8_18
4. Chapman, P., Roberts, W.: Towards diagram-based editing of ontologies. In: Chapman, P., Stapleton, G., Moktefi, A., Perez-Kriz, S., Bellucci, F. (eds.) Diagrams 2018. LNCS (LNAI), vol. 10871, pp. 699–703. Springer, Cham (2018). https://doi.org/10.1007/978-3-319-91376-6_62
5. Chapman, P., Stapleton, G., Rodgers, P.: PaL diagrams: a linear diagram-based visual language. J. Vis. Lang. Comput. 25(6), 945–954 (2014)
6. Dix, A., Ellis, G.: Starting simple: adding value to static visualisation through simple interaction. In: AVI 1998, pp. 124–134. ACM (1998)
7. Elmqvist, N., Moere, A.V., Jetter, H.C., Cernea, D., Reiterer, H., Jankun-Kelly, T.: Fluid interaction for information visualization. Inf. Vis. 10(4), 327–340 (2011)
8. Galbraith, S., Daniel, J.A., Vissel, B.: A study of clustered data and approaches to its analysis. J. Neurosci. 30(32), 10601–10608 (2010)
9. Halekoh, U., Højsgaard, S., Yan, J., et al.: The R package GEEPACK for generalized estimating equations. J. Stat. Softw. 15(2), 1–11 (2006)
10. Harrell, F.E., Jr., Lee, K.L., Mark, D.B.: Multivariable prognostic models: issues in developing models, evaluating assumptions and adequacy, and measuring and reducing errors. Stat. Med. 15(4), 361–387 (1996)

11. Langer, E.J.: The illusion of control. J. Pers. Soc. Psychol. **32**(2), 311 (1975)
12. Misue, K., Eades, P., Lai, W., Sugiyama, K.: Layout adjustment and the mental map. J. Vis. Lang. Comput. **6**(2), 183–210 (1995)
13. Munzner, T.: Visualization Analysis and Design. CRC Press, Boca Raton (2014)
14. Purchase, H.: Experimental Human Computer Interaction: A Practical Guide with Visual Examples. Cambridge University Press, Cambridge (2012)
15. Rodgers, P., Stapleton, G., Chapman, P.: Visualizing sets with linear diagrams. ACM Trans. Comput. Hum. Interact. (TOCHI) **22**(6), 27 (2015)
16. Shneiderman, B.: The eyes have it: a task by data type taxonomy for information visualizations. In: IEEE Symposium Visual Languages, pp. 336–343 (1996)
17. Stapleton, G., Rodgers, P., Touloumis, A., Blake, A.: Well-matchedness in Euler and linear diagrams. In: Pietarinen, A.-V., Chapman, P., Bosveld-de Smet, L., Giardino, V., Corter, J., Linker, S. (eds.) Diagrams 2020. LNCS (LNAI), vol. 12169, pp. 247–263. Springer, Cham (2020). https://doi.org/10.1007/978-3-030-54249-8_20
18. Yi, J.S., Ah Kang, Y., Stasko, J., Jacko, J.A.: Toward a deeper understanding of the role of interaction in information visualization. IEEE TVCG **13**(6), 1224–1231 (2007)

Diagrammatic Representations of Uncertainty in Meteorological Forecasting

Brandon Boesch[✉] [iD]

Morningside University, Sioux City, IA 51106, USA
boeschb@morningside.edu

Abstract. Meteorological prediction is complex, but an important scientific practice to help keep people safe. In the past few decades, forecasts have become much more accurate. However, this accuracy has not removed uncertainty from forecasts. Indeed, it is widely acknowledged that weather forecasting is an inherently uncertain practice. Given that the aims of weather prediction are practical (to help keep people and property safe), it is important that weather forecasts be effectively communicated to the broader public. But, there is an interesting question here about how the uncertainty (itself an interesting epistemological concept) of the forecasts is communicated. One of the ways this is done by the National Weather Service in the United States is to use various forms of graphical and diagrammatic representations to identify and communicate the uncertainty that remains as part of a forecast. In this essay, I will explore some of these diagrammatic and graphical representations to identify the ways in which uncertainty is expressed. I argue that there is an interesting and important way in which the aims of meteorological practice constrain the representations—such that they do not aim at perfect (or true) representations, but rather at products which will help change behavior to save lives.

Keywords: Meteorology · Weather · Diagrammatic and graphical representations · Uncertainty

1 Introduction

Meteorological prediction is complex [2]. Changes in the structure or path of a low-pressure system near Seattle can impact the weather in Charleston days later. Temperature changes of a degree or two can significantly alter the practical impacts of a winter weather system. A shift of fifty miles of a hurricane can cause or prevent billions of dollars in damage. Given the effects of weather on human life, meteorological prediction is an important scientific endeavor. There are, of course, the familiar and minor ways in which weather prediction impacts our daily life—decisions about what clothes to wear, whether to pack an umbrella, and so on. But weather forecasting of more extreme events has a large impact on a society. The United States averages around "26,000 severe storms, 1,300 tornadoes, 12 Atlantic basin tropical storms, 5,000 floods, 69,000 fires, and dozens of heavy snowstorms and blizzards" each year [15]. According to the National Safety Center, there were 787 weather-related deaths and 1,797 weather-related injuries

© Springer Nature Switzerland AG 2021
A. Basu et al. (Eds.): Diagrams 2021, LNAI 12909, pp. 466–479, 2021.
https://doi.org/10.1007/978-3-030-86062-2_48

in 2018 [10]. There is an economic impact, too—with about 3.4% of yearly economic variability (around $400 billion) attributable to weather events [7]. Given that the ability to successfully predict weather events can help save lives and property, it is important that any predictions are as accurate as they can be.

Weather forecasting has improved substantially in the past decades [18]. Much of this is attributable to developments related to Numerical Weather Prediction (NWP)—the use of computational simulations of the weather for the next several weeks. Comparing the skill of a prediction made by a NWP system now to that made in the early 1980s, there has been about a two-day improvement in forecasts: a five-day weather prediction today is more accurate than a three-day weather forecast from the early 1980s. Similarly, the seven-day and ten-day forecasts today are better than the five-day and seven-day forecasts from the 1980s, respectively. Further, predictions for the southern hemisphere used to be substantially worse than those for the northern hemisphere—a gap which has nearly vanished [18]. Similar success can be seen in the prediction of hurricane tracks in the Atlantic basin, where the error of predicted track of a hurricane three days out today is similar to that of the prediction one day out in the 1980s [19]. Despite these impressive improvements, it is widely accepted that there is a limit to predictive capabilities of NWP, and therefore of forecasting in general [8, 9, 18].

Practically, this means that meteorological prediction—which has massive practical impacts on the day-to-day lives of individuals and on the functioning of a society— is inherently uncertain. Whether it be the variability of the effects of a winter storm based on a snow-rain line that shifts by thirty miles or the impacts of stronger hurricane winds from a track error of fifty miles, weather forecasting is imprecise. Indeed, this is something we have all experienced: the snowstorm which turned out to be just a few flurries; the hurricane bearing down which turns last minute; and so on.

There is an interesting and important set of philosophical considerations here related to this uncertainty. One question is epistemological: in virtue of what do meteorologists know that their predictions are uncertain? Another question is methodological: what are the methods and techniques by which uncertainty is reduced for meteorological prediction. And yet another question, and the one upon which I will focus in this essay, is the way in which this uncertainty is communicated with the people who can actively and effectively use it, i.e. the broader public. One common method (though not the only method) used by the National Weather Service in the United States involves the use of diagrammatic representations of uncertainty to help the public better understand what, in particular, is uncertain and how things might change. I'll focus especially on visual representations produced and released by various United States governmental meteorological organizations including the National Weather Service, the Storm Prediction Center, and the National Hurricane Center.

After analyzing some examples, I will move on to draw some conclusions. I will primarily focus on the interesting way in which these diagrammatic representations do not aim at accuracy, but rather aim at helping people to take meaningful action steps to help protect themselves, their livestock and pets, and their property. Since the goal is not perfect prediction of the meteorological event, but rather behavior change, it is more important that people take the risks seriously than that the prediction perfectly verifies. Thus, some uncertainty is built not only in the forecasting process but also in the representations themselves.

2 Philosophy and Meteorology: Some Background

Though there has not been much written from a philosophical perspective on meteorological prediction, Wendy Parker [11–14] has explored the ways in which meteorological prediction relies upon a particular method (ensembles) in numerical weather models. Numerical weather models are computational simulations of the weather for the next several weeks. These simulations model the way in which weather systems may unfold, given a particular set of inputs—showing temperature changes, low pressure systems, anticipated precipitation, etc.

In order to make these models useful, it is important that the initial conditions used by the model are as close to the real conditions as possible. Realistic starting conditions are important because, as Edward Lorenz famously showed [8, 9], errors can multiply quickly—doubling every few days. Therefore, any errors in the initial data can quickly lead to inaccurate forecasts.

Improving data can help, but it's noteworthy that even if the data were perfect—there are other potential sources of error arising from idealization of the atmosphere or a lack of complete theoretical understanding of a particular phenomenon [11]. Indeed, as Fuqing Zhang and colleagues [18] demonstrate, even under nearly perfect initial conditions, there will still be some time point beyond which our predictions are inaccurate. Given the liability of errors to multiply quickly, these are significant hurdles to meaningful weather predictions.

To combat these effects, meteorologists use ensemble methods. Rather than using one set of initial data, numerical weather models run several times with slight variations of that initial data. If there is general agreement among the many different runs, then there is agreement in the ensemble, which leads (at least in theory, see below) to a higher degree of confidence of the outcome demonstrated in the model—because even if the data was off or the assimilation process skewed the data, the same solution is showing up again and again. When there is not agreement among the different forecasts, then there is low confidence in any particular outcome. However, the different forecasts still give a sense of what possibilities may exist. This approach is an interesting example of what Michael Weisberg calls "multiple models idealization" [17]—the use of several different models, each of which has distinct outputs, for the sake of understanding or predicting some distinct phenomenon.

Parker [11] describes some of the variations to the ensemble approach practiced in different settings. The National Center for Environmental Prediction (NCEP) uses an "ensemble transform bred vector approach" in which the variations in initial conditions mimic the variations in other recent forecasts for the initialization time. The European Center for Medium-Range Weather Forecasting (ECMWF) uses a different method of "singular value decomposition"—identifying variations that are most likely to multiply. A different approach (used by the Meteorological Service of Canada) is to use different initial conditions as well as different models (sets of equations).

Parker's analysis suggests that we can give a limited inductive argument in favor of the skill of numerical weather models (especially if we include post-processing). Assuming that the model and relevant atmospheric causes are not significantly different from previous years, she argues that we are justified in expecting a particular level of skill with regard to the forecast of some weather event. Parker's example is helpful: "Suppose that in each of the last eight summers a particular ensemble system has delivered daily forecasts of the probability of E: The temperature in San Diego will exceed 30 °C on at least one of the next three days. Suppose further that in each of those summers the forecasts of E had a Brier Skill Score between 0.44 and 0.48, with a mean of 0.46" [11]. Under these conditions, and assuming that the models and the atmospheric causes are not significantly different, then we are justified in believing that any forecasts of E will have approximately the same skill score.

This is a fairly modest conclusion. It is not the conclusion that weather models, generally speaking, are sufficiently predictive. Nor is it the claim that at least one member of an ensemble will be in a certain skill range with respect to some forecasted element. It claims that weather models can be expected to be consistent in their skill, provided that the atmosphere isn't significantly changing (and we know the opposite to be the case, due to climate change). Whatever the case, it is clear that the use of weather models alone will be insufficient to generate perfect accuracy. And, in reality, meteorologists rely on a wide range of data beyond just NWP to make predictions, meaning that much of their forecast will depend upon their own judgments—something that can help but not fully remove uncertainty.

Practically this means that the uncertainty in weather forecasting never completely goes away. Oftentimes, the uncertainty will have limited practical impacts on the predictions that are made—perhaps uncertainty about the precise location of some high pressure system, or of the actual mid-day 850 mb temperature. But there are other times when the uncertainty has great practical impacts—a sharp gradient between rain and snow, or the precise location of a hurricane in five days, and so on. In these cases, it is the challenge of meteorologists to decide how to make (or withhold) predictions. If predictions are made, some thought must be given to explore how to also share and communicate the uncertainty about the forecast.

3   Meteorological Prediction in the United States

In the United States, the NWS (a division of the National Oceanic and Atmospheric Administration, itself a division of the U.S. Department of Commerce) is tasked with developing weather forecasts and advisories which people can use to prepare for weather events. The NWS is split into six regions of local weather forecast offices in addition to several river forecast centers and national centers, e.g. the NHC and the SPC [16]. The 122 local weather forecasts offices are each tasked with collecting and disseminating data, developing local forecasts, and issuing advisories, watches, and warnings for a particular region.

Local forecasts often involve graphical information—e.g. about the expected depth of snowfall for particular regions. There are also textual forecasts for about ten days out and statements about potential severe weather. The advisories, watches, and warnings issued by a local weather forecast office relate to a wide range of weather phenomena—temperature (excessive heat, freeze, frost, wind chill, etc.), wind (high wind, extreme wind), fires (red flag warning) severe weather (severe thunderstorm, tornado, winter storm, blizzard, etc.), and flooding (flash flood, flood, river flood), among others. Though their forecast period is much more immediate, advisories, watches, and warnings are themselves a particular kind of forecast (about expected significant weather effects) with a shorter time span—the next few hours or minutes, depending on the length of the advisory. There is also a wide range of data collected at these offices—information about the atmosphere via radiosondes on weather balloons, information from weather stations on the ground, rainfall data, etc. Finally, these offices are also responsible for conducting storm reports, including tornado surveys—wherein teams work to survey damage and write a report about the storm and its damage.

4   Diagrammatic Representations of Uncertainty in Graphical Weather Products

Because of the inherent uncertainty of weather forecasting, examples of the representation of uncertainty could be made for any entity which makes meteorological predictions. In this essay, I will focus on three examples coming from three different parts of NOAA—the Storm Prediction Center (SPC), the National Hurricane Center (NHC), and a local office of the National Weather Service (NWS).

4.1   The Storm Prediction Center (SPC)

The SPC is located in Norman, Oklahoma and is responsible for issuing several products related to severe weather, including weather watches, mesoscale discussions, and convective outlooks in addition to several fire weather products. I will here focus on the products related to severe weather. This is a particularly interesting context because it is not possible to predict precisely where some severe weather event (e.g. a tornado, damaging hail, or damaging wind). Instead, meteorologists are able to identify the places where there is increased probability of such an event. This is visible at all the stages of severe weather products.

The first product issued by the SPC which identifies the possibility of severe weather are "Convective outlooks." Convective outlooks are first issued as early as one week from a potential severe weather event. The outlooks are fairly vague for days 4–8, and more specific for days 1 through 3. A given outlook will identify regions in which there is a particular likelihood of a severe weather event using five categories: marginal, slight, enhanced, moderate, and high. Depending upon the likelihood of different severe weather events, different categories are used, according to the chart below (Fig. 1):

Day 1 Outlook Probability	TORN	WIND	HAIL
2%	MRGL	Not Used	Not Used
5%	SLGT	MRGL	MRGL
10%	ENH	Not Used	Not Used
10% with Significant Severe	ENH	Not Used	Not Used
15%	ENH	SLGT	SLGT
15% with Significant Severe	MOT	SLGT	SLGT
30%	MDT	ENH	ENH
30% with Significant Severe	HIGH	ENH	ENH
45%	HIGH	ENH	ENH
45% with Significant Severe	HIGH	MOT	MOT
60%	HIGH	MOT	MOT
60% with Significant Severe	HIGH	HIGH	MOT

Fig. 1. Day 1 Outlook Probability to Category Conversion. https://www.spc.noaa.gov/misc/about.htm

So, the general way this works is that a forecaster or group of forecasters will make judgments about the likelihood of some particular severe weather events. If they judge there to be a 30% chance of tornadoes occurring then they'd issue an outlook of "moderate" overall. They also give the event-specific probabilities for tornadoes, wind, and hail. Thus, each day will have four products. The following is an example of the day 1 outlook for April 12, 2021, in which a large area of moderate risk was identified in the Southeastern United States.

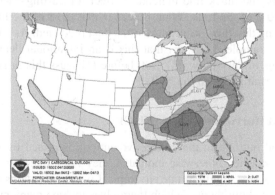

Additional graphics are released which show the risk of particular weather events, from left to right below, tornadoes, wind, and hail for the same day:

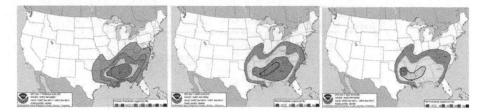

A few remarks are necessary. One is that the probability is based on the "best estimate of a severe weather event occurring within 25 miles of a point" (SPC Products 2021). So, for example, a 10% chance of a tornado indicates that there is a 10% chance that a tornado will occur within 25 miles of an enhanced tornado risk area. It's also worth noting that these are subjective judgments made with meteorological expertise.

Studies of the verification of these events has shown that the convective outlooks tend to be pretty accurate, though they tend to underpredict [4–6]. By this, I mean to say that they tend to give a lower probability of events happening than they actually happen. This is perhaps best understood diagrammatically, as made by Herman, Nielsen, and Schumacher [4]:

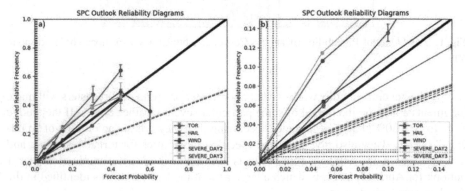

In these graphs, the black line indicates perfect forecasting—where a 20% event happens exactly 20% of the time. What is seen is that most of the lines are above the black line, indicating that they occur more often than predicted. As I'll explore in more detail in the next example, I suspect that this is an intentional decision so that they can avoid criticisms of "crying wolf." But, before I discuss this in greater detail, let's consider a similar example from the NHC.

4.2 The National Hurricane Center (NHC)

A similar graphical outlook to the convective outlook is issued by the NHC. During hurricane season, or when otherwise indicated, they issue five-day graphical tropical weather outlooks. These highlight current tropical weather systems (with links to get updated information about them—which themselves include many graphical representations) as well as the potential for development in the coming days. Below is an example from August 27th, 2017, which indicates the likelihood of the development of a wave of energy coming off the West African coast into the Atlantic. This would eventually go on to be Hurricane Irma—a strong Category 5 hurricane with the second-strongest wind speeds of a hurricane at landfall on record (285 km/h).

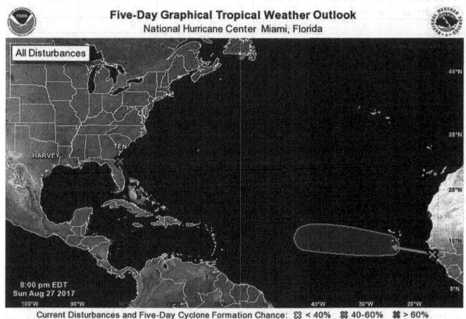

These are similar in many ways to the outlooks offered by the SPC—they serve as a sort of initial alert to be attentive to the weather conditions going forward and begin considering preparations. Information is given probabilistically (at intervals of ten percent, though color coding puts chances into three categories) with percentages being shared when the area is hovered over with a cursor. Additionally, graphical information is given about where that genesis is likely to occur.

Each year, the NHC releases a season report including information about how their five- and two-day graphical outlooks performed. Below is a summary of compiled data from 2015–2019 (Fig. 2).

There are two interesting remarks to make about this data. The first is that, at 120-h, the forecasters are reticent to make predictions of a probability of 0 genesis (with only 212 forecasts over the five year time span of 0, about 6% of all forecasts made) compared to at 48-h, where they are fairly likely to make such a prediction (with 1,474 forecasts over the five-year time span, about 48% of all forecasts made). This is reasonable even when looking at the verification data, given that the forecasts of 0 probability of genesis were much more accurate at 48-h than 120-h. There's also the fact that forecaster ability to predict at five days is much lower than the ability to predict at two days (Zhang and Weng 2014).

48-h Predicted Probability of Genesis	Actual Percentage of Genesis at 48-h	Number of Forecasts	120-h Predicted Probability of Genesis	Actual Percentage of Genesis at 120-h	Number of Forecasts
0	3.6	1474	0	8	212
10	15.6	624	10	20.7	964
20	24.3	313	20	34.4	545
30	35.8	190	30	50.1	304
40	47.9	134	40	42.8	263
50	51	109	50	51.3	235
60	61.8	76	60	66.4	180
70	85.2	88	70	76.4	144
80	90.7	54	80	82.5	126
90	91.6	36	90	96.5	118
100	100	2	100	100	2

Fig. 2. Summary of Atlantic Basin Genesis Verification Rates. Forecast Reliability from 2015–2019 at 48-h and 120-h. Data compiled using data from the annual NHC verification reports, available at: https://www.nhc.noaa.gov/verification/verify3.shtml.

The other interesting feature to note is, as was seen in the data from the SPC, the NHC outlooks systematically underestimate the probability of genesis. Nearly all probabilities given at both 48-h and 120-h are all underestimations of the likelihood (with the exception of 100% probability at both time periods, for which there were only four predictions and for which the actual genesis rate was equal to the prediction rate).

If it is intentional (as I suspect it may be), this approach seems to be in line with the general philosophy of the NHC, which is deeply concerned with its credibility. The concern about credibility is evident in the graphical five-day forecasts made about tropical cyclone development and track. Even when the data from NWP models or other information indicates a change in forecast is necessary, the NHC makes only modest changes with each forecast. So, for example, if the newest models have shown a track much further west than previous runs, then the official forecast will also shift west, but to a lesser degree. According to Michael Brennan, the Branch Chief of the Hurricane Specialist Unit, the reasoning behind this is mostly sociological: "Credibility can be damaged by making big changes from one forecast to the next, and then having to go back" [3]. The reasoning seems to be that they'd rather be a bit wrong in actual path than all over the place, since flip-flopping predictions will lead to a lack of confidence in the ability of the NHC to predict the location of the hurricane.

It seems that a similar approach is present both here and in the SPC forecasts: by systematically trying to underplay threats (by a little bit), both the SPC and NHC can ensure that they avoid the real-world risks associated with "crying wolf". This same attitude in also important in graphical products put out by local offices of the NWS.

4.3 The National Weather Service

For the SPC and NHC, I examined standard graphical representations that are issued frequently, in the same format, and at predictable times. Similar products are also available from the NWS local offices, but I want to focus on some of the one-off diagrams and figures that are made by these offices which are meant to help the audience understand the uncertainty associated with this particular weather event. For this purpose, I will select a couple of graphical representations shared by local NWS offices near to me on social media (specifically on Twitter). Consider the following series of tweets made by the NWS Omaha regarding a winter storm that occurred from January 30, 2021 to January 31, 2021. The first tweet is from January 27, three days before the start of the event[1]:

At this early stage, they show a vague graphical representation of probabilities of "impactful winter weather" careful to note the possibility of freezing rain to impact the forecast outcome. The next tweet, issued a day later (January 28) gets a bit more specific about the types and locations of snow—but does not yet issue a traditional forecast map with the total amount of expected snow or ice.

---

[1] All tweets come from the social media account of the NWS in Omaha (@NWSOmaha).

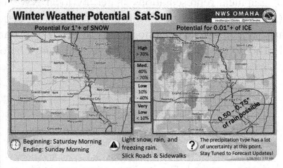

At the next social media update offered later that same day, the NWS office issued its first prediction for the amount of snow and ice expected at different areas. Note the region where the greatest amount of uncertainty persists highlighted:

In the next update, they modified their predictions and continued to identify and note that there was persistent uncertainty about the snowfall totals in a region where temperature would make a big difference.

Once again, we see an attempt being made to carefully and slowly make specific predictions (to avoid "going on the record" before there is sufficient certainty) and the way in which representing that uncertainty is important to help individuals understand the potential for impacts.

5 Conclusion

There are a few worthwhile conclusions to make about these examples which reveal how diagrams function within the landscape of uncertainty management within which these organizations function.

5.1 Practical Orientation of Diagrammatic Representations

The mission of the NWS (which oversees both the SPC and NHC) is to "Provide weather, water, and climate data, forecasts and warnings for the protection of life and property and enhancement of the national economy" [1]. Representing and communicating uncertainty is an essential part of that project, given that uncertainty is an unavoidable feature of weather prediction.

The diagrams here show valuable benefits for this purpose. The map format of many of these diagrams is a useful element to help the public locate themselves visually within different levels of risk while also allowing for comparative judgments about risk level. They are thus well-suited to the practical aims and goals at hand, encouraging individuals to make practical change to behavior to avoid or mitigate risks.

5.2 Credibility Maintenance

In part due to these practical concerns, the NWS, SPC, and NHC are clearly committed to maintaining their credibility with the public, so that they will continue to take any forecasted threats seriously. In order to maintain that credibility, these organizations seem to frequently underestimate the risks at hand. This is to avoid "crying wolf". Here, we can see that the uncertainty of forecasts might be overemphasized, not for epistemic reasons, but for non-epistemic, value-oriented reasons. That is to say that the judgments made within this scientific discipline are governed not only by epistemic norms of truth or prediction, but also by non-epistemic norms of saving lives and mitigating damage to property and livestock. This is a noteworthy example of value-judgments in science. In this context, they are affecting judgments of uncertainty.

5.3 The Nature of Uncertainty

Uncertainty is an interesting concept and comes in both continuous and discrete or binary forms. In the context of meteorological prediction, binary uncertainty occurs in the context of long-term forecasts, for which there is insufficient information to make any particular prediction. The other form is continuous--and occurs in contexts where there is reason to support some particular event happening, but there is uncertainty about where, precisely, it will happen or the likelihood of it happning.

The diagrams that I examined in this essay are examples of continuous uncertainty. These examples show that this form of uncertainty is well-suited to diagrammatic representation, given the way it is spread across location and often admits of degree. This is potentially useful for understanding how this form of uncertainty functions in other domains (perhaps in medicine or public health, as the recent Covid-19 crisis has revealed, with variable risks in different locales).

References

1. "About the NWS": 2021. 2021. https://www.weather.gov/about/
2. Alley, R.B., Emanuel, K.A., Zhang, F.: Advances in weather prediction. Science **363**(6425), 342–344 (2019). https://doi.org/10.1126/science.aav7274
3. Brennan, M.: Tropical cyclone track: overview, challenges, and forecast philosophy. In: In: Presented at the national hurricane conference, national hurricane center, April 20 (2017). https://www.nhc.noaa.gov/outreach/presentations/NHC2017_TrackForecasting.pdf
4. Herman, G.R., Nielsen, E.R., Schumacher, R.S.: Probabilistic verification of storm prediction center convective outlooks. Weather Forecast. **33**(1), 161–184 (2018). https://doi.org/10.1175/WAF-D-17-0104.1
5. Hitchens, N.M., Brooks, H.E.: Evaluation of the storm prediction center's day 1 convective outlooks. Weather and Forecasting **27**(6), 1580–1585 (2012)
6. Hitchens, N.M., Brooks, H.E.: Evaluation of the storm prediction center's convective outlooks from day 3 through day 1. Weather and Forecasting **29**(5), 1134–1142 (2014)
7. Lazo, J.K., Lawson, M., Larsen, P.H., Waldman, D.M.: US economic sensitivity to weather variability. Bull. Am. Meteor. Soc. **92**(6), 709–720 (2011)
8. Lorenz, E.N.: A study of the predictability of a 28-variable atmospheric model. Tellus **17**(3), 321–333 (1965)
9. Lorenz, E.N.: The predictability of a flow which possesses many scales of motion. Tellus **21**(3), 289–307 (1969). https://doi.org/10.1111/j.2153-3490.1969.tb00444.x
10. National Safety Council: n.d. Weather-Related Deaths and Injuries. Injury Facts (blog). https://injuryfacts.nsc.org/home-and-community/safety-topics/weather-related-deaths-and-injuries/. Accessed 26 June 2020
11. Parker, W.S.: Predicting weather and climate: uncertainty, ensembles and probability. Studies in History and Philosophy of Science Part B: Studies in History and Philosophy of Modern Physics **41**(3), 263–272 (2010)
12. Parker, W.S.: Whose probabilities? Predicting climate change with ensembles of models. Philosophy of Science **77**(5), 985–997 (2010)
13. Parker, W.S.: Ensemble modeling, uncertainty and robust predictions. Wiley Interdisciplinary Reviews: Climate Change **4**(3), 213–223 (2013)
14. Parker, W.S.: Simulation and understanding in the study of weather and climate. Perspect. Sci. **22**(3), 336–356 (2014)
15. Uccellini, L.W., Ten Hoeve, J.E.: Evolving the national weather service to build a weather-ready nation: connecting observations, forecasts, and warnings to decision-makers through impact-based decision support services. Bull. Am. Meteor. Soc. **100**(10), 1923–1942 (2019). https://doi.org/10.1175/BAMS-D-18-0159.1
16. US Department of Commerce, NOAA. n.d. "We Are the National Weather Service." Accessed July 2, 2020. https://www.weather.gov/about/nws.
17. Weisberg, M.: Three Kinds of Idealization. J. Philos. **104**(12), 639–659 (2007)

18. Zhang, F., et al.: What is the predictability limit of midlatitude weather? J. Atmospheric Sci. **76**(4), 1077–1091 (2019)
19. Zhang, F., Weng, Y.: Predicting hurricane intensity and associated hazards: a five-year real-time forecast experiment with assimilation of airborne doppler radar observations. Bull. Am. Meteor. Soc. **96**(1), 25–33 (2014). https://doi.org/10.1175/BAMS-D-13-00231.1

Structuralist Analysis for Neural Network System Diagrams

Guy Clarke Marshall[1]([✉]) [iD], Caroline Jay[1] [iD], and André Freitas[1,2] [iD]

[1] Department of Computer Science, University of Manchester, Manchester, UK
`guy.marshall@postgrad.manchester.ac.uk,`
`{caroline.jay,andre.freitas}@manchester.ac.uk`
[2] Idiap Research Institute, Martigny, Switzerland

Abstract. This paper examines diagrams describing neural network systems in academic conference proceedings. Many aspects of scholarly communication are controlled, particularly with relation to text and formatting, but often diagrams are not centrally curated beyond a peer review. Using a corpus-based approach, we argue that the heterogeneous diagrammatic notations used for neural network systems has implications for signification in this domain. We divide this into two questions (i) what content is being represented and (ii) how relations are encoded. Using a novel structuralist framework, we use a corpus analysis to quantitatively cluster diagrams according to the author's representational choices. This quantitative diagram classification in a heterogeneous domain may provide a foundation for categorising representational properties of diagrams.

Keywords: Neural networks · Scholarly diagrams · Empirical · Semiotics

1 Introduction

Currently, there is no consistent model for visually or formally representing the architecture of neural networks (NNs). This lack of representation brings interpretability, correctness and completeness challenges in the description of existing models and systems engineering. In the context of scientific communication, most approaches and systems today are described by a combination of arbitrary diagrammatic elements, algorithms, formulae and natural language descriptions. In this paradigm, there is little consistency on abstraction levels or notation. From the perspective of scientific practice, these limitations challenge dialogue, transparency and reproducibility.

We show that authors of applied NN systems papers could be getting more communicative value from their diagrams. Focusing on the diagrammatic representations used, it seems the author's notion of relevance is skewed, resulting in missing context and content. Diagrams often feature partial information on, for example, the key novelty or contribution the paper is articulating. This is unlike

© Springer Nature Switzerland AG 2021
A. Basu et al. (Eds.): Diagrams 2021, LNAI 12909, pp. 480–487, 2021.
https://doi.org/10.1007/978-3-030-86062-2_49

other disciplines such as mechanical, electronic, or structural engineering, where standardised and complete diagrams are utilised.

This short paper is an empirical analysis based on semiotic theory, introducing a semiotic viewpoint to the analysis of NN diagrams used at leading Natural Language Processing (NLP) conferences, applying a novel structuralist approach, inspired by structural linguistics [8] to survey a corpus of NN diagrammatic representations. Our contributions are:

- To introduce diagrammatic structuralism as a method for analysing diagrammatic semiotics.
- To show the heterogeneity of neural network diagrams through an empirical corpus-based analysis.
- To identify clusters of diagrams through quantitative analysis.

2   Related Work

2.1   Scientific Diagrams

In an interview study involving academics about NN system diagrams, Marshall et al. [11] found differing views on whether precision was meaningful: "If you have four, it means you have four" contrasts with "it cannot be four dimensions, it must be much more than that."

Specific semiotic principles for diagrams can be found in other technical applications. In the domain of Business Process Mapping, it was found that the shape of the graphic symbols improved ease of understanding more effectively than colour or number of symbols [3]. In a small study of mechanical aviation diagrams, Kim et al. [9] found that highlighting/shading to identify key components was more usable and useful, compared to other visual techniques such as "zooming in". Additionally, Heisner and Tversky [6] showed in mechanical diagrams that arrows added to static structural (parts and relations) diagrams increase the capability of the representation to convey function (temporal, dynamic, or causal process).

2.2   Diagrammatic Representations

According to Peirce [15], the concept of a diagram as an icon is malleable to Kant's concept of Schema (the procedural rule by which a Kantian category is associated with a sense impression). Applying Saussurian semiotic thinking beyond the linguistic domain, in a diagram we might wish to minimise the distance between the signifier and the signified. Tylén et al. [19] concisely summarise that diagrams:

- are external representational support to cognitive processes [2].
- make abstract properties and relations accessible [7].
- can be in a public space, therefore enabling collective and temporally distributed forms of thinking [15].

– are manipulated in order to profile known information in an optimal fashion.

Diagrams and text, even if together in the same media, are different modalities. Due to the prevalence of their combined usage, when referencing diagrams we assume dual coding of text and graphics as part of our working definition of a diagram. We build on Morris [14] by partitioning semiotics, the field examining the process by which something functions as a sign, into the following categories:

- *Semantics*: relation between signs and the things to which they refer, or their meaning
- *Syntactics*: relations among or between signs
- *Pragmatics*: relation between signs and sign-users [14]
- *Empirics*: the statistical analysis of repeated use of signs, adapted from the Information Systems domain [18]

Recent research into diagrammatic empirics include those by Lechner [10] and Marshall et al. [13].

3   Neural Network Diagram Empirics

3.1   Context

Neural approaches are popular in AI research, and many of the diagrammatic design decisions are centred on the machine-learning mechanisms. We performed a corpus-based analysis on the use of diagrams to represent NNs, with NLP as an application domain.

This study applies an empirical approach to diagrammatic representations, inspired by structuralist linguistics, in order to uncover the potential implicit semiotic processes and laws. This approach has a potential shortcoming in revealing only partial truths [16], but allows us to access a complex conceptual space. By analogy to Saussure's [17] linguistic concept of *parole* (speech as concrete instances of the uses of language), the proposed *diagrammatic structuralism* purely considers the sign itself and does not include the act of using it. To assess the similarity of diagrams from a visual perception standpoint, we should consider the overall structure rather than substructures [4]. Following Saussure, the American Structuralist school (e.g. Bloomfield [1]) advocate a mechanistic approach to the analysis of linguistics, which we apply to NN diagrams. We believe this approach to be expedient and not excessively reductionist, as the establishment of common graphical elements is beneficial for understanding what is being articulated.

3.2   Method

In order to map the current state of diagrammatic representations of NNs, we conducted systematic literature analysis, categorising NN diagrams based on Diagrammatic and Content features (see [12] for a complete list, including observed frequencies). This partition is inspired by Jakobson's [8] "axis of

Table 1. Surveyed Papers

Source	Papers	Sampled	Diagrams
COLING 2016	338	16	6
COLING 2018	330	16	8
EMNLP 2016	275	9	4
EMNLP 2017	323	14	6
NAACL 2016	181	8	4
NAACL 2018	332	14	7
NIPS 2016	569	14	2
NIPS 2017	679	23	3

selection" and "axis of combination" applied in structural linguistics. Prior to the sampling, we identified repeated concepts, divided into "Content" and "Diagrammatic" primitives, a diagrammatic equivalent of Jakobson's "selection" and "combination." The categorisation is in this way because the sets of symbols and their relationships together can also form a symbol themselves, in the same way that a set of low-level symbols come together to form a combined diagram symbol. This intertwining means it is not possible to distill semantic and syntactic information independently, as argued by Jakobson for linguistics.

We then sampled and applied our coding based on content and diagrammatic features (see Table 1 for content, diagrammatic features table omitted for brevity), analysing the results.

With a corpus-based approach, we randomly sampled papers and manually extracted 40 diagrams (see Table 1). Where multiple diagrams were identified, we have taken the diagram including the most features. We excluded Computer Vision venues to avoid complication with the visual nature of their input and high specificity of their architectures, and as such this is not representative of all scholarly NN diagrams.

The diagrams were assessed by one NN practitioner according to pre-defined categories covering Content (what features have been selected for diagrammatic representation) and Diagrammatic concepts (their visual and relational encoding).

3.3   Analysis

The low occurrence of data-centric content features such as operations indicates that these diagrams are not describing transformations the data undergoes, but instead prioritise the NN system, particularly the layers, dimensionality, input and output.

In order to extract 40 diagrams, we sampled 114 papers. Table 1 details the paper counts from each conference. From papers at these recent top neural network-centric venues, we expect that at least 25% will contain NN diagrams

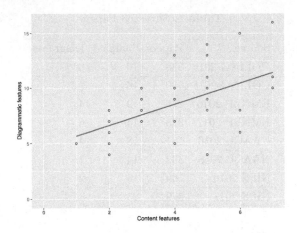

Fig. 1. Surveyed semiotic content of diagrams

(with 99% confidence, due to our sampling of 40 diagrams from 114 sampled papers from a total population of 3027). Note that the sampling was not filtered for NN content, and the prevalence of NN diagrams underscores the community requirement for effective NN diagrams.

Most diagrams are representing the same content. In 82.5% of papers, the role of NN diagrams in the sample was to communicate the function of the NN, rather than context, the whole system, or a component. In the sample, *no diagram included identical Diagrammatic features*. Two of the sampled diagrams shared the same Content features (lower semantic content, NN-only diagrams, with an output and an explicit example). Figure 2 shows the quantity of semiotic elements according to our schema, which were found to be normally distributed (13.07) and have a large standard deviation (3.96). Details are available [12]. Some highlights:

- 92.5% used directional arrows, and 42.5% dotted arrows. Given the utility of arrows to provide a sense of functional composition, this uniformity is as expected.
- 60% had a caption over one sentence long, perhaps suggesting the authors struggled to get the relevant content into the diagram.
- 60% used an ellipsis in order to indicate "incomplete sets of objects."
- 47.5% included labels on neural layers.
- 42.5% made use of an explicit example input.
- 52.5% did not indicate dimensionality.
- 27.5% indicated the key features by boundary lines, shading or labeling.
- 10% indicated data resources of any kind, and a different 10% utilised symbolic mathematical tensor operation.
- None used UML, SysML, OML or other existing formal diagrammatic language.

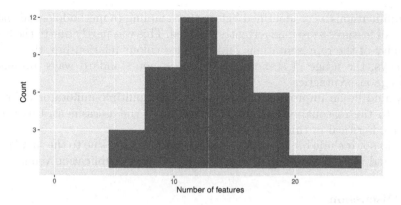

Fig. 2. Distribution of the total number of features in each diagram

Figure 1 shows the relationship between NN Content and Diagrammatic features. Diagrams with additional feature content are more likely to have additional diagrammatic features. It may be that there is variable author diagramming efficiency, or that these systems are intrinsically more complex in both dimensions.

The study found little evidence of standardisation. In the language of Peirce, the level of heterogeneity in representing such similar *objects* implies that it is more appropriate to consider the components of NN diagrams as *symbols* rather than *icons*, since conventions appear to be utilised over physical resemblance of system or data structures. Despite this heterogeneity, common themes have been identified in order to classify and group diagram types. Based on *k-means clustering* on the 69-dimensional space created by applying diagrammatic structuralism (24 content and 45 diagrammatic), together with manual adjustment, we identified four potential NN diagram classes:

1. *Visual* using coloured circles to represent tensors, mostly left-to-right shape, mostly including an input and output example, a short caption, and featuring an embedding.
2. *Mathematical* with more mathematical notation, often with an upward linear shape.
3. *Lightweight* with less content and fewer visual encoding mechanisms, relying on text, often with labelled layers, and sometimes in a Block diagram format.
4. *Unorthodox* not fitting into the above groups.

Applying this classification manually, we identified 8 Visual, 4 Mathematical, 12 Lightweight and 16 Unorthodox diagrams. This classification may help structure future research, with specific studies for each diagram class.

3.4 Limitations

– The dimensions were chosen to have a high coverage of representative features in the space. Whilst this was intended to be comprehensive, because the dimensions were identified before the sampling (a methodological choice), several features were inadvertently omitted. This was partly due to the heterogeneity of the representations and included colour relationship, the presence of keys, the usage of standard notation in non-standard ways, presence of ambiguous syntactics.

– It would be an improvement in rigor to have multiple annotators. However, due to the reasonably unambiguous nature of the assessment, this was not deemed critical in this early stage of analysis.

– We do not include bibliometrics such as citation count due to the small sample size and the sensitivity in adjustment for different publication venues.

3.5 Discussion

The consequence of variable signification methods within the same domain is that readers are required to switch modalities even between very similar papers. Readers may also be unable to access the information they require, as the diagrammatic signifier may not contain some important content-related aspects of the signified system.

We found that additional visual objects were correlated with additional diagrammatic grammar features. This suggests that either the expressions were more complex and required both types of encoding, or that some author were more likely to use a larger variety of encoding objects generally (not discriminating between objects and relations). We have not shown that this heterogeneity is a problem per se, but it does differ from many other Computer Science domains where diagrammatic representation of systems is more standardised.

Grice [5] distinguishes between "what is said" and "what is implicated." In this domain, the figure caption text sometimes refers to a system rather than a scholarly contribution. It may be that in these diagrams we are observing an ambiguity between "what is drawn" and "what is implicated," even for the author's own mind. The system diagram may be functioning as a Peircian index for a scholarly contribution, rather than a system, the functioning of that system, or the manipulation of data. This ambiguity introduces complexity in identifying the object of the diagram. Despite this ambiguity, a semiotic lens has allowed quantitative exploration of diagram components and poses further questions about the intent of the authors.

4   Conclusion

Using a structuralist framework we gathered, sampled and analysed neural network diagrams from a variety of recent neural network-centric conferences, using this to demonstrate aspects of the heterogeneity of diagrammatic representations employed. Using a k-means clustering algorithm on the high-dimensional visual semiotic space, we derived clusters of NN diagram types.

Beyond diagram classification and heterogeneity quantitification, this work aims to encourage increase attention on scholarly diagramming practices, which may help authors to better fulfil their communicative intent.

References

1. Bloomfield, L.: Language. University of Chicago Press (1984)
2. Clark, A., Chalmers, D.: The extended mind. Analysis **58**(1), 7–19 (1 1998). https://doi.org/10.1093/analys/58.1.7
3. Gabryelczyk, R., Jurczuk, A.: Does experience matter? Factors affecting the understandability of the business process modelling notation. Procedia Eng. **182**, 198–205 (2017). https://doi.org/10.1016/j.proeng.2017.03.164
4. Goldmeier, E.: Similarity in visually perceived forms. Psychol. Issues **8**(1), 1–136 (1972). http://www.ncbi.nlm.nih.gov/pubmed/4661205
5. Grice, P.: Studies in the Way of Words. Harvard University Press, Cambridge (1989)
6. Heiser, J., Tversky, B.: Arrows in comprehending and producing mechanical diagrams. Technical report, Department of Psychology, Stanford University (2006). https://doi.org/10.1207/s15516709cog0000_70
7. Hutchins, E.: How a cockpit remembers its speeds. Cogn. Sci. **19**(3), 265–288 (7 1995). https://doi.org/10.1207/s15516709cog1903_1
8. Jakobson, R., Halle, M.: Fundamentals of Language, vol. 1. Walter de Gruyter (2010)
9. Kim, S., Woo, I., Maciejewski, R., Ebert, D.S., Ropp, T.D., Thomas, K.: Evaluating the effectiveness of visualization techniques for schematic diagrams in maintenance tasks. In: Proceedings of the 7th Symposium on Applied Perception in Graphics and Visualization - APGV 2010, New York, p. 33 (2010)
10. Lechner, V.E.: Modality and uncertainty in data visualizations: a corpus approach to the use of connecting lines. In: Pietarinen, A.-V., Chapman, P., Bosveld-de Smet, L., Giardino, V., Corter, J., Linker, S. (eds.) Diagrams 2020. LNCS (LNAI), vol. 12169, pp. 110–127. Springer, Cham (2020). https://doi.org/10.1007/978-3-030-54249-8_9
11. Marshall, G., Freitas, A., Jay, C.: How researchers use diagrams in communicating neural network systems. arXiv preprint arXiv:2008.12566 (2020)
12. Marshall, G., Jay, C., Freitas, A.: Data supporting "Structuralist analysis for neural network system diagrams" (2021). https://doi.org/10.6084/m9.figshare.14813025
13. Marshall, G.C., Jay, C., Freitas, A.: Number and quality of diagrams in scholarly publications is associated with number of citations. In: Basu, A., et al. (ed.) Diagrams 2021, LNAI 12909, pp. 512–519. Springer, Heidelberg (2021)
14. Morris, C.W.: Foundations of the theory of signs. J. Symbolic Logic **3**(04), 158 (1938). https://doi.org/10.2307/2267781
15. Peirce, C.S.: The collected papers of Charles S. Peirce. Harvard University Press (1931–1966)
16. Rutherford, D.: How structuralism matters. J. Ethnographic Theory **6**(3), 61–77 (2016). https://doi.org/10.14318/hau6.3.008
17. Saussure, F.D., Baskin, W., Meisel, P., Saussy, H.: Course in General Linguistics. Columbia University Press (2011)
18. La Mantia, F.: Semantics. In: Vercellone, F., Tedesco, S. (eds.) Glossary of Morphology. LNM, pp. 459–463. Springer, Cham (2020). https://doi.org/10.1007/978-3-030-51324-5_106
19. Tylén, K., Fusaroli, R., Bjørndahl, J.S., Rączaszek-Leonardi, J., Østergaard, S., Stjernfelt, F.: Diagrammatic reasoning: abstraction, interaction, and insight. Pragmatics Cogn. **22**(2), 264–283 (2014). https://doi.org/10.1075/pc.22.2.06tyl

Modeling Multimodal Interactions and Feedback for Embodied Geovisualization

Markus Berger(✉)

University of Rostock, Rostock, Germany
markus.berger@uni-rostock.de

Abstract. Modern wearable devices, especially those aimed at enabling Augmented and Virtual Reality, permit an increasing number of ways to translate data about us and our surroundings into sensory impressions. This is especially useful for geospatial data – data that already has a location in the environment. In the same way that a modern weather map translates simulation results into red, blue, and green pixels, we could now, for example, translate air quality data in a user's vicinity into an audio rhythm, creating a sort of virtual Geiger counter not limited to ambient ionizing radiation. However, there are currently few conceptual tools that help with designing these kinds of multisensory representations. This poster shows preliminary work on a type of diagram specifically aimed at modelling the interactions and feedback loops between these interactive, embodied visualizations and the human body.

Keywords: Geospatial data · Multisensory visualization · Embodiment · Interaction diagram

1 Introduction

In the earth sciences and other disciplines dealing with geospatial data, diagrams and data visualization are closely related. Behind map-generating services we often find UML diagrams describing their underlying data structures. The same goes for immersive (AR&VR) visualizations, where UML can not only describe structure, but also behaviour. However, these diagrams rarely explicate the larger concept of the modelled system, instead focusing on specific programming choices and subsystems. This can become an issue for visualization systems that don't make use of established methodologies, as the very technical diagrams might mask some of the novelty.

Recently, some immersive visualizations have started to move beyond the traditionally visual-only means of showing information and began incorporating other sensory modalities, for example through "sonification", which has been a very active area of research for close to three decades [1].

Such multisensory systems are often "embodied", i.e. they "involve a user's body in a natural and significant way, such as by using gestures" [3]. If we now immersively visualize geospatial data in this way, we put a user's body into

© Springer Nature Switzerland AG 2021
A. Basu et al. (Eds.): Diagrams 2021, LNAI 12909, pp. 488–491, 2021.
https://doi.org/10.1007/978-3-030-86062-2_50

a virtual environment that *is* the data, which they then experience through multiple senses, while simultaneously manipulating it with the same body that is also "carrying" the aforementioned senses. In this way, the body is made the center of a multivariate feedback loop.

Developing such systems requires careful planning and modelling. This poster presents preliminary efforts to develop a diagram that can help us document and reason about these immersive, multisensory geovisualizations.

2   Results

A popular concept in data visualization is the visualization pipeline. It exists in several incarnations, but usually includes one or multiple stages of data analysis, filtering or enrichment, a mapping step in which the data is transformed into a representation that can be interpreted visually, a rendering step to show the representations on a computer screen, and "backwards" interaction steps [2].

McCormack et al. [4] adapt this pipeline for immersive, multisensory systems. They extend the traditionally visual mapping step to a "sensorial mapping", which is defined as "a mapping from data elements and data attributes to sensory channels (sight, hearing, touch, proprioception, smell and taste) and their respective sensorial variables (color, pitch, roughness, etc.)". The rendering stage is extended into a device-focused stage, where a device enacts every one of the sensory channels onto the human body, and a body-focused stage, where the involved human senses perceive and body movements trigger interactions. This pipeline will be the starting point for our diagram.

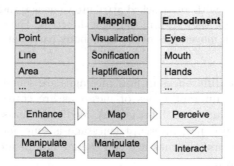

Fig. 1. The multisensory visualization and interaction pipeline, adapted from [4].

For our purposes, there is no way to present a digital sensory channel to a body part if there is no device that can do so. Thus, the device and body stages are combined into an "embodiment" stage. Interactions also need to be able to change both mapping and the data itself. The initial adaptation of the pipeline can be seen in Fig. 1.

To more explicitly model how senses, interactions, and data sets are related, we make use of the concept of "messages", as is also common in UML diagrams.

Here, these messages will run from data sets to body parts (sensory mapping) and back (interactions). Certain messages trigger others in specific ways, modeling the embodied feedback a system provides. Because we are working with geospatial data, we also need to keep in mind what types of "spatiality" different senses can accommodate - we can visually see very complex types of geometry, but only if they are in front of us. Hearing on the other hand works in all directions, but can generally only resolve sound emitter positions as points.

The steps of the pipeline in Fig. 1 are thus translated into more specific message lines connecting a data set (or more specifically one of its properties) with a body part. On the way the data spatiality needs to be mapped to a sensory spatiality, i.e. there need to be one or multiple spatial transformations in the mapping stage. One example for such a transformation is an interpolation over a collection of points, which yields a continuous surface. This might be useful for a haptic application, as we can only touch surfaces, not points. An abstracted sample graph for this message flow is shown in Fig. 2.

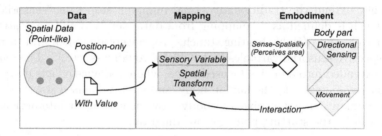

Fig. 2. A point data set with an attached attribute value is transformed into a surface that the user can sense if they turn in the right direction. The surface representation can be changed through movements of the same body part that did the sensing.

There can also be differences in how messages are sent to the sensory channels. Currently, four different types of messages are considered here: queried (message is sent once upon an interaction), feedback (message is sent continually as interaction happens), interrupt (message is sent once a specific state is reached) and continuous (the message can change independently of interactions and is sent constantly).

As an example for a possible real world application, we imagine a user wearing a pair of augmented reality glasses with hand interaction and vibration functionality. Running on the glasses is an augmented reality weather system. Users would be walking around in the real world and could see weather maps directly projected into the sky. An approaching rainstorm would regularly emit pitched warning sounds as it approaches. If a user looks up, more detailed textual information about a cloud could be displayed, and a rumble in the AR glasses could signify that a cloud the user is currently looking at is a storm center. With a hand swipe the user can fast forward through the predicted weather. A possible diagram for this system is shown in Fig. 3.

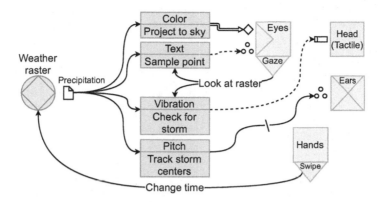

Fig. 3. An example diagram for an immersive weather visualization. A dotted line represents feedback, a crossed line an interrupt, and the double line a continuous message.

What the diagram makes clear is that all these disparate interactions and sensory channels depend on only one data set, that most of the sensory load is on the head, and that most of the input is fully user-controllable or otherwise selective (i.e. feedback and interrupt messages), making it less likely to sensorily overwhelm the user, but also requiring them to learn how to control the system.

In conclusion, the diagram can model a system with several modalities and interaction types, and it can promote thinking about specific steps and necessary transformations. In future iterations, it should also include information about the temporality of the data and time-based interactions, as well as a way to move parts of the visualization from world-space into UI space, for example in the form of a floating mini map. The diagram could also be applied to certain non-geospatial visualization problems, for example when abstract data is turned into "data landscapes".

The usefulness of this approach should also be assessed through a formal literature review. This review could focus on whether existing applications can be appropriately modeled and whether all possible message types and channels are covered, similar to the review methodology used by Dubus and Bresin [1].

References

1. Dubus, G., Bresin, R.: A systematic review of mapping strategies for the sonification of physical quantities. PloS One **8**(12), e82491 (2013)
2. Haber, R.B., McNabb, D.A.: Visualization idioms: a conceptual model for scientific visualization systems. Visual. Sci. Comput. **74**, 93 (1990)
3. Hartson, R., Pyla, P.S.: The UX Book: Agile UX Design for a Quality User Experience. Morgan Kaufmann, Amsterdam (2018)
4. McCormack, J., et al.: Multisensory immersive analytics. In: Immersive Analytics. LNCS, vol. 11190, pp. 57–94. Springer, Cham (2018). https://doi.org/10.1007/978-3-030-01388-2_3

Formal Diagrams

On the Cognitive Potential of Derivative Meaning in Aristotelian Diagrams

Hans Smessaert[1]([✉])[iD], Atsushi Shimojima[2][iD], and Lorenz Demey[3][iD]

[1] Department of Linguistics, KU Leuven, Blijde-Inkomststraat 21,
3000 Leuven, Belgium
`hans.smessaert@kuleuven.be`
[2] Faculty of Culture and Information Science, Doshisha University,
1-3 Tatara-Miyakodani, Kyotanabe 610-0394, Japan
`ashimoji@mail.doshisha.ac.jp`
[3] Center for Logic and Philosophy of Science, KU Leuven, Kardinaal Mercierplein 2,
3000 Leuven, Belgium
`lorenz.demey@kuleuven.be`
`https://www.logicalgeometry.org`

Abstract. In this paper we investigate the cognitive potential of Derivative Meaning—defined in terms of Abstraction Tracking (Shimojima 2015)—in order to characterise various families of Aristotelian diagrams. In a first part we consider the notion of subdiagrams—i.e. uniform triangles of contrariety relations and implication relations—inside Aristotelian hexagons and octagons. In a second part we look at different strategies for embedding complete Aristotelian diagrams—i.e. classical and degenerate squares—into Aristotelian hexagons and octagons.

Keywords: Cognitive potential · Derivative meaning · Aristotelian diagram · Abstraction tracking · Informational/computational equivalence · Logical geometry · Classical/degenerate Aristotelian square

1 Introduction

The overall aim of this paper is to apply the general semantic and cognitive framework for the analysis of diagrams proposed by Shimojima [13] to the analysis of Aristotelian diagrams developed by Demey and Smessaert in the framework of Logical Geometry. In our joint Diagrams 2020 paper [16] we demonstrated the relevance of Shimojima's first cognitive potential—namely that for Free Ride in Inference—in drawing the distinction between consequential constraint tracking

The first author acknowledges the financial support from the Research Foundation–Flanders (FWO) of his research stay at Doshisha University with the second author. The third author holds a Research Professorship (BOFZAP) from KU Leuven. The research was partially funded through the KU Leuven ID-N project 'Bitstring Semantics for Human and Artificial Reasoning' (BITSHARE).

© Springer Nature Switzerland AG 2021
A. Basu et al. (Eds.): Diagrams 2021, LNAI 12909, pp. 495–511, 2021.
https://doi.org/10.1007/978-3-030-86062-2_51

by consequence with Logical Space Diagrams and consequential constraint tracking by correlation with Aristotelian diagrams. In the present paper we focus on Shimojima's fourth cognitive potential—namely that for Derivative Meaning—in order to characterise various families of Aristotelian diagrams. In this introductory section we lay out the basic ingredients of the two frameworks. On the one hand, we present the phenomenon of Derivative Meaning and its technical analysis in terms of Abstraction Tracking. On the other hand, we introduce the Aristotelian relations, diagrams and subdiagrams.

Derivative Meaning. In the realm of visualising quantitative information, tables with numerical data are standardly 'translated' into statistical graphs such as scatter plots, line graphs or bar charts. The fundamental constraints in these graphical representation systems concern 'point-wise' facts: if a dot appears at X-coordinate m and Y-coordinate n in a scatter plot or line graph, then there is an instance in the data with the X-value m and the Y-value n. These semantic constraints are not 'natural' but CONVENTIONAL or arbitrary: they hold because a group of people started to conform to them at some time in history and kept respecting them for the common interest of effective communication [13, p. 103]. We call these historically prior conventions BASIC SEMANTIC CONVENTIONS, to distinguish them from additional conventions that are logically derivable in the way described later in this paper. In this limited point-wise sense, the statistical graphs have the same informational content as their corresponding tables.

Nevertheless, there is a clear sense in which each of these graphs expresses more information than its corresponding table. With scatter plots, the particular shapes of dot configurations indicate different types of correlations—e.g. linear versus quadratic—between the variables X and Y. With line diagrams, differences in degree or direction of the inclines made by the various lines indicate differences in speed, trend or intensity of the changes in the data. In both cases, the observation of general trends or overall shapes yields additional informational relations which are *not* part of the basic semantic conventions of the representation system. Instead, these patterns indicate more abstract or general facts about the represented data. It is important to stress that these 'new' facts have a different logical and historical status. In particular, it is *not* necessary for the establishment of these more abstract relations that the relevant group of people be conforming to a new basic semantic convention. These constraints, by contrast, seem to be holding *naturally*—or logically—once the basic conventions are established, and in this sense can be considered *derived* constraints. We will therefore refer to these additional informational relations as instances of DERIVATIVE MEANING [13, p. 103]. It is precisely these derivative informational relations which significantly contribute to the *informational utility* of statistical graphs: their existence is often the very reason why a given type of graphs is more effective than others as a method of displaying certain information. Furthermore, the distinction between basic semantic conventions and derivative meaning has led researchers to distinguish 'levels of meaning', and to consider derivative or higher-level meanings as the 'main messages' of statistical graphs.

The phenomenon of derivative meaning is by no means restricted to the representation of numerical data in statistical charts. First of all, it is also relevant when displaying spatial information in the form of topographic contour maps or meteorological maps. With such maps, the basic semantic conventions are concerned with individual points on individual contour lines or isobars, whereas overall patterns formed by several proximate contour lines or isobars also carry important information about the mapped topographic or meteorological reality. Such more abstract constraints also illustrate another crucial property of this type of derivative meaning relations, namely the importance of expertise, reading skills and recognition memory in appreciating perceptual patterns [13, p. 108]. Secondly, derivative meaning also plays a crucial role in graphical representation systems for more abstract, symbolic—i.e. non-numerical—data. With node-edge graphs, for instance—such as route maps for underground or subway systems in large cities—the basic semantic convention is that if a (single) edge connects two (adjacent) nodes, it means that the objects denoted by the two nodes stand in the particular relationship represented by the edge. Derivative meaning in these graphs, by contrast, gives rise to 'overview effects' when we observe the presence of a *path* consisting of multiple, consecutive edges connecting two (non-adjacent) nodes, or when we derive the hub-like nature of a particular node from the number of nodes directly connected to it. To sum up: graphical systems can support derivative meaning relations that go beyond their basic semantic conventions. This enables us to read off a richer set of informational relations directly from graphics, expanding the expressive coverage of a system with a relatively simple set of basic semantic conventions.

General Framework for the Analysis of Diagrams. In order to characterise the semantic content of a diagrammatic representation, the framework adopted in this paper [13, p. 23ff.] has a two-tier semantics. It draws a distinction between a TOKEN level at the bottom of Fig. 1(a)—with a REPRESENTATION relation $\rightsquigarrow$ from a representation s to represented object t—and a TYPE level at the top of Fig. 1(a)—with an INDICATION relation $\Rightarrow$ from a source type σ to a target type θ. In the case of a street map, for instance, the representation s is a particular sheet of paper (token) and the arrangement of lines and symbols is the source type σ or property holding of (or 'being supported' by) that s. The actual streets and buildings constitute the represented object t (token) and their overall arrangement is the target type θ or property holding of that t. A representation s represents an object or situation t as being of target type θ if s represents t and s supports a source type σ that indicates θ. Since the notions of Derivative Meaning and Abstraction Tracking will be defined in terms of source and target types, this paper will focus on the type level and the indication relation established by the semantic conventions for the representational practice. A set Γ of source types COLLECTIVELY INDICATES a set Δ of target types ($\Gamma \Rightarrow \Delta$) if Γ and Δ stand in a one-to-one correspondence under the indication relation $\Rightarrow$.

Abstraction Tracking. In order to characterise the notion of Derivative Meaning in a more technical manner in terms of relations between source types and target types, we take the simple example of a so-called Round-Robin table in

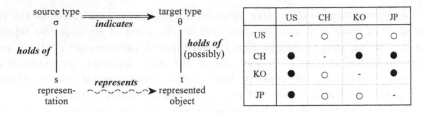

Fig. 1. (a) General framework [13, Figure 21] (b) Round-Robin Table

Table 1. Sets of source types Γ_n and target types Δ_n.

$\Gamma_1 = \{\circ(\text{JP,CH}), \circ(\text{JP,KO}), \bullet(\text{JP,US})\}$ $\Delta_1 = \{W(JP,CH),\ W(JP,KO),\ L(JP,US)\}$

$\Gamma_2 = \{\circ(\text{JP,CH}), \circ(\text{JP,US}), \bullet(\text{JP,KO})\}$ $\Delta_2 = \{W(JP,CH),\ W(JP,US),\ L(JP,KO)\}$

$\Gamma_3 = \{\circ(\text{JP,KO}), \circ(\text{JP,US}), \bullet(\text{JP,CH})\}$ $\Delta_3 = \{W(JP,KO),\ W(JP,US),\ L(JP,CH)\}$

Fig. 1(b) as our starting point.[1] Such a table is used to represent the results of a (sports) competition in which each contestant or team—in this case the US, China, Korea and Japan—meets all other contestants or teams in turn. The basic semantic conventions of a Round-Robin table are as follows: if a white (resp. black) circle appears at the intersection of a row headed by name X and a column headed by name Y, then this means that the team with name X has won against (resp. lost to) the team with name Y. The two source types (characterising the graphical representation) will be abbreviated as $\circ(X,Y)$ and $\bullet(X,Y)$ respectively, whereas the two target types (characterising the represented situation) will be abbreviated as $W(X,Y)$ and $L(X,Y)$, assuming that the italicised X and Y are the teams denoted by the names X and Y. This allows us to reformulate the basic semantic conventions as two general CONSTRAINTS between source and target types, namely (i) if $\circ(X,Y)$ holds, then $W(X,Y)$ holds, and (ii) if $\bullet(X,Y)$ holds, then $L(X,Y)$ holds [13, p. 115]. These constraints are concerned with individual symbols in individual cells, i.e. with the results of individual games. Derivative meaning relations, by contrast, are concerned with an entire row or column, or multiple rows or columns, specifying the information carried by the particular *distributions or patterns* of white and black circles on them. From the two white circles on the final row in Fig. 1(b), for instance, we can read off that the Japanese team won two of its three games (namely against China and Korea). This kind of derivative meaning relation can also be characterised in terms of a relation between a source type—n white circles appear in a row with name X, abbreviated as $\circ_n$(X-row)—and a target type—team X has won against n teams, abbreviated as $W_n(X)$. The crucial question now will be how we can account for the valid informational relation in Fig. 1(b) going from the source type $\circ_2$(JP-row) to the target type $W_2(JP)$, since this is *not* directly specified by the system's basic semantic conventions [13, p. 115].

---

[1] The figure is a considerably simplified/modified version of [13, p. 112, Fig. 84(a)].

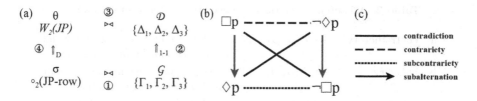

Fig. 2. (a) Abstraction tracking (b) Aristotelian diagram (c) coding conventions.

First of all, it is important to note that both the source type $\circ_2(\text{JP-row})$ and the target type $W_2(JP)$ are *abstract* conditions in the sense that there are several alternative ways in which they are true. Table 1 lists some sets Γ_n of source types, and Δ_n of target types (for $1 \leq n \leq 3$). When we say that Γ_n is a way in which $\circ_2(\text{JP-row})$ holds, we mean that there is a *constraint* holding on possible properties of Round-Robin tables stating that 'if all members of Γ_n hold, then $\circ_2(\text{JP-row})$ holds'. Notice that Γ_1 is satisfied by the particular table in Fig. 1(b), whereas Γ_2 and Γ_3 are two alternative ways in which $\circ_2(\text{JP-row})$ holds. We now obtain the collection $\mathcal{G}$ of sets of source types—i.e. $\mathcal{G} = \{\Gamma_1, \Gamma_2, \Gamma_3\}$—which exhausts all alternative ways in which $\circ_2(\text{JP-row})$ holds (since $_3C_2 = 3$). Intuitively, $\circ_2(\text{JP-row})$ captures a piece of information commonly implied by the members of $\mathcal{G}$, *abstracting away* the specific information particular to individual members. We therefore say that $\circ_2(\text{JP-row})$ is an ABSTRACTION OVER $\mathcal{G}$—written as $\circ_2(\text{JP-row}) \bowtie \mathcal{G}$—iff (i) for every set Γ_n in $\mathcal{G}$, if all members of Γ_n hold, then $\circ_2(\text{JP-row})$ holds, and (ii) if $\circ_2(\text{JP-row})$ holds, then there is some set Γ_n in $\mathcal{G}$ whose members all hold. Completely analogously we can obtain the collection $\mathcal{D}$ of sets of target types—i.e. $\mathcal{D} = \{\Delta_1, \Delta_2, \Delta_3\}$—which exhausts all alternative ways in which target type $W_2(JP)$ holds (since $_3C_2 = 3$), with Δ_1 being indicated by the table in Fig. 1(b). As a consequence, we can say that $W_2(JP)$ is an ABSTRACTION OVER $\mathcal{D}$—written as $W_2(JP) \bowtie \mathcal{D}$—since (i) for every set Δ_n in $\mathcal{D}$, if all members of Δ_n hold, then $W_2(JP)$ holds, and (ii) if $W_2(JP)$ holds, then there is some set Δ_n in $\mathcal{D}$ whose members all hold [13, pp. 116–8]. Given (i) the two general constraints capturing the basic semantic conventions for individual cells of the Round-Robin table and (ii) the relation of collective indication between sets of source and target types, Table 1 shows that each Γ_n collectively indicates its Δ_n counterpart. This relation straightforwardly carries over to the entire collections $\mathcal{G}$ and $\mathcal{D}$. In other words, $\mathcal{G}$ and $\mathcal{D}$ are in a one-to-one correspondence under the collective indication relation by the system's basic semantic conventions.

Now we can finally characterise the derivative meaning relation between source type $\circ_2(\text{JP-row})$ and target type $W_2(JP)$ in terms of the notion of ABSTRACTION TRACKING. A source type σ is said to *track* a target type θ *in abstraction* if there are a collection $\mathcal{G}$ of sets of source types and a collection $\mathcal{D}$ of sets of target types such that (i) $\sigma \bowtie \mathcal{G}$, (ii) $\theta \bowtie \mathcal{D}$ and (iii) $\mathcal{G}$ and $\mathcal{D}$ are in a one-to-one correspondence. As is shown in Fig. 2(a), $\circ_2(\text{JP-row})$ abstracts over collection $\mathcal{G}$ (bottom $\bowtie$) and this abstraction 'tracks' the abstraction

Table 2. Aristotelian relations between two propositions α and β.

a	contradictory	$CD(\alpha, \beta)$	iff α and β cannot be true together and α and β cannot be false together
b.	contrary	$CR(\alpha, \beta)$	iff α and β cannot be true together but α and β can be false together
c	subcontrary	$SCR(\alpha, \beta)$	iff α and β can be true together but α and β cannot be false together
d.	in subalternation	$SA(\alpha, \beta)$	iff α entails β but β doesn't entail α

relation between $W_2(JP)$ and $\mathcal{D}$ (top $\bowtie$). This tracking of abstraction is mediated by the one-to-one correspondence between $\mathcal{G}$ and $\mathcal{D}$ by the collective indication relation $\Rightarrow_{1-1}$. The source type σ and the target type θ are thus taken to stand in a derivative indication relation $\Rightarrow_D$ whenever σ and θ stand in an abstraction tracking relation. Figuratively speaking, the information relation in Fig. 2(a) goes (1) from $\circ_2$(JP-row) to $\mathcal{G}$, then (2) from $\mathcal{G}$ to $\mathcal{D}$ and then (3) from $\mathcal{D}$ to $W_2(JP)$. Therefore, it goes (4) directly from $\circ_2$(JP-row) to $W_2(JP)$. This is precisely the sense in which the latter is a *derivative* relation, based on the system's basic semantic conventions, plus the two instances of abstraction relations holding within the source and the target domains.

Aristotelian Relations and Diagrams. In the research programme of Logical Geometry [6, 15] a central object of investigation is the so-called 'Aristotelian square' or 'square of opposition', which visualises logical relations of opposition and implication. Table 2 gives an informal definition and abbreviations for the different ARISTOTELIAN RELATIONS in which two propositions α and β can stand. In order to draw an ARISTOTELIAN DIAGRAM (AD for short), we first of all need a (non-empty) fragment F of a language L, i.e. a subset of formulas of that language. The formulas in the fragment F are typically assumed to be contingent and pairwise non-equivalent, and the fragment is standardly closed under negation: if formula φ belongs to F, then its negation $\neg\varphi$ also belongs to F. For the language of the modal logic S5 (with operators $\Box$ for necessity and $\Diamond$ for possibility), for instance, such a fragment F could be $\{\Box p, \neg\Box p, \Diamond p, \neg\Diamond p\}$. An Aristotelian diagram AD for F is then defined as a diagram that visualises an edge-labeled graph G. Fig. 2(b) presents the AD for the modal fragment $\{\Box p, \neg\Box p, \Diamond p, \neg\Diamond p\}$. The vertices of G are the elements of F, whereas the edges of G are labeled by *all* the Aristotelian relations holding between those elements, using the coding conventions in Fig. 2(c) and abbreviations in Table 2: full line for CD, dashed line for CR, dotted line for SCR, and arrow for SA. In this respect, it is worth observing that the systematic study of Aristotelian diagrams is a very recent and emerging field of research, and that an international research community is very much in the process of being established.[2] Hence, it should

---

[2] See the *World Congress on the Square of Opposition* (Montreux 2007, Corsica 2010, Beirut 2012, Vatican 2014, Easter Island 2016, Crete 2018, Leuven 2022).

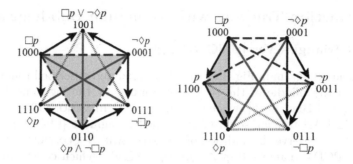

Fig. 3. (a) Jacoby-Sesmat-Blanché hexagon (b) Sherwood-Czeżowski hexagon.

come as no surprise that the actual basic semantic conventions—in the sense of a community's graphical 'practice'—are themselves also still being established.[3]

Derivative Meaning and Abstraction Tracking in ADs. In the next two parts of this paper we will explore the phenomenon of Derivative Meaning and its technical analysis in terms of Abstraction Tracking in the realm of Aristotelian diagrams. Whereas the basic semantic conventions in ADs are concerned with *individual* Aristotelian relations between two formulas (i.e. individual edges between two vertices), Derivative Meaning mainly arises when we consider larger *patterns* or constellations of such relations in order to identify or distinguish different families of ADs beyond the basic square. Such Derivative Meaning patterns—the recognition of which clearly involves expert knowledge and training—hence play an important heuristic and methodological role in establishing a systematic and exhaustive typology of Aristotelian families. In Sect. 2 we consider 'uniform' triangular shapes consisting of three contrariety relations or three subalternation relations. These two patterns first of all allow us to distinguish between the so-called Aristotelian family of Jacoby-Sesmat-Blanché hexagons (JSB for short) and that of the Sherwood-Czeżowski (SC for short) hexagons (Sect. 2.1). Secondly, they allow us to connect the 'arbitrariness' of graphical conventions to the distinction between informational and computational equivalence of diagrams (Sect. 2.2). In Sect. 3 we go into a second mechanism which plays a central role in drawing up a typology of Aristotelian families, namely the way in which smaller ADs are systematically embedded into larger ADs. On the one hand we look at the embedding of the so-called classical Aristotelian squares in JSB hexagons and SC hexagons (Sect. 3.1). On the other hand, the embedding of the so-called degenerate square inside two types of Aristotelian octagons—namely the Buridan and the Béziau octagon—nicely illustrates the idea that the observation of a pattern can also be based on the *absence* of certain (individual) relations (Sect. 3.2).

---

[3] The basic colour code (red ≈ CD, blue ≈ CR, green ≈ SCR and black ≈ SA) has become the de facto standard in the Square community. As for black and white line style counterparts, full line ≈ CD and full line arrow ≈ SA are default, but some variation remains as to dotted/dashed/... lines for CR and SCR.

2   Abstraction Tracking with Triangular Subdiagrams

2.1   JSB Triangles Versus SC Triangles

The hexagonal diagram in Fig. 3(a) was discovered and described roughly simultaneously in the middle of the twentieth century by the logicians Jacoby [7], Sesmat [12] and Blanché [2], whence its name JSB HEXAGON. If we compare it to the original Aristotelian square for the modal fragment $\{\Box p, \neg\Box p, \Diamond p, \neg\Diamond p\}$ in Fig. 2(b), we observe the addition of an extra PAIR OF CONTRADICTORY FORMULAS (or PCD)—namely $\{\Box p \vee \neg\Diamond p, \Diamond p \wedge \neg\Box p\}$—which constitutes a third diagonal, crossing the original square vertically. Furthermore, we have added the BITSTRING encoding of the six formulas involved since that will greatly facilitate the systematic description and comparison with diagrams further on. For this particular fragment, these bitstrings consist of four bitpositions β_n, which have the value 1 or 0, and which correspond to four anchor formulas α_n—together constituting a partition Π of logical space.[4] Formulas can now be classified as level one (L1), level two (L2) or level three (L3) according to the number of values 1 in their bitstring. Thus the two PCDs constituting the diagonals of the original square in Fig. 2(b) connect a L1 and a L3 formula, whereas the extra vertical diagonal in the JSB hexagon of Fig. 3(a) connects two L2 formulas. The addition of this third PCD gives rise to the typical downward pointing equilateral triangular shape (marked in grey) which is defined by three interconnected contrariety relations (marked by the dashed lines). Such a triangle thus serves as a diagnostic for identifying the hexagon as a member of the Aristotelian family of JSB hexagons. It is important to stress that this shape—which we refer to as a triangular Aristotelian SUBDIAGRAM (or AsD)—is *not* itself an Aristotelian diagram: since ADs are standardly closed under negation, they consist of an even number of vertices/formulas (i.e. they are built out of PCDs), and therefore triangles are excluded in principle.[5]

The observation of such an overall shape—over and beyond that of the individual Aristotelian relations that it consists of—can now straightforwardly be captured in terms of Derivative Meaning and Abstraction Tracking. On the level of source types we first of all need the set of individual graphical components—i.e. the dashed lines between the pairs of vertices. If we represent the presence of a dashed line between vertex β and β' as $\|(\beta, \beta')$, then the grey triangle in Fig. 3(a) corresponds to the set of source types $\Gamma_1 = \{\|(1000, 0001), \|(0001, 0110), \|(0110, 1000)\}$. An alternative JSB hexagon could

---

[4] For the technical details see [6]. In this particular case $\Pi = \{\alpha_1, \alpha_2, \alpha_3, \alpha_4\} = \{\Box p, \neg\Box p \wedge p, \Diamond p \wedge \neg p, \neg\Diamond p\}$ and for every formula φ, its bitstring representation $\beta(\varphi) = \beta_1\beta_2\beta_3\beta_4$ is such that $\beta_n = 1$ iff $\models \alpha_n \to \varphi$. Thus, $\beta(\Box p) = 1000$, since only $\models \alpha_1 \to \Box p$ and $\beta(\Diamond p) = 1110$, since $\models \alpha_1 \to \Diamond p$, $\models \alpha_2 \to \Diamond p$ and $\models \alpha_3 \to \Diamond p$ but $\not\models \alpha_4 \to \Diamond p$. Strictly speaking, bitstrings of length 3 suffice for the JSB hexagon in Fig. 3(a). However, for the sake of uniformity with the SC hexagon in Fig. 3(b)— which does require length 4—and the octagons later on, we stick to length 4.

[5] The contrariety triangle in the JSB hexagon thus closely resembles the four Aristotelian subdiagrams—left/right triangle, hour glass and bow tie—in [16].

be characterised as $\Gamma_2 = \{\| (1000, 0001), \| (0001, 0100), \| (0100, 1000)\}$. It can easily be shown that, with bitstrings of length 4, exactly ten of these sets of source types Γ_n can be built.[6] Taking these ten Γ_n sets together, we get the source type collection $\mathcal{G}$. Their common property—namely that they constitute (upside down) dashed line triangles—can be captured by means of the source type $\sigma = \|\triangledown$. This σ counts as an *abstraction over* $\mathcal{G}$—i.e. $\|\triangledown \bowtie \mathcal{G}$—as defined in Sect. 1. On the level of target types, the individual contrariety relations between two formulas φ and φ' can be represented as $CR(\varphi, \varphi')$. The grey triangle in Fig. 3(a) then corresponds to the set of target types $\Delta_1 = \{ CR(1000,0001), CR(0001,0110), CR(0110,1000)\}$. The target level counterpart of the alternative JSB hexagon with the Γ_2 source type set could be characterised as $\Delta_2 = \{ CR(1000,0001), CR(0001,0100), CR(0100,1000)\}$. Obviously, all ten Γ_n source type sets defined above have Δ_n counterparts, which, taken together, yield the target type collection $\mathcal{D}$. Its defining property—namely that all its members yield a JSB contrariety constellation—can be captured by means of the target type $\theta = CR(JSB)$ which counts as an *abstraction over* $\mathcal{D}$—i.e. $CR(JSB) \bowtie \mathcal{D}$. The overall result is a constellation of *abstraction tracking* similar to the one depicted in Fig. 2(a): by virtue of the one-to-one correspondence between $\mathcal{G}$ and $\mathcal{D}$ under the basic semantic conventions in Fig. 2(c) and Table 2, the source level abstraction $\|\triangledown \bowtie \mathcal{G}$ 'tracks' the target level abstraction $CR(JSB) \bowtie \mathcal{D}$. As a consequence, the source type $\|\triangledown$ is said to stand in a derivative meaning relation with the target type $CR(JSB)$.

As illustrated in the top hexagon of Fig. 4(a), whenever we observe an upside down (dark grey) CR triangle ($\|\triangledown$), the logic of the basic semantic conventions—and in particular the central symmetry of the contradiction diagonals—predicts the existence of additional meaningful objects, i.e. an overlapping or intertwined (light grey) SCR triangle ($\dagger\triangle$), and alternating SA arrows constituting the outer edges of the hexagon ($\downarrow\bigcirc$). We can think of these three basic shapes as 'first-order' source type abstractions which can be combined into the more complex shape $\Gamma_n = \{\|\triangledown, \dagger\triangle, \downarrow\bigcirc\}$. The resulting source type collection $\mathcal{G}$ is characterised by the 'higher-order' source type abstraction $\sigma = \triangledown\triangle\bigcirc \bowtie \mathcal{G}$, 'fusing' the two triangles and the arrow edges. Analogously, we combine the three first-order target type abstractions into $\Delta_n = \{ CR(JSB), SCR(JSB), SA(JSB)\}$. The resulting target type collection $\mathcal{D}$ is then characterised by the higher-order target type abstraction denoting the complete JSB hexagon, i.e. $\theta = HEX(JSB) \bowtie \mathcal{D}$. Figure 4(a) thus illustrates the mechanism of HIGHER-ORDER ABSTRACTION TRACKING: the source type $\triangledown\triangle\bigcirc$ stands in a HIGHER-ORDER DERIVATIVE MEANING RELATION with the target type $HEX(JSB)$.[7]

---

[6] First of all, there are six so-called *strong* JSBs that form their contrariety triangle like Γ_1, i.e. by first choosing two L1 formulas ($_4C_2 = 6$) and then adding the one 'complementary' L2 formula. In addition, there are four so-called *weak* JSBs that form their contrariety triangle like Γ_2, i.e. by means of 3 L1 formulas ($_4C_3 = 4$).

[7] A very similar process occurs with spatial information in topographic contour maps, where the basic, first-order abstraction shape of concentric contour lines indicates a single mountain top, whereas a series of adjacent basic shapes indicates the higher-order abstraction shape of a mountain range.

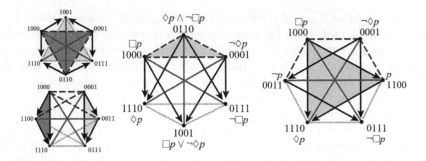

Fig. 4. (a) Double JSB/SC triangles (b) alternative JSB (c) alternative SC.

We can now turn to the second family of Aristotelian hexagons, namely the so-called Sherwood-Czeżowski (SC) hexagon illustrated in Fig. 3(b). It was described in the mid twentieth century by Czeżowski [3], but Khomskii [8] convincingly demonstrated that the figure already occurs in the work of the mediaeval logician William of Sherwood [10]. If we again compare this hexagon to the original Aristotelian square for the modal fragment $\{\Box p, \neg\Box p, \Diamond p, \neg\Diamond p\}$ in Fig. 2(b), we observe that the extra PCD—namely the third diagonal $\{p, \neg p\}$— now crosses the original square horizontally, as opposed to vertically in the JSB of Fig. 3(a). The addition of this third PCD yields an obtuse triangular shape (marked in grey) which is defined by three interconnected subalternation relations (marked by the arrows). Such a triangle thus serves as a diagnostic for identifying the hexagon as a member of the Aristotelian family of SC hexagons.[8] Within the framework of Logical Geometry, identifying which of the five logically possible Aristotelian families a particular hexagon belongs to, provides crucial (typo)logical information, in particular concerning the bitstring length required to encode the fragment involved.[9] The crucial difference between the symmetric relation of contrariety—$CR(\alpha, \beta) \Leftrightarrow CR(\beta, \alpha)$—and the asymmetric relation of subalternation—$SA(\alpha, \beta) \Leftrightarrow \neg SA(\beta, \alpha)$—is graphically reflected in the 'undirectedness' of the JSB triangle in Fig. 3(a) as opposed to the top down 'directedness' of the subalternation triangle in Fig. 3(b). With all its arrows pointing downwards, the latter represents the 'transitive closure' of the subalternation relation: if a first SA arrow gets you from α to β and a second SA arrow gets you from β to γ, then a third one gets you directly from α to γ.

The observation of such an overall triangular shape—in addition to that of the individual arrows that it consists of—can again straightforwardly be captured in terms of Derivative Meaning and Abstraction Tracking. On the level of source types, we first of all represent the presence of an arrow from vertex β to β' as $\downarrow (\beta, \beta')$. The grey triangle in Fig. 3(b) thus corresponds to the set of source

---

[8] As with the contrariety triangle in Fig. 3(a), the subalternation triangle in Fig. 3(b) is not itself an AD, but an AsD, since it is not closed under negation.

[9] For an analysis of the Boolean differences between JSB and SC hexagons—the two most common and well-studied Aristotelian families of hexagons—see [14].

types $\Gamma_1 = \{\downarrow(1000, 1100), \downarrow(1100, 1110), \downarrow(1000, 1110)\}$. It can easily be shown that, with bitstrings of length 4, exactly 24 of these sets of source types Γ_n can be built.[10] Taking these 24 Γ_n sets together, we get the source type collection $\mathcal{G}$. Their common property—namely that they constitute obtuse arrow triangles— can be captured by means of the source type $\sigma = \downarrow\lhd$. This σ counts as an *abstraction over* $\mathcal{G}$—i.e. $\downarrow\lhd \bowtie \mathcal{G}$. On the level of target types, the individual subalternation relations from formulas φ to φ', can be represented as $SA(\varphi, \varphi')$. The grey triangle in Fig. 3(b) then corresponds to the set of target types $\Delta_1 = \{SA(1000,1100), SA(1100,1110), SA(1000,1110)\}$. Obviously, all 24 Γ_n source type sets defined above have Δ_n counterparts, which, taken together, yield the target type collection $\mathcal{D}$. Its defining property—namely that all its members yield a SC subalternation constellation—can be captured by means of the target type $\theta = SA(SC)$ which counts as an *abstraction over* $\mathcal{D}$—i.e. $SA(SC) \bowtie \mathcal{D}$. The overall result is again a constellation of *abstraction tracking* (see Fig. 2(a)): by virtue of the one-to-one correspondence between $\mathcal{G}$ and $\mathcal{D}$ under the basic semantic conventions in Fig. 2(c) and Table 2, the source level abstraction $\downarrow\lhd \bowtie \mathcal{G}$ 'tracks' the target level abstraction $SA(SC) \bowtie \mathcal{D}$. Hence, source type $\downarrow\lhd$ is said to stand in a derivative meaning relation with target type $SA(SC)$.

As illustrated in the bottom hexagon of Fig. 4(a), whenever we observe an obtuse (dark grey) SA triangle on the left side of a SC hexagon ($\downarrow\lhd$), the basic semantic conventions and the centrally symmetric CD diagonals predict the existence of additional meaningful objects, in particular a non-overlapping (light grey) SA triangle on the right side of the hexagon ($\downarrow\rhd$). Furthermore, the relations of (sub)contrariety also yield easily recognizable 'hour glass' patterns, where the $\overline{\times}$ shaped constellation for CR at the top is the mirror image of the $\times$ shape for SCR at the bottom. As argued above for the JSB hexagon, this idea of combining basic shapes into more complex shapes can be analysed in terms of HIGHER-ORDER ABSTRACTION TRACKING and derivative meaning.[11]

2.2 Alternative JSB Versus SC Triangles

In this subsection, we briefly go into the aspect of 'arbitrariness' in the way in which basic semantic conventions—and graphical practice in general—come about within a given community and at a given point in history. As we argued above, both the JSB hexagon and the SC hexagon start out from the same original Aristotelian square and then add a third PCD diagonal: with the JSB hexagon the latter is inserted vertically, with the SC hexagon it is inserted horizontally. Notice that—exceptionally—an alternative visualisation shows up in which a minimal change takes place at the moment of inserting the respective

---

[10] For each of the four L1 starting points α, three L3 end points γ can be chosen (the contradictory L3 being excluded), and for each of these twelve L1-L3 pairs two L2 intermediate steps β can be chosen: $_4C_1 \times _3C_1 \times _2C_1 = 4 \times 3 \times 2 = 24$.

[11] In particular, the higher-order source type set $\Gamma_n = \{\downarrow\lhd, \downarrow\rhd, \|\overline{\times}, \dagger\times\}$ corresponds to the higher-order target type set $\Delta_n = \{SA_1(SC), SA_2(SC), CR(SC), SCR(SC)\}$ and the higher-order source type abstraction $\sigma = \lhd\rhd\overline{\times}\times \bowtie \mathcal{G}$ tracks the higher-order target type abstraction of the complete SC hexagon $\theta = HEX(SC) \bowtie \mathcal{D}$.

third PCD, i.e. not so much by changing its fundamental orientation (vertical vs horizontal) but simply by switching around its two formulas [5]. In the case of the alternative JSB in Fig. 4(b), this means that the 1001-0110 vertices for $\{\Box p \lor \neg \Diamond p, \Diamond p \land \neg \Box p\}$ are switched from top to bottom, as compared to the original JSB hexagon in Fig. 3(a). In the case of the alternative SC in Fig. 4(c) this means that the 1100-0011 vertices for $\{p, \neg p\}$ are switched from left to right, as compared to the original SC hexagon in Fig. 3(b).

The two JSB hexagons in Fig. 3(a) and Fig. 4(b) represent exactly the same Aristotelian relations between exactly the same formulas, and the same holds for the two SC hexagons in Fig. 3(b) and Fig. 4(c). In the terminology of Larkin & Simon [11], the two variants for both types of hexagons stand in a relation of INFORMATIONAL EQUIVALENCE, which need not coincide with that of COMPUTATIONAL EQUIVALENCE. In other words, two variants representing exactly the same information may still differ in terms of cognitive processing requirements. From the point of view of Derivative Meaning in the previous subsection, the alternative JSB hexagon in Fig. 4(b) resembles the original SC hexagon in Fig. 3(b) in that the CR and SCR triangles have become obtuse (instead of equilateral) and are no longer intertwined. At the same time, all subalternation arrows between the top CR triangle and the bottom SCR triangle are now pointing downwards, thus resembling to some extent an upside down Hasse diagram. One could argue that this property is more in line with the Congruence Principle [17]—the structure of the visualisation should match the represented logical structure as closely as possible—at least if one prefers to focus on implication relations instead of opposition relations [15]. Conversely, the alternative SC hexagon in Fig. 4(c) resembles the original JSB hexagon in Fig. 3(a) in that the two SA triangles—the one in grey pointing to the right, the other one pointing to the left—have now become equilateral (instead of obtuse) as well as intertwined. At the same time the CR and SCR relations have now become two 'semicircles'—the top and the bottom half respectively—along the outer edges of the hexagon. From the perspective of the Congruence Principle, one disadvantage of this alternative SC hexagon in Fig. 4(c) could be that the iconicity for the transitivity closure of the SA arrows is lost. The L1-L3 arrow from 1000 to 1110 has the same length as the two 'intermediate' arrows from 1000 to 1100 (L1-L2) and from 1100 to 1110 (L2-L3), whereas from a logical point of view it has the combined effect of the two intermediate SA relations. Obviously, much more research is needed in order to determine how a representational variant of a given Aristotelian family has obtained 'canonical' status at a given point in history and to what extent cognitive principles played a role in that process (see [5] for some initial observations). But apart from that, the analysis of Derivative Meaning in terms of Abstraction Tracking—which was spelt out in full detail in § 2.1—straightforwardly carries over to the alternative representations of the JSB and SC hexagons in Fig. 4(b-c).

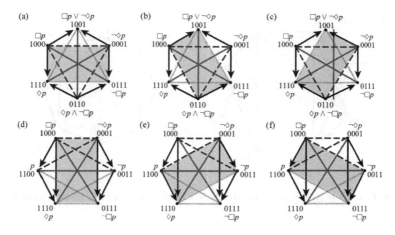

Fig. 5. (a-b-c) classical squares in JSB hexagon (d-e-f) classical squares in SC hexagon.

3   Abstraction Tracking with Embedded Squares

In addition to the identification of triangular AsD shapes illustrated in Sect. 2, a second important heuristic and diagnostic technique for distinguishing families of ADs concerns the identification of the number and position of smaller ADs EMBEDDED in larger ADs. Aristotelian squares—the smallest non-trivial ADs to be embedded—come in two families, namely the CLASSICAL SQUARE and the DEGENERATE SQUARE. The classical square—represented in Fig. 2(b)—has six Aristotelian relations between four vertices, whereas the degenerate square only has two Aristotelian relations left, namely the two diagonals for contradiction. We consider the embedding of classical squares in two families of hexagons (Sect. 3.1), and that of degenerate squares in two families of octagons (Sect. 3.2).

3.1   Classical Squares Inside JSB Versus SC Hexagons

In addition to the 'basic' embedding of the classical square in Fig. 5(a), two more classical squares are embedded in a JSB hexagon, with 120° (counter)clockwise rotations in Fig. 5(b-c). Completely analogously, Fig. 5(d) represents the 'basic' embedding of a classical square in a SC hexagon, whereas Fig. 5(e-f) have embedded squares with 30° (counter)clockwise rotations.[12]

The technical analysis in terms of abstraction tracking illustrated in Sect. 2 for the triangular AsD shapes straightforwardly generalises to embedded squares. The grey square in Fig. 5(a) yields the source type set $\Gamma_1 = \{\|((1000, 0001),$ $\dagger(1110, 0111), \downarrow (1000, 1110), \downarrow (0001, 0111)\}$. With bitstrings of length 4, the

---

[12] As to the Apprehension Principle [17]—the structure/content of the visualisation should be readily/accurately perceived/comprehended—the three SC squares in Fig. 5(d-f) are basically 'upright', whereas with the 120° rotations in Fig. 5(b-c) the two JSB squares are almost upside down, and thus less easily perceivable.

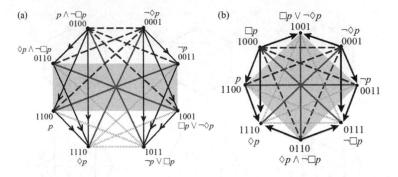

Fig. 6. Degenerate square inside (a) Buridan octagon (b) Béziau octagon.

source type collection $\mathcal{G}$ contains 18 of these Γ_n sets.[13] The source type $\sigma = {\downarrow}\boxtimes{\downarrow}$ counts as an *abstraction over* $\mathcal{G}$—i.e. ${\downarrow}\boxtimes{\downarrow}\bowtie\mathcal{G}$. Γ_1 collectively indicates target type set $\Delta_1 = \{CR(1000,0001),\ SCR(1110,0111),\ SA(1000,1110),\ SA(0001,0111)\}$, resulting in a one-to-one correspondence between source and target type collections $\mathcal{G}$ and $\mathcal{D}$. The target type $\theta = CL(JSB)$ refers to the embedding of a classical square in a JSB hexagon, and abstracts over $\mathcal{D}$—i.e. $CL(JSB)\bowtie\mathcal{D}$. The derivative meaning relation between σ and θ holds, since the source level abstraction 'tracks' the target level abstraction. The embedding of a classical square in the SC hexagon in Fig. 5(d) receives a perfectly analogous treatment. The generalisation—visualised in Fig. 5—that every JSB and SC hexagon contains exactly three classical squares, could then be analysed in terms of the so-called 'higher-order' abstraction tracking and derivative meaning introduced in Sect. 2. The details of such an analysis would take us too far here, mainly because it involves the simultaneous consideration of three different diagrams, or at least of three different perspectives on the same diagram.[14]

3.2 Degenerate Squares in Buridan and Béziau octagons

In the last step in this paper we will consider a very special type of Derivative Meaning. In the examples studied so far—the contrariety and subalternation triangles and the embedded classical squares—Derivative Meaning arose when we shifted the focus from individual Aristotelian relations between pairs of formulas to the observation of shapes or constellations of interconnected relations between three or four formulas inside an Aristotelian diagram. We have characterised the DEGENERATE ARISTOTELIAN SQUARE above as a constellation of four formulas which only consists of the two diagonal relations of contradiction. In other words,

---

[13] Six of them have a L1-L1 CR relation ($_4C_2 = 6$) and twelve of them a L1-L2 CR relation ($_4C_1 \times {_3}C_2 = 4 \times 3 = 12$).

[14] See [13, Ch. 6] for the closely related notion of ASPECT SHIFTING, an important method of mathematical discovery: diagrammatic proofs of mathematical theorems often involve (constraints between) two decomposition types of the same figure.

both the two horizontal relations of CR and SCR and the two vertical arrows of SA are missing. So when we look for degenerate squares embedded in bigger ADs, we are actually identifying a pattern, not by the *presence* of a number of relations, but by their *absence*. Observing such embedded degenerate squares as constellations of 'missing links' thus—in a figurative sense—boils down to 'seeing invisible squares'.

One very well studied family of ADs which contain an embedded degenerate square is that of the so-called BURIDAN OCTAGON [4,9], named after the mediaeval logician John Buridan and illustrated in Fig. 6(a). The latter can be seen as the 'superimposition' of two SC hexagons, in the sense that—in between the 0100 L1 vertex at the top left and the 1110 L3 vertex at the bottom left—not just one L2 vertex is inserted, but two, namely 0110 and 1100, thus interlocking two of the triangular SA arrow shapes that were discussed in full detail in Sect. 2.1.[15] The crucial thing to observe now is that these two L2 formulas themselves do not stand in any Aristotelian relation whatsoever. In [15] these formulas are said to be UNCONNECTED (Buridan himself calls them 'disparatae'). In Fig. 6(a) all four pairs of L2 formulas turn out to be unconnected, as indicated by the grey shaded area identifying the degenerate square embedded in the Buridan octagon.

Transferring the abstraction tracking analysis of the embedded squares in Sect. 3.1, we represent the absence of any line or arrow between vertex β and β' as $\varnothing(\beta, \beta')$. The grey square in Fig. 6(a) corresponds to the source type set Γ_1 = $\{\varnothing(0110, 0011), \varnothing(0011, 1001), \varnothing(1001, 1100), \varnothing(1100, 0110)\}$. With bitstrings of length 4, the source type collection $\mathcal{G}$ contains 3 of these sets Γ_n.[16] The source type $\sigma = \varnothing\boxtimes$ captures the idea that the latter constitute an 'invisible square' and counts as an *abstraction over* $\mathcal{G}$—i.e. $\varnothing\boxtimes \bowtie \mathcal{G}$. On the target level, the absence of any Aristotelian relation between 'unconnected' formulas φ and φ' is represented as $UN(\varphi, \varphi')$. Γ_1 collectively indicates target type set Δ_1 = $\{UN(0110,0011), UN(0011,1001), UN(1001,1100), UN(1100,0110)\}$, resulting in a one-to-one correspondence between source and target type collections $\mathcal{G}$ and $\mathcal{D}$. The target type $\theta = UN(BUR)$ refers to the embedding of a degenerate square in a Buridan octagon, and abstracts over $\mathcal{D}$—i.e. $UN(BUR) \bowtie \mathcal{D}$. The derivative meaning relation between σ and θ holds, since the source level abstraction $\varnothing\boxtimes$ $\bowtie \mathcal{G}$ 'tracks' the target level abstraction $UN(BUR) \bowtie \mathcal{D}$.

A second Aristotelian family of octagons in which a degenerate square is embedded is the so-called BÉZIAU OCTAGON [1], illustrated in Fig. 6(b). This octagon can be characterised as the 'superimposition' of the JSB hexagon in Fig. 3(a) and the SC hexagon in Fig. 3(b), in the sense that both the horizontal L2-L2 diagonal 1100-0011 and the vertical L2-L2 diagonal 1001-0110 are inserted into the basic square simultaneously. As a consequence, the Béziau octagon contains both the two interlocking equilateral triangles for the CR and SCR relations and the two non-interlocking obtuse triangles for the SA relations. The co-occurrence of two L2-L2 diagonals by definition results in the embedding of a degenerate square in a Béziau octagon, as indicated by the grey shaded area

---

[15] The analysis of some Buridan octagons requires bitstrings of length 5 and 6 [4].

[16] Each degenerate square consists of two out of the three L2-L2 PCDs ($_3C_2 = 3$).

in Fig. 6(b). Notice that—in contrast to the Buridan octagon in Fig. 6(a)—the Béziau octagon does not have any adjacent L2 vertices. In other words, the four L2 vertices are laid out alternatingly around the outer edges of the octagon, resulting in the 'tilted' degenerate square standing on one of its corners. Obviously, the technical analysis in terms of Abstraction Tracking provided above for the degenerate square embedded in the Buridan octagon carries over straightforwardly to the analogous constellation in the Béziau octagon.

4 Conclusion

In this paper we have used the mechanism of Abstraction Tracking [13] in order to describe Derivative Meaning arising in various Aristotelian diagrams (ADs). This collaborative enterprise has not only turned out to be fruitful and relevant for a deeper understanding of ADs in the framework of Logical Geometry but has also yielded a deeper understanding of the Cognitive Potential of Derivative Meaning itself, in particular w.r.t. the patterns of so-called higher-order Abstraction Tracking and Derivative Meaning.

References

1. Béziau, J.: The new rising of the square of opposition. In: Beziau, J., Jacquette, D. (eds.) Around and Beyond the Square of Opposition, pp. 3–19. Birkhäuser (2012). Basel
2. Blanché, R.: Structures Intellectuelles. J. Vrin, Paris (1969)
3. Czeżowski, T.: On certain peculiarities of singular propositions. Mind **64**(255), 392–395 (1955)
4. Demey, L.: Boolean considerations on John Buridan's octagons of opposition. Hist. Philos. Logic **40**(2), 116–134 (2019)
5. Demey, L., Smessaert, H.: The interaction between logic and geometry in Aristotelian diagrams. In: Jamnik, M., Uesaka, Y., Elzer Schwartz, S. (eds.) Diagrams 2016. LNCS (LNAI), vol. 9781, pp. 67–82. Springer, Cham (2016). https://doi.org/10.1007/978-3-319-42333-3_6
6. Demey, L., Smessaert, H.: Combinatorial bitstring semantics for arbitrary logical fragments. J. Philos. Log. **47**, 325–363 (2018)
7. Jacoby, P.: A triangle of opposites for types of propositions in Aristotelian logic. New Scholasticism **24**, 32–56 (1950)
8. Khomskii, Y.: William of Sherwood, singular propositions and the hexagon of opposition. In: Béziau, J.Y., Payette, G. (eds.) New Perspectives on the Square of Opposition. Peter Lang, Bern (2011)
9. Klima, G. (ed.): John Buridan, Summulae de Dialectica. Yale UP. New Haven, CT (2001)
10. Kretzmann, N.: William of Sherwood's Introduction to Logic. Minnesota Archive Editions, Minneapolis (1966)
11. Larkin, J., Simon, H.: Why a diagram is (sometimes) worth ten thousand words. Cogn. Sci. **11**, 65–99 (1987)
12. Sesmat, A.: Logique II. Hermann, Paris (1951)

13. Shimojima, A.: Semantic Properties of Diagrams and Their Cognitive Potentials. CSLI Publications. Stanford, CA (2015)
14. Smessaert, H.: Boolean differences between two hexagonal extensions of the logical square of oppositions. In: Cox, P., Plimmer, B., Rodgers, P. (eds.) Diagrams 2012. LNCS (LNAI), vol. 7352, pp. 193–199. Springer, Heidelberg (2012). https://doi.org/10.1007/978-3-642-31223-6_21
15. Smessaert, H., Demey, L.: Logical geometries and information in the square of opposition. J. Logic Lang. Inform. **23**, 527–565 (2014)
16. Smessaert, H., Shimojima, A., Demey, L.: Free rides in logical space diagrams versus Aristotelian diagrams. In: Pietarinen, A.-V., Chapman, P., Bosveld-de Smet, L., Giardino, V., Corter, J., Linker, S. (eds.) Diagrams 2020. LNCS (LNAI), vol. 12169, pp. 419–435. Springer, Cham (2020). https://doi.org/10.1007/978-3-030-54249-8_33
17. Tversky, B.: Visualizing thought. Top. Cogn. Sci. **3**, 499–535 (2011)

Number and Quality of Diagrams
in Scholarly Publications is Associated
with Number of Citations

Guy Clarke Marshall[1]([⊠]) [iD], Caroline Jay[1] [iD], and André Freitas[1,2] [iD]

[1] Department of Computer Science, University of Manchester, Manchester, UK
`guy.marshall@postgrad.manchester.ac.uk,`
`{caroline.jay,andre.freitas}@manchester.ac.uk`
[2] Idiap Research Institute, Martigny, Switzerland

Abstract. Diagrams are often used in scholarly communication. We analyse a corpus of diagrams found in scholarly computational linguistics conference proceedings (ACL 2017), and find inclusion of a system diagram to be correlated with higher numbers of citations after three years. Inclusion of more than three diagrams in this 8-page limit conference was found to correlate with a lower citation count. Focusing on neural network system diagrams, we find a correlation between highly cited papers and "good diagramming practice" quantified by level of compliance with a set of diagramming guidelines. This study suggests that diagrams may be a useful source of quality data for predicting citations, and that "graphicacy" is a key skill for scholars with insufficient support at present.

Keywords: Neural networks · Scholarly diagrams · Corpus analysis · Bibliometrics · Graphicacy

1 Introduction

Diagrams form a part of communications about Artificial Intelligence (AI) systems, such as papers published at the Association of Computational Linguistics (ACL), a top natural language processing (NLP) conference. We argue that system diagrams are an important source of data about scholarly authorship practices in computer science, specifically neural networks (NN) for natural language processing, and have insufficient attention in many academic writing guides. Using "Transactions of ACL 2017" as a corpus, we show that system diagrams are prevalent. We find that papers containing a system diagram are more likely to have a higher number of citations, perhaps indicating that their authors are effective science communicators, or that they write papers about systems, which are more highly cited. Further, papers containing more than two diagrams are found to be more likely to have a lower number of citations, and possible reasons for this are explored.

© Springer Nature Switzerland AG 2021
A. Basu et al. (Eds.): Diagrams 2021, LNAI 12909, pp. 512–519, 2021.
https://doi.org/10.1007/978-3-030-86062-2_52

Corpus analysis of diagrams is nascent, with recent analysis into connecting lines in data visualisations [8]. We use a corpus-based approach to examine diagrams within a wider social context, and have designed our approach to leverage existing document-level citation metrics, allowing quantitative analysis. Our main contribution is to test for compliance with an existing set of neural network system diagram guidelines, using a corpus-based approach. In summary, we find system diagrams are prevalent, occurring in 82% of papers at ACL 2017, and that diagrams in highly cited papers are more likely to contain "good diagrams" in the sense of conforming to an existing set of guidelines [10].

2 Background

Natural Language Processing is a discipline within Computer Science, and is concerned with creating systems that solve tasks relating to natural language interpretation. NLP systems take a text input, go through data manipulation steps, and create an output that is usually a classification, ranking, regression or prediction, such as what the next word in a sequence is likely to be. The state-of-the-art systems are technically complex, requiring application of mathematical and algorithmic techniques. These NLP systems are often described through diagrams. We have chosen to examine scholarly neural network systems, described in diagrams within NLP conference proceedings.

Contemporary NLP systems are often based on neural networks, and it is these systems we focus on. A neural network takes an input (in NLP, text), and then processes this via a series of *layers*, to arrive at an output (classification/prediction). Within each layer are a number of *nodes* which are attributes of the representation, and there is a connection between them. Specific mathematical functions or operations are also used in these systems, such as sigmoid, concatenate, softmax, max pooling, and loss. The *system architecture* describes the way in which the components are arranged. Different architectures are used for different types of activities. For example Convolutional Neural Networks (CNN), inspired by the human visual system, are commonly used for processing images. Long Short Term Memory networks (LSTM), a type of Recurrent Neural Network (RNN) which are designed for processing sequences, are often used for text.

These neural networks "learn" a function, but have to be trained to do so. Training consists of providing inputs and expected outputs, allowing the system to develop a representation which can be used for interpretation. The system is then tested with unseen inputs, to verify for generalisation. Diagrams almost always depict the training process. A more detailed introduction to LSTM architectures, including schematics, is provided by Olah [16].

3 Method

We use ACL 2017 scholarly papers as a corpus from which to extract diagrams, because it is an appropriate size for analysis (195 long papers), is distributed with a CC-BY licence, and is recent enough to be relevant whilst allowing for short-term (3 year) citation analysis. Web of Science contains statistics of peer reviewed citations which provides additional robustness to the measure which we use ("Times cited, WoS Core"). Using a chi-squared test we found this metric highly correlated with the less curated "Times cited, All". Our method follows Lee et al. [9], adapted to use a manual extraction process in order to reduce systematic omission and make use of the validity ensuring method of Lechner [8].

1. Using Web of Science, publication metadata was manually extracted from all long papers from ACL 2017, including number of citations.
2. Every figure that displayed a diagram was manually extracted, except figures in the Results section. We added diagram count as additional paper metadata. The term "diagram" is used to describe a conceptual diagram, usually a figure, which is not reporting results, displayed as a table, nor describing an algorithm. In practice, it encompasses system diagrams, parts of systems, graphical representations of algorithms, concept maps, flow charts of methods or systems and parse trees.
3. Diagrams were stored as separate image files, labelled according to which paper they were extracted from.
4. In each paper, at most one diagram was identified as the primary system diagram. Where multiple system diagrams were found, the one with the largest number of graphical elements was used. Additional metadata was captured, including conformity to each individual guideline, and whether the diagram was colour or monochrome.
5. Following the method of Lechner [8], inter-rater reliability was measured to validate scoring of guidelines compliance, on a subset of 15% of the resulting NN system diagram (17/119). This resulted in 204 pairs of pieces of metadata scored as "true", "false", or "not applicable". Using this, Gwet's AC_1 coefficient was calculated [7], finding "good" reliability when considering the guidelines as a set. Individual guideline conformity was variable, with Guidelines 2, 4, 7, 10 and 12 (in the ordering presented in [10]) scoring a less than "good" Gwet's AC_1, and required further clarification beyond the guideline text alone to agree scoring. Subsequent assessment was done with a single coder. This manual coding resulted in the addition of over 1,600 pieces of diagram metadata, together with 400 additional paper metadata items (diagram count, and system diagram inclusion).
6. The conference area of each paper was manually extracted, as defined by ACL organisers [3].
7. Data were analysed in R [17], using ggplot2 to create graphics [19].

4 Results

4.1 Diagrams in Context

Figure 1 shows the frequency of diagrams in ACL 2017 proceedings. The large number of papers, particularly highly cited papers, which include system diagrams demonstrates the importance of system diagrams in communicating at ACL 2017.

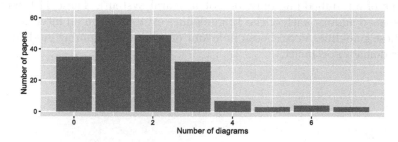

Fig. 1. Number of (non-results) diagrams in ACL 2017 papers is normally distributed, with inclusion of one or two diagrams most common. Most papers include a diagram.

To summarise the key insights, with correlations using chi squared test:

- 160/195 (82%) of all ACL 2017 papers included diagrams to represent system conceptualisations (not including results or algorithms).
- 124/195 (64%) of all ACL 2017 papers included at least one system diagram.
- Including 1–2 diagrams, of which at least one is a system diagram, is correlated with a 250% higher number of citations.
- Having more than two diagrams is correlated with lower number of citations. In a linear model each additional diagram is correlated with 5.6 fewer citations ($p = 0.02$). In the subset of papers which include a system diagram, this effect increases to 7 fewer citations per additional diagram.
- 82/119 (69%) of NN diagrams used colour, which may affect accessibility.
- Diagrams may be a valuable source of data for modeling number of citations. See Sect. 6.2.

4.2   Conference Areas

In an attempt to remove some of the effect of the content of paper, we analysed whether there was a relationship between the 17 conference areas [3] and (i) citations (ii) inclusion of a system diagram (iii) number of diagrams (iv) usage of examples. We found no significant difference between pairs of these attributes using chi-squared tests, using the entire dataset.

To further investigate any potential paper-content-related cause, we found 21 papers contain the word "architecture" and 18 of those contain a system diagram. Number of citations and the abstract containing the word "architecture" are correlated ($p < 0.01$), with those containing "architecture" having on average 20.4 more citations (than 15.9). As would be expected, the abstract containing the word "architecture" and including a system diagram are not independent: There is a significant relationship ($p < 0.05$). Causality is therefore ambiguous, as to whether architectural papers are more likely to be highly cited, or whether it is due to the presence of the diagrams.

4.3   Neural Network System Diagram Guideline Conformity

119 of 124 system diagrams described neural network systems (the others being diagrams of an embedding only, or not a neural system). These 119 diagrams were assessed against each of the 12 guidelines established in an interview study [10]. These guidelines were chosen in favour of other diagramming guidelines due to their domain specificity.

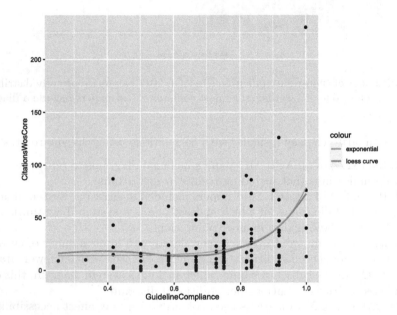

Fig. 2. Scatter plot of number of citations versus NN system diagram guideline compliance (as a quantitative proxy for "how good the diagram is"). LOESS curve for locally weighted smoothing is in blue, and the function $y = e^{10(x-7/12)} + 14$ is in red. (Color figure online)

In this exploratory analysis, we found a correlation between number of citations and "specific" ($p < 0.05$), and also "self contained" ($p < 0.05$) guidelines. The other guidelines alone did not correlate with a significant difference in number of citations. However, the best correlation was found with an average of the guideline compliance. The LOESS curve in Fig. 2 can be approximated by an exponential function, $citations = e^{10(compliance-7/12)} + 14$, where "7/12" captures the increase in citations observed from higher levels of compliance, "14" captures the asymptotic average number of citations for low compliance papers, and the multiplier "10" fits the curve. The only independent variable is average guideline compliance of each diagram. We do not aim to model citations accurately, arguing instead that the guidelines capture aspects of diagramming behaviours of effective communicators. Data has been made available [11].

5   Related Work

Much attention is given to the automated extraction of information from scholarly figures, including the classification of charts into bar charts, pie charts, etc. Roy et al. [18] recently created a classification system for neural network system diagrams. Their system classifies deep learning architectures into six categories, (e.g. 2D boxes, pipeline) based on how the layers are visually represented. This, and many other scholarly processing systems, rely on pdffigures 2.0 [4] for diagram extraction, which has known limitations and edge-cases. In particular, some types of figure (such as those with an L-shape) are systematically omitted [5]. Manual classification of NN system diagrams has conducted based on mental model categories [13] and semiotics [14], and VisDNA has been applied to neural network system diagrams [15]. Marshall et al. [10] conducted an interview study on the role of diagrams in scholarly AI papers, which reported 12/12 participants using diagrams to get a summary of the paper, and found some participants (3/12) used the diagram before any text in the paper. The potential role of NN diagrams in improving scholarly communication has also been explored [12].

6   Discussion

6.1   Limitations

- Our corpus analysis is based on one year of one venue, and cannot be generalised.
- The manual data extraction process does not scale well.
- Number of citations can be affected by many other factors, including author institution, author name, twitter presence, and so on. We mitigate venue by restricting to one venue. We do not take action to reduce the impact of other factors, focusing analysis on features of diagrams.
- Our inter-rater reliability covered only guideline conformity, not the diagram extraction or classification of figures.
- Whilst using the guidelines alone provided "good" inter-rater reliability, raters needed to make subjective judgements, and required more than the guidelines alone to ensure replicability.
- Unlike Lee et al. [9], we examine only diagrams, not all figures.

6.2   Using Diagrams in Models to Predict Number of Citations

A simple linear model for number of citations in the ACL 2017 corpus can be made using only (i) whether the abstract contains the word "architecture" and (ii) the level of conformity of the system diagram to a set of guidelines (zero if a system diagram is absent). This model has an R-squared of 0.132, suggesting that 13.2% of the variation in number of citations can be explained by these two factors alone. This simple model performs comparatively to existing citation predictions based on the entire text of the paper [20], which report 0.13 R-squared on a different scholarly corpus. Richer state-of-the-art models using social variables have R-squared around 0.4 at the 3rd year, again on a different scholarly corpus [1]. This supports the claim of the utility of diagrams in the medical-centric "viziometrics" research agenda of Lee et al. [9] and suggests figures may be a underutilised data source more broadly for scientometrics.

6.3   The Potential of Scholarly Diagrams

Our results provide further motivation for improved scientific scholarly graphicacy, the benefits of which to are often pedagogically focused. "Drawing to learn" is an active research area [2], and studies have been conducted concerning benefits of drawing for scientific thinking specifically (see Fan [6]). Our findings support the centrality of diagrams in scholarly communication previously identified in Medical Science [21], and lends weight to the reported primacy for some users of diagrams within the AI scholarly context [10].

7   Conclusion

Diagrams are an important, prevalent, and neglected component of scholarly communication about neural network systems, and diagramming is not proportionately discussed in many scholarly writing guides. At ACL 2017, high quantities of diagrams were found to be correlated with lower numbers of citations. Usage of system diagrams was found to be correlated with higher numbers of citations, suggesting this is a good scholarly communication practice in this domain. We have shown good domain-specific diagramming practices, quantified by compliance with a set of guidelines, to be correlated with a higher number of citations for ACL 2017 papers. This study demonstrates diagrams are important for communicating about scholarly neural network systems, and may be an underutilised tool for understanding and improving scholarly communication.

References

1. Abrishami, A., Aliakbary, S.: Predicting citation counts based on deep neural network learning techniques. J. Informetr. **13**(2), 485–499 (2019)
2. Ainsworth, S.E., Scheiter, K.: Learning by drawing visual representations: potential, purposes, and practical implications. Curr. Dir. Psychol. Sci. **30**(1), 61–67 (2021)
3. Barzilay, R., Kan, M.Y.: Accepted papers, demonstrations and TACL articles for ACL 2017 (2017). https://acl2017.wordpress.com/2017/04/05/accepted-papers-and-demonstrations/. Accessed 08 Jan 2021
4. Clark, C., Divvala, S.: Pdffigures 2.0: mining figures from research papers. In: 2016 IEEE/ACM Joint Conference on Digital Libraries (JCDL), pp. 143–152. IEEE (2016)
5. Clark, C., Divvala, S.: Pdffigures 2.0 readme (2016). https://github.com/allenai/pdffigures2/blob/master/README.md. Accessed 05 Mar 2021

6. Fan, J.E.: Drawing to learn: how producing graphical representations enhances scientific thinking. Transl. Issues Psychol. Sci. **1**(2), 170 (2015)
7. Gwet, K.L.: Handbook of inter-rater reliability: the definitive guide to measuring the extent of agreement among raters. Advanced Analytics, LLC (2014)
8. Lechner, V.E.: Modality and uncertainty in data visualizations: a corpus approach to the use of connecting lines. In: Pietarinen, A.-V., Chapman, P., Bosveld-de Smet, L., Giardino, V., Corter, J., Linker, S. (eds.) Diagrams 2020. LNCS (LNAI), vol. 12169, pp. 110–127. Springer, Cham (2020). https://doi.org/10.1007/978-3-030-54249-8_9
9. Lee, P.S., West, J.D., Howe, B.: Viziometrics: analyzing visual information in the scientific literature. IEEE Trans. Big Data **4**(1), 117–129 (2017)
10. Marshall, G., Freitas, A., Jay, C.: How researchers use diagrams in communicating neural network systems. arXiv preprint arXiv:2008.12566 (2020)
11. Marshall, G., Jay, C., Freitas, A.: Data supporting "number and quality of diagrams in scholarly publications is associated with number of citations" (2021). http://dx.doi.org/10.6084/m9.figshare.14812959
12. Marshall, G.C., Jay, C., Freitas, A.: Diagrammatic summaries for neural architectures. In: Beyond Static Papers: Rethinking how we Share Scientific Understanding in ML-ICLR 2021 Workshop (2021)
13. Marshall, G.C., Jay, C., Freitas, A.: Scholarly AI system diagrams as an access point to mental models. arXiv preprint arXiv:2104.14811 (2021)
14. Marshall, G.C., Jay, C., Freitas, A.: Structuralist analysis for neural network system diagrams. In: International Conference on Theory and Application of Diagrams. Springer (2021)
15. Marshall, G.C., Jay, C., Freitas, A.: Understanding scholarly neural network system diagrams through application of VisDNA. In: International Conference on Theory and Application of Diagrams. Springer (2021)
16. Olah, C.: Understanding LSTM networks (2015). https://colah.github.io/posts/2015-08-Understanding-LSTMs/. Accessed 22 May 2020
17. R Core Team: R: A Language and Environment for Statistical Computing. R Foundation for Statistical Computing, Vienna, Austria (2020) https://www.R project.org/
18. Roy, A., et al.: Diag2graph: representing deep learning diagrams in research papers as knowledge graphs. In: 2020 IEEE International Conference on Image Processing (ICIP), pp. 2581–2585. IEEE (2020)
19. Wickham, H.: ggplot2: Elegant Graphics for Data Analysis. Springer-Verlag, New York (2016). https://doi.org/10.1007/978-0-387-98141-3, https://ggplot2.tidyverse.org
20. Yan, R., Huang, C., Tang, J., Zhang, Y., Li, X.: To better stand on the shoulder of giants. In: Proceedings of the 12th ACM/IEEE-CS Joint Conference on Digital Libraries, pp. 51–60 (2012)
21. Yang, S.T., et al.: Identifying the central figure of a scientific paper. In: 2019 International Conference on Document Analysis and Recognition (ICDAR), pp. 1063–1070. IEEE (2019)

How Can Numerals Be Iconic?
More Varieties of Iconicity

Dirk Schlimm[(✉)]

McGill University, Montreal, QC H3A 2T7, Canada
dirk.schlimm@mcgill.ca

Abstract. The standard notion of iconicity, which is based on degrees of similarity or resemblance, does not provide a satisfactory account of the iconic character of some representations of abstract entities when those entities do not exhibit any imitable internal structure. Individual numbers are paradigmatic examples of such structureless entities. Nevertheless, numerals are frequently described as iconic or symbolic; for example, we say that the number three is represented symbolically by '3', but iconically by '|||'. To address this difficulty, I discuss various alternative notions of iconicity that have been presented in the literature, and I propose two novel accounts.

Keywords: Philosophy of notation · Iconicity · Numeral systems

1 Introduction

One of the most influential distinctions in the philosophy of notation is that between *icons* and *symbols*. It was originally introduced by C. S. Peirce as part of his tripartite characterization of representations as 'indices', 'symbols', and 'likenesses', where the latter are characterized by 'a mere community in some quality' of a representation and its object [10, p. 294]. Later, Peirce changed the terminology from 'likeness' to 'icon', for a sign that 'stands for something merely because it resembles it', and, as examples, he discussed geometric diagrams, Euler circles, and algebraic formulas and rules [11, p. 181]. The relation between a symbol and its object, on the other hand, is arbitrary, established only by convention. It is generally accepted that the terminology of 'icons' and 'symbols' is concerned with aspects of representations that come in degrees and are not mutually exclusive. In this paper, we restrict our attention to written representations. For the sake of simplicity, we also set aside Peirce's notion of an interpretant, but we follow his terminology and consider a representation to be a relation between a *sign* and an *object*; alternatively, one could also speak of a 'vehicle of representation' and its 'content' [4].

The general problem that is addressed in this paper is: How can we make sense of speaking of iconic representations when we are dealing with abstract objects that do not exhibit any internal features? As paradigmatic examples, I

© Springer Nature Switzerland AG 2021
A. Basu et al. (Eds.): Diagrams 2021, LNAI 12909, pp. 520–528, 2021.
https://doi.org/10.1007/978-3-030-86062-2_53

will use numbers, or to be more exact, the positive integers, as they are conceived from the contemporary structuralist perspective [14,17]. In this account, numbers are positions or places in a structure without any further organization, which means each individual integer has no internal structure or parts.[1] To focus the discussion, I will consider two different representational systems for numbers, namely the tally system, consisting of the signs '|', '||', '|||', ..., and the familiar Indo-Arabic decimal place-value system, consisting of the numerals '1', '2', '3', and so on. Intuitively, I take it that the tally numerals more iconic than the decimal place-value numerals (see, e. g., [1, p. 18]), so that the central question of this investigation becomes: What notion of iconicity can support this intuition?

2 Operational Iconicity

Before turning to various ways of explicating the notion of resemblance that was alluded to in the previous quotations, I'd first like to discuss an alternative characterization of iconicity, also put forward by Peirce, that does not rely on resemblance at all. According to it, 'a great distinguishing property of the icon is that by the direct observation of it other truths concerning its object can be discovered than those which suffice to determine its construction' [13, 2.279]. This understanding has been called *operational iconicity* by Stjernfelt, as it provides us with a criterion for iconicity by testing 'whether it is possible to manipulate or develop the sign so that new information as to its object appears' [19, p. 397]. However, the direct comparison of two different representations in terms of this criterion is, in practice, often less than clear-cut. As Stjernfelt's discussion of the comparison between an algebraic and a geometric representation of a mathematical function shows, whether one representation is considered to be *more* iconic than another depends on what kind of information one is interested in (e. g., exact values vs. qualitative properties about overall structure) and what means one uses to make the information 'appear' (e. g., calculus vs. direct representation) [19, pp. 413–415].

What can the criterion of operational iconicity tell us with regard to the representations of numbers by tallies and decimal place-values? Just by itself, the symbol '3' does not offer any new observations or invite any manipulations. By counting the individual strokes in '|||', on the other hand, we arrive at the numerical value it represents. This, however, was presumably part of the rule of its construction in the first place, so it should not count as a 'new' truth according to Peirce's criterion. However, we can easily separate the strokes, either mentally or by drawing them on paper, into two groups of '|' and '||', which yields the interpretation $3 = 1 + 2$. Other properties of the represented number can also be determined by some simple manipulations of the tally representation: if the tallies can be arranged into n rows and m columns such that none is left, we can infer that the represented number is the product of n and m; if the tallies can be arranged in a series of rows, starting with a row of a single tally

---

[1] This differs from other conceptions of numbers, e. g., Euclid's [6, Bk. 7, Def. 2].

and increasing the number of tallies by one for each subsequent row, we know that the represented number is 'triangular'. At first glance, these considerations seem to indicate that the tally representation is more iconic than the decimal place-value representation. However, once we start considering representations of numbers greater than nine, also the decimal place-value notation gives us some immediate insights: a closer inspection of '1230', for example, reveals how the represented number can be written as a product of powers of ten, that it is even and divisible by ten (because of the trailing '0'), and that it is also divisible by 3 (because the cross sum is evenly divisible by 3). Now, this information is also obtainable by manipulating the tally notation, but only with much more effort and a higher possibility of making errors, as it requires keeping track of 1230 different strokes. In turn, by consulting addition and multiplication tables and using the familiar paper-and-pencil algorithms for arithmetical operations, the information obtained by manipulating the tally notation can also be extracted from the corresponding representations in the decimal notation. To summarize, both numeral systems under consideration are operationally iconic, but which one of them is more iconic than the other depends on what information we consider most pertinent, on what means of manipulation and additional information we are willing to employ, and, in particular, on what range of numbers we are interested in representing. Let us now turn to various ways of explicating iconicity in terms of resemblance.

3   Iconicity as Resemblance

The standard notion of iconicity is based on degrees of similarity or resemblance between a sign and the object it denotes. Setting aside well-known general criticisms of this idea by Goodman and Eco (see [18, pp. 53–66] for a discussion), let me briefly present three ways this notion of resemblance may be understood. I will thereby introduce different ways in which a representation can be said to be iconic, and then discuss whether they can be fruitfully applied to systems of numerals.

3.1   Resemblance in Qualities

As hinted at in the first quote in the Introduction, one way in which a sign Σ and an object Ω can resemble each other is if they have a 'quality', say Q, in common: $Q(\Sigma) \leftrightarrow Q(\Omega)$. In the case of 'simple' qualities, such as color, shape, or form, Peirce referred to such icons as 'images' [12, 2.277]. Examples are '•' as a sign for coal (both are colored black) and the common astronomical sign for the sun, '⊙', which shares the round shape with the object it denotes. This criterion for resemblance could be generalized by requiring the sign and the object to have not one and the same quality, but rather two different qualities Q and Q' that are *naturally* related. Following Giardino and Greenberg, 'naturally' can be explicated in such a way that a representational system is natural 'to the degree to which human *nature*—including relatively universal aspects of cognition, physiology, social behavior, and environmental interaction—rather than enculturation, makes that system easy to internalize and use' [4, p. 8]. Examples from mathematics of qualities that are related in this way are the proximity of signs and their binding strengths [8], and vertical symmetry of signs and the commutativity of the associated operations [20].

However, regardless of whether the relation between the qualities Q and Q' is that of identity or of some other kind, this notion of resemblance does not help in addressing the problem of numerals: mathematical properties of numbers (such as being even or being a successor) are only definable in relation to other numbers, and are thus not isolated qualities of individual objects.

3.2   Resemblance in Structure

The most important analysis of the resemblance between a sign and an object is that of a correspondence between the components of a sign $(s_1, s_2, \dots)$ and the parts of the object $(o_1, o_2, \dots)$, such that for relations $(R_1, R_2, \dots)$ that hold between the components, corresponding relations $(R'_1, R'_2, \dots)$ hold between the associated parts: if $R_x(s_y, s_z)$, then $R'_x(o_y, o_z)$. Formally, resemblance is explicated here as a homomorphism (or isomorphism). Peirce speaks in this case of 'analogous' relations[2] and calls icons that are based on this notion of resemblance 'diagrams' [12, 2.277]. Norman distinguishes further between iconicity that is based on visual appearance ('VA-iconicity'), such that the relations $R_1, R_2, \dots$ can be observed directly without any conscious process of inference, and iconicity that is purely structural ('S-iconicity'). Geometric diagrams are examples of the former; maps of the latter [9, p. 111].

In addition to the relation-based correspondence between components of signs and parts of objects, this kind of resemblance can be augmented by allowing similarity in qualities between the components and parts (i. e., not just between the signs and objects, as was the case in the previous section). This is done, for example, when bigger dots are used on a map to represent cities with larger populations.

While it is possible to distinguish different components in the numeral signs under consideration, namely the individual strokes in a tally numeral and the individual digits in a decimal place-value numeral, which are themselves spatially related, no analogous decomposition of a number into parts is possible. Independent of any particular representation, an abstract number simply does not have any parts. Therefore, the various attempts to characterize the resemblance between a sign and the object it denotes in terms of similarity in structure do not apply to the relation between a numeral and a number.

The compound nature of icons and objects that is necessary for a structural account of resemblance also underlies the notion of *optimal iconicity*, which Stjernfelt attributes to Peirce, according to which a representation is more iconic when there is a one-to-one correspondence between the components of the sign and the parts of the object; thus, the icons represent the objects *like they really are* [19, p. 406; italics in original]. Because each number is only a single entity,

---

[2] This view of analogy also underlies the structure-mapping theory of analogy [3].

the applicability of this notion to systems of numerals is very restricted: in the tally notation, only '|' represents a number with a single character, and in the Indo-Arabic notation, this is the case only for the one-digit numerals, '0' to '9'.

3.3 Indirect Resemblance

A third variety of icons in addition to images and diagrams, which are based on resemblance of qualities and resemblance of structure respectively, are what Peirce calls 'metaphors' [12, 2.277]. Here, the relation is not between a sign and the object it denotes, but rather indirect: both are 'parallel' to a third entity. Such entities can be the words that are used to denote the object in question. For example, to denote logical conjunction, which is expressed in Latin by 'et', Hilbert and Ackermann use the sign '&', which originates from the ligature of 'et' [7, p. 1]; similarly, Peano frequently used word mnemonics to motivate his choice of logical signs (see [16, p. 154]).

Due to the lack of any obvious relation of numerals to other entities that are themselves related in some way or other to numbers, the notion of indirect resemblance does not help in explicating the iconicity of numerals.

4 Two New Notions of Iconicity

We have seen in the previous two sections that the various notions of iconicity that have been discussed in the literature in relation to Peirce are not suited to analyze the relation between numerals and numbers. I will now present two novel notions of iconicity that fare better in this regard: *exemplar iconicity*, which relates signs to concepts instead of objects, and *systematic iconicity*, which does not consider individual signs but rather entire systems of signs.

4.1 Exemplar Iconicity

The main difficulty with the various notions of iconicity discussed above lies in the fact that we considered numbers to be individual objects that cannot be further decomposed into parts, and thus do not have any internal structure that could serve to establish a relation to the numeral signs. Based on the distinction between ordinal and cardinal numbers, however, numbers can also be conceived of as being associated with particular classes or concepts. For example, Russell identified the cardinal number two as the class of all possible couples: 'The number of a class is the class of all those classes that are similar [i. e., one-to-one] to it' [15, p. 18]. This allows us to rigorously establish a relation between the sign '||' and the number two: the components of the sign form a class that is contained in the class that is the number two. Formulated in terms of concepts, one can say that every couple falls under the concept of two. In other words, '||' is an exemplar or instance of the number two. The details of an account of numbers that allows such parlance might vary considerably, but that need not interest us here, since what matters is that there is such an account at all.[3]

It is worth emphasizing at this point that exemplar iconicity is not based on any underlying notion of resemblance between the signs (numerals) and the objects (numbers). Despite the possibly misleading terminology, Russell's class of classes that are similar (in Russell's sense) to a class with two elements is not similar (in the ordinary sense) in any way to '||'. Rather, the latter, understood as the class of two individual strokes, is an element of Russell's class. For this reason, one might wonder whether this relation should be considered to be iconic at all. However, the relation is certainly not purely conventional (i.e., symbolic) and the lack of resemblance between sign and object is similar to the case of indirect or metaphorical resemblance, discussed above.

As we have seen, the tallies turn out to be exemplar iconic, as every sign consists of exactly as many strokes as the number it stands for. The signs of the decimal-place value notation, on the other hand, are not exemplar iconic in general. In fact, only the numeral '1' is exemplar iconic. Thus, using the notion of exemplar iconicity, we can account for the iconic character of the tally system as well as for the non-iconic character of the decimal place-value notation.

4.2 Systematic Iconicity

To arrive at the notion of exemplar iconicity, we replaced the objects that the signs stood for with concepts. To arrive at the notion of *systematic iconicity*, we consider the relations between signs within a system and objects within a system instead of the relations between individual signs and individual objects. Following Gamkrelidze, we call the latter 'vertical' and the former 'horizontal' [2], as illustrated by the following figure:

$$
\begin{array}{ccccccccc}
 & & \text{succ.} & & \text{succ.} & & \text{succ.} & & \cdots \\
\text{Numbers:} & \text{one} & \curvearrowright & \text{two} & \curvearrowright & \text{three} & \curvearrowright & \text{four} & \curvearrowright \cdots \\
\hline
\text{Numerals:} & | & \rightsquigarrow & || & \rightsquigarrow & ||| & \rightsquigarrow & |||| & \rightsquigarrow \cdots \\
 & \text{add } | & & \text{add } | & & \text{add } | & & \cdots
\end{array}
$$

The relation between the individual signs and their corresponding objects, in this case between the numerals in the tally system and the numbers (e.g., between '|' and the number one), is vertical. But we also notice a horizontal pattern among the numerals themselves, regardless of their relation to numbers: Starting from '|', we can arrive at another numeral by simply adding one stroke, and so on. These horizontal relations are indicated by the curved arrows in the figure. We can see how the sequence of numerals generated by adding a stroke ('|', '||', '|||', ...) mirrors the sequence of numbers related by the successor function. This resemblance between the entire systems of numerals and numbers (formally, an isomorphism) underlies the notion of systematic iconicity.

---

[3] See [5, pp. 52–57] for a general discussion of exemplification in the context of representations.

Let us now consider the system of decimal place-value numerals:

		succ.		succ.		succ.		...	
Numbers:	one	⌢	two	⌢	three	⌢	four	⌢	...
Numerals:	1	⤳	2	⤳	3	⤳	4	⤳	...
		?		?		?		...	

Prima facie, one might be inclined to say that we have the same systematic iconicity in this case. However, this is misleading, because the operation '?' in the figure is *not* a simple manipulation of the signs as was the case for adding a stroke in the tally system. Knowing that one needs to replace the sign '1' with the sign '2' in the first step is no help for knowing what to do in the second step. We are so familiar with manipulating numerals in the decimal place-value system that we easily overlook the nature and complexity of these operations! Due to the recursiveness of the decimal-place value numerals, it is possible to formulate rules for their generation, but these impose an additional, arbitrary structure on the numerals than that based on the successor operation, namely one that is based on the decomposition of the number into multiples of powers of ten. The main point is, however, that while there is a pattern also in the decimal place-value system, the syntactic operations that generate this pattern are much more complex than those that generate the sequence in the tally system.[4] Thus, the degree of systematic iconicity of the tally system is much greater than that of the decimal place-value system.[5]

Similarly to the other notions of iconicity, systematic iconicity should be understood as an aspect of representational systems which comes in degrees. For example, while the structures of the tally system and of the natural number system match up perfectly with regard to the successor operation, the system of Roman numerals does so only for certain sub-sequences of numerals. Thus, with regard to the successor operations, the Roman numeral system is more systematically iconic than the decimal-place value system, but less so than the tally system.

5   Conclusion

In the first, negative part of the paper, I have argued that the intuitive characterizations of the tally numerals as more iconic than the decimal place-value numerals cannot be adequately accounted for by the main notions of iconicity that emerge in Peirce's writings: operational iconicity and iconicity based on resemblance (in qualities, in structure, and indirect). To overcome this problem, I have presented, in the second, positive part of the paper, two new notions of iconicity: one based on the idea that the signs are exemplars of the concepts

---

[4] The exact measure of simplicity that is applied here is difficult to make precise, but it might correlate with the notion of naturalness invoked in Sect. 3.1.

[5] Note that some of the caveats about the assessment of operational iconicity discussed at the end of Sect. 2 also apply to the notion of systematic iconicity.

they signify and one based on the idea of a resemblance between the system of signs and the system of objects they denote.

While the notion of exemplar iconicity might not have many applications to written representations, the notion of systematic iconicity has potential for opening up further fruitful investigations in the philosophy of mathematics and in the study of notational systems. In particular, it might be employed to shed more light into Peirce's own discussions of the diagrammatic character of systems of algebraic and logical formulas, and to compare different notational systems according to their degrees of systematic iconicity with regard to specific structural aspects of the represented systems.

Acknowledgments. I would like to thank Viviane Fairbank, David Waszek, Jessica Carter, and two anonymous reviewers for fruitful discussions and helpful comments on a draft of this paper. This research was supported by the Social Sciences and Humanities Research Council of Canada.

References

1. Chrisomalis, S.: Numerical Notation: A Comparative History. Cambridge University Press, Cambridge (2010)
2. Gamkrelidze, T.V.: The problem of "l'arbitraire du signe". Language **50**(1), 102–111 (1974)
3. Gentner, D.: Structure-mapping: a theoretical framework for analogy. Cogn. Sci. **7**(2), 155–170 (1983)
4. Giardino, V., Greenberg, G.: Introduction: varieties of iconicity. Rev. Philos. Psychol. **6**(1), 1–25 (2015)
5. Goodman, N.: Languages of Art. An Approach to a Theory of Symbols. Bobbs-Merrill Co., Indianapolis (1968)
6. Heath, T.: The Thirteen Books of Euclid's Elements. Dover, Chicago (1953)
7. Hilbert, D., Ackermann, W.: Grundzüge der theoretischen Logik. Springer, Heidelberg (1928)
8. Landy, D., Goldstone, R.: How abstract is symbolic thought? J. Exp. Psychol. Learn. Mem. Cogn. **33**(4), 720–733 (2007)
9. Norman, J.: Iconicity and "direct interpretation". In: Malcolm, G. (ed.) Studies in Multidisciplinarity, Chap. 8, pp. 99–113. Elsevier (2004)
10. Peirce, C.S.: On an improvement of Boole's calculus of logic. In: Proceedings of the American Academy of Arts and Sciences, vol. 7, pp. 250–261. Welch, Bigelow, and Co. (1867). Repr. in [13], 3.1–19
11. Peirce, C.S.: On the algebra of logic: a contribution to the philosophy of notation. Am. J. Math. **7**(2–3), 180–202 (1885). Repr. in [13], 3.154–403
12. Peirce, C.S.: Syllabus. Manuscript (1902). Repr. in [13] 2.274–277, 283–4, 292–4
13. Peirce, C.S.: Collected Papers. Harvard University Press, Cambridge (1932–1958). Eight volumes. Edited by Charles Hartshorne and Paul Weiss
14. Resnik, M.D.: Mathematics as a Science of Patterns. Claredon Press, Oxford (1997)
15. Russell, B.: Introduction to Mathematical Philosophy. George Allen & Unwin, London (1919). Reprinted by Dover Publications, New York (1993)
16. Schlimm, D.: Peano on symbolization, design principles for notations, and the dot notation. Philosophia Scientiæ **25**(1), 139–170 (2021)

17. Shapiro, S.: Philosophy of Mathematics. Structure and Ontology. Oxford University Press, Oxford (1997)
18. Stjernfelt, F.: Diagrammatology. An Investigation of the Borderlines of Phenomenology, Ontology, and Semiotics. Springer, Dordrecht (2007). https://doi.org/10.1007/978-1-4020-5652-9
19. Stjernfelt, F.: On operational and optimal iconicity in Peirce's diagrammatology. Semiotica **186**(1/4), 395–419 (2011)
20. Wege, T., Batchelor, S., Inglis, M., Mistry, H., Schlimm, D.: Iconicity in mathematical notation: commutativity and symmetry. J. Numer. Cognit. **6**(3), 378–392 (2020)

Natural Deduction for Intuitionistic Euler-Venn Diagrams

Sven Linker$^{(\boxtimes)}$ (iD)

Lancaster University in Leipzig, Leipzig, Germany
`s.linker@lancaster.ac.uk`

Abstract. We present preliminary results for a proof system based on Natural Deduction for intuitionistic Euler-Venn diagrams. These diagrams are our building blocks to visualise intuitionistic arguments, that is, arguments avoiding classical Boolean arguments, for example "proof by contradiction" or the "law of the excluded middle". We have previously presented semantics for these diagrams and a proof system in the style of Sequent Calculus. Within this proof system, proof search consists of gradually decomposing the diagram to prove. While such a style is easy to automate, it is not very intuitive for human reasoners. Since formalising human reasoning was the main intention behind Natural Deduction, we choose this style for a proof system that is easier to apply and to understand than the existing system.

Keywords: Intuitionistic logic · Euler-Venn diagrams · Proof theory

1 Introduction

While the original intention of intuitionistic logic was only to formalise purely constructive reasoning (i.e., avoiding arguments that relied on principles of classical logic, like the law of the excluded middle and proofs by contradiction), its connection to type systems and program execution make it highly relevant in more applied fields as well. However, intuitionistic reasoning is often (in contrast to its name) unintuitive to untrained people. This makes it important to find good visualisations, that may make it easier to grasp the ideas of constructive reasoning and get a feel for the type of arguments involved.

Visualisations of intuitionistic reasoning are sparse and were only recently defined [1,3,4]. In this work, we consider our recent definition of intuitionistic Euler-Venn diagrams, for which we presented a sound, complete and cut-free sequent calculus. This style of proof systems is well suited for automated reasoning, since at each step, the diagrams allow only for a subset of the rules to be applied. The rules decompose the diagrams into smaller diagrams, until only literals remain. Then, if all leaves of the deduction tree are axioms, it is a proof.

When humans reason mathematically, however, they typically work in a different way. Reasoners generally make auxiliary assumptions, from which they may deduce the validity of other assertions. For example, if we have a disjunctive assertion, $F \vee G$, and want to show that H follows, we typically assume first

© Springer Nature Switzerland AG 2021
A. Basu et al. (Eds.): Diagrams 2021, LNAI 12909, pp. 529–533, 2021.
https://doi.org/10.1007/978-3-030-86062-2_54

F and deduce that H is true, and then assume G and show that H is true. While this style of reasoning is only indirectly represented in a sequent calculus, Natural Deduction is intended to mimic this reasoning. In fact, Natural Deduction was originally defined explicitly to formalise such arguments [2]. This may make it a better choice for human reasoners. In this work, we present diagrammatic rules for such a proof system. Instead of decomposing the diagrams as much as possible, the intention behind the rules is to reflect the meaning of the diagrams.

2   Intutionistic Euler-Venn Diagrams

We have previously presented syntax and semantics for intuitionistic Euler-Venn diagrams [3] based on Heyting algebras. A main property of Heyting algebras is that, in contrast to Boolean algebras, neither the lattice operations *meet* and *join*, nor the *implication* can be defined as abbreviations.[1] So, we need distinct syntactic elements to represent meets, joins and implications. For this purpose, we distinguished three types of diagrams: pure Venn, pure Euler, and general Euler-Venn diagrams. Pure Venn diagrams contain all possible zones for a set of contours, where some of the zones may be shaded. Pure Euler diagrams may not contain any shading, but some of the possible zones may be missing. General Euler-Venn diagrams may both have missing zones, and contain shading. To distinguish pure Euler-diagrams from general Euler-Venn diagrams without shading, we use dashed contours for the former.

The semantics of the diagrams are defined with respect to Heyting algebras $\mathcal{H}$. Each contour c is associated with an element $\mathbf{c} \in \mathcal{H}$. A shaded zone z in a pure Venn-diagram d_V denotes the meet of all contours z is contained in and of the complements of the contours z is not contained in. For example, the shaded zone in the premiss of the application of er_V in Fig. 2a denotes $\mathbf{a} \sqcap \mathbf{b} \sqcap -\mathbf{c}$. The whole semantics of d_V is then the join of all semantics of its shaded zones. This means that our notion of shading is constructive: it expresses the existence of a certain element of the Heyting algebra. In contrast, in classical Venn diagrams, shading is used to denote that the corresponding set is empty. The semantics of a missing zone z in a pure Euler diagram denotes the implication $x \mapsto y$, where x is the meet of all the contours z would be contained in, while y is the join of all contours z would not be in. For example, consider the pure Euler diagram in the assumptions of (1). It only misses one zone: the zone contained in a but not in b. The semantics of this zone would be the implication $\mathbf{a} \mapsto \mathbf{b}$. Finally, the semantics of a general Euler-Venn diagram consists of an implication, where the antecedent is given by the Euler-part of the diagram (i.e., its missing zones), while the consequent is given by the Venn-part (i.e., the shaded zones). For example, the goal diagram in (1) represents $((\mathbf{b} \sqcap \mathbf{c}) \mapsto 0) \mapsto (\mathbf{b} \sqcap -\mathbf{c})$. The Euler-part acts as a constraint: if we can prove the topological constraints, then we can deduce the information in the shaded zones.

---

[1] The lattice operations reflect the logical operators: the meet $\sqcap$ is the algebraic equivalent to conjunction $\wedge$, the join $\sqcup$ to disjunction $\vee$, and the algebraic implication $\mapsto$ corresponds to logical implication $\rightarrow$. The complement $-\mathbf{a}$ is defined by $\mathbf{a} \mapsto 0$ (where 0 is the bottom element of the algebra), and corresponds to negation.

3   Proof Rules

We present some proof rules on diagrams by two examples of derivations. A derivation consists of a tree, where the diagram to prove is at the root, and the branches are given by the applications of proof rules. The leaves of the tree denote either open assumptions, or *discharged* assumptions that have been used as temporary assumptions, and are no longer necessary for the derivation. Discharged assumptions are enclosed in brackets, and may be annotated by a symbol that identifies the rule application, where they were discharged. For example, in Fig. 2a, the pure Euler diagram consisting of the disjoint contours b and c has been eliminated by the application of the last rule in the tree, with the label att.

We will use the following two derivations to explain how the rules operate:[2]

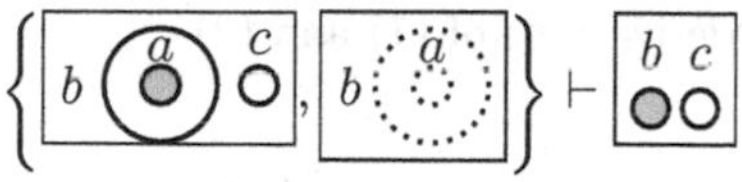

$$\left\{ b\,\overset{c}{\textcircled{a}}\,\circ,\; b\,\overset{\cdots a}{\vdots} \right\} \vdash \overset{b\; c}{\bullet\bullet} \quad (1) \qquad \left\{ \overset{b\; c}{\bullet\bullet},\; a\,\bullet \right\} \vdash b\,\overset{a}{\textcircled{a}}\,\overset{c}{\circ} \quad (2)$$

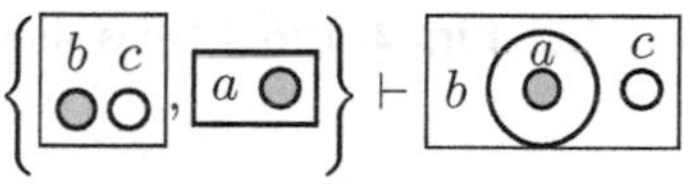

The derivation trees showing that these statements are true can be found in Fig. 2. In these derivations, the rules er_V and er_E erase contours. In the application of er_V, it can be seen that, in contrast to rules for classical Venn diagrams, shading is not erased, but proliferates into zones that were split. Consider the intersection of the contours a and b outside of c in the application of er_V.

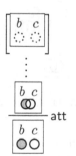

In the result, the zone inside of b and outside of c is shaded. If we erase a contour in a pure Euler diagram, using the rule er_E, all topological relations not concerning this contour are preserved (see the uppermost step in Fig. 2b). We also have rules for pure Venn and Euler diagrams that allow us to combine the information contained within two diagrams into one. For pure Euler diagrams, the rule is the *topological combination rule* co_T, an application of which can be seen in Fig. 2a in the topmost step. For pure Venn diagrams, the rule is the *consolidation rule* cons, shown in the middle of Fig. 2b. In the conclusion of this rule, we shade the zones that are allowed combinations of the shadings in the premises. For example, we

Fig. 1. Attachment

shade the zone within a and b but outside of c, since the right premiss states that the element $\mathbf{b} \sqcap -\mathbf{c}$ can be constructed, and the left premiss states that we can construct $\mathbf{a}$. So, taking both of these in conjunction means that we can construct $\mathbf{a} \sqcap \mathbf{b} \sqcap -\mathbf{c}$. For general Euler-Venn diagrams, we only employ rules that either *de-* or *attach* the Euler part from/to the Venn part. In the derivations, we have two examples of the *attachment rule* att. It allows us to deduce a diagram d, if we can derive a pure Venn diagram containing the shaded zones of d,

---

[2] $\Gamma \vdash d$ means that using the diagrams in Γ as assumptions, we can prove d.

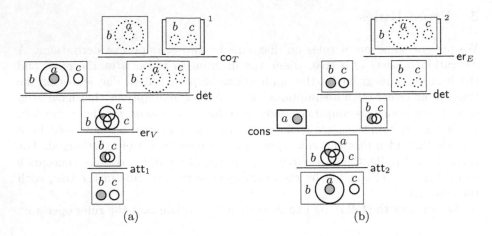

Fig. 2. Proof trees showing the derivations of (1) and (2)

while assuming its Euler part. We may discharge the assumed Euler-part, since its information is now contained in d itself. Figure 1 extracts the application of this rule from Fig. 2a. As explained in Sect. 2, the semantics of the discharged assumption is $(\mathbf{b} \sqcap \mathbf{c}) \mapsto 0$, the semantics of the derived pure Venn diagram is $\mathbf{b} \sqcap -\mathbf{c}$ and the semantics of the conclusion is $((\mathbf{b} \sqcap \mathbf{c}) \mapsto 0) \mapsto (\mathbf{b} \sqcap -\mathbf{c})$. Thus, the rule att is similar to the sentential rule of *implication introduction* [2].

We have proven all of the rules presented in this work (and some more) sound, i.e., if the premisses are valid, then so is the conclusion. However, there are still rules missing that are desirable. For example, we conjecture that we need a rule to transfer information from a pure Euler diagram directly to a pure Venn diagram. Consider the derivation in Fig. 3. Intuitively, such a rule should hold: the left premiss denotes $\mathbf{c} \mapsto$ $\mathbf{a}$, the right $\mathbf{c} \sqcap -\mathbf{e}$, and the conclusion $\mathbf{a} \sqcap -\mathbf{e}$. This

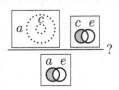

Fig. 3. Desired rule

is easily provable in intuitionistic logic. While this instance of the diagrammatic rule is clearly true, the general conditions under which such a rule should hold are not at all obvious.

References

1. Bellucci, F., Chiffi, D., Pietarinen, A.V.: Assertive graphs. J. Appl. Non-Classical Logics **28**(1), 72–91 (2018)
2. Gentzen, G.: Untersuchungen über das logische Schließen I. Math. Z. **39**, 176–210 (1935)

3. Linker, S.: Intuitionistic Euler-Venn diagrams. In: Pietarinen, A.-V., Chapman, P., Bosveld-de Smet, L., Giardino, V., Corter, J., Linker, S. (eds.) Diagrams 2020. LNCS (LNAI), vol. 12169, pp. 264–280. Springer, Cham (2020). https://doi.org/10.1007/978-3-030-54249-8_21
4. Ma, M., Pietarinen, A.V.: A graphical deep inference system for intuitionistic logic. Logique et Anal. (N.S.) **245**, 73–114 (2019)

Understanding Thought Processes

Observing Strategies of Drawing Data Representations

Fiorenzo Colarusso[1] ✉ ⓘ, Peter C.-H. Cheng[1] ✉ ⓘ, Grecia Garcia Garcia[1] ✉ ⓘ, Daniel Raggi[2] ✉ ⓘ, and Mateja Jamnik[2] ✉ ⓘ

[1] University of Sussex, Brighton, UK
{f.colarusso,p.c.h.cheng,g-garcia-garcia}@sussex.ac.uk
[2] University of Cambridge, Cambridge, UK
{daniel.raggi,mateja.jamnik}@cl.cam.ac.uk

Abstract. New methods for the assessment of drawing strategies are examined that focus on the analysis of perceptual chunking. The methods are demonstrated with four diverse participants as they copied a line-graph and a bar-chart. Video recordings of the transcriptions were analysed stroke by stroke. Diverse global drawing strategies were used for the line graph whereas all four participants used a similar approach on the bar-chart, but with local differences. Performance fluency varied substantially, particularly in stimuli viewing frequency. Differences in behaviours can be explained in terms of how they perceptually chunked the stimuli. Sample GOMS models were constructed in order to verify that chunking explains the drawing strategies. The potential of using drawing transcription tasks to assess users' competence with graphs and charts is discussed.

Keywords: Chunking · Competence measurement · Diagrams · Bar chart · Line graph · Graph comprehension · Task analysis · GOMS · Representations

1 Introduction

How people draw diagrams has been rather neglected in the diagrams research literature, but it is worth studying for many reasons. It is an interesting complex cognitive phenomenon (von Sommers 1984). A drawing individual's approach depends on their familiarity with the diagram and whether it is reproduced from long term memory, copied or traced (Obaidellah and Cheng 2015). Different drawing strategies can reflect learners' understanding of technical topics (Roller and Cheng 2014). Our particular reason for studying the nature of drawing is motivated by whether individual's diagram drawing behaviours reflect their familiarity with the diagrams being produced. Are there signals in drawing behaviour that can be used to assess an individual's competence with a particular class of diagrams? The focus here is on data charts and graphs.

Our interest in assessing people's competence with particular diagrams, and representations more generally, is motivated by the *Rep2Rep* project that is attempting to build an automated system to selected appropriate representations for individual as they attempt to solve specific problems in some target domain (Jamnik and Cheng 2021).

© Springer Nature Switzerland AG 2021
A. Basu et al. (Eds.): Diagrams 2021, LNAI 12909, pp. 537–552, 2021.
https://doi.org/10.1007/978-3-030-86062-2_55

Representation selection is essential because the choice of representations substantially determines the ease and success of problem solving and learning. The Rep2Rep framework involves two aspects: (a) selecting representations that are formally adequate for the problem using a formal AI system and (b) picking representations that are cognitively suited to individuals. This paper focuses on the second. A key aspect of assessing the cognitive suitability of a representation for an individual is knowledge of their level of familiarity with the given representation, that is, how well the representation is understood. So, we need quick and reliable means of evaluating people's familiarity with different representations. We contend that when individuals reproduce a representation, or copy a diagram, their behaviours provide signals (i.e., pause latency between successive pen-strokes) that reflect their familiarity from which measures of their competence can be derived. This paper proposes such an approach and uses it to examine the variety of people's drawing behaviours as they reproduce data diagrams. We first consider some theoretical background on: (i) how the graphical structure of graphs and charts are critical to how they are understood, in general; (ii) how the role of chunking in comprehension and their analysis provides means to assess competence; (iii) existing approaches to assessing familiarity.

1.1 Interpreting and Comprehending Graphs

As our goal is to assess users' familiarity with data diagrams, we consider what it means for someone to understand them. In general, line graphs are employed to depict x-y trends while bar charts support comparisons between data values as bars. Wu et al. (2010) argue that the tendency to associate lines with trends is cognitively natural and an easy perceptual process. Shah et al. (1999) found that format and scale influence the graph interpretation: bar charts should be used when two independent variables are equally important, whereas line graphs should be used when a particular trend is more relevant. According to the model for graph interpretation (Shah and Carpenter 1995; Shah et al. 1999), three processes are particularly relevant: encoding, transition of visual features to conceptual relations, and referential processes. An accurate encoding of the major visual pattern in a graph, such as whether there is a straight or jagged line, is essential for its correct comprehension. The transition of visual features to conceptual relations requires the retrieval of quantitative knowledge associated with the visual pattern, such as the fact that a downward curve represents a decreasing function. Therefore, when the visual pattern evokes familiar quantitative concepts, the comprehension is relatively effortless. Moreover, Peebles and Cheng (2003) found that changes to a graph's design affect users' performance in graph reading task in terms of the visual pattern to find the required information. Thus, it is likely that competence in graph comprehension is closely linked to users' familiarity with the perceptual patterns or visual features of graphs and charts.

Processing models of graph comprehension also imply the importance of grasping the organisation and processing structure of graphs and charts. Freedman and Shah (2002) applied Kintsch's (1988) Construction-Integration (CI) model to graph comprehension. The CI model states that comprehension can be subdivided into two subphases: a construction phase and a comprehension phase. Moreover, three pools of information are included in the model: visual features, domain knowledge, and interpretation propositions (Freedman and Shah 2002). During the construction phase, textual information, prior knowledge, and goals interact to form a coherent representation of the available information. During the comprehension phase, when the information is depicted in the graph by visual features and can be linked with prior knowledge without making inferences, the comprehension is effortless while, if inferences are needed the process becomes effortful. Similarly, Hegarty (2005) proposed a Model of Comprehension of visual displays to explain how people construct a mental model starting from the display visualisation. The model claims that bottom-up information (design features) interacts with top-down processes (prior knowledge). Thus, familiarity and background knowledge may influence the manner in which attention is directed to the external display and how information is perceived, interpreted, and modelled internally (Kriz and Hegarty 2007). As explained by Freedman et al. (2001), taking as an example a line graph, an expert can integrate into a coherent mental representation the visual features and the interpretation of data while a novice, lacking prior knowledge of the graph, can't explicitly represent information, thus inferences are necessary and the comprehension effortful. Thus, an expert automatically forms a link between the visual features (the shape of the line) and the theoretical interpretation of the data. When a perceiver lacks the relevant prior knowledge, or the display does not explicitly represent information that must then be inferred, comprehension is effortful. When diagrams do not contain all the information that a user needs, the prior knowledge associated with a specific diagrammatic representation allows experts a better interaction with the diagram itself. Thus, users with high levels of expertise will have higher performance in information processing and inference processes, generating new knowledge and awareness concerning the depicted information in the diagram through the interaction process itself (Cheng et al. 2001).

1.2 Chunking in Competence Measurement

As graphical features and structure underpin comprehension theories from psychology and cognitive science, our new approach to assessing competence focuses on how people process such features and structures. Of particular relevance are schema (Bartlett 1932) and chunking (Miller 1956) theories. Learning increases the size of our chunks in memory, so someone who is more familiar with a topic has chunks with greater content. According to Gobet et al. (2001) the chunking process has a dual nature based on two opposite assumptions: the first which defines chunking as deliberate and conscious (goal-oriented chunking), while the second as an automatic and continuous process that occurs during the perception (perceptual-chunking). Thus, tasks that involve perceptual processing and deliberate processing of chunks may provide rich behavioural signals that reveal an individual's personal organisation of chunks in their memory (Gobet 2005, Gobet and Simon 1996, Gobet et al. 2001, Holding 1985). For instance, the time between successive task activities (i.e., pauses) varies depending on where in the hierarchy of chunks processing is occurring. They reveal the total amount of cognition to perform a specific action, so pauses between the production of intra-chunk elements (within a chunk) will be shorter, whereas the pauses between actions spanning inter-chunk boundaries (between chunks) will be relatively longer. Further, the higher in the hierarchy of chunks an inter-chunk transition occurs, the greater will be its pause.

Based on these ideas about pause analysis of chunk structures, Cheng (2014) and colleagues (Cheng and Rojas-Anaya 2007; Albehaijan and Cheng 2019) developed an approach to assess competences using transcription tasks, in which stimuli, such as mathematical formulas or program code, are copied. The measure of competence in those tasks exploits pauses between successive written characters that are sensitive to chunk structure in an individual's memory. The shape of the distribution of pauses varies with the competence of the transcriber. The potential of assessing competence using temporal chunk signals in transcription tasks is demonstrated by measures that are well correlated with independent measures of competence.

The issue is now whether this approach can be used to assess individuals' competence with diagrammatic representations. The previous work mainly focused on linear symbolic notations and natural language. Will the technique and measures be applicable to diagrams? As diagrams are 2D, they do not have an obvious linear format to follow during transcription, but some previous work on chunking in diagram drawing suggests that there is some potential (Cheng et al. 2001; Obaidellah and Cheng 2015; Roller and Cheng 2014). That work did not include data graphs and charts, so the question for this paper is whether clear signs of chunking are manifested in the transcription of charts and graphs. Specifically: (a) Will the transcription of these representations show temporal signals, patterns of pauses between drawing actions, that reflect the structure of chunks? (b) Will those signals vary between individuals in ways that suggest they possess different chunk structures?

These questions will be addressed empirically and theoretically. In the next main section, a small-scale study of four participants transcribing diagrams is presented, which answers the questions affirmatively. In the third main section, task analysis – with GOMS – is used to model the differences in the observed behavioural strategies in order to show that the distribution of pauses can be attributed to participants' possession of different hierarchical chunk structures.

1.3   Existing Methods for the Assessment of Graph Familiarity

To end the introduction, two previous approaches to the assessment of familiarity with graphs should be acknowledged. Xi (2005) assessed competence for line graphs and bar charts using a Graph Familiarity questionnaire. It has verbal statements, which are judged on a 6-point scale, in groups concerning: participants' prior experience using graphs, their ability to read graphs, and their typical reactions to graphs. Moreover, Xi found that planning time affects the accuracy of graph descriptions as participants captured the major points of the graph and described more elements.

The other approach by Cox and Grawemeyer (2003) assesses how people organise their knowledge of external representations (ERs) through a card-sorting task. They found that expert ability to use ERs in reasoning and problem solving was associated with high performance in semantic distinctions and accurate naming of ERs, thus high competence participants produced few categories in the ER card-sorting task as they had better mental representations of ER knowledge, and perceived the semantic commonality between visually different ERs.

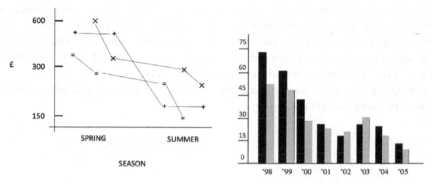

Fig. 1. Study stimuli: (a) Line graph (left), (b) bar chart (right)

2   Observing Strategies of Drawing Graphs and Charts

It is imaginable that participants might differ little in how they transcribe graphs and charts, because such diagrams have been designed to make particular visual features and structures particularly salient. This might mask any effects of familiarity with these representations and the behavioural signals due to chunking. This would contrast with linear sentential notations in which the structure of expressions depends heavily on the content of the expressions. Thus, it is essential to show that the transcription behaviours of these 2D representations actually reveal signs of chunking. To this end, a small-scale observational study was conducted.

2.1   Study

Participants. Four right-handed participants with Master's degrees in different subjects were recruited. All completed Xi's (2005) graph familiarity questionnaire (on a scale of 1 to 6; 6 is high). Their scores (and subjects) are: P1 = 4.7 (Finance); P2 = 4.3 (Engineering); P3 = 2.9 (Literature); P4 = 2.3 (Law). The scores are clearly dichotomous and consistent with the participants' educational speciality.

Materials. Figure 1 shows the two stimuli used. A grouped bar graph (Fig. 1b) from the Wall Street Journal, "Auto Industry, at a Crossroads, Finds Itself Stalled by History", January 2, 2006 was used. We designed the line-graph (Fig. 1a) especially so that it has two sets of points that might be perceived as corners of two hexagons, as potential distractor to the three data lines. Each was accompanied by a summary of their general meaning. To show the stimuli, we used a laptop computer running a logging program especially written in our lab. We recorded the participants' drawing from above with a video-camera. All drawing actions, pen strokes, were coded using the ELAN video analysis software (Sloetjes and Wittenburg 2008) and the duration of pauses between strokes computed with milliseconds (ms) accuracy.

Procedure. The study consisted of two trials where participants copied the stimulus on to a blank sheet of paper using the participant-driven "hide-show" interaction method (Albehaijan and Cheng 2019), in which the stimulus only appears on the computer screen when the participant holds down a special key. To write on the sheet, participants must release the key and the stimulus is hidden. This method allowed us to record various measures in addition to pauses: (a) view-numbers – the total number of views of the stimulus in a trial; (b) view-times – the duration of each look at the stimulus; (c) writing-times – the time spent writing between two successive views.

The results for each measure are presented and discussed separately and then followed by some general discussion.

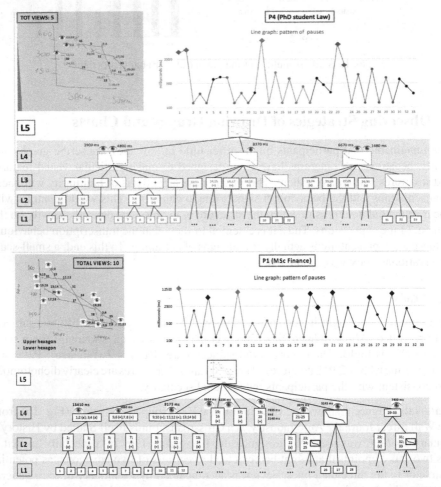

Fig. 2. Data plots for (a) P4 (top panel) and (b) P1 (bottom panel). Each panel comprises: the participant's drawing (top left), with strokes numbered in sequence; a log axis pause profile graph (top right) for each stroke; a chunk processing tree (bottom), where each leaf is a pen stroke and nodes are chunks or sub-chunks of lines. The eye symbols in the chunk process tree and the diamond data points in the pause graph are participant's views of the stimulus.

2.2 Line Graph: Results

Observing the drawing behaviours, we found various approaches across the participants, including: P2 & P3 – reproduced each set of data, switching continuously between dots and lines; P4 – reproduced each set of data in turn, with a tendency to do all the data points first followed by the connecting lines; P1 – drew all data points first, for the two hexagons, then fill in the connecting lines. Consistent with previous studies, all participants had distributions of pauses (times between strokes) that appear to reflect hierarchal organisation of in chunks memory (Cheng and Rojas-Anaya 2007; Roller and Cheng 2014; Thompson et al. 2017). Specifically, the pauses for the first stoke of meaningful groups of elements is longer than the pauses within those groups: longer pauses seem to reveal within the transitions between chunk or sub-chunk boundaries and may reflect the extra cognition required to make inter chunk switches.

Take, for example, the strategies employed by P4 and P1 (Fig. 2a and 2b), as representative of those with high and low (P1) familiarity, respectively. The chunk processing tree is navigated in a depth-first manner. Level L5 is for the whole drawing, L4 is for the chunk(s) acquired in a view(s) of the stimulus, and L3 is a sub-chunk level, where sub-chunks are defined by a pause threshold of 500 ms (Obaidellah and Cheng 2015; Roller and Cheng 2014). The overall structure of the trees changes little with reasonable variations of the threshold. L2 and L1 are symbols and strokes.

P4's drawing, Fig. 2a, appears to be organised by a graphical schema that separates datapoints and connecting lines, particularly in the second and third chunks, where all the datapoints are produced before the line connecting them is completed. Each schema is acquired in one view (or two consecutive views) and the connecting lines appear to be treated as sub-chunks. In contrast, P1's production, Fig. 2b, shows a different strategy that starts with datapoints at the extremes of the plot, then completes points within each hexagon, and finally the three sets of connecting lines. The profile of pauses shows less evidence of large chunks, but still includes signs of chunks. Consequently, the process hierarchy is shallower as the sub-chunk level is absent (L3). P1 took more views than P4, and initially appears to be treating the three sets of data points as a single field.

P2 and P3 present a similar overall approach to P1, focusing on each data set in turn, but they took approximately twice the number of views (P2 = 8; P3 = 9). At a lower level, however, they broke down each dataset into groups of a few points and lines, each associated with a view. Thus, it appears they did not use a high-level schema for each dataset but were nevertheless chunking.

2.3 Line Graph: Discussion

Diverse strategies were employed by the participants during the task as expected for such a heavily perceptually oriented task (van Sommers 1984). The strategies vary overall in relation to the datasets and also locally in relation to subgroups of elements within a data set. Despite these differences, it is clear from the pause profiles and structure of the derived processing hierarchies – which incorporates information about the occurrence of views beyond the pause threshold – that chunking is being used in the transcription of the line graph.

Contrary to expectations, the two participants with the greatest familiarity with graphs did not exploit chunking processes the most in their drawings. Only P4, the lowest scorer on the questionnaire, can be characterised as showing processes with a clear pattern of chunking and sub-chunking. However, P1, who scored highest on the questionnaire, adopted an approach that might be considered as an attempt to use an overly sophisticated strategy, because P1 ignored the three distinct groups of data; faced with a large field of points, P1 appears to have tried to demark their overall shape and then fill in the individual points. This may be linked to Roller and Cheng's (2014) and Obaidellah and Cheng's (2015) observations that the drawing of complex diagrams may follow a decomposition strategy in which an overall frame is first produced and followed by details within. Even so, there is evidence of chunking, albeit not consistently associated with the sets of data and the intended meaning of the line graph.

2.4 Bar Chart: Results and Discussion

Unlike the line graph, where evident differences occur in the drawings, for the bar chart all the participants shared similarities in terms of their drawing sequences. The overall strategy adopted by all was to reproduce the bars from left to right. Also, the black bar was always drawn first in each pair of bars.

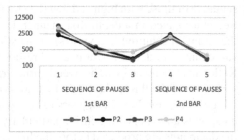

Fig. 3. Participants' pause profiles (log scale in ms) for one pair of bars in the bar chart.

At a lower level, evidence of chunking is apparent in the number of views required by each participant and their pause profiles. The number of views varied markedly between participants: P1 = 10, P2 = 16, P3 = 6, P4 = 8. As there are 8 pairs of bars (Fig. 1b), P1, P3 and P4 require approximately one view per pair, whereas P2 dealt with one bar at a time. Given the similarity of the overall strategy, we were able to derive a general pause profile per each participant, shown in Fig. 3 (numbers 1 to 5 indicate the sequence of strokes). All participants' values within chunks and sub-chunks are means, except the 1st of each participant and the 4th of P2 for which the first quartile (Q1) of the view time was used for the pause duration, as appropriate to their specific local strategy. The pause at the start of each bar is longer than that for strokes within a bar, and the pause for the start of the second bar is shorter than the first's, but longer than the pauses within the first bar. Thus, all participants are showing signs of treating each pair of bars as a chunk and each bar within as a sub-chunk.

It is odd that P2, who scored highly on the familiarity questionnaire, did not group two bars as one chunk. After the trial P2 explained that their overall goal was to accurately represent the values of the bars, which accounts for P2's one bar at a time approach. As P2 was attending to extra information, it is possible that this may have sufficiently loaded their working memory that no capacity was spare for the second bar. The approach is reflected in P2's generally longer durations of pauses for the first line of each bar compared to the other participants. Despite this difference in strategy, it is noteworthy that P2's pause profile has the same overall shape as the others.

The pause profiles are independent of the precise order of the drawing of lines within the bars. P1, P2 and P3 started the black bar drawing: (i) left line, (ii) top line and (iii) right line. However, they differ for the direction from which they start the left line: P2 drew the line from the top to the x-axis whereas P3-P4 did the opposite. P4 drew bars differently producing in sequence: (i) top line, (ii) left line and (iii) right line. The production of the second bar was consistent for all participants. Despite the marked difference in the specific strategy of line production, it is clear that this is secondary to the role of chunking in their performance as shown by the profile of pauses in Fig. 4.

2.5 Overall Discussion of the Study

The purpose of the study was to investigate the possibility that the task of transcribing graphs and charts could be used as a method to assess users' familiarity or competence with data charts. In particular, did chunking have a major role in the production process, so that measures of chunking can be engaged, such as the number of views and distributions of pauses (Albehaijan and Cheng 2019; Cheng 2014)? Overall, there is clear evidence of chunking in participants' transcriptions of both line graph and bar chart stimuli. This is demonstrated by the pause profiles and also the coherence of derived chunk process hierarchies, that is, putative chunks correspond to meaningful groups of elements (Figs. 2 and 3). Chunking provides a good explanation of the participants' performance despite the wide variety of strategies they used, at global and local levels. Further, the size of the observed chunks (2–4 sub-chunks) is in line with chunking theory for complex tasks. In this respect, the transcription of graphs and charts may have potential as method for competence measurement.

However, the diverse drawing strategies are problematic as they may not be associated with chunks that are related to the meaning of the target representation, but encode superficial visual features. As Kriz and Hegarty (2007) noted, the interaction between prior knowledge and the bottom-up features presented by a stimulus influences the perceptual processing and therefore the way chunks are drawn. The sequence of production may be affected by the specific design of our stimulus, which invites a specific order of production (Van Sommers 1984). Strong visual patterns, such as those highlighted by gestalt principles of visual perception, may determine a drawing strategy. This is a particular concern, because the behaviours associated with such superficial features may appear similar to the signals of behaviours associated with meaningful chunks and so mask the target signals. An implication is that methods must be developed so that the strategies adopted during transcription are closely tied only to whatever meaningful chunk the participants have of target stimuli; that is, chunks must reflect the way in which information is encoded when transcribing a representation rather than accidental perceptually salient patterns. At minimum, participants must be instructed that precise values of data points are not of concern, in order to prevent behaviours such as P2's narrow precision goal on the bar chart. We might, for instance, instruct participants to focus on the meaning and communicative intent embodied by the representations.

Although only four participants contributed transcriptions for each representation, it is noteworthy that despite the clear difference in the familiarity of the two pairs of participants, there was no indication of difference in competence. One explanation is that any effect of familiarity may have been masked by the issue of diverse drawing strategies. Another explanation is that the selected stimuli (Fig. 1) are too simple for the chosen participants. In future work we will test more complex line graphs and bar charts.

3   CPM-GOMS Verification of the Line Graph Chunking

The aim of this second part of the study is to obtain converging evidence that the behaviours in the transcription of the graph and chart were largely determined by chunking processes. We will use cognitive modelling, specifically task analysis, for this and focus specifically on the line graph. The idea is to generalise the strategies used by participants on the line graph to produce an ideal chunk hierarchy that shares the common features of the individual approaches, see Fig. 4 top. Each line in the graph is treated as a chunk that is composed of sub-chunks consisting of different combinations of data points and lines produced in sequence. For an ideal chunk hierarchy we built a task analytic model composed of a typical sequence of cognitive processes for such tasks with standard values of timings for basic cognitive operations. The chunk hierarchy determines the sequencing of operators in the model from which predictions of pause values for each pen-stroke can be derived. If the predicted pauses of the simulation match the typical distribution of the participants' pauses well, this will support the idea that chunking processes are also responsible for the participants' pauses.

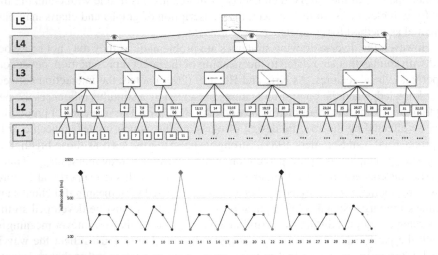

Fig. 4. Idealised chunk hierarchy (top) and predicted pause profile (bottom).

3.1   CPM-GOMS Modelling

We adopted the GOMS approach to task analysis. GOMS is a family of modelling techniques that analyses user performance in interactive systems (John and Kieras 1996). Each type of GOMS task analysis consists of a hierarchical task decomposition based on Goals, Operators, Methods, and Selection rules (Card et al. 1983). The Goal is what the user is trying to accomplish. Operators are atomic elements that generally exhibit a fixed execution time. Methods consist of sets of operators commonly applied together to achieve a goal. Selection rules choose between methods. The *Cognitive, Perceptual, Motor* GOMS (CPM-GOMS) technique, using the *Model Human Processor* (MHP) as a framework (Card et al. 1983), is the most suitable for drawing transcription tasks, because it can deal with parallel execution of visual perception, cognitive and motor operations. In CPM-GOMS, the perceptual processor is responsible for transforming external information into a form that the cognitive system can process; the cognitive processor uses contents of WM and LTM to make decisions and schedule actions with the motor system; and, the motor processor translates thoughts into actions. The CPM-GOMS architecture employs PERT/Gantt-like charts to represent relations between the operators and to derive a critical path that estimates the total time required for the task execution. We used the software Cogulator (Estes 2016) to build the models.

Figure 5 shows a section of the Cogulator model for the ideal chunk hierarchy, derived in Fig. 4, top. Figure 5 is a template used for our CPM-GOMS model. The model has three principal types of statements: `GOAL`, `.Goal` and `.Also`. The `GOAL` statements represent the main goals required to perform the transcription task, specifically a perceptual goal and a drawing goal. The `.Goal` statements are sub-goals included within the main goal and deal with the different items that must be drawn. `.Also` is used to represent the parallel processes that occur during the pauses when the pen is moving or hovering between the inscriptions. Each operation has a separate line in the code. The number of full stops before a code word indicates the nesting level of the operator.

Most of the values employed for the operators are provided by the literature (John and Newell 1989; Gray and Boehm-Davis 2000). At the beginning of each chunk within the drawing `GOAL`, we assumed a `recall` operator of 1200 ms (John and Newell 1989; Lee 1995) to retrieve information from WM. For the `pen_stroke` motor operators, which execute the pen strokes, we picked values based on the average time (ms) to draw the single components of the line graph. The `jump` motor operators, which correspond to the pauses between the inscriptions, were predicted by summing the values for all cognitive operators that we assumed occur in parallel during the pause. Moreover, as pauses between symbol inscriptions are assumed to be automatic processes, they are not included in the model.

As diagram drawing is not typically modelled by GOMS, we decided to deal with the spatial information separately as the spatial coordinates' values are fundamental both in drawing and graph comprehension. So, some non-standard operators are defined: `verify_location`, `shifting` and `updating`. We decided to assign 50 ms to `verify_location`, to match that of the standard GOMS `verify_information`. Furthermore, as the task is complex, it likely involves executive functions (EF) (Miyake et al. 2000, 2012, Morra and Panesi 2016). EF operators comprise those mental capacities necessary for formulating goals, planning how to achieve them, and carrying out the plans effectively (Lezak 1982), they differ from the cognitive functions (CF) as they explain how and whether a person goes about doing something, rather than what and how much. Thus, we also define EF cognitive operators: `ignore`, `shifting`, and `updating`. The `shifting` operator is a main component of cognitive flexibility. It is an ability

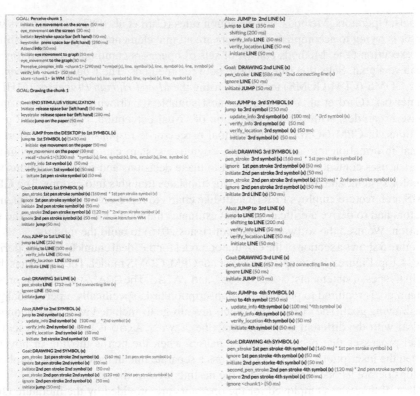

Fig. 5. Sample Cogulator code for a CPM-GOMS for the drawing of the first chunk of the idealized model in Fig. 4. Two main GOALs are required, one for perceiving the stimulus and one for drawing the chunks. The second is always broken down into several sub-goals (i.e. .Goal statements) necessary for drawing the symbols and connecting lines in the chunk. All the pause durations between the inscriptions represented by the .jump motor operators were obtained summing the cognitive operators values, which occur in parallel within .Also statements.

used by people to represent their knowledge about a task and the possible strategies in which to engage (Cañas et al. 2006). A shifting operator is required before drawing each connecting line, however its value varies depending on the hierarchical position of the connecting lines. Thus, it has a value of 100 ms whenever a connecting line within sub-chunk must be drawn. Its value increases to 200 ms during the switching between sub-chunks as an upper level item in the hierarchy needs to be picked (e.g., the transition from pen stroke 5 to 6 in Fig. 4). The updating operator of 100 ms, updates and monitors the working memory contents (Miyake et al. 2000, 2012) before drawing each point or top line which represent the numerical values, respectively, in the line graph and bar chart. Otherwise, the .ignore operator correspond to the inhibitory executive function which suppresses inappropriate responses (Miyake et al. 2000, 2012), so it occurs whenever a pen stroke is made before starting the subsequent cognitive operation.

3.2 Modelling Results

The idealized chunk hierarchy has three chunks with two or three sub-chunks, Fig. 4. Applying standard sequences of CMP-GOMS operators to this hierarchy, with the values given above, the full series of operations needed for the task and their timings were assembled. Pauses between the end of each stroke and beginning of each stroke were computed (Fig. 5) and the pause profile graph plotted (Fig. 4).

The overall shape of the profile resembles the profiles for the participants (e.g., Fig. 2). There are long pauses for the views, very short pauses for strokes within symbols, but critically medium and short pauses for sub-chunks, which are comparable to the participants' durations of pauses. When the idealised chunk structure is modified, for example, to more closely reflect a specific participant, the precise pattern of the pause profile also changes, but the overall nature of the distribution remains the same. Thus, the match between the CPM-GOMS models and that of the participants, suggests that chunking is primarily responsible for the shape of the profiles, and hence chunking is critical in these drawing transcription tasks.

4 Discussion

Chunking is a well-studied phenomenon in cognitive science due to its ubiquity in learning and information processing. We aim to produce a method to assess competence in representations that goes beyond current tests that only indirectly assess familiarity using questionnaires, verbal descriptions or simple tasks (Cox and Grawemeyer 2003; Xi 2005). Critical for our approach was to show that chunking occurs in transcription-based drawing of diagrams, following previously established measures of chunk-based assessment with linear notation (Cheng 2014; Cheng and Rojas-Anaya 2007; Albehaijan and Cheng 2019).

In the observations of drawings of the four participants on line graph and bar chart, evidence was found of chunking. Pauses between strokes had distributions typical of tasks involving chunking, and values typical of chunking. Longer pauses occur for inter-chunk transition at higher level, and shorter pauses for intra-chunk transitions at lower levels. From the pause profiles, putative chunk hierarchies were systematically derived that exhibit structures typical of chunking processes.

Generalising over the predicted chunk hierarchies an idealised chunk hierarchy was constructed and used as the foundation of a CMP-GOMS task analytic model. The good correspondence between the model and participants pause profiles, particularly in the levels and magnitudes of pauses, adds weight to the claim that chunking was central in the transcription processes. This suggests, at least in principle, that such drawing transcription task may have potential as a measure of competence.

However, the diversity of strategies and the actual patterns of drawn elements suggest that the chunks may often reflect obvious perceptual patterns and conventions, rather than chunks and schemas that underpin the participants' underlying knowledge of the two representations. Thus, recommendations for refinements to the method have been suggested (in Sect. 2) to ensure that meaningful chunks are most likely to be probed.

Acknowledgements. We thank Gem Stapleton for her comments and suggestions for this paper. This work was supported by the EPSRC grants EP/R030650/1, EP/T019603/1, EP/R030642/1, and EP/T019034/1.

References

Albehaijan, N., Cheng, P.C.-H.: Measuring programming competence by assessing chunk structures in a code transcription task. In: Goel, A., Seifert, C., Freksa, C. (eds.) Proceedings of the 41st Annual Conference of the Cognitive Science Society, pp. 76–82. Cognitive Science Society, Austin, TX (2019)

Bartlett, F.C.: Remembering: A Study in Experimental and Social Psychology, p. 329. Cambridge University of Press, Cambridge (1932)

Cañas, J.J., Fajardo, I., Salmeron, L.: Cognitive flexibility. In: Karwowski, W. (ed.) International encyclopedia of ergonomics and human factors, 2nd edn., pp. 297–301. CRC Press, Boca Raton (2006)

Card, S., Moran, T.P., Newell, A.: The Psychology of Human-Computer Interaction. Lawrence Erlbaum Associates, Hillsdale (1983)

Cheng, P.C.H., Lowe, R.K., Scaife, M.: Cognitive science approaches to understanding diagrammatic representations. Artif. Intell. Rev. **15**(1–2), 79–94 (2001)

Cheng, P.C.-H.: Copying equations to assess mathematical competence: an evaluation of pause measures using graphical protocol analysis. In: Bello, P., Guarini, M., McShane, M., Scassellati, B. (eds.) Proceedings of the 36th Annual Meeting of the Cognitive Science Society, pp. 319–324. Cognitive Science Society, Austin, TX (2014)

Cheng, P., Rojas-Anaya, H.: Measuring mathematic formula writing competence: an application of graphical protocol analysis. In: Proceedings of the Thirtieth Annual Conference of the Cognitive science Society (2007)

Cheng, P., McFadzean, J., Copeland, L.: Drawing out the temporal signature of induced perceptual chunks. In: Proceedings of the Twenty-Third Annual Conference of the Cognitive Science Society, pp. 200–205 (2001)

Cox, R., Grawemeyer, B.: The mental organisation of external representations. Proc. Eurocogsci **03**, 91–96 (2003)

Estes, S.: Introduction to simple workload models using cogulator (2016)

Freedman, E.G., Shah, P.: Toward a model of knowledge-based graph comprehension. In: Hegarty, M., Meyer, B., Narayanan, N.H. (eds.) Diagrams 2002. LNCS (LNAI), vol. 2317, pp. 18–30. Springer, Heidelberg (2002). https://doi.org/10.1007/3-540-46037-3_3

Freedman, E.G., Shah, P.S.: Individual differences in domain knowledge, graph reading skills, and explanatory skills during graph comprehension. Paper presented at the 42nd annual meeting of the psychonomic society, Orlando, FL, November, 2001

Gobet, F.: Chunking models of expertise: implications for education. Appl. Cogn. Psychol. **19**, 183–204 (2005)

Gobet, F., Simon, H.A.: Templates in chess memory: a mechanism for recalling several boards. Cogn. Psychol. **31**, 1–40 (1996)

Gobet, F., et al.: Chunking mechanisms in human learning. Trends Cogn. Sci. (2001). https://doi.org/10.1016/S1364-6613(00)01662-4

Gray, W.D., Boehm-Davis, D.A.: Milliseconds matter: an introduction to microstrategies and to their use in describing and predicting interactive behavior. J. Exp. Psychol. Appl. **6**(4), 322–335 (2000)

Hegarty, M.: Multimedia learning about physical systems. In: Mayer, R.E. (ed.), The Cambridge Handbook of Multimedia Learning, pp. 447–465. Cambridge University Press, Cambridge (2005)

Holding, D.H.: The Psychology of Chess Skill. Erlbaum, Hillsdale (1985)

Jamnik, M., Cheng, P.C.-H.: Endowing machines with the expert human ability to select representations: why and how. In: Muggleton, S., Chater, N. (eds.) Human-Like Machine Intelligence. Oxford University Press, Oxford (2021)

John, B.E., Kieras, D.E.: The GOMS family of user interface analysis techniques. ACM Trans. Comput. Hum. Interact. 3(4), 320–351 (1996)

John, B.E., Newell, A.: Cumulating the science of HCI: from S-R compatibility to transcription typing. In: Proceedings of the Conference on Human Factors in Computing Systems, May, pp. 109–114 (1989). https://doi.org/10.1145/67449.67472

Kintsch, W.: The role of knowledge in discourse comprehension. A construction- integration model. Psychol. Rev. 95, 163–182 (1988)

Kriz, S., Hegarty, M.: Top-down and bottom-up influences on learning from animations. Int. J. Hum. Comput. Stud. 65, 911–930 (2007)

Lee, A.: Exploring user effort involved in using history tools through MHP/GOMS: results and experiences. In: Nordby, K., Helmersen, P., Gilmore, D.J., Arnesen, S.A. (eds.) Human—Computer Interaction. IFIP Advances in Information and Communication Technology. Springer, Boston (1995). https://doi.org/10.1007/978-1-5041-2896-4_18

Lezak, M.D.: The problem of assessing executive functions. Int. J. Psychol. 17, 281–297 (1982)

Miller, G.A.: The magical number seven, plus or minus two: Some limits on our capacity for processing information. Psychol. Rev. 63(2), 81–97 (1956)

Miyake, A., Friedman, N.P., Emerson, M.J., Witzki, A.H., Howerter, A., Wager, T.D.: The Unity and Diversity of Executive Functions and Their Contributions to Complex "Frontal Lobe" Tasks: A Latent Variable Analysis. Cognitive Psychol. 41(1), 49–100 (2000). https://doi.org/10.1006/cogp.1999.0734

Miyake, A., Friedman, N.P.: The nature and organization of individual differences in executive functions: four general conclusions. Curr. Dir. Psychol. Sci. 21(1), 8–14 (2012). https://doi.org/10.1177/0963721411429458

Obaidellah, U.H., Cheng, P.C.H.: The role of chunking in drawing Rey complex figure. Percept. Mot. Skills 120(2), 535–555 (2015)

Panesi, S., Morra, S.: Drawing a dog: The role of working memory and executive function. Journal of Experimental Child Psychology 152, 1–11 (2016). https://doi.org/10.1016/j.jecp.2016.06.015

Peebles, D., Cheng, P.C.H.: Modeling the effect of task and graphical representation on response latency in a graph reading task. Hum. Factors 45(1), 28–46 (2003)

Roller, R., Cheng, P.C.-H.: Observed strategies in the freehand drawing of complex hierarchical diagrams. In: Bello, P., Guarini, M., McShane, M., Scassellati, B. (eds.) Proceedings of the 36th Annual Meeting of the Cognitive Science Society, pp. 2020–2025. Cognitive Science Society, Austin, TX (2014)

Shah, P., Carpenter, P.A.: Conceptual limitations in comprehending line graphs. J. Exp. Psychol. Gen. 124(1), 43–61 (1995)

Shah, P., Mayer, R.E., Hegarty, M.: Graphs as aids to knowledge construction: signaling techniques for guiding the process of graph comprehension. J. Educ. Psychol. 91(4), 690–702 (1999)

Sloetjes, H., Wittenburg, P.: Annotation by category - ELAN and ISO DCR. In: Proceedings of the 6th International Conference on Language Resources and Evaluation (LREC 2008) (2008)

Thompson, J.J., McColeman, C.M., Stepanova, E.R., Blair, M.R.: Using video game telemetry data to research motor chunking, action latencies, and complex cognitive-motor skill learning. Top. Cogn. Sci. 9(2), 467–484 (2017). https://doi.org/10.1111/tops.12254

van Sommers, P.: Drawing and Cognition: Descriptive and Experimental Studies of Graphic Production Processes. Cambridge University Press, Cambridge (1984)

Wu, P., Carberry, S., Elzer, S., Chester, D.: Recognizing the intended message of line graphs. In: Goel, A.K., Jamnik, M., Narayanan, N.H. (eds.) Diagrams 2010. LNCS (LNAI), vol. 6170, pp. 220–234. Springer, Heidelberg (2010). https://doi.org/10.1007/978-3-642-14600-8_21

Xi, X.: Do visual chunks and planning impact performance on the graph description task in the SPEAK exam? Lang. Test. 22(4), 463–508 (2005)

Diagrams in Essays: Exploring the Kinds of Diagrams Students Generate and How Well They Work

Emmanuel Manalo[1](✉) and Mari Fukuda[2]

[1] Graduate School of Education, Kyoto University, Kyoto, Japan
manalo.emmanuel.3z@kyoto-u.ac.jp
[2] Graduate School of Education, The University of Tokyo, Tokyo, Japan
mari_fukuda@p.u-tokyo.ac.jp

Abstract. Using appropriate diagrams is generally considered efficacious in communication. However, although diagrams are extensively used in printed and digital media, people in general rarely construct diagrams to use in common everyday communication. Furthermore, instruction on diagram use for communicative purposes is uncommon in formal education and, when students are required to communicate what they have learned, the usual expectation is they will use words – not diagrams. Requiring diagram inclusion in essays, for example, would be almost unheard of. Consequently, current understanding about student capabilities in this area is very limited. The aim of this study therefore was to contribute to addressing this gap: it comprised a qualitative exploration of 12 undergraduate students' diagram use in two essays (in which they were asked to include at least one diagram). Analysis focused on identifying the kinds of diagrams produced, and the effectiveness with which those diagrams were used. Useful functions that the diagrams served included clarification, summarization, integration of points, and provision of additional information and/or perspectives in visual form. However, there were also redundancies, as well as unclear, schematically erroneous, and overly complicated representations in some of the diagrams that the students constructed. These findings are discussed in terms of needs, opportunities, and challenges in instructional provision.

Keywords: Self-constructed diagrams · Essay writing · Effective communication · Student instructional needs

1 Introduction

Alongside problem solving and thinking, communication is one of the areas of human activity where diagram use is considered to be beneficial. When appropriately used, diagrams can clarify and/or complement verbal information presented in speech or text, so that both verbal and visual channels of working memory are utilized, thus facilitating more efficient cognitive processing [1, 2].

Diagrams can contribute to both message encoding and decoding (i.e., the production and the comprehension of communication), thus being of value to both the communicator and the communication receiver. Especially in contexts where there are

© The Author(s) 2021
A. Basu et al. (Eds.): Diagrams 2021, LNAI 12909, pp. 553–561, 2021.
https://doi.org/10.1007/978-3-030-86062-2_56

some constraints or limitations to conveying the message through verbal means, diagrams can be indispensable. They can supplement speech or text by providing complementary or alternative means of conveying the intended message. Examples of such contexts include communicating complicated procedures, like furniture assembly [3], and communicating with people who speak a different language [4].

However, despite the apparent usefulness of diagrams in communication, its actual use remains very limited. Pictures and various kinds of diagrams, including illustrations, are regularly used in books, magazines, websites, and various forms of printed and digital media, but most of those visual representations are commercially or professionally created. They are not generated by regular people in everyday communication contexts. Regular people are often only receivers of such visual representations. In most communication contexts, they do not generate their own diagrams: they rely almost exclusively on written or spoken words. In formal education provided in modern societies, diagram use for communicative purposes is rarely taught. Despite the recognition in research and policy documents of the value of being able to use multiple forms of representation [5, 6], students seldom receive explicit instruction about how to create and use diagrams. In both school and higher education, when students are asked to communicate what they have learned and what they think (e.g., in essays, which are focused pieces of writing intended to inform or persuade), the general expectation is that they will express that information in words – without the use of any diagrams [7].

Considering that much of the knowledge and ideas that students have to engage with, learn, *and then communicate* are quite complex, and diagrams have the capacity of representing complex ideas effectively [8], the general lack of attention in education to cultivating skills in diagram use is troubling. Like words, diagrams can be used effectively or ineffectively [9], so the question of the extent to which instruction or guidance may be necessary would appear important to address. In tasks like problem solving and information organization in subjects like mathematics and science, the kinds of diagrams that students generate and use have previously been investigated [10–12]. However, very few studies have examined diagram use in communicating information in the social sciences [13], where traditionally a greater emphasis has been placed on the quality of language that is used. In fact, the present authors are not aware of any studies that have examined students' diagram use in *essays*. Our current understanding of student capabilities in using diagrams in such contexts is very limited, including what we know of the potential benefits that such use might afford.

The present study was motivated by this knowledge gap, and it comprised a qualitative exploration of student diagram use in two essays they produced for an undergraduate-level introductory course in educational psychology. In the two essays, the students were asked to include at least one diagram to portray processes or mechanisms of moderately complex ideas. Both essays were expository-type essays, hence requiring the students to demonstrate not only knowledge of the topic, but also the ability to communicate information clearly – which the appropriate use of diagrams is supposed to facilitate. The following were the main questions we addressed:

(i) What kinds of diagrams would students use to portray processes/mechanisms?
(ii) In what ways do students use diagrams effectively in their essays?
(iii) In what ways do they not use diagrams effectively?

2 Method

This investigation comprised analysis of the contents of two essays that students produced as part of their coursework. No experimental manipulation was involved. The analysis was conducted following completion of the course, so it had no bearing on the students' grading. Permission was obtained from all the students for use of their essays.

There were 12 students in the course (females = 4), 7 of whom were in their first year of study, while the remaining 5 were in their second year or higher. Nine of the students were Japanese, and 3 were international students from other East Asian countries. All had English as a foreign language, but were adequately proficient in that language (a requirement for acceptance to the university). The course was conducted entirely in English, and all assignments (including the essays) had to be written in English.

The essays were each worth 20% of the students' final grade, and they dealt with topics covered in the course. However, both essays required students to seek additional information (beyond what was covered in class), and to provide explanations that could not simply be obtained from the instructions provided in the course. The first essay required the students to research and then explain one theory about how young children develop their understanding of the world around them. The second required them to undertake research on formative assessment and explain how it can promote more successful learning. In both essays, the students were asked to include at least one diagram, which was allotted 3 points (out of 20) in the grading rubrics (in this case, for "demonstrating a clear understanding of the mechanisms or processes that it illustrates"). Diagrams can assist in clarifying ideas [8] and so, from a pedagogical perspective, one purpose of asking the students to include the diagram was for them to better understand key processes covered in the course. The diagram had to be self-constructed (i.e., not copied-and-pasted from some other source). In the first essay, the instruction given to the students indicated that the diagram was to "help in explaining *the progress in understanding that children develop*", while in the second essay the diagram was to "help in clarifying *how formative assessment facilitates learning*". No other instruction was provided on what form the diagram should take or how they should construct it, and *no diagrams relating to those or other similar mechanisms/processes were shown during instructions provided in class*. During grading of the essays, apart from the score out of 3 on the grading rubrics (see above), no explicit comment or feedback was provided on the type, content, or quality of construction of the diagrams the students included.

In the analysis, firstly the number and kinds of diagrams included in the essays were determined with the use of a coding schema comprising categories from previous research [10, 11, 13]. Apart from the first author's coding, the second author, who initially was not involved in this research, also independently coded the diagrams. Initial inter-coder agreement was 75%. Differences were then discussed and subsequently agreed upon. Second, the diagrams were evaluated in terms of how effectively they were used. For this, key questions asked were: Does the diagram contribute to clarifying the process it refers to – and, if so, how? Apart from clarification, does it serve other useful functions? When diagrams did not appear to work well, the reasons were also carefully considered. Again, both authors independently coded the diagrams (initial inter-coder agreement was 92%), and then discussed differences to reach agreement.

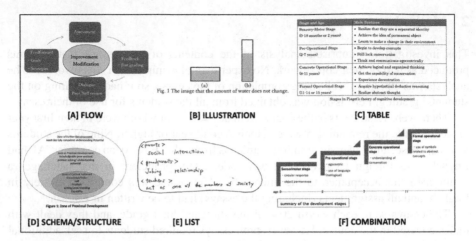

Fig. 1. Examples of diagrams belonging to each of the categories that were identified

3   Results and Discussion

3.1   Kinds of Diagrams Used to Portray Processes/Mechanisms

Table 1 shows the kinds/categories and corresponding frequencies of diagrams the students included in their first and second essays, and in total, while Fig. 1 shows examples of the diagrams belonging to each of those categories. In the first essay four of the 12 students included more than one diagram (two student with 3 diagrams, and two with 2 diagrams), and in the second essay two students included more than one diagram (both with 2 diagrams). The frequencies shown in Table 1 include all the diagrams the students generated.

All except one student included self-constructed diagrams (as the assignments required). We were fairly confident about this because the language use in and appearance of the diagrams included suggested non-native and/or non-professional creators. The one student who included diagrams that obviously came from some Internet source (they were both water-marked) did so in both the first and second essays. Both diagrams were in the category of illustrations.

Table 1. Kinds of diagrams and the frequencies with which they were used

Kinds of diagram	Essay 1	Essay 2	Total
Flow	3	7	10
Illustration	5	2	7
Table	1	2	3
Schema/Structure	2	0	2
List	1	0	1
Combinations	6	3	9

Although a "list" on its own does not – technically speaking – count as a diagram, it has been included in the categories because one student erroneously included a list as one of his 3 diagrams for the first essay, and three other students included lists as part of their "combination" diagram.

The kind of diagram most frequently used was a flow diagram: apart from the total of 10 flow diagrams shown in Table 1, 7 of the 9 combination diagrams comprised a flow diagram with another kind of diagram. This is probably understandable given that flow diagrams (also known as "flow charts") are meant to depict processes, procedures or sequence of steps, and cause-and-effect relationships. An interesting point to note is how the number of flow diagrams increased from the first to the second essay. No instruction or hint was given to the students about what diagram to use, so this increase could have been due to a number of other possible reasons, including differences between the two essays in the procedures/mechanisms that needed to be represented, the students seeing other diagrams their peers have generated (although there were no indications of copying), and development in the students' understanding of what works well (or not) in using diagrams to communicate particular kinds of information.

It is also worth noting that although we often consider flow diagrams as being most appropriate for representing processes and mechanisms, other forms of diagrams can work as effectively when designed well to match their intended purposes. For example, Panels C and F in Fig. 1 show two examples depicting the progression through the stages of Piaget's theory of cognitive development, using a table in C, and a combination of illustration (of steps) and line diagram in F. Although they differ in appearance, the diagrams can be considered as working equally well not only in showing the proposed stages of the theory, but also in conveying the incremental progression through key cognitive abilities with increasing age (corresponding to those stages).

3.2   Ways that Diagrams Were Used Effectively

The majority of the students did not refer to their diagrams in the text of their essays: only two students did in both essays. Thus, this is perhaps an academic writing method that undergraduate students (like these students) could usefully be instructed to do. However, in general, the students placed their diagrams appropriately, following the text where they deal with the information that is portrayed in the diagram – thereby making the connection between text information and the diagram more apparent.

Concerning the question of whether the diagrams that the students constructed contributed to clarifying any of the processes or mechanisms they were explaining: in the first essay, 7 of the students were considered to have satisfactorily achieved this with at least one of their diagrams, while 8 of them were considered to have done so in the second essay. In each of these cases, the diagrams served a useful function in the essay, to the extent that if they were not included, something sufficiently important in the essay would have been lost, not achieved, or not conveyed as adequately. In most of the cases, the diagrams clarified how the stages or processes referred to in the essay text connect or relate to each other and progress through particular sequences: Panels A and D in Fig. 1 are good examples of this. However, in a few cases, the diagrams also made clearer concepts that – to those unfamiliar with them – could be difficult to understand, such as what is involved in developing the ability of conservation (Panel B of Fig. 1).

In some cases, the diagrams also showed or clarified the connections to other components, such as children's abilities in connection to the progression of developmental stages in the previously referred to Panels C and F in Fig. 1. In a way, some of

the diagrams that worked well served a summarizing function: they visually represented key components and showed more saliently how they were related to each other
– which were not as easy to apprehend in sentences because of temporal/sequential
separation. This is one of the reasons diagrams are considered effective: they integrate
all information that is used together, reducing the need and effort for searching [14].

In the majority of cases where the diagrams worked well, they visually represented
content that was already represented in words in the text – albeit with some
enhancements like integration, as noted above. However, in a few exceptional cases,
the diagrams also introduced content that was not present in the text of the essays.
Figure 2 provides two examples of this. In Panel A, the diagram includes details in the
lower part about unsuccessful (arrows with x) and successful (arrows with o) outcomes
which require different responses. These details were not explicitly provided in the text
but they enable readers to better understand how formative feedback is used in the
example of solving story problems. Likewise, in Panel B, the diagram shows details not
duplicated in the text about how different categories of complexes are formed, leading
eventually to the formation of concepts [15]. The illustrations of different object
combinations make the categories of complexes easier to grasp and distinguish from
each other.

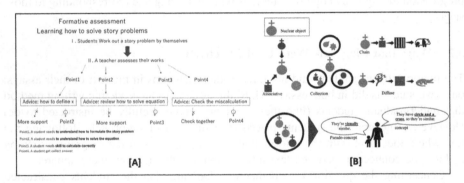

Fig. 2. Examples of diagrams that introduced content or elaborations not present in the text

3.3   Ways that Diagrams Were Not Used Effectively

There were also numerous instances when the diagrams the students included did not
appear to serve any useful function in the essay. In a couple of those cases, the
diagrams were redundant: they showed images that portrayed information from the text
that was simple enough not to require visual clarification. The illustration in Panel A in
Fig. 3 is an example of this. Another ineffective use manifested was when the schematic structure of the diagram was unclear or erroneous. Examples of this are shown in
Panels B and C of Fig. 3. In Panel B, both the intended message and the connections
between the components shown are unclear. In Panel C, the meaning of the arrows, and
therefore what process might be depicted by the diagram, is unclear. In addition, there

were a few diagrams, like the one shown in Panel D of Fig. 3, which were quite complicated and therefore hard to understand. The contents of Panel D were also referred to in the text of the student's essay, but the relationships shown in the diagram are new configurations that are not obvious and not explained explicitly in the text. It is therefore difficult to grasp its possible contribution to explaining, in this case, children's development of understanding of the world around them.

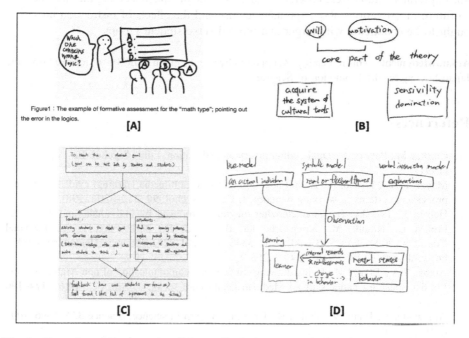

Fig. 3. Examples of diagrams that did not effectively serve their intended purpose in the essay

3.4 Implications for Theory, Research, and Practice

The many purposes that diagrams can serve in enhancing communication have been established in previous research [2, 5, 8], and the findings of the present study provide additional evidence for those in the area of student essay writing at the tertiary level. The findings also provide support for the idea that the same communicative purpose can be served by different kinds of diagrams [14]. Thus, for example, conveying the progression of a particular process can effectively be achieved using a flow diagram – or a table ... or an illustration. This means that, in the same way that different words can convey the same meaning, different diagrams – if used/constructed appropriately – can communicate the same meaning. However, in future research, it would be useful to examine the *range* of communicative purposes that different kinds of diagrams can serve as there are probably important limitations to it. For example, if the communicative purpose is to *describe what something looks like*, other kinds of diagrams may not be quite as effective as an illustration (e.g., a table or a flow chart would be quite limited in conveying qualities pertaining to appearance).

The findings of the present research also suggest that many students, even at the tertiary level, would likely benefit from receiving some instruction or guidance on the use of diagrams not only in essays but also other forms of communication. While some students were able to generate diagrams that served useful functions in their essays, there were also quite a few who did not manage to do this. However, the findings of this exploratory study were based on a small sample of students taking the same course, so future investigations about spontaneous use and the effects of instruction provision ought to be conducted with larger and more diverse student groups.

Acknowledgment. This research was supported by a grant-in-aid (20K20516) received from the Japan Society for the Promotion of Science.

References

1. Clark, J.M., Paivio, A.: Dual coding theory and education. Educ. Psychol. Rev. **3**, 149–210 (1991)
2. Mayer, R.E., Moreno, R.: A split-attention effect in multimedia learning: evidence for dual processing systems in working memory. J. Educ. Psychol. **90**, 312–320 (1998)
3. Heiser, J., Tversky, B.: Characterizing diagrams produced by individuals and dyads. In: Freksa, C., Knauff, M., Krieg-Brückner, B., Nebel, B., Barkowsky, T. (eds.) Spatial Cognition 2004. LNCS (LNAI), vol. 3343, pp. 214–226. Springer, Heidelberg (2005). https://doi.org/10.1007/978-3-540-32255-9_13
4. Alabsi, T.A., Taha, I.M.: Using drawings to facilitate communication of non Arabic and non English speaking patients in Al Medinah health care sector. J. Am. Sci. **10**(6), 174–190 (2014)
5. Ainsworth, S., Prain, V., Tytler, R.: Drawing to learn in science. Science **333**, 1096–1097 (2011)
6. National Research Council: Education for Life and Work: Developing Transferable Knowledge and Skills in the 21st Century. National Academies Press, Washington, DC (2012)
7. Manalo, E., Ueaska, Y., Kriz, S., Kato, M., Fukaya, T.: Science and engineering students' use of diagrams during note taking versus explanation. Educ. Stud. **39**, 118–123 (2013)
8. Tversky, B.: Visualizing thought. Top. Cogn. Sci. **3**, 499–535 (2011)
9. Hegarty, M., Kozhevnikov, M.: Types of visual-spatial representations and mathematical problem solving. J. Educ. Psychol. **91**, 684–689 (1999)
10. Novick, L.R., Hurley, S.M.: To matrix, network, or hierarchy: that is the question. Cognitive Psychol. **42**, 158–216 (2001)
11. Zahner, D., Corter, J.E.: The process of probability problem solving: use of external visual representations. Math. Think. Learn. **12**, 177–204 (2010)
12. Manalo, E., Uesaka, Y.: Students' spontaneous use of diagrams in written communication: understanding variations according to purpose and cognitive cost entailed. In: Dwyer, T., Purchase, H., Delaney, A. (eds.) Diagrams 2014. LNCS (LNAI), vol. 8578, pp. 78–92. Springer, Heidelberg (2014). https://doi.org/10.1007/978-3-662-44043-8_13

13. Manalo, E., Uesaka, Y.: Hint, instruction, and practice: the necessary components for promoting spontaneous diagram use in students' written work? In: Jamnik, M., Uesaka, Y., Elzer Schwartz, S. (eds.) Diagrams 2016. LNCS (LNAI), vol. 9781, pp. 157–171. Springer, Cham (2016). https://doi.org/10.1007/978-3-319-42333-3_12
14. Larkin, J.H., Simon, H.A.: Why a diagram is (sometimes) worth ten thousand words. Cognitive Sci. **11**, 65–99 (1987)
15. Vygotsky, L.: Studies in communication. In: Hanfmann, E., Vakar, G. (eds.) Thought and Language. MIT Press, Cambridge (1962)

Open Access This chapter is licensed under the terms of the Creative Commons Attribution 4.0 International License (http://creativecommons.org/licenses/by/4.0/), which permits use, sharing, adaptation, distribution and reproduction in any medium or format, as long as you give appropriate credit to the original author(s) and the source, provide a link to the Creative Commons license and indicate if changes were made.

The images or other third party material in this chapter are included in the chapter's Creative Commons license, unless indicated otherwise in a credit line to the material. If material is not included in the chapter's Creative Commons license and your intended use is not permitted by statutory regulation or exceeds the permitted use, you will need to obtain permission directly from the copyright holder.

How Can We Statistically Analyze the Achievement of Diagrammatic Competency from High School Regular Tests?

Yuri Uesaka[1(✉)], Shun Saso[1], and Takeshi Akisawa[2]

[1] Graduate School of Education, The University of Tokyo, Tokyo, Japan
yuri.uesaka@ct.u-tokyo.ac.jp
[2] Tsurumine Senior High School, Chigasaki, Kanagawa, Japan

Abstract. Owing to the recent global changes in education goals, students nowadays need to achieve 'key competencies' in school. 'Diagrammatic competency' is an essential part of such competencies. To cultivate diagrammatic competency, it is necessary to evaluate teachers and students and provide feedback on the students' degree of achieving diagrammatic competency. Regular school tests can provide useful opportunities for assessing such achievement. However, in such tests, Japanese high schools mainly focus on evaluating the understanding of learning contents rather than the development of competencies (such as diagrammatic competency). The current study was a collaboration between educational psychologists and a high school mathematics teacher. Together they modified a regular school test to incorporate tasks that require diagrammatic competency to solve them, thus enabling the assessment of such achievement. The study was conducted in an actual high school. The students' performance was analyzed using cognitive diagnostic models [1], which statistically estimate how well students have mastered the elements of cognitive abilities and skills required to solve problems, generating 'attribute mastery probabilities'. The attribute mastery probabilities obtained demonstrated that students' achievement of diagrammatic competency was insufficient, indicating a need for cultivating such competency in subject learning instruction provided in schools.

1 Necessity of Assessing Diagrammatic Competency in School

Although traditional school instruction emphasizes the acquisition of the contents of various subjects, due to the recent global changes in educational goals, developing students' learning competencies are now increasingly considered an essential objective of school education. This perspective is embodied in the idea of 'key competencies' proposed by the Organization for Economic Cooperation and Development (OECD). The ability to use external resources, such as diagrams, has been considered one of the most critical among those key competencies.

An essential aspect that needs to be cultivated in school is using diagrams effectively, which can be referred to as 'diagrammatic competency'. For example, Kragten et al. [2] emphasized that "diagrammatic literacy" is vital in secondary science education. Furthermore, based on findings about the efficacy of self-constructed diagrams

© The Author(s) 2021
A. Basu et al. (Eds.): Diagrams 2021, LNAI 12909, pp. 562–566, 2021.
https://doi.org/10.1007/978-3-030-86062-2_57

(e.g., [3]), the ability to construct diagrams has also been integrated into the concept of diagrammatic competency.

To cultivate diagrammatic competency, it is necessary to evaluate and provide feedback to students and teachers on the students' extent of achieving it. Regular school tests can provide valuable opportunities for assessing such achievement. However, such tests in Japanese high schools mainly focus on evaluating the understanding of learning contents rather than the development of competencies like diagrammatic competency. Thus, this study proposes a framework that can evaluate diagrammatic competency in school students statistically, with the use of regular school tests.

In this study, the students' performance was analyzed using cognitive diagnostic models (CDMs) [1]. As described later in more detail, this analysis assumes that students can solve problems if they master the elements of cognitive abilities and the skills required for problem-solving (i.e., 'attributes'), and it estimates how well students have mastered attributes (i.e., 'attribute mastery probabilities'). The current study incorporated diagrammatic competency into the analysis as one of the attributes and attempted to statistically calculate how students master diagrammatic competency using CDMs.

2 Development of Tasks to Assess Diagrammatic Competency

Educational psychologists and a high-school mathematics teacher collaborated for this study; they worked together to modify a regular school test to incorporate tasks requiring diagrammatic competency to solve problems.

The problems assessing diagrammatic competency essentially were of two types: the first ones were tasks that were also included in the traditional regular test; however, it could also assess the diagrammatic competency (e.g., item 1 in Fig. 1). The second type required diagrammatic competency for problem-solving, which was newly incorporated in this regular test (e.g., item 2 in Fig. 1). As reflected by these problems, the current study defined diagrammatic competency operationally as the ability to imagine a necessary diagrammatic representation when solving a given problem. The regular test was conducted in an actual high school, and 40 students took the test.

In this study, the attributes representing diagrammatic competency along with four other attributes were identified. In addition to 'diagrammatic competency (A1)', 'comprehension of math terminology (A2)', 'application of a mathematical formula (A3)', 'understanding relations between numerical expressions (A4)', and 'computational skills (A5)' were incorporated as attributes in this study.

To conduct the analysis of CDMs, it was necessary to specify the item-attribute relationship in the form of a matrix, called a 'Q-matrix'. The Q-matrix for this study was specified by the math teacher and two educational psychologists. Inter-rater agreement between the math teacher and one of the educational psychologists was confirmed as substantially equivalent (Cohen's $\kappa = .66$, 95% confidence interval, .54–.78). Inconsistencies in the Q-matrix were discussed between the two educational psychologists.

An example of the Q-matrix for the tasks in Fig. 1 is presented in Table 1. It indicates, for instance, that to solve item 1 in Fig. 1, other attributes such as 'A4', 'A5' are necessary, together with diagrammatic competency (A1). A total of 40 students took the regular test that was administered.

	Item 1	Item 2
Task	Let D be the domain represented by the three inequalities $y - x \leq 0, y + 2x \geq 0, 2x - y - 4 \leq 0$. Find the maximum and minimum values of $x + y$ when point (x, y) moves within this domain.	Let C be a parabola $y = f(x)$ and take a point $P(a, 2a^2)$ on C, where $a > 0$. Let the tangent at the P on C be l. Let point Q be the intersection of line l and the x-axis, line m be perpendicular to l through Q, and point A be the intersection of m and the y-axis. Choose one of the following (1) to (4) as the most appropriate representation of $T = \int_0^a \{2x^2 - (4ax - 2a^2)\}dx$ (1) Area of the square $OQPA$ (2) Area of the figure bounded by curve C and line l (3) Area of the figure enclosed by curve C, line l, and y-axis (4) Area of the figure enclosed by curve C, line l, and the x-axis
Nec-essary Dia-gram		

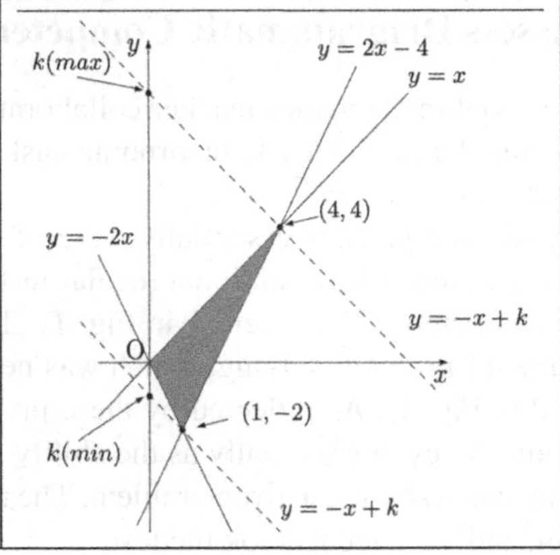

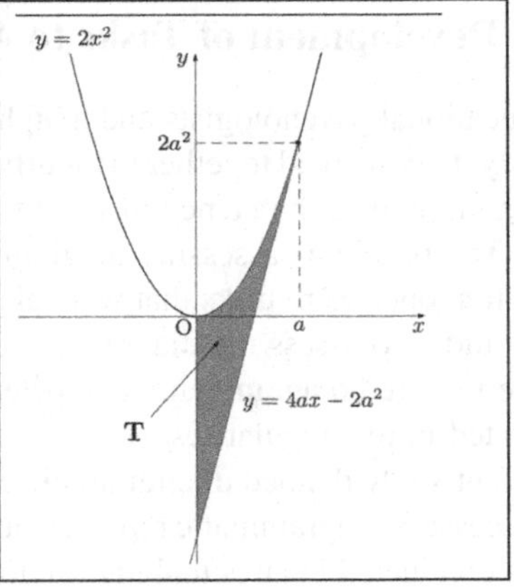

Fig. 1. Example tasks assessing diagrammatic competency

Table 1. Example of Q-matrix

Attribute / Task	A1:Diagrammatic Competency	A2:Comprehension of Math Terminology	A3:Application of a Mathematical Formula	A4:Understanding Relations between Numerical Expressions	A5:Computational Skills
Item1	1	0	0	1	1
Item2	1	0	1	1	1

3 Results and Discussion

In the current study, DINA (the deterministic inputs, noisy "and" gate) model was applied for analysis. The R2jags package in the R program was used with five chains, 30,000 iterations, and 10,000 burn-in. All the model parameters were confirmed to converge. Attribute mastery probabilities for each student were calculated. Table 2 shows examples of outputs. It also illustrates that the analysis with CDMs makes it possible to empirically demonstrate unique patterns of students' acquisition of each attribute and their acquisition of diagrammatic competency.

Following previous studies, in the current study, 0.5 was used to decide whether students were judged as mastering the attribute. The result of the analysis revealed the percentages of students who mastered each attribute: acquisition of 'diagrammatic competency' was 52.5%, 'comprehension of math terminology' was 52.5%, 'application of a mathematical formula' was 67.5%, 'understanding relations between numerical expressions' was 45.0%, and 'computational skills' was 60.0%. This suggests that almost half of the students in a class did not sufficiently master diagrammatic competency.

Table 2. Example of output: attribute mastery probabilities (excerpt)

Student ID	A1:Diagrammatic Competency	A2:Comprehension of Math Terminology	A3:Application of a Mathematical Formula	A4:Understanding Relations between Numerical Expressions	A5:Computational Skills
1	0.19	0.55	0.90	0.54	0.80
17	0.97	0.90	1.00	0.05	0.99
23	0.09	0.89	0.93	0.31	0.92
37	0.44	0.83	0.71	0.59	0.60

In this study, cluster analysis was conducted to categorise the students into several groups. Four clusters were obtained, as listed in Table 3. The students belonging to the two larger clusters did not achieve diagrammatic competency. The results indicate the necessity of cultivating diagrammatic competency in school practices.

Table 3. Results of cluster analysis (ward's method) of students

Cluster ID	N	A1:Diagrammatic Competency	A2:Comprehension of Math Terminology	A3:Application of a Mathematical Formula	A4:Understanding Relation between Numerical Expressions	A5:Computational Skills
1	12	0.30	0.62	0.87	0.59	0.79
2	17	0.46	0.41	0.31	0.46	0.41
3	8	0.92	0.64	1.00	0.05	0.99
4	3	0.98	0.93	1.00	0.03	1.00

References

1. Rupp, A.A., Templin, J., Henson, R.A.: Diagnostic Measurement: Theory. Methods and Applications. Guilford Press, New York (2010)
2. Kragten, M., Admiraal, W., Rijlaarsdam, G.: Diagrammatic literacy in secondary science education. Res. Sci. Educ. **43**, 1785–1800 (2012)
3. Ainsworth, S., Prain, V., Tytler, R.: Drawing to learn in science. Science **26**, 1096–1097 (2011)

566   Y. Uesaka et al.

Open Access This chapter is licensed under the terms of the Creative Commons Attribution 4.0 International License (http://creativecommons.org/licenses/by/4.0/), which permits use, sharing, adaptation, distribution and reproduction in any medium or format, as long as you give appropriate credit to the original author(s) and the source, provide a link to the Creative Commons license and indicate if changes were made.

The images or other third party material in this chapter are included in the chapter's Creative Commons license, unless indicated otherwise in a credit line to the material. If material is not included in the chapter's Creative Commons license and your intended use is not permitted by statutory regulation or exceeds the permitted use, you will need to obtain permission directly from the copyright holder.

Author Index

Printed in the United States
by Baker & Taylor Publisher Services